OT 50
Operator Theory: Advances and Applications
Vol. 50

Birkhäuser Verlag
Basel · Boston · Berlin

Topics in Matrix and Operator Theory

Workshop on Matrix
and Operator Theory
Rotterdam (The Netherlands),
June 26–29, 1989.

Edited by

H. Bart
I. Gohberg
M. A. Kaashoek

1991

Birkhäuser Verlag
Basel · Boston · Berlin

Volume Editorial Office:

Econometric Institute
Erasmus University Rotterdam
Postbus 1738
3000 DR Rotterdam
The Netherlands

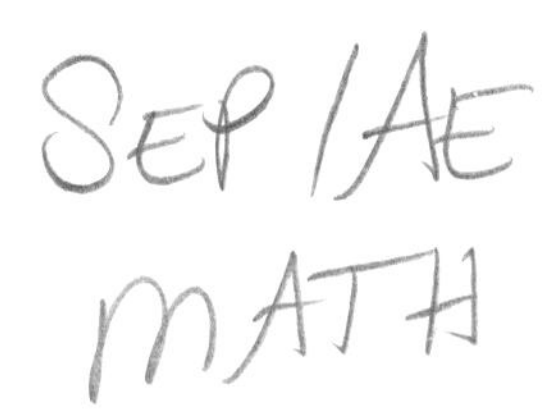

Library of Congress Cataloging-in-Publication Data

Workshop on Matrix and Operator Theory (1989 : Rotterdam, Netherlands)
Topics in matrix and operator theory : [proceedings] of the Workshop on Matrix and Operator Theory, Rotterdam (The Netherlands), June 26–29, 1989 / edited by H. Bart, I. Gohberg, M. A. Kaashoek
p. cm. – (Operator theory, advances and applications ; v. 50)
Includes bibliographical references.
ISBN 3-7643-2570-4 (alk. paper). – ISBN 0-8176-2570-4 (alk. paper)
1. Matrices-Congresses. 2. Operator theory-Congresses.
I. Bart, H. (Harm), 1942– . II. Gohberg, I. (Israel), 1928– .
III. Kaashoek, M. A. IV. Title. V. Series.
QA188.W67 1989
512.9' 434 – dc20

Deutsche Bibliothek Cataloging-in-Publication Data

Topics in matrix and operator theory / Workshop on Matrix and Operator Theory, Rotterdam (The Netherlands), June 26–29, 1989. Ed. by H. Bart ... – Basel ; Boston ; Berlin : Birkhäuser, 1991
(Operator theory ; Vol. 50)
ISBN 3-7643-2570-4 (Basel ...)
ISBN 0-8176-2570-4 (Boston)
NE: Bart, Harm [Hrsg.]; Workshop on Matrix and Operator Theory
<1989, Rotterdam>; GT

Printed in Germany on acid-free paper
ISBN 3-7643-2570-4
ISBN 0-8176-2570-4

Contents

EDITORIAL INTRODUCTION

This volume contains the proceedings of the Workshop on Matrix and Operator Theory held at the Erasmus University in Rotterdam, June 26–29, 1989. The workshop, which followed the international symposium on Mathematical Theory of Networks and Systems (Amsterdam, June 19–23, 1989), was the fifth of its kind; the previous four were held at Santa Monica, California (1981), Rehovot (1983), Amsterdam (1985) and Mesa, Arizona (1987). The next conference in this series is planned for June 11–14, 1991, Hokkaido, Japan.

The papers in these proceedings give a cross-section of the lectures presented at the workshop. They review recent advances and discuss problems in matrix theory, the theory of linear operators, mathematical physics and other areas of analysis. The concrete classes of operators that are analysed concern differential and difference operators, Toeplitz matrices (both finite and infinite), and Wiener–Hopf integral operators. Various interpolation problems for matrix functions, matrix completion problems, and inversion and perturbation problems are investigated. Model theory and its connection with mathematical systems theory is reviewed. System theoretical methods also appear in the applications of the state space method.

The Workshop on Matrix and Operator Theory was made possible through the generous support of

- *Faculteit der Economische Wetenschappen, Erasmus Universiteit*
- *IBM Nederland N.V.*
- *Koninklijke Nederlandse Academie van Wetenschappen*
- *Stichting Universiteitfonds Rotterdam*
- *Wiskundig Genootschap (Vertrouwenscommissie).*

The Econometric Institute of the Erasmus University Rotterdam provided most valuable administrative assistance.

October 1990, *H. Bart, I. Gohberg, M.A. Kaashoek*

Operator Theory:
Advances and Applications, Vol. 50

WIENER-HOPF FACTORIZATION IN THE INVERSE SCATTERING THEORY FOR THE n-D SCHRÖDINGER EQUATION

Tuncay Aktosun and Cornelis van der Mee

We study the n-dimensional Schrödinger equation, $n \geq 2$, with a nonspherically symmetric potential in the class of Agmon's short range potentials without any positive energy bound states. We give sufficient conditions that guarantee the existence of a Wiener-Hopf factorization of the corresponding scattering operator. We show that the potential can be recovered from the scattering operator by solving a related Riemann-Hilbert problem utilizing the Wiener-Hopf factors of the scattering operator. We also study the properties of the scattering operator and show that it is a trace class perturbation of the identity when the potential is also integrable.

1. INTRODUCTION

In this article we study the inverse scattering problem for the n-D Schrödinger equation

$$\nabla_x^2 \psi(k,x,\theta) + k^2 \psi(k,x,\theta) = V(x)\psi(k,x,\theta),$$

where $n \geq 2$, ∇_x^2 is the Laplacian, $x \in \mathbf{R}^n$ is the spatial coordinate, $\theta \in S^{n-1}$ is a unit vector in $\mathbf{R}^n$, and $k^2 \in \mathbf{R}$ is energy. The potential $V(x)$ is assumed to decrease to zero sufficiently fast as $|x| \to \infty$, but need not be spherically symmetric. Then, as $|x| \to \infty$, the wave function $\psi(k,x,\theta)$ behaves as

$$\psi(k,x,\theta) = e^{ik\theta\cdot x} + ie^{-\frac{\pi}{4}i(n-1)} \frac{e^{ik|x|}}{|x|^{\frac{n-1}{2}}} A(k, \frac{x}{|x|}, \theta) + o\left(\frac{1}{|x|^{\frac{n-1}{2}}}\right)$$

where $A(k,\theta,\theta')$ is the scattering amplitude. The scattering operator $S(k,\theta,\theta')$ is then defined as

$$S(k,\theta,\theta') = \delta(\theta-\theta') + i\left(\frac{k}{2\pi}\right)^{\frac{n-1}{2}} A(k,\theta,\theta'),$$

where δ is the Dirac delta distribution on S^{n-1}. The scattering operator acts on $L^2(S^{n-1})$, the Hilbert space of square-integrable complex-valued functions with respect to the surface Lebesgue measure on S^{n-1}. Then, in operator notation, the above equation becomes

$$S(k) = \mathbf{I} + i\left(\frac{k}{2\pi}\right)^{\frac{n-1}{2}} A(k).$$

Here **I** denotes the identity operator. $S(k)$ is known to be unitary and to satisfy

$$S(-k) = QS(k)^{-1}Q, \tag{1.1}$$

where Q is the signature operator on $L^2(S^{n-1})$ defined by $(Qf)(\theta) = f(-\theta)$.

The inverse scattering problem consists of retrieving the potential $V(x)$ from the scattering matrix $S(k)$. For one-dimensional and radial Schrödinger equations the inverse scattering problem is fairly well understood [**CS89**]. In higher dimensions the methods available to solve the inverse scattering problem have not yet led to a complete and satisfactory solution. These methods include the Newton-Marchenko method [**Ne80, Ne81, Ne82**], the generalized Gel'fand-Levitan method [**Ne74, Ne80, Ne81, Ne82**], the $\bar{\partial}$-method [**NA84, BC85, BC86, NH87**], the generalized Jost-Kohn method [**Pr69, Pr76, Pr80, Pr82**], a method based on the Green's function of Faddeev [**Fa65, Fa74, Ne85**], and the generalized Muskhelishvili-Vekua method [**AV89b**]. A comprehensive review of the methods and related open problems in 3-D inverse scattering prior to 1989 can be found in Newton's forthcoming book [**Ne89b**] and in Chapter XIV of [**CS89**].

The basic idea behind the Newton-Marchenko, Gel'fand-Levitan, and Muskhelishvili-Vekua methods is to formulate the inverse scattering problem as a Riemann-Hilbert boundary value problem and to use the Fourier transform to obtain a vector-valued integral equation on the half-line (the Newton-Marchenko method), or to use the solution of the Riemann-Hilbert problem in the kernel of an integral equation (the generalized Gel'fand-Levitan method), or to transform the Riemann-Hilbert problem into a Fredholm integral equation with a weakly singular kernel (the generalized Muskhelishvili-Vekua method). The key Riemann-Hilbert problem in n-D inverse scattering theory is given by (3.3), where the operator $G(k)$ is the x-dependent unitary transform of the scattering operator defined by

$$G(k) = U_x(k)QS(k)QU_x(k)^{-1}, \tag{1.2}$$

where $(U_x(k)f)(\theta) = e^{-ik\theta\cdot x}f(\theta)$. Note that we suppress the x-dependence of $G(k)$. The spectra of the three integral operators mentioned above are closely related to the partial indices of $G(k)$. Hence, the study of the Wiener-Hopf factorization of $G(k)$ not only leads to a direct solution of the Riemann-Hilbert problem (3.3) but also helps us to study the solvability of the integral equations in these three inversion methods.

This paper is organized as follows. In Section 2 we establish the Hölder continuity of the scattering operator by using the limiting absorption principle for the free Hamiltonian [**Ag75, Ku80**] and using the estimates given by Weder [**We90**]. In Section 3 using the Hölder continuity of the scattering operator and the results by Gohberg and Leiterer [**GL73**], we prove the existence of the Wiener-Hopf factorization for $G(k)$. In this section we also study the properties of the partial indices of $G(k)$, solve the Riemann-Hilbert problem (3.3) in terms of the Wiener-Hopf factors of $G(k)$, and show that the potential of the n-dimensional Schrödinger equation can be recovered from the scattering operator. Hence, the results in this paper generalize those in [**AV89a**] from 3-D to n-D. Note also that the generalized Muskhelishvili-Vekua method in 3-D given in [**AV89b**] is now seen to be valid also for n-D because the Hölder continuity of $G(k)$ is basically all that is needed in that method. In Section 4 we prove that the scattering operator $S(k)$ is a trace class perturbation of the identity and evaluate the trace of $S(k) - \mathbf{I}$ as $k \to \pm\infty$.

Throughout we will use the following notation. $\mathbf{C}$ is the complex plane, $\mathbf{C}^{\pm} = \{z \in \mathbf{C} : \pm\text{Im} z > 0\}$, $\mathbf{R}^{+} = (0, \infty)$, $\mathbf{R}_{\infty} = \mathbf{R} \cup \{\pm\infty\}$, $\mathbf{T} = \{z \in \mathbf{C} : |z| = 1\}$, $\mathbf{T}^{+} = \{z \in \mathbf{C} : |z| < 1\}$, $\mathbf{T}^{-} = \{z \in \mathbf{C} : |z| > 1\} \cup \{\infty\}$ and $\mathbf{D}^{\pm} = \mathbf{C}^{\pm} \cup \mathbf{R}^{+}$. The closure of a set $\mathbf{P}$ in the Riemann sphere $\mathbf{C}_{\infty} = \mathbf{C} \cup \{\infty\}$ will be denoted by $\overline{\mathbf{P}}$.

The domain, kernel, range, and spectrum of a linear operator T will be denoted by $\mathcal{D}(T)$, $\mathcal{N}(T)$, $\mathcal{R}(T)$, and $\sigma(T)$, respectively. $\mathcal{L}(X;Y)$ will denote the set of bounded linear operators from the Banach space X into the Banach space Y, while $\mathcal{L}(X)$ will stand for $\mathcal{L}(X;X)$. The adjoint of T on a Hilbert space will be denoted by $T^{\dagger}$.

$\hat{u}(\xi)$ will denote the Fourier transform of $u \in L^2(\mathbf{R}^n)$; i.e.,

$$\hat{u}(\xi) = \lim_{N \to \infty} \frac{1}{(2\pi)^{n/2}} \int_{|x| \leq N} e^{-i\xi \cdot x} u(x) dx,$$

and hence $\|u\|_2 = \|\hat{u}\|_2$, where $\|\cdot\|_2$ is the norm in $L^2(\mathbf{R}^n)$. We will use $L^2_s(\mathbf{R}^n)$ to denote the Hilbert space of all complex measurable functions $u(x)$ on $\mathbf{R}^n$ such that $(1 + |x|^2)^{s/2} u(x) \in L^2(\mathbf{R}^n)$, endowed with the norm

$$\|u\|_s = \|(1 + |x|^2)^{s/2} u(x)\|_2. \tag{1.3}$$

By $C_0^{\infty}(\mathbf{R}^n)$ we will denote the linear space of all C^{∞}-functions on $\mathbf{R}^n$ of compact support. Then $H^{\alpha}(\mathbf{R}^n)$ will denote the Sobolev space of order α, which is the completion of $C_0^{\infty}(\mathbf{R}^n)$ in the norm

$$\|u\|_{H^{\alpha}(\mathbf{R}^n)} = \|(1 + |\xi|^2)^{\alpha/2} \hat{u}(\xi)\|_2.$$

By $H^{\alpha}_s(\mathbf{R}^n)$ we denote the weighted Sobolev space of order (α, s), which is defined as

$$H^{\alpha}_s(\mathbf{R}^n) = \{u(x) : (1 + |x|^2)^{s/2} u(x) \in H^{\alpha}(\mathbf{R}^n)\}$$

with norm $\|u\|_{H^{\alpha}_s(\mathbf{R}^n)} = \|(1 + |x|^2)^{s/2} u(x)\|_{H^{\alpha}(\mathbf{R}^n)}$. Note that we have $H^0_s(\mathbf{R}^n) = L^2_s(\mathbf{R}^n)$, $H^{\alpha}_0(\mathbf{R}^n) = H^{\alpha}(\mathbf{R}^n)$, and $H^0_0(\mathbf{R}^n) = L^2(\mathbf{R}^n)$.

For any Banach space $\mathcal{X}$, Borel set $\mathcal{J}$ in $\mathbf{C}_{\infty}$, and $\gamma \in (0, 1]$, we will denote by $\mathcal{C}(\mathcal{J}; \mathcal{X})$ the Banach space of all bounded and continuous functions $\psi : \mathcal{J} \to \mathcal{X}$ endowed with the norm

$$\|\psi\|_{\mathcal{C}(\mathcal{J};\mathcal{X})} = \sup_{t \in \mathcal{J}} \|\psi(t)\|_{\mathcal{X}},$$

and by $\mathcal{H}_{\gamma}(\mathcal{J}; \mathcal{X})$ the Banach space of all uniformly Hölder continuous functions $\psi : \mathcal{J} \to \mathcal{X}$ endowed with the norm

$$\|\psi\|_{\mathcal{H}_{\gamma}(\mathcal{J};\mathcal{X})} = \sup_{t \in \mathcal{J}} \|\psi(t)\|_{\mathcal{X}} + \sup_{t \neq s \in \mathcal{J}} \frac{\|\psi(t) - \psi(s)\|_{\mathcal{X}}}{|t - s|^{\gamma}}.$$

Here the continuity pertains to the strong topology of $\mathcal{X}$. If $\mathcal{X}$ is a Banach algebra, so are $\mathcal{C}(\mathcal{J}; \mathcal{X})$ and $\mathcal{H}_{\gamma}(\mathcal{J}; \mathcal{X})$. For $\gamma \in (0, 1)$, the closed subspace of $\mathcal{H}_{\gamma}(\mathcal{J}; \mathcal{X})$ consisting of those $\psi : \mathcal{J} \to \mathcal{X}$ such that $\|\psi(t) - \psi(s)\|_{\mathcal{X}} = o(|t-s|^{\gamma})$ as $t \to s$, will be denoted by $\mathcal{H}^0_\gamma(\mathcal{J};\mathcal{X})$. Note also that we will use the norm $\|\cdot\|$ without a subscript in order to denote the operator norm in $\mathcal{L}(L^2(S^{n-1}))$.

Acknowledgement. The research leading to this article was supported in part by the National Science Foundation under grant No. DMS 8823102.

2. ESTIMATES ON THE SCATTERING OPERATOR

In this section, starting with the representation given in (2.3), we prove that the scattering operator $S(k)$ and its unitary transform $G(k)$ defined in (1.2) are both Hölder continuous.

For $\alpha \geq 0$, a real function $V(x) \in L^2_{loc}(\mathbf{R}^n)$ is said to belong to the class $\mathbf{B}_\alpha$ if, for some $s > \frac{1}{2}$, the multiplication by $(1+|x|^2)^s V(x)$ represents a bounded linear operator from $H^\alpha(\mathbf{R}^n)$ into $L^2(\mathbf{R}^n)$. (Our definition of $\mathbf{B}_\alpha$ differs from the one used in [**We90**] in that we require $V(x)$ to be locally L^2). For such potentials, the multiplication by $V(x)$ represents a bounded operator from $H^\alpha_{-t}(\mathbf{R}^n)$ into $L^2_t(\mathbf{R}^n)$ for all $t \in (\frac{1}{2}, s]$ with s as in the definition of $\mathbf{B}_\alpha$. It follows from results in Chapter 6 of [**Sc71**] that $V(x) \in \mathbf{B}_\alpha$ if $\exists \epsilon > 0, \beta \in (0, 2\alpha)$ such that

$$\sup_{x \in \mathbf{R}^n} (1+|x|^2)^{1+\epsilon} \int_{|x-y|\leq 1} dy \, \frac{|V(y)|^2}{|x-y|^{n-\beta}} < +\infty.$$

If $V(x) \in \mathbf{B}_0$; i.e., if $\exists c > 0, s > \frac{1}{2}$ such that

$$|V(x)| \leq \frac{c}{(1+|x|^2)^s}, \quad x \in \mathbf{R}^n,$$

then $V(x) \in \mathbf{B}_\alpha$ for every $\alpha \geq 0$.

A real function $V(x) \in L^2_{loc}(\mathbf{R}^n)$ is said to be **short range** or to belong to the class **SR** [**Ag75**] if for some $s > \frac{1}{2}$ the multiplication by $(1+|x|^2)^s V(x)$ represents a compact operator from $H^2(\mathbf{R}^n)$ into $L^2(\mathbf{R}^n)$. Since $H^2(\mathbf{R}^n)$ is compactly imbedded in any of the spaces $H^\alpha(\mathbf{R}^n)$ for $0 \leq \alpha < 2$, we have

$$\mathbf{B}_\alpha \subset \mathbf{SR}, \quad 0 \leq \alpha < 2.$$

Now let $\mathbf{K}_2$ denote the set of all real functions $V(x) \in L^2_{loc}(\mathbf{R}^n)$ such that the multiplication by $V(x)$ represents a compact operator from $H^2(\mathbf{R}^n)$ into $L^2(\mathbf{R}^n)$ [**We90**]. Then $\mathbf{SR} \subset \mathbf{K}_2$ and hence

$$\mathbf{B}_\alpha \subset \mathbf{SR} \subset \mathbf{K}_2, \quad 0 \leq \alpha < 2.$$

Let $H_0 = -\nabla^2_x$ be the free Hamiltonian with domain $\mathcal{D}(H_0) = H^2(\mathbf{R}^n)$. Then H_0 is a selfadjoint operator on $L^2(\mathbf{R}^n)$ with absolutely continuous spectrum $[0,\infty]$. Define

$$R_0^{\pm}(\lambda) = \lim_{z \to \lambda} (H_0 - z)^{-1}, \quad \lambda \in \mathbf{R}^+, \, z \in \mathbf{C}^{\pm}. \tag{2.1}$$

According to the limiting absorption principle, the limit in (2.1) exists in the uniform operator topology of $\mathcal{L}(L_s^2(\mathbf{R}^n); H_{-s}^2(\mathbf{R}^n))$ for $s > \frac{1}{2}$ [**Ag75, Ku80**]. $R_0^{\pm}(\lambda)$ can be extended to $\mathbf{C}^{\pm}$ by defining

$$R_0^{\pm}(z) = (H_0 - z)^{-1}, \quad z \in \mathbf{C}^{\pm}. \tag{2.2}$$

Then from (2.1) and (2.2) it follows that $R_0^{\pm}(z)$ is a bounded linear operator from $L_s^2(\mathbf{R}^n)$ into $H_{-s}^2(\mathbf{R}^n)$, which is continuous (in operator norm) in $z \in \mathbf{D}^{\pm}$, bounded and analytic for $z \in \mathbf{C}^{\pm}$.

If $V(x) \in \mathbf{SR}$, then the Hamiltonian $H = H_0 + V(x)$, whose domain is $\mathcal{D}(H) = \mathcal{D}(H_0) = H^2(\mathbf{R}^n)$, is selfadjoint and bounded below with essential spectrum $[0, \infty)$, while its negative eigenvalues have finite multiplicity and can only accumulate at zero. Moreover [**Ag75, Ku80**], the set, $\sigma_+(H)$, of positive eigenvalues of H consists only of eigenvalues of finite multiplicity, which can only accumulate at zero and at infinity. Apart from that, H may have a bound state or half-bound state at zero energy. If $V(x) = O(\frac{1}{|x|})$ as $x \to \infty$, the set $\sigma_+(H)$ is empty [**Ka59**]. As in (2.1) we define

$$R^{\pm}(\lambda) = \lim_{z \to \lambda} (H - z)^{-1}, \quad \lambda \in \mathbf{R}^+ \setminus \sigma_+(H),\ z \in \mathbf{C}^{\pm}.$$

The operator $R^{\pm}(\lambda)$ exists in the norm topology of $\mathcal{L}(L_s^2(\mathbf{R}^n); H_{-s}^2(\mathbf{R}^n))$ for all $s > \frac{1}{2}$ and it can be extended to $\mathbf{C}^{\pm}$, by defining $R^{\pm}(z) = (H - z)^{-1}$ for $z \in \mathbf{C}^{\pm}$. Then $R^{\pm}(z)$ becomes a continuous (in operator norm) function on $\mathbf{D}^{\pm} \setminus \sigma_+(H)$ and analytic for $z \in \mathbf{C}^{\pm}$.

If $V(x) \in \mathbf{K}_2$, then $V(x)R_0^{\pm}(z)$ is compact on $L_s^2(\mathbf{R}^n)$ for all $z \in \mathbf{D}^{\pm}$ and $s > \frac{1}{2}$, while $[\mathbf{I} + V(x)R_0^{\pm}(z)]$ and $[\mathbf{I} + R_0^{\pm}(z)V(x)]$ are invertible on $L_s^2(\mathbf{R}^n)$ and $H_{-s}^2(\mathbf{R}^n)$, respectively, for all $z \in \mathbf{D}^{\pm} \setminus \sigma_+(H)$ and satisfy the identity [**Ag75, Ku80**]

$$R^{\pm}(z) = R_0^{\pm}(z)[\mathbf{I} + V(x)R_0^{\pm}(z)]^{-1} = [\mathbf{I} + R_0^{\pm}(z)V(x)]^{-1}R_0^{\pm}(z), \quad z \in \mathbf{D}^{\pm} \setminus \sigma_+(H).$$

The scattering matrix may then be represented in the form [**Ku80**]

$$S(k) = \mathbf{I} - \frac{ik^{n-2}}{2(2\pi)^{n-1}}\sigma(k)[\mathbf{I} + V(x)R_0^{+}(k^2)]^{-1}V(x)\sigma^{\dagger}(k), \quad k^2 \in \mathbf{R}^+ \setminus \sigma_+(H), \tag{2.3}$$

where

$$(\sigma(k)f)(\theta) = \int_{\mathbf{R}^n} dx\ e^{-ik\theta\cdot x}f(x), \quad \theta \in S^{n-1},$$

$$(\sigma^{\dagger}(k)g)(x) = \int_{S^{n-1}} d\theta\ e^{ik\theta\cdot x}g(\theta), \quad x \in \mathbf{R}^n. \tag{2.4}$$

We will use the multiplicative representation (2.3) in order to prove the Hölder continuity of $S(k)$, but we first need a few propositions concerning the Hölder continuity of the individual factors appearing in (2.3).

PROPOSITION 2.1. *Let $s > \frac{1}{2}$. Then, for every compact subset $\mathcal{J}$ of $\mathbf{R}_\infty \setminus \{0\}$, $\sigma^\dagger(k)$ is a uniformly Hölder continuous function from $\mathcal{J}$ into $\mathcal{L}(L^2(S^{n-1}); L^2_{-s}(\mathbf{R}^n))$ of exponent γ where $0 < \gamma < s - \frac{1}{2}$.*

Proof: It is known ([**Ag75**], Lemma 5.2; [**We90**], Lemma 2.4) that $\sigma^\dagger(k)$ belongs to $\mathcal{L}(L^2(S^{n-1}); L^2_{-s}(\mathbf{R}^n))$ for all $s > \frac{1}{2}$ and $k \in \mathbf{R} \setminus \{0\}$; i.e., for all $g \in L^2(S^{n-1})$

$$\|\sigma^\dagger(k)g\|_{-s} \leq C_{k,s}\|g\|_{L^2(S^{n-1})}. \tag{2.5}$$

More details on $C_{k,s}$ are given in Appendix A. For $s > \frac{3}{2}$ we have

$$\|\frac{d}{dk}(\sigma^\dagger(k)g)\|^2_{-s} \leq \sum_{j=1}^n \|\sigma^\dagger(k)(\theta_j g)\|^2_{-(s-1)} \leq nC^2_{k,s-1}\|g\|^2_{L^2(S^{n-1})},$$

so that for $k_1 \neq k_2 \in \mathbf{R} \setminus \{0\}$

$$\|[\sigma^\dagger(k_1) - \sigma^\dagger(k_2)]g\|_{-s} \leq \sqrt{n}\; C_{\mathcal{J},s-1}|k_1 - k_2|\; \|g\|_{L^2(S^{n-1})},$$

where $C_{\mathcal{J},s} = \sup_{k\in\mathcal{J}}\; C_{k,s}$. So, if $s > \frac{3}{2}$, $\sigma^\dagger(k)$ belongs to $\mathcal{H}_1(\mathcal{J}; \mathcal{L}(L^2(S^{n-1}); L^2_{-s}(\mathbf{R}^n)))$ with its norm bounded above by $(1 + \sqrt{n})C_{\mathcal{J},s-1}$. Now note that, by Hölder's inequality, for $s_1, s_2 > 0$ and $\epsilon \in (0, 1)$

$$\|f\|_{-s} \leq \|f\|^\epsilon_{-s_1}\|f\|^{1-\epsilon}_{-s_2} \tag{2.6}$$

whenever $s = \epsilon s_1 + (1 - \epsilon)s_2$. Thus, for $s_1 > \frac{3}{2}$, $s_2 > \frac{1}{2}$, $\epsilon \in (0, 1)$ and $s = \epsilon s_1 + (1 - \epsilon)s_2$, using (2.6) we obtain

$$\begin{aligned}\|[\sigma^\dagger(k_1) - \sigma^\dagger(k_2)]g\|_{-s} \leq \\ \leq \left(2C_{\mathcal{J},s_2}\|g\|_{L^2(S^{n-1})}\right)^{1-\epsilon}\left((1 + \sqrt{n})C_{\mathcal{J},s_1-1}\|g\|_{L^2(S^{n-1})}|k_1 - k_2|\right)^\epsilon = \\ = 2^{1-\epsilon}(1 + \sqrt{n})^\epsilon(C_{\mathcal{J},s_1})^{1-\epsilon}(C_{\mathcal{J},s_2-1})^\epsilon|k_1 - k_2|^\epsilon\|g\|_{L^2(S^{n-1})},\end{aligned}$$

so that $\sigma^\dagger(k)$ belongs to $\mathcal{H}^0_\epsilon(\mathcal{J}; \mathcal{L}(L^2(S^{n-1}); L^2_{-s}(\mathbf{R}^n)))$. In fact, its Hölder norm of exponent ϵ is bounded above by $C_{\mathcal{J},s} + 2^{1-\epsilon}(1 + \sqrt{n})^\epsilon(C_{\mathcal{J},s_1})^{1-\epsilon}(C_{\mathcal{J},s_2-1})^\epsilon$, which is in turn bounded above by

$$(2 + \sqrt{n})(C_{\mathcal{J},s_1})^{1-\epsilon}(C_{\mathcal{J},s_2-1})^\epsilon. \tag{2.7}$$

Here we have used (2.6) and the inequality $1 + 2^{1-\epsilon}(1 + \sqrt{n})^\epsilon \leq 2 + \sqrt{n}$. ∎

PROPOSITION 2.2. *Let $s > \frac{1}{2}$ and $\alpha \geq 0$. Then, for every compact subset $\mathcal{J}$ of $\mathbf{R}_\infty \setminus \{0\}$, $\sigma^\dagger(k)$ is a uniformly Hölder continuous function from $\mathcal{J}$ into $\mathcal{L}(L^2(S^{n-1}); H^\alpha_{-s}(\mathbf{R}^n))$ of exponent γ where $0 < \gamma < s - \frac{1}{2}$.*

Proof: Clearly,

$$(\nabla^2_x\sigma^\dagger(k)g)(x) = -k^2 \int_{S^{n-1}} d\theta\; e^{-ik\theta\cdot x}g(\theta), \tag{2.8}$$

so that for $s > \frac{1}{2}$

$$\|\sigma^\dagger(k)g\|_{2,-s} \leq \left(\|\sigma^\dagger(k)g\|_{-s}^2 + \|\nabla_x^2\sigma^\dagger(k)g\|_{-s}^2\right)^{\frac{1}{2}} \leq$$

$$\leq (1+k^2)^{\frac{1}{2}}\|\sigma^\dagger(k)g\|_{-s} \leq (1+k^2)^{\frac{1}{2}} C_{k,s}\|g\|_{L^2(S^{n-1})}.$$

Repeated application of the above process leads to the estimate

$$\|\sigma^\dagger(k)g\|_{\alpha,-s} \leq (1+k^2)^{\alpha/4} C_{k,s}\|g\|_{L^2(S^{n-1})}$$

for $\alpha = 0, 2, 4, 6, \cdots$ and hence, by interpolation, this inequality remains valid for all $\alpha \geq 0$.

We may now repeat the proof of Proposition 2.1 using (2.8) instead of (2.5). The result is the Hölder continuity of $\sigma^\dagger(k)$ from $\mathcal{J}$ into $\mathcal{L}(L^2(S^{n-1}); H^\alpha_{-s}(\mathbf{R}^n))$ with its Hölder norm bounded above by

$$(2+\sqrt{n})\left(D_{\mathcal{J},s_1}\right)^{1-\epsilon}\left(D_{\mathcal{J},s_2-1}\right)^{\epsilon},$$

where $D_{\mathcal{J},s} = \sup_{k\in\mathcal{J}}(1+k^2)^{\alpha/4}C_{k,s}$ and $s = \epsilon s_1 + (1-\epsilon)s_2$ for some $s_1 > \frac{1}{2}$, $s_2 > \frac{3}{2}$ and $\epsilon \in (0,1)$. ∎

PROPOSITION 2.3. *Let $s > \frac{1}{2}$. Then, for every compact subset $\mathcal{J}$ of $\mathbf{R}_\infty \setminus \{0\}$, $\sigma(k)$ is a uniformly Hölder continuous function from $\mathcal{J}$ into $\mathcal{L}(L^2_s(\mathbf{R}^n); L^2(S^{n-1}))$ of exponent γ where $\gamma \in (0, s-\frac{1}{2})$.*

Proof: By duality from Proposition 2.1. The Hölder norm again is bounded by the quantity in (2.7). ∎

PROPOSITION 2.4. *Let $s > \frac{1}{2}$. Then, for every compact subset $\mathcal{J}$ of $\mathbf{R}_\infty \setminus \{0\}$, $R_0^+(k^2)$ is a uniformly Hölder continuous function from $\mathcal{J}$ into $\mathcal{L}(L^2_s(\mathbf{R}^n); H^2_{-s}(\mathbf{R}^n))$ of exponent γ where $\gamma \in (0, s-\frac{1}{2})$.*

Proof: According to [**Ag75**], Eq. (4.7),

$$< R_0^\pm(k^2)f, g > = \pm\pi i\Phi_{f,g}(k) + CPV\int_0^\infty dt\, \frac{t^{(n-2)/2}}{t-k^2}\Phi_{f,g}(t), \tag{2.9}$$

where CPV stands for Cauchy's principal value, $\Phi_{f,g}$ is defined as

$$\Phi_{f,g}(k) = \frac{1}{2|k|}\int_{|k|S^{n-1}} d\theta\, \hat{f}(\theta)\overline{\hat{g}(\theta)} = \frac{1}{2} < \sigma(k)f, \sigma(k)g >_{L^2(S^{n-1})}$$

and $f, g \in L^2_s(\mathbf{R}^n)$. According to Proposition 2.3, $\Phi_{f,g}$ is a Hölder continuous complex function on $\mathcal{J}$ of exponent $\gamma \in (0, s-\frac{1}{2})$ and the Hölder norm is bounded by

$$\|\Phi_{f,g}\|_{\mathcal{H}_\gamma} \leq M_\gamma\|f\|_s\|g\|_s,$$

where M_γ is a constant independent of f and g. Then (2.9) implies that $< R_0^\pm(k^2)f, g >$ is a Hölder continuous complex function on $\mathcal{J}$ of exponent $\gamma \in (0, s-\frac{1}{2})$ with its Hölder norm bounded above by $N_{\gamma,\mathcal{J}}\|f\|_s\|g\|_s$ for some constant $N_{\gamma,\mathcal{J}}$ not depending on f and g. Thus we obtain

$$|<[R_0^{\pm}(k_1^2)-R_0^{\pm}(k_2^2)]f,g>|\leq N_{\gamma,\mathcal{J}}|k_1-k_2|^{\gamma}\|f\|_s\|g\|_s,$$

which completes the proof. ∎.

PROPOSITION 2.5. *Let $V(x)\in \mathbf{B}_\alpha$ for some $\alpha\in[0,2)$, and let $s>\frac{1}{2}$ be the constant such that multiplication by $(1+|x|^2)^s V(x)$ is a bounded linear operator from $H^\alpha(\mathbf{R}^n)$ into $L^2(\mathbf{R}^n)$. Then the scattering operator $S(k)$ is a uniformly Hölder continuous function from $\mathbf{R}\setminus[\sigma_+(H)^{1/2}\cup\{0\}]$ into $\mathcal{L}(L^2(S^{n-1}))$ satisfying*

$$\|S(k)-\mathbf{I}\|\leq\frac{C_\delta}{(1+|k|^2)^{1-\delta}}\tag{2.10}$$

for all $0<\delta<\frac{1}{2}$. Here $\sigma_+(H)^{1/2}=\{z\in\mathbf{R}:z^2\in\sigma_+(H)\}$ and C_δ is a constant.

Proof: According to (2.3) and Propositions 2.1-2.4, we have the following commutative diagram of bounded linear operators for all nonzero k such that $k^2\notin\sigma_+(H)$:

$$\begin{array}{ccccc}
L^2(S^{n-1}) & \xrightarrow{S(k)-\mathbf{I}} & L^2(S^{n-1}) & \xleftarrow{\frac{-ik^{n-2}}{2(2\pi)^{n-1}}\mathbf{I}} & L^2(S^{n-1})\\
\sigma^\dagger(k)\big\downarrow & & & & \big\uparrow\sigma(k)\\
H^\alpha_{-s}(\mathbf{R}^n) & \xrightarrow{(\mathbf{I}+R_0^+(k^2)V)^{-1}} & H^\alpha_{-s}(\mathbf{R}^n) & \xrightarrow{V} & L^2_s(\mathbf{R}^n)
\end{array}$$

Then, for each compact subset $\mathcal{J}$ of $\mathbf{R}\setminus[\sigma_+(H)^{1/2}\cup\{0\}]$, every operator in the diagram is uniformly Hölder continuous in k as a function from $\mathcal{J}$ into $\mathcal{L}(\mathcal{X};\mathcal{Y})$ where $\mathcal{X}$ and $\mathcal{Y}$ are the spaces at the beginning and the end of the corresponding arrow, respectively. Hence, $S(k)$ is a uniformly Hölder continuous function from any such $\mathcal{J}$ into $\mathcal{L}(L^2(S^{n-1}))$. Moreover, the Hölder index γ may be chosen to satisfy $0<\gamma<s-\frac{1}{2}$.

Using the estimate (A.4) in Appendix A we obtain, for some constant C_1,

$$\|S(k)-\mathbf{I}\|\leq C_1|k|^{n-2}\left(|k|^{s-\frac{1}{2}n}\right)^2,$$

which implies (2.10) on all compact subsets $\mathcal{J}$ of $\mathbf{R}_\infty\setminus[\sigma_+(H)^{1/2}\cup\{0\}]$. ∎

Suppose $V(x)\in\mathbf{K}_2$ and $[\mathbf{I}+V(x)R_0^+(k^2)]$ has a limit in the operator norm of $L^2_s(\mathbf{R}^n)$ as $k\to 0$. Then this limit is a compact perturbation of the identity. We denote it by $\mathbf{I}+V(x)R_0^+(0)$. We call $k=0$ an **exceptional point [Ne89b, We90]** if $[\mathbf{I}+V(x)R_0^+(0)]$ is not boundedly invertible on $L^2_s(\mathbf{R}^n)$. In that case there exists $0\neq\varphi\in L^2_s(\mathbf{R}^n)$ such that $\varphi=-VR_0^+(0)\varphi$.

The estimates obtained so far must be refined in order to deal with the case $k \to 0$. This will lead us to the additional assumption that $s > \frac{3}{2} - \frac{1}{n}$ in the definition of $\mathbf{B}_\alpha$. First of all, for all $s > \frac{3}{2} - \frac{1}{n}$, $k \in \mathbf{R} \setminus \{0\}$ and $g \in L^2(S^{n-1})$, we have

$$\|k^{\frac{1}{2}(n-2)}\sigma^\dagger(k)g\|_{-s} \le D_{n,s}|k|^{\frac{(2s-3)n+2}{2(n-1)}}\|g\|_{L^2(S^{n-1})},$$

as a result of (A.4). As in the proof of Proposition 2.1, we obtain for $s > \frac{5}{2} - \frac{1}{n}$

$$\|\frac{d}{dk}(k^{\frac{1}{2}(n-2)}\sigma^\dagger(k)g)\|_{-s} \le D'_{n,s}|k|^{\frac{(2s-5)n+2}{2(n-1)}}\|g\|_{L^2(S^{n-1})}.$$

Here $D_{n,s}$ and $D'_{n,s}$ are constants which do not depend on k. Then through interpolation it follows that, for $s > \frac{3}{2} - \frac{1}{n}$, the operator $k^{\frac{1}{2}(n-2)}\sigma^\dagger(k)$ is a uniformly Hölder continuous function from $[-1,1]$ into $\mathcal{L}(L^2(S^{n-1}); L^2_{-s}(\mathbf{R}^n))$ of exponent γ where $0 < \gamma < s - \frac{3}{2} + \frac{1}{n}$. Next, converting the integral on the right-hand side of (2.9) to a CPV-integral on all of $\mathbf{R}$ we obtain the Hölder continuity of $R_0^+(k^2)$ as a function from $[-1,1]$ into $\mathcal{L}(L^2_{-s}(\mathbf{R}^n); H^2_{-s}(\mathbf{R}^n))$ of exponent γ where $0 < \gamma < s - \frac{3}{2} + \frac{1}{n}$. Thus, in the absence of an exceptional point at $k = 0$, using (2.3) we conclude that $S(k)$ is uniformly Hölder continuous from any compact subset of $[-1,1] \setminus \sigma_+(H)^{1/2}$ into $\mathcal{L}(L^2(S^{n-1}))$ of exponent γ where $0 < \gamma < s - \frac{3}{2} + \frac{1}{n}$. We then readily obtain the following result.

THEOREM 2.6. *Let $V(x) \in \mathbf{B}_\alpha$ for some $\alpha \in (0,2)$, and let $s > \frac{3}{2} - \frac{1}{n}$ be the constant such that multiplication by $(1+|x|^2)^s V(x)$ is a bounded linear operator from $H^\alpha(\mathbf{R}^n)$ into $L^2(\mathbf{R}^n)$. Suppose $\sigma_+(H) = \emptyset$ while $k = 0$ is not an exceptional point. Then $S(k)$ is a uniformly Hölder continuous function from $\mathbf{R}$ into $\mathcal{L}(L^2(S^{n-1}))$ satisfying (2.10) for all $0 < \delta < \frac{1}{2}$.*

In order to prove the existence of a Wiener-Hopf factorization of $S(k)$, we transform Theorem 2.6 to the unit circle $\mathbf{T}$. Let us define

$$\widetilde{S}(\xi) = S(i\frac{1+\xi}{1-\xi}), \quad \xi \in \mathbf{T}. \tag{2.11}$$

Throughout $\widetilde{\ }$ will denote the Möbius transform of a function on the real line to the unit circle, according to the rule (2.11). The next theorem shows that $\widetilde{S}(\xi)$ is also Hölder continuous.

THEOREM 2.7. *Let $V(x) \in \mathbf{B}_\alpha$ for some $\alpha \in (0,2)$, and let $s > \frac{3}{2} - \frac{1}{n}$ be the constant such that multiplication by $(1+|x|^2)^s V(x)$ is a bounded linear operator from $H^\alpha(\mathbf{R}^n)$ into $L^2(\mathbf{R}^n)$. Suppose $\sigma_+(H) = \emptyset$ while $k = 0$ is not an exceptional point. Then $\widetilde{S}(\xi)$ is a uniformly Hölder continuous function from $\mathbf{T}$ into $\mathcal{L}(L^2(S^{n-1}))$ satisfying $\widetilde{S}(1) = \mathbf{I}$. The Hölder exponent can be any ζ satisfying $0 < \zeta < \min\{\frac{1}{2}, 1 - (s - \frac{1}{2} + \frac{1}{n})^{-1}\}$.*

Proof: From Proposition 2.5 we have

$$\|S(k_1) - S(k_2)\| \le M_1|k_1 - k_2|^{\epsilon_1}$$

for every $\epsilon_1 \in (0, \min\{1, s - \frac{3}{2} + \frac{1}{n}\})$, and

$$\|S(k) - \mathbf{I}\| \leq \frac{M_2}{(1+|k|)^{2(1-\epsilon_2)}}$$

for every $\epsilon_2 \in (0,1)$. Here M_1 and M_2 are constants independent of k. Now put

$$\lambda(k,\delta) = (k^2+1)^{\zeta/2}[(k+\delta)^2+1]^{\zeta/2}\delta^{-\zeta}\|S(k+\delta) - S(k)\|,$$

where $\zeta \in (0,1)$ is to be determined later. Then for $\xi = \frac{k-i}{k+i}$ and $\eta = \frac{k+\delta+i}{k+\delta-i}$ we have

$$\frac{\|\tilde{S}(\xi) - \tilde{S}(\eta)\|}{|\xi - \eta|^{\zeta}} = 2^{-\zeta}\lambda(k,\delta),$$

so that the theorem follows if we can prove the boundedness of $\lambda(k,\delta)$.

Indeed, for $|k| \leq 1 \leq \delta$ we have $\lambda(k,\delta) \leq 2^{\zeta/2}(5\delta^2)^{\zeta/2}\delta^{-\zeta}\|S(k+\delta) - S(k)\| \leq 2 \cdot 10^{\zeta/2}$. For $\max(|k|,\delta) \leq 1$ we have $\lambda(k,\delta) \leq 2^{\zeta/2}5^{\zeta/2}\delta^{\epsilon_1 - \zeta}M_1$. For $1 \leq |k| \leq \delta$ we get $\lambda(k,\delta) \leq (2k^2)^{\zeta/2}(5k^2)^{\zeta/2}2M_2(1+|k|)^{-2(1-\epsilon_2)}$, since $\delta^{-\zeta} \leq 1$. Finally, for $\delta \leq 1 \leq |k|$ we have for $0 < \zeta \leq \epsilon_1$

$$\lambda(k,\delta) \leq (2k^2)^{\zeta/2}(5k^2)^{\zeta/2}M_1^{\zeta/\epsilon_1}\left[\frac{2M_2}{(1+|k|)^{2(1-\epsilon_2)}}\right]^{1-\frac{\zeta}{\epsilon_1}}.$$

Hence, to have $\lambda(k,\delta)$ bounded we must choose $0 < \zeta \leq \epsilon_1(1-\epsilon_2)/(1+\epsilon_1-\epsilon_2)$. Taking the maximum over $\epsilon_2 \in (0,1)$ with $0 < \epsilon_1 < \min\{1, s - \frac{3}{2} + \frac{1}{n}\}$, we get $\zeta \in (0, \min\{\frac{1}{2}, 1 - (s - \frac{1}{2} + \frac{1}{n})^{-1}\})$. ∎

3. WIENER-HOPF FACTORIZATION OF THE SCATTERING OPERATOR

The incoming and outgoing scattering solutions $\psi(k,x,\theta)$ and $\psi(-k,x,\theta)$ of the n-D Schrödinger equation are related to each other, as in the 3-D case [**Ne80**], as

$$\psi(k,x,\theta) = \int_{S^{n-1}} d\theta' \; S(k,-\theta,\theta')\psi(-k,x,\theta'),$$

where $x \in \mathbf{R}^n$, $k \in \mathbf{R}$ and $\theta \in S^{n-1}$. Defining

$$f(k,x,\theta) = e^{-ik\theta\cdot x}\psi(k,x,\theta), \tag{3.1}$$

we obtain the Riemann-Hilbert problem

$$f(k,x,\theta) = \int_{S^{n-1}} d\theta' \; e^{-ik\theta\cdot x}S(k,-\theta,\theta')e^{-ik\theta'\cdot x}f(-k,x,\theta'),$$

where $f(k,x,\theta) = 1 + O(\frac{1}{|k|})$ as $|k| \to \infty$ from $\mathbf{C}^+ \cup \mathbf{R}$. In the absence of bound states, $f(k,x,\theta)$ has an analytic continuation in k to $\mathbf{C}^+$. Using $G(k)$ given in (1.2) and defining the sectionally analytic functions

$$X_{\pm}(k,x,\theta) = f(\pm k,x,\pm\theta) - 1, \tag{3.2}$$

we obtain the vector Riemann-Hilbert problem

$$X_+(k) = G(k)X_-(k) + [G(k) - \mathbf{I}]\hat{1} \tag{3.3}$$

on $L^2(S^{n-1})$, where $\hat{1}(\theta) \equiv 1$ and the x-dependence has been suppressed. In general, $f(k, x, \theta)$ is meromorphic on $\mathbf{C}^+$ with simple poles at $k = i\gamma$ where $-\gamma^2$ are the bound state energies. It is possible to remove these simple poles from the Riemann-Hilbert problem by a reduction method [**Ne82**] and to obtain a Riemann-Hilbert problem of the form (3.3) where $X_\pm(k)$ are continuous on $\mathbf{C}^\pm \cup \mathbf{R}$, are analytic on $\mathbf{C}^\pm$, and vanish as $k \to \infty$ from $\mathbf{C}^\pm \cup \mathbf{R}$. For $n = 3$, we refer the reader to [**Ne89b, AV89a**] for details. Note that once (3.3) is solved, the solution of the Schrödinger equation can be obtained using (3.2) and (3.1).

We will solve the Riemann-Hilbert problem (3.3) by using the Wiener-Hopf factors [**Go64**] of the operator function of $G(k)$. By a (**left**) **Wiener-Hopf factorization** of an operator function $G : \mathbf{R}_\infty \to \mathcal{L}(L^2(S^{n-1}))$, we mean a representation of $G(k)$ in the form

$$G(k) = G_+(k)D(k)G_-(k), \quad k \in \mathbf{R}_\infty, \tag{3.4}$$

with

$$D(k) = P_0 + \sum_{j=1}^{m} (\frac{k-i}{k+i})^{\rho_j} P_j,$$

where
1. $G_+(k)$ is continuous on $\mathbf{C}^+ \cup \mathbf{R}$ in the operator norm of $\mathcal{L}(L^2(S^{n-1}))$ and is boundedly invertible there. Similarly, $G_-(k)$ is continuous on $\mathbf{C}^- \cup \mathbf{R}$ in the operator norm and is boundedly invertible there.
2. $G_+(k)$ is analytic on $\mathbf{C}^+$ and $G_-(k)$ is analytic on $\mathbf{C}^-$.
3. $G_+(\pm\infty) = G_-(\pm\infty) = \mathbf{I}$.

The projections $P_1, \cdots, P_m$ are finite in number, are mutually disjoint, have rank one, and $P_0 = \mathbf{I} - \sum_{j=1}^m P_j$. The (**left**) **partial indices** $\rho_1, \cdots, \rho_m$ are nonzero integers. In the absence of partial indices, we have $D(k) = \mathbf{I}$, in which case the Wiener-Hopf factorization is called (**left**) **canonical**. The partial indices of $G(k)$ depend neither on the choice of the factors $G_+(k)$ and $G_-(k)$ nor on the choice of the projections $P_1, \cdots, P_m$. If the factorization is (left) canonical, the factors $G_+(k)$ and $G_-(k)$ are unique as a result of Liouville's theorem.

In the same way we define a right Wiener-Hopf factorization, right partial indices, and a right canonical factorization by interchanging the roles of $G_+(k)$ and $G_-(k)$ in (3.4). The right indices may be different, both in number and in value, from the left indices, but the sum of the left indices coincides with the sum of the right indices. This sum is called the **sum index** of $G(k)$.

By using the Möbius transformation defined in (2.11), we can define the left and right Wiener-Hopf factorizations of operator functions on $\mathbf{T}$. The left and right partial indices are invariant under this Möbius transformation. For details, we refer the reader to [**AV89a**].

REMARK 3.1. The scattering operator $S(k)$ is unitary, and hence from (1.2) it follows that $G(k)$ is also unitary. As a consequence [**AV89a**], the sets of left and right partial indices of $G(k)$ coincide. Moreover, the projections and factors appearing in the right and left Wiener-Hopf factorizations of $G(k)$ are related by

$$\hat{P}_j = (P_j)^\dagger \quad \text{for} \quad j = 1, \cdots, m; \quad \text{and} \quad \hat{G}_\pm(k)^{-1} = G_\mp(\overline{k})^\dagger.$$

Here the quantities with ^ pertain to the right Wiener-Hopf factorization of $G(k)$.

THEOREM 3.2. *Let $V(x) \in \mathbf{B}_\alpha$ for some $\alpha \in (0,2)$, and let $s > \frac{3}{2} - \frac{1}{n}$ be the constant such that multiplication by $(1+|x|^2)^s V(x)$ is a bounded linear operator from $H^\alpha(\mathbf{R}^n)$ into $L^2(\mathbf{R}^n)$. Suppose $\sigma_+(H) = \emptyset$ while $k = 0$ is not an exceptional point. Then the operator function $G(k)$ defined by (1.2) has right and left Wiener-Hopf factorizations. The sets of left and right indices coincide, and the sum index is independent of the choice of $x \in \mathbf{R}^n$.*
Proof: According to Theorem 6.1 (or 6.2) of [**GL73**], it is sufficient to show the following:
1. $G(k)$ is boundedly invertible for every $k \in \mathbf{R}_\infty$.
2. $G(k)$ is a compact perturbation of the identity for every $k \in \mathbf{R}_\infty$.
3. $\widetilde{G}(\xi) \in \mathcal{H}_\alpha(\mathbf{T}; \mathcal{L}(L^2(S^{n-1})))$ for some $\alpha \in (0,1)$, where $\widetilde{G}(\xi)$ is the Möbius transform of $G(k)$, as defined in (2.11).

Under these conditions there exists a left Wiener-Hopf factorization of $\widetilde{G}(\xi)$ with respect to $\mathbf{T}$ which is given by

$$\widetilde{G}(\xi) = \widetilde{G}_+(\xi)\widetilde{D}(\xi)\widetilde{G}_-(\xi),$$

where

$$\widetilde{D}(\xi) = P_0 + \sum_{j=1}^m \xi^{\rho_j} P_j,$$

$\widetilde{G}_+(\xi) \in \mathcal{H}_\alpha(\mathbf{T}^+; \mathcal{L}(L^2(S^{n-1})))$ and is invertible there, $\widetilde{G}_-(\xi) \in \mathcal{H}_\alpha(\mathbf{T}^-; \mathcal{L}(L^2(S^{n-1})))$ and is invertible there, $\widetilde{G}_+(\xi)$ and $\widetilde{G}_-(\xi)$ are analytic on $\mathbf{T}^+$ and on $\mathbf{T}^-$, respectively. The inverse of the Möbius transformation given by (2.11) then yields a left Wiener-Hopf factorization for $G(k)$ of the type (3.4) where the Möbius transformed factors $\widetilde{G}_+(\xi)$ and $\widetilde{G}_-(\xi)$ as well as their inverses are Hölder continuous of exponent α in operator norm on $\mathbf{T}^+$ and $\mathbf{T}^-$, respectively. These properties are easily seen to extend to $G(k)$ for all $x \in \mathbf{R}^n$, since the operator function $U_x(k)$ appearing in (1.2) is continuously differentiable with respect to k and the norm of $G(k) - \mathbf{I}$ does not depend on x. As mentioned in Remark 3.1, the coincidence of the sets of left and right indices of $G(k)$ for every $x \in \mathbf{R}^n$ is a direct consequence of the unitarity of this operator function. The x-independence of the sum index is due to the fact that $\widetilde{G}(\xi)$ depends continuously on x in $\mathcal{H}_\alpha(\mathbf{T}; \mathcal{L}(L^2(S^{n-1})))$ [cf. [**GL73**], Section 7]. ∎

The Riemann-Hilbert problem (3.3) can be solved in terms of the Wiener-Hopf factors of $G(k)$ as in the case of 3-D [**AV89a**] to obtain

$$X_+(k) = [G_+(k)^{-1} - \mathbf{I}]\hat{1} + G_+(k) \sum_{\rho_j > 0} \frac{\phi_j(k)}{(k+i)^{\rho_j}} \pi_j$$

and

$$X_-(k) = [G_-(k)^{-1} - \mathbf{I}]\hat{1} + G_-(k)^{-1} \sum_{\rho_j > 0} \frac{\phi_j(k)\pi_j + [(k+i)^{\rho_j} - (k-i)^{\rho_j}]P_j\hat{1}}{(k-i)^{\rho_j}}$$

provided $P_j\hat{1} = 0$ whenever $\rho_j < 0$. Here π_j is a fixed nonzero vector in the range of P_j and $\phi_j(k)$ is an arbitrary polynomial of degree less than ρ_j. Using the Schrödinger equation the potential is obtained as

$$V(x) = \frac{[\nabla_x^2 + 2ik\theta \cdot \nabla_x]X_+(k)}{1 + X_+(k)}$$

provided the right-hand side is independent of θ and k; as in the case of 3-D [**AV89a**], this θ- and k-independence is equivalent to the "miracle" condition of Newton [**Ne89b**].

From (1.1) and (1.2), we obtain

$$G(-k) = QG(k)^{-1}Q.$$

It can be shown that the factors $G_+(k)$ and $G_-(k)$ in (3.4) can be chosen in such a way that

$$G_\pm(-k) = QG_\mp(k)^{-1}Q.$$

For details we refer the reader to Remark 4.3 of [**AV89a**].

In the special case where $S(k)$ is a meromorphic function on $\mathbf{C}$ with only finitely many poles and zeros, we can obtain additional information on the partial indices of $G(k)$. If $\sigma_-(H) = \emptyset$ and $k = 0$ is not an exceptional point, these poles and zeros are nonreal. Using that $\lim_{k\to\pm\infty} \|S(k) - \mathbf{I}\| = 0$, we may represent $G(k)$ in the form [**BGK79**]

$$G(k) = E_x(k)\left[\mathbf{I} + C_x(k\mathbf{I} - A_x)^{-1}B_x\right]F_x(k), \tag{3.5}$$

where $E_x(k)$, $E_x(k)^{-1}$, $F_x(k)$ and $F_x(k)^{-1}$ are entire operator functions satisfying the identities $\lim_{k\to\pm\infty} \|E_x(k) - \mathbf{I}\| = 0$ and $\lim_{k\to\pm\infty} \|F_x(k) - \mathbf{I}\| = 0$ and A_x, B_x and C_x are bounded linear operators. In that case

$$G(k)^{-1} = F_x(k)^{-1}\left[\mathbf{I} - C_x(k\mathbf{I} - A_x^\times)^{-1}B_x\right]E_x(k)^{-1},$$

where $A_x^\times = A_x - B_xC_x$. One may choose A_x, B_x, and C_x in such a way that A_x acts on a finite-dimensional space $\mathcal{X}$, and B_x and C_x act between $L^2(S^{n-1})$ and $\mathcal{X}$, and that the spectra of A_x and $A_x^\times$ coincide with the sets of poles and zeros of $S(k)$, respectively. Moreover, the representation (3.5) may be chosen in such a way that it is minimal [i.e., $\cap_{i=0}^\infty \mathcal{N}(C_xA_x^i) = \{0\}$ and $\sum_{i=0}^\infty \mathcal{R}(A_x^iB_x) = \mathcal{X}$] and there exists a signature operator J [i.e., $J = J^\dagger = J^{-1}$] such that $A^\times$ coincides with $(-A)$ and iA is J-selfadjoint. Since a zero of $S(k)$ is a pole of $S(k)^{-1}$, the poles of $G(k)$ in $\mathbf{C}^+$ correspond exactly with the (negative) bound states and hence these poles are simple and located on the imaginary axis.

THEOREM 3.3. *Let $V(x) \in \mathbf{B}_\alpha$ for some $\alpha \in (0,2)$, and let $s > \frac{3}{2} - \frac{1}{n}$ be the constant such that multiplication by $(1+|x|^2)^s V(x)$ is a bounded linear operator from $H^\alpha(\mathbf{R}^n)$ into $L^2(\mathbf{R}^n)$. Suppose $\sigma_+(H) = \emptyset$ while $k=0$ is not an exceptional point. Finally, assume $S(k)$ extends to a meromorphic operator function on the entire complex plane with only finitely many poles and zeros. Then the partial indices of $G(k)$ are nonnegative.*

Proof: Let us construct the representation (3.5) with the properties mentioned above and let us suppress the x-dependence. If T is a bounded linear operator without real spectrum and $\Gamma_\pm$ is a simple, positively oriented rectifiable Jordan contour in $\mathbf{C}^\pm$ enclosing the spectrum of T in $\mathbf{C}^\pm$, we write $\mathcal{M}_\pm(T)$ for the ranges of the complementary bounded projections $\frac{1}{2\pi i}\int_{\Gamma_\pm} dz\ (z\mathbf{I}-T)^{-1}$. Then, according to the main result of [**BGK86**], the sum of the positive (resp. negative) indices ρ_j of $G(k)$ satisfies

$$\sum_{\pm\rho_j>0} |\rho_j| = \dim[\mathcal{M}_\mp(A)\cap\mathcal{M}_\pm(A^\times)].$$

Here we have used the fact that the sets of left and right partial indices of $G(k)$ coincide. If $^\perp$ denotes the orthogonal complement with respect to the indefinite scalar product $[\cdot,\cdot] = < J\cdot,\cdot >$, then

$$\sum_{\pm\rho_j>0} |\rho_j| = \dim\{\mathcal{M}_\mp(A)\cap\mathcal{M}_\mp(A)^\perp\},$$

which is the J-neutral part of $\mathcal{M}_\mp(A)$; i.e., the subspace of those vectors u of $\mathcal{M}_\mp(A)$ such that $[u,u]=0$. Now let $i\kappa_1,\cdots,i\kappa_q$ be the *different* positive imaginary eigenvalues of A. Then in terms of a J-orthogonal direct sum we have

$$\mathcal{M}_+(A) = \bigoplus_{j=1}^{q}\bigoplus_{l=1}^{\infty} \mathcal{N}(A-i\kappa_j)^l.$$

Since there are no generalized eigenvectors associated with these eigenvalues, we obtain

$$\mathcal{M}_+(A) = \bigoplus_{j=1}^{q} \mathcal{N}(A-i\kappa_j).$$

Hence,

$$\mathcal{M}_+(A)\cap\mathcal{M}_+(A)^\perp = \bigoplus_{j=1}^{q}\{\mathcal{N}(A-i\kappa_j)\cap\mathcal{R}(-A+i\kappa_j)\} = \{0\},$$

the last equality being clear from the absence of generalized eigenvectors. Thus, the sum of the negative indices is zero, and hence there are no negative indices, as claimed. ∎

Even if $S(k)$ does not have the analyticity properties of Theorem 3.3, it is possible to obtain expressions for the sum of the positive indices and the sum of the negative indices of $G(k)$. Let $\mathcal{X}$ be one of the Banach spaces $\mathcal{H}_\gamma(\mathbf{T};L^2(S^{n-1}))$ where $0<\gamma<1$, and let $\mathcal{X}^{\pm}$ be the subspaces of $\mathcal{X}$ consisting of those $F \in \mathcal{X}$ which extend to a function in $\mathcal{H}_{\gamma}(\mathbf{T}^{\pm} \cup \mathbf{T}; L^2(S^{n-1}))$ that are analytic on $\mathbf{T}^{\pm}$ and, in the case of $\mathcal{X}^-$, satisfy $F(\infty) = 0$. Then $\mathcal{X}^+ \oplus \mathcal{X}^- = \mathcal{X}$ ([**Mu46**], extended to the vector-valued case). Then (cf. [**AV89b**])

$$\sum_{\rho_j > 0} \rho_j = \dim\left\{\tilde{G}[\mathcal{X}^-] \cap \mathcal{X}^+\right\}, \quad -\sum_{\rho_j < 0} \rho_j = \dim\left\{\mathcal{X}^- \cap \tilde{G}[\mathcal{X}^+]\right\}. \tag{3.6}$$

Indeed, if $f_+ \in \tilde{G}[\mathcal{X}^-] \cap \mathcal{X}^+$, then $f_+ = \tilde{G} f_-$ for some $f_- \in \mathcal{X}^-$. Representing $G(k)$ as in (3.4), we obtain

$$\left[P_0 + \sum_{j=1}^{m} \xi^{\rho_j} P_j\right] \tilde{G}_-(\xi) f_-(\xi) = \tilde{G}_+(\xi)^{-1} f_+(\xi).$$

Premultiplication by P_0 gives $P_0 \tilde{G}_-(\xi) f_-(\xi) = P_0 \tilde{G}_+(\xi)^{-1} f_+(\xi)$; both sides vanish, because they belong to $\mathcal{X}^+ \cap \mathcal{X}^-$. Similarly, we get $\xi^{\rho_j} P_j \tilde{G}_-(\xi) f_-(\xi) = P_j \tilde{G}_+(\xi)^{-1} f_+(\xi)$. For $\rho_j < 0$ both sides vanish. For $\rho_j > 0$, however, Liouville's theorem implies that the two sides equal a scalar polynomial $\varphi_j(\xi)$ of degree at most $(\rho_j - 1)$ multiplied by P_j. But then, by adding the contributions of P_0 and the various P_j,

$$f_+(\xi) = \tilde{G}_+(\xi) \sum_{\rho_j > 0} \varphi_j(\xi) P_j$$

and hence the first identity in (3.6) follows. The second identity in (3.6) can be proven by employing a right Wiener-Hopf factorization of $G(k)$ and using the fact that the left and right indices of $G(k)$ coincide.

4. TRACE CLASS PROPERTIES OF THE SCATTERING OPERATOR

In this section we will prove that $S(k)$ is a trace class operator on $L^2(S^{n-1})$ and study its effect on its Wiener-Hopf factorization.

THEOREM 4.1. *Let $V(x) \in \mathbf{B}_\alpha$ for some $\alpha \in (0,1)$, and let $s > 1$ be the constant such that multiplication by $(1+|x|^2)^s V(x)$ is a bounded linear operator from $H^\alpha(\mathbf{R}^n)$ into $L^2(\mathbf{R}^n)$. Suppose $0 \neq k \notin \sigma_+(H)^{1/2}$ and $V(x) \in L^1(\mathbf{R}^n)$. Then $S(k) - \mathbf{I}$ is a trace class operator on $L^2(S^{n-1})$ and*

$$\lim_{k \to \pm\infty} k^{2-n} \mathrm{tr}[S(k) - \mathbf{I}] = 2i(2\pi)^{n-1} \Sigma_n < V >, \tag{4.1}$$

where $< V > = \int_{\mathbf{R}^n} dx\, V(x)$ and Σ_n is the surface area of S^{n-1}.

Proof: Observe that $V[H^\alpha_{-t}(\mathbf{R}^n)] \subset L^2_t(\mathbf{R}^n)$ for all $t \in [0, s]$, and put $V^{1/2} = |V|^{1/2}\mathrm{sgn}(V)$. Note that

$$\left|< u, v >_{H^\alpha_t(\mathbf{R}^n)}\right| \leq \|u\|_{H^{\alpha_1}_{t_1}(\mathbf{R}^n)} \|v\|_{H^{\alpha_2}_{t_2}(\mathbf{R}^n)},$$

where $\alpha_1 + \alpha_2 = 2\alpha$ and $t_1 + t_2 = 2t$. Hence, since $V(x) \in \mathbf{B}_\alpha \subset \mathbf{B}_{2\alpha}$, we have

$$\|V^{1/2} u\|^2_{H^\alpha(\mathbf{R}^n)} \leq \|Vu\|_{L^2_s(\mathbf{R}^n)} \|u\|_{H^{2\alpha}_{-s}(\mathbf{R}^n)}.$$

Furthermore,

$$\|V^{1/2}u\|^2_{H^{\frac{1}{2}\alpha}_{\frac{1}{2}s}(\mathbf{R}^n)} \leq \|Vu\|_{L^2_s(\mathbf{R}^n)}\|u\|_{H^\alpha(\mathbf{R}^n)}.$$

Since $\frac{1}{2}s > 1$, the following diagram consists of bounded linear operators:

$$H^\alpha(\mathbf{R}^n) \xrightarrow{|V|^{1/2}} H^{\frac{1}{2}\alpha}_{\frac{1}{2}s}(\mathbf{R}^n) \xrightarrow{\text{imbedding}} L^2_{\frac{1}{2}s}(\mathbf{R}^n) \xrightarrow{R_0^+(k^2)} H^{2\alpha}_{-s}(k^2) \xrightarrow{V^{1/2}} H^\alpha(\mathbf{R}^n).$$

Then $0 \neq k \notin \sigma_+(H)^{\frac{1}{2}}$ implies that $\left[\mathbf{I} + V^{1/2}R_0^+(k^2)|V|^{1/2}\right]$ is boundedly invertible on $H^\alpha(\mathbf{R}^n)$ and the following diagram is commutative:

$$\begin{array}{ccc} H^{2\alpha}_{-s}(\mathbf{R}^n) & \xrightarrow{[\mathbf{I}+R_0^+(k^2)V]^{-1}} & H^{2\alpha}_{-s}(\mathbf{R}^n) \\ {\scriptstyle V^{1/2}}\downarrow & & \downarrow{\scriptstyle V^{1/2}} \\ H^\alpha(\mathbf{R}^n) & \xrightarrow{[\mathbf{I}+V^{1/2}R_0^+(k^2)|V|^{1/2}]^{-1}} & H^\alpha(\mathbf{R}^n) \end{array}$$

Here we have used that $\alpha \in [0,1]$ so that $V(x) \in \mathbf{B}_{2\alpha} \in \mathbf{SR}$. Also, defining

$$\mathbf{S}(k) = \mathbf{I} - \frac{ik^{n-2}}{2(2\pi)^{n-1}}[\mathbf{I} + V^{1/2}R_0^+(k^2)|V|^{1/2}]^{-1}V^{1/2}\sigma^\dagger(k)\sigma(k)|V|^{1/2},$$

we have the commutative diagram

$$\begin{array}{ccccc} H^\alpha(\mathbf{R}^n) & \xrightarrow{\mathbf{S}(k)-\mathbf{I}} & H^\alpha(\mathbf{R}^n) & \xleftarrow{\frac{-ik^{n-2}}{2(2\pi)^{n-1}}N(k)} & H^\alpha(\mathbf{R}^n) \\ {\scriptstyle |V|^{1/2}}\downarrow & & & & \uparrow{\scriptstyle V^{1/2}} \\ L^2_{\frac{1}{2}s}(\mathbf{R}^n) & \xrightarrow{\sigma(k)} & L^2(S^{n-1}) & \xrightarrow{\sigma^\dagger(k)} & H^{2\alpha}_{-s}(\mathbf{R}^n) \end{array}$$

where

$$N(k) = [\mathbf{I} + V^{1/2}R_0^+(k^2)|V|^{1/2}]^{-1}.$$

A swift comparison with the diagram in the proof of Proposition 2.5 yields that

$$S(k) = \mathbf{I} + \sigma(k)|V|^{1/2}T(k), \quad \mathbf{S}(k) = \mathbf{I} + T(k)\sigma(k)|V|^{1/2},$$

for bounded operators $\sigma(k)|V|^{1/2} : H^\alpha(\mathbf{R}^n) \to L^2(S^{n-1})$ and $T(k) : L^2(S^{n-1}) \to H^\alpha(\mathbf{R}^n)$. As a result, the nonzero spectra of $S(k) - \mathbf{I}$ and $\mathbf{S}(k) - \mathbf{I}$ coincide. Since $S(k) - \mathbf{I}$ is also a normal compact operator, the nonzero spectra of $S(k) - \mathbf{I}$ and $\mathbf{S}(k) - \mathbf{I}$ consist of **the same** discrete set of eigenvalues without associated generalized eigenvectors. Even the multiplicities of the nonzero eigenvalues coincide. Hence, it suffices to prove that

$$\Sigma(k) = |V|^{1/2}\sigma^\dagger(k)\sigma(k)|V|^{1/2}$$

is a trace class operator on $H^\alpha(\mathbf{R}^n)$. This will immediately imply that

$$\mathbf{S}(k) - \mathbf{I} = \frac{-ik^{n-2}}{2(2\pi)^{n-1}}[\mathbf{I} + V^{1/2}R_0^+(k^2)|V|^{1/2}]^{-1}\mathrm{sgn}(V)\Sigma(k)$$

is a trace class operator on $H^\alpha(\mathbf{R}^n)$. The approximation numbers $\{s_n(\mathbf{S}(k)-\mathbf{I})\}_{n=1}^\infty$ [i.e. the non-increasing sequence of eigenvalues of $\{[\mathbf{S}(k)-\mathbf{I}]^\dagger[\mathbf{S}(k)-\mathbf{I}]\}^{\frac{1}{2}}$ (cf. [**GK65**])] form a sequence in ℓ^1. As a result, the eigenvalues $\{\lambda_n(k)\}_{n=1}^\infty$ of $S(k)-\mathbf{I}$ [or of $\mathbf{S}(k)-\mathbf{I}$] satisfy

$$\sum_{n=1}^\infty |\lambda_n(k)| \le \sum_{n=1}^\infty s_n(\mathbf{S}(k)-\mathbf{I}) \le \|[\mathbf{I}+V^{1/2}R_0^+(k^2)|V|^{1/2}]^{-1}\| \sum_{n=1}^\infty s_n(\Sigma(k)) < +\infty.$$

However, due to the unitarity of $S(k)$ the operator $S(k)-\mathbf{I}$ does not have a Volterra part and hence the trace norm of $S(k)-\mathbf{I}$ satisfies

$$\|S(k)-\mathbf{I}\|_{S_1} = \sum_{n=1}^\infty |\lambda_n(k)| \le \|[\mathbf{I}+V^{1/2}R_0^+(k^2)|V|^{1/2}]^{-1}\|\|\Sigma(k)\|_{S_1} < +\infty,$$

which proves $S(k)-\mathbf{I}$ to be a trace class operator on $L^2(S^{n-1})$.

Note that the kernel of the integral operator $\Sigma(k)$ is given by

$$\Sigma(k;x,y) = \int_{S^{n-1}} d\theta\ e^{ik\theta\cdot(x-y)}|V(x)V(y)|^{\frac{1}{2}}.$$

Now let us first consider the case $\alpha = 0$ with $V(x)\in\mathbf{B}_0$. Then $\Sigma(k)$ is a positive self-adjoint operator on $L^2(\mathbf{R}^n)$, the space that takes the place of $H^\alpha(\mathbf{R}^n)$. Thus, if $V(x)\in L^1(\mathbf{R}^n)$ and $V(x)$ is continuous on $\mathbf{R}^n$, $\Sigma(k)$ is a trace-class operator and

$$\mathrm{tr}(\mathrm{sgn}(V)\Sigma(k)) = \int_{\mathbf{R}^n} dx\ \mathrm{sgn}(V(x))\Sigma(k;x,x) = \Sigma_n\int_{\mathbf{R}^n} dx\ V(x),$$

where Σ_n is the surface area of S^{n-1}. If $V(x)\in L^1(\mathbf{R}^n)$ and $V(x)$ is not necessarily continuous on $\mathbf{R}^n$, we put for every $h>0$

$$V_h(x) = \frac{1}{V_n}h^{-n}\int_{|x-y|\le h} dy\ V(y),$$

where V_n is the volume of the unit sphere in $\mathbf{R}^n$. Then $\|V_h\|_1\le\|V\|_1$, $\lim_{h\to 0}\|V_h-V\|_1 = 0$ and V_h is a bounded continuous function on $\mathbf{R}^n$. If the original $V(x)\in\mathbf{B}_0$ and $s>1$ is the constant such that multiplication by $(1+|x|^2)^sV(x)$ is a bounded operator from $L^2_{-s}(\mathbf{R}^n)$, the space that stands for $H^\alpha_{-s}(\mathbf{R}^n)$ if $\alpha=0$, into $L^2_s(\mathbf{R}^n)$, then $|V(x)|\le C(1+|x|^2)^{-s}$ for some constant C. Using the estimates $\frac{1+|x|^2}{1+|y|^2}\le 1+|x|^2\le 1+h^2\le(h+1)^2$ if $|x|\le h$ and $\frac{1+|x|^2}{1+|y|^2}\le 1+|x|^2\le\frac{1+(h+1)^2}{2}$ if $|x|\ge h$, we obtain

$$|V_h(x)| \le \frac{C}{(1+|x|^2)^s}\frac{(h+1)^{2s}}{V_nh^n}\int_{|x-y|\le h} dy = \frac{C(h+1)^{2s}}{(1+|x|^2)^s},$$

so that $V_h(x)$ belongs to $\mathbf{B}_0$ if $V(x)$ does. However, then $S(k) - \mathbf{I}$ with $V(x)$ replaced by $V_h(x)$ is trace class with a trace norm which is $O(1)$ as $h \to 0$. Hence, $S(k) - \mathbf{I}$ is trace class for the original $V(x)$ and the trace of $S(k) - \mathbf{I}$ with the original $V(x)$ is obtained from the trace of $S(k) - \mathbf{I}$ with $V(x)$ replaced by $V_h(x)$ by taking $h \to 0$. On the other hand, since $\mathrm{tr}(\mathrm{sgn}(V)\Sigma(k)) = \Sigma_n < V >$ for continuous $V(x)$, this must also be the case for discontinuous $V \in L^1(\mathbf{R}^n)$. From the special form of $\mathbf{S}(k) - \mathbf{I}$, the fact that its trace is the sum of its eigenvalues and its eigenvalues coincide with those of $S(k) - \mathbf{I}$, we eventually get (4.1), where we have also used that $[\mathbf{I} + V^{1/2} R_0^+(k^2)|V|^{1/2}]^{-1}$ approaches $\mathbf{I}$ in the norm as $k \to \pm\infty$.

Next, consider arbitrary $\alpha \in [0,1)$, but $V(x) \in \mathbf{B}_0$. Then a simple compactness argument yields that $S(k) - \mathbf{I}$ has the same eigenvalues with the same multiplicities, and hence the same trace, as an operator on either $L^2(\mathbf{R}^n)$ or $H^\alpha(\mathbf{R}^n)$. Thus (4.1) is immediate. More generally, if $V(x) \in \mathbf{B}_\alpha \cap L^1(\mathbf{R}^n)$ for some $\alpha \in [0,1)$, we can always approximate it by potentials in $\mathbf{B}_0 \cap L^1(\mathbf{R}^n)$ in the L^1-norm. Then the expression for the trace of $\mathrm{sgn}(V)\Sigma(k)$ will extend to these more general potentials and hence (4.1) will apply to them. ∎

For $V(x) \in \mathbf{B}_0$ we simply have the diagram of bounded operators

$$L^2_{-s}(\mathbf{R}^n) \xrightarrow{V^{\frac{1}{2}}} L^2(\mathbf{R}^n) \xrightarrow{|V|^{1/2}} L^2_s(\mathbf{R}^n)$$

with $\sigma(k) : L^2_s(\mathbf{R}^n) \to L^2(S^{n-1})$, $\sigma^\dagger(k) : L^2(S^{n-1}) \to L^2_{-s}(\mathbf{R}^n)$ and $R_0^+(k^2) : L^2_s(\mathbf{R}^n) \to L^2_{-s}(\mathbf{R}^n)$ bounded. Hence, in that case Theorem 4.1 is valid if the constant s satisfies $s > \frac{1}{2}$.

If, in addition to the hypotheses of Theorem 4.1, zero is not an exceptional point, the trace norm of $S(k) - \mathbf{I}$ is easily seen to be $O(k^{n-2})$ as $k \to 0$, due to the boundedness of $[\mathbf{I} + R_0^+(k^2)]^{-1}$ on a neighborhood of $k = 0$. Hence, in that case $S(0) - \mathbf{I}$ is a trace class operator on $L^2(S^{n-1})$ if $n \geq 3$. For $n = 2$ and zero not an exceptional point, $s > 1$ implies that $\sigma(0)$ and $\sigma^\dagger(0)$ are bounded operators between suitable spaces [cf. (A.3)]. Hence, we may then repeat the entire proof of Theorem 4.1 and prove that $S(0) - \mathbf{I}$ is trace class on $L^2(S^{n-1})$ if $n = 2$ and if zero is not an exceptional point. Summarizing, if zero is not an exceptional point, $S(0) - \mathbf{I}$ is trace class on $L^2(S^{n-1})$ if $V(x) \in \mathbf{B}_\alpha$ for some $\alpha \in [0,1)$ and $s > 1$ is the constant such that multiplication by $(1 + |x|^2)^s V(x)$ is a bounded operator from $H^\alpha_{-s}(\mathbf{R}^n)$ into $L^2_s(\mathbf{R}^n)$.

APPENDIX A: NORM ESTIMATES FOR $\sigma^\dagger(k)$

From (2.4) we have $(\sigma^\dagger(k)g)(x) = (\sigma^\dagger(1)g)(kx)$. Letting $y = kx$ and using the identity $\frac{1+|y|^2}{k^2+|y|^2} \leq \max(1, \frac{1}{k^2})$, from (2.4) we obtain

$$\|\sigma^\dagger(k)g\|^2_{-s} \leq |k|^{2s-n} \max(1, |k|^{-2s}) \|\sigma^\dagger(1)g\|^2_{-s}. \tag{A.1}$$

where the norm $\|\cdot\|_{-s}$ is the norm defined in (1.3). From the paragraph following the proof of Theorem 4.1, we have $\sigma^\dagger(1) \in \mathcal{L}(L^2(S^{n-1}); L^2_s(\mathbf{R}^n))$ for $s > \frac{1}{2}$. Hence, from (A.1) and the definition of $C_{k,s}$ given in (2.5), it follows that

$$C_{k,s} \leq C_{1,s} \max\left(|k|^{s-\frac{1}{2}n}, |k|^{-\frac{1}{2}n}\right). \tag{A.2}$$

Note also that from (1.3) and (2.4) we obtain, for $s > \frac{1}{2}n$,

$$\|\sigma^\dagger(k)g\|_{-s}^2 \leq \int_{\mathbf{R}^n} \frac{dx}{(1+|x|^2)^s}\Big(\int_{S^{n-1}} d\theta\, |g(\theta)|\Big)^2 \leq (\Sigma_n)^2 \int_0^\infty dr\, \frac{r^{n-1}}{(1+r^2)^s}\|g\|_{L^2(S^{n-1})}^2,$$

where Σ_n is the surface area of S^{n-1}. Hence, a comparison with (2.5) shows that

$$C_{k,s} \leq \Sigma_n\Big[\int_0^\infty dr\, \frac{r^{n-1}}{(1+r^2)^s}\Big]^{\frac{1}{2}}, \quad s > \frac{1}{2}n. \tag{A.3}$$

It is possible to improve the estimates in (A.2) and (A.3) as follows. Using (2.6) we obtain

$$C_{k,s} \leq C\max(|k|^{\epsilon(s_1-\frac{1}{2}n)}, |k|^{-\frac{1}{2}n\epsilon})$$

for some constant C if $s = \epsilon s_1 + (1-\epsilon)s_2$ with $\epsilon \in [0,1]$, $s_1 > \frac{1}{2}$ and $s_2 > \frac{1}{2}n$, which restricts ϵ to $\frac{n-2s}{n-1} < \epsilon \leq 1$. Maximizing $-\frac{1}{2}n\epsilon$ and minimizing $\epsilon(s_1 - \frac{1}{2}n)$ under these constraints we get

$$C_{k,s} = \begin{cases} O(|k|^{\min(0,s-\frac{1}{2}n)}) & (k \to \pm\infty) \\ O(|k|^{-\frac{n(n-2s)}{2(n-1)}+\delta}) & (k \to 0, \forall\delta > 0), \end{cases} \tag{A.4}$$

which for $k \to \pm\infty$ corresponds with [**We90**].

LITERATURE

[**Ag75**] S. Agmon, *Spectral Properties of Schrödinger Operators and Scattering Theory*, Ann. Scuola Norm. Sup. Pisa **2**, 151-218 (1975).

[**AV89a**] T. Aktosun and C. van der Mee, *Solution of the Inverse Scattering Problem for the 3-D Schrödinger Equation by Wiener-Hopf Factorization of the Scattering Operator*, J. Math. Phys., to appear.

[**AV89b**] T. Aktosun and C. van der Mee, *Solution of the Inverse Scattering Problem for the 3-D Schrödinger Equation using a Fredholm Integral Equation*, Preprint.

[**BGK79**] H. Bart, I. Gohberg and M.A. Kaashoek, *Minimal Factorization of Matrix and Operator Functions*, Birkhäuser OT **1**, Basel and Boston, 1979.

[**BGK86**] H. Bart, I. Gohberg and M.A. Kaashoek, *Explicit Wiener-Hopf Factorization and Realization.* In: I. Gohberg and M.A. Kaashoek, *Constructive Methods of Wiener-Hopf Factorization*, Birkhäuser OT **21**, Basel and Boston, 1986, pp. 235-316.

[**BC85**] R. Beals and R.R. Coifman, *Multidimensional Inverse Scattering and Nonlinear P.D.E.'s*, Proc. Symp. Pure Math. **43**, 45-70 (1985).

[**BC86**] R. Beals and R.R. Coifman, *The D-bar Approach to Inverse Scattering and Nonlinear Evolutions*, Physica D **18**, 242-249 (1986).

[**CS89**] K. Chadan and P.C. Sabatier, *Inverse Problems in Quantum Scattering Theory*, Second Edition, Springer, New York, 1989.

[**Fa65**] L.D. Faddeev, *Increasing Solutions of the Schrödinger Equation*, Sov. Phys. Dokl. **10**, 1033-1035 (1965) [Dokl. Akad. Nauk SSSR **165**, 514-517 (1965) (Russian)].

[**Fa74**] L.D. Faddeev, *Three-dimensional Inverse Problem in the Quantum Theory of Scattering*, J. Sov. Math. **5**, 334-396 (1976) [Itogi Nauki i Tekhniki **3**, 93-180 (1974) (Russian)].

[**Go64**] I.C. Gohberg, *The Factorization Problem for Operator Functions*, Amer. Math. Soc. Transl., Series **2**, **49**, 130-161 (1966) [Izvestiya Akad. Nauk SSSR, Ser. Matem., **28**, 1055-1082 (1964) (Russian)].

[**GK65**] I.C. Gohberg and M.G. Krein, *Introduction to the Theory of Linear Nonselfadjoint Operators*, Transl. Math. Monographs, Vol. **18**, A.M.S., Providence, 1969 [Nauka, Moscow, 1965 (Russian)].

[**GL73**] I.C. Gohberg and J. Leiterer, *Factorization of Operator Functions with respect to a Contour. III. Factorization in Algebras*, Math. Nachrichten **55**, 33-61 (1973) (Russian).

[**Ka59**] T. Kato, *Growth Properties of Solutions of the Reduced Wave Equation with a Variable Coefficient*, Comm. Pure Appl. Math. **12**, 403-425 (1959).

[**Ku80**] S. Kuroda, *An Introduction to Scattering Theory*, Lecture Notes Series, Vol. **51**, Math. Inst., Univ. of Aarhus, 1980.

[**Mu46**] N.I. Muskhelishvili, *Singular Integral Equations*, Noordhoff, Groningen, 1953 [Nauka, Moscow, 1946 (Russian)].

[**NA84**] A.I. Nachman and M.J. Ablowitz, *A Multidimensional Inverse Scattering Method*, Studies in Appl. Math. **71**, 243-250 (1984).

[**Ne74**] R.G. Newton, *The Gel'fand-Levitan Method in the Inverse Scattering Problem in Quantum Mechanics*. In: J.A. Lavita and J.-P. Marchand (Eds.), *Scattering Theory in Mathematical Physics*, Reidel, Dordrecht, 1974, pp. 193-225.

[**Ne80**] R.G. Newton, *Inverse Scattering. II. Three Dimensions*, J. Math. Phys. **21**, 1698-1715 (1980); **22**, 631 (1981); **23**, 693 (1982).

[**Ne81**] R.G. Newton, *Inverse Scattering. III. Three Dimensions, Continued*, J. Math. Phys. **22**, 2191-2200 (1981); **23**, 693 (1982).

[**Ne82**] R.G. Newton, *Inverse Scattering. IV. Three Dimensions: Generalized Marchenko Construction with Bound States*, J. Math. Phys. **23**, 2257-2265 (1982).

[**Ne85**] R.G. Newton, *A Faddeev-Marchenko Method for Inverse Scattering in Three Dimensions*, Inverse Problems **1**, 371-380 (1985).

[**Ne89a**] R.G. Newton, *Eigenvalues of the S-matrix*, Phys. Rev. Lett. **62**, 1811-1812 (1989).

[**Ne89b**] R.G. Newton, *Inverse Schrödinger Scattering in Three Dimensions*, Springer, New York, 1989.

[**NH87**] R.G. Novikov and G.M. Henkin, *Solution of a Multidimensional Inverse Scattering Problem on the Basis of Generalized Dispersion Relations*, Sov. Math. Dokl. **35**, 153-157 (1987) [Dokl. Akad. Nauk SSSR **292**, 814-818 (1987) (Russian)].

[**Pr69**] R.T. Prosser, *Formal Solution of Inverse Scattering Problems*, J. Math. Phys. **10**, 1819-1822 (1969).

[**Pr76**] R.T. Prosser, *Formal Solution of Inverse Scattering Problems. II*, J. Math. Phys. **17**, 1775-1779 (1976).

[**Pr80**] R.T. Prosser, *Formal Solution of Inverse Scattering Problems. III*, J. Math. Phys. **21**, 2648-2653 (1980).

[**Pr82**] R.T. Prosser, *Formal Solution of Inverse Scattering Problems. IV*, J. Math. Phys. **23**, 2127-2130 (1982).

[**Sc71**] M. Schechter, *Spectra of Partial Differential Operators*, North-Holland, Amsterdam, 1971.

[**We90**] R. Weder, *Multidimensional Inverse Scattering Theory*, Inverse Problems, in press (April 1990).

Tuncay Aktosun
Dept. of Mathematical Sciences
University of Texas at Dallas
Richardson, TX 75083

Cornelis van der Mee
Dept. of Mathematical Sciences
University of Delaware
Newark, DE 19716

Operator Theory:
Advances and Applications, Vol. 50

EXISTENCE AND REGULARITY OF SOLUTIONS OF CAUCHY PROBLEMS FOR INHOMOGENEOUS WAVE EQUATIONS WITH INTERACTION[1]

Felix ALI MEHMETI

0 INTRODUCTION

The main aim of this paper is a nonrecursive formula for the compatibility conditions ensuring the regularity of solutions of abstract inhomogeneous linear wave equations, which we derive using the theory of T. Kato [11]. We apply it to interaction problems for wave equations (cf. [3]), generalizing regularity results of Lions-Magenes [12].

Important special cases of interaction problems are transmission problems on ramified spaces (i.e. families of domains, where certain parts of the boundaries are identified), which have been treated by G. Lumer [13], S. Nicaise [14], J.v. Below [5] (mainly for parabolic equations) and B. Gramsch [9] and in many other contributions of these authors. The hyperbolic case has been investigated e.g. in [1], ..., [4]. There, interaction problems are introduced as natural extensions of transmission problems.

To explain the nature of this extension, let us consider first the following classical transmission problem for the wave equation on two half planes $\Omega_1 := \{x \in I\!R^n : x_n > 0\}$, $\Omega_2 := \{x \in I\!R^n : x_n < 0\}$, and the interface $\Sigma := \{x \in I\!R^n : x_n = 0\}$:
find $u_i : [0,T] \times \Omega_i \to I\!R$, $i = 1,2$ with

$$\begin{array}{ll} (W) & \partial_t^2 u_i(t,x) - a_i \Delta u_i(t,x) + c_i u_i(t,x) = \theta_i(t,x) \ \forall t,x \\ (I) & u_i(0,x) = x_{0,i}(x),\ \partial_t u_i(0,x) = y_{0,i}(x) \ \forall x \\ (T_0) & u_1(t,\cdot)\big|_\Sigma = u_2(t,\cdot)\big|_\Sigma \ \forall t \\ (T_1) & a_1 \partial_n u_1(t,\cdot)\big|_\Sigma = a_2 \partial_n u_2(t,\cdot)\big|_\Sigma \ \forall t\ , \end{array} \tag{1}$$

[1]This research was supported by the Deutsche Forschungsgemeinschaft under grant Al 289/1-1

where $a_i, c_i > 0$ are constants and $\partial_n := \frac{\partial}{\partial x_n}$. This is a model for scalar wave propagation in the Ω_i with reflection and refraction phenomena along Σ.

(1) can be reformulated abstractly (cf. [3]):

$$\begin{cases} \ddot{x}(t) + Ax(t) = \theta(t) \ \forall t \in [0,T] \\ x(0) = x_0, \quad \dot{x}(0) = y_0 \\ x(t) \in D(A) \ \forall t \in [0,T] \end{cases} \tag{2}$$

where $x(t) := (u_i(t,\cdot))_{i=1,2}$ and $A : D(A) \to H$ is a selfadjoint operator in the Hilbert space $H := \prod_{i=1}^2 L^2(\Omega_i)$ with

$$D(A^{1/2}) = V := \{(u_i) \in \prod_{i=1}^{2} H^1(\Omega_i) : (u_i) \text{ satisfies } (T_0)\} \ .$$

The formalism works even if V is any closed subspace of $\prod_{i=1}^2 H^1(\Omega_i)$ (cf. [3], A is then the socalled interaction operator, V the interaction space). For example replace (T_0) by

$$(I_0) \qquad (F_1 u_1(t,\cdot))\Big|_\Sigma = (F_2 u_2(t,\cdot))\Big|_\Sigma$$

where F_i are suitable Fourier integral operators on Ω_i which can be restricted to Σ and are continuous on $\prod_{i=1}^2 H^1(\Omega_i)$ (cf. [7]). (2) corresponds then to (1), where (T_0) is replaced by (I_0) and (T_1) by

$$(I_1) \qquad (F_1 a_1 \partial_n u_1(t,\cdot))\Big|_\Sigma = (F_2 a_2 \partial_n u_2(t,\cdot))\Big|_\Sigma \ .$$

Such a system arises naturally as result of a coordinate transformation of a transmission problem or as a model of somehow 'active' interfaces (with modified laws of reflection and refraction).

Semigroup theory gives regular solutions if $\theta \equiv 0$ and $x_0, y_0 \in D(A^\infty)$. If θ is not identically zero, the situation is more complicated. In 2.3 we transform (2) into a homogeneous time dependent evolution equation (following [11]). In 1.2 we recall the recursive formula given in [11] for the general case, which describes the compatibility conditions ensuring the regularity of the solution.

As a main concern of this paper, we derive from 1.2 a nonrecursive formula for the compatibility conditions for abstract wave equations (theorem 1.4) and apply it to wave equations with interaction (sections 2 and 3). Specializing to the Dirichlet problem for a wave equation on a single domain with a smooth boundary, we get an extension of the regularity result in Lions-Magenes [12].

These results are useful for the study of propagation of singularities for certain interaction problems following [15] and [10].

Some straightforward calculations have been omitted in the following.

I am grateful to Prof. B. Gramsch for valuable support. Further I thank Dr.

S. Nicaise for valuable discussions.

1 A NONRECURSIVE FORMULA FOR THE COMPATIBILITY CONDITIONS FOR CERTAIN EVOLUTION EQUATIONS

Let us recall a result of T.Kato on time dependent evolution equations in [11]:

ASSUMPTIONS 1.1

(i) Consider a fixed $m \in I\!N$ and a scale of Banach spaces $(X_j, |\cdot|_j)$, $j = 0, \ldots, m$ with $X_m \hookrightarrow X_{m-1} \hookrightarrow \ldots \hookrightarrow X_0 = X$ and $|\cdot|_0 \leq |\cdot|_1 \leq \ldots \leq |\cdot|_m$.
Let $Y \subseteq X_0$ be a closed subspace of X_1 and $Y_j := Y \cap X_j$, $j = 1, \ldots, m$.

(ii) Let $T, M, \beta, \nu > 0$ be fixed and $\mathcal{A} := \{\mathcal{A}(t) : D(\mathcal{A}(t)) \longrightarrow X : 0 \leq t \leq T\}$ be a family of generators of strongly continuous semigroups in X with the properties (as in theorem 2.13 in [11]):

(P1) Stability: $\mathcal{A} \in \mathcal{G}(X, M, \beta)^2$

(P2) Smoothness:
$D(\mathcal{A}(t)) \cap X_1 = Y \;\; \forall t \in [0,T]$ and
$$d^k\mathcal{A} \in L^\infty_*([0,T], \mathcal{L}(Y_{j+k}, X_j)) \quad \text{for } 0 \leq j \leq m-k,\ 1 \leq k \leq m.$$
($d^k\mathcal{A}$ means the $k-th$ derivative of the operator valued function $\mathcal{A}(\cdot)$ on $[0,T]$.
$L^\infty_*([0,T], \mathcal{L}(M,N))$ means the space of essentially bounded, strongly measurable functions $\mathcal{A} : [0,T] \rightarrow \mathcal{L}(M,N)$.)

(P3) Ellipticity: $\Phi \in D(\mathcal{A}(t))$ and $\mathcal{A}(t)\Phi \in X_j \implies \Phi \in X_{j+1}$ and
$|\Phi|_{j+1} \leq \nu(|\mathcal{A}(t)\Phi|_j + |\Phi|_0) \;\; \forall t \in [0,T],\ 0 \leq j \leq m-1$
(ν is called constant of ellipticity).

In 2.15 in [11], T. Kato gives the following recursion formula for the compatibility conditions ensuring the regularity of evolution equations associated to $\mathcal{A}$:

DEFINITION 1.2 For $j = 0, \ldots, m$ and $t \in [0,T]$ we define recursively $S^j(t) : D^j(t) \longrightarrow X$ by $D^0(t) := X$, $S^0(t) := Id$ and
$$D^{j+1}(t) := \{\Phi \in D^j(t) : S^k(t)\Phi \in Y_{j+1-k};\ 0 \leq k \leq j\},$$
$$S^{j+1}(t)\Phi := -\sum_{k=0}^{j} \binom{j}{k} (d^{j-k}\mathcal{A}(t)) S^k(t)\Phi, \qquad \forall \Phi \in D^{j+1}(t)$$

[2] i.e. $\mathcal{A}$ is a stable family of generators $\mathcal{A}(t) \in G(X, M, \beta)$ of strongly continuous semigroups; for the notion of stability cf. [11], section 1.

(this expression makes sense due to assumption (P1) and (P2); note that $D^j(t) = D(\mathcal{A}^j)\ \forall t \in [0,T]$, if $\mathcal{A}(t) =$ const).

Then holds the following statement (theorem 2.13 in [11]):

THEOREM 1.3 *Consider a family of generators $\{\mathcal{A}(t) : 0 \leq t \leq T\}$ satisfying (P1), (P2) and (P3).*
Conclusion:
For $j \in \{1, \ldots, m\}$ we have:
for all $\Phi \in D^j(0)$ exists exactly one solution $u(\cdot)$ of

$$\begin{cases} \dot{u}(t) \ + \ \mathcal{A}(t)u(t) \ = \ 0 \ \ \forall t \geq 0 \\ u(0) \ = \ \Phi \end{cases} \tag{3}$$

with

$$\begin{cases} d^k u \ \in \ C([s,T], Y_{j-k}), \quad \textit{für}\ 0 \leq k \leq j-1 \\ d^j u \ \in \ C([s,T], X). \\ u(t) \ \in D^j(t)\ \forall t \in [0,T] \ . \end{cases} \tag{4}$$

In applications to inhomogeneous wave-equations appears the following special case, where we can give a nonrecursive formula for the $S^k(t)$:

THEOREM 1.4 *Assumptions:*
(i) Assume for every $t \in [0,T]$ the existence of a (continuous) operator $\mathcal{A}_f(t) : X_1 \to X_0$ coinciding with $\mathcal{A}(t)$ on $D(\mathcal{A}(t)) \cap X_1 \ = \ Y_1$ and

$$d^k \mathcal{A}_f(\cdot) \in L^\infty_*([0,T], \mathcal{L}(X_{j+k}, X_j))\ \textit{for}\ 0 \ \leq \ j \ \leq m-k; \ 1 \ \leq \ k \ \leq m;$$

$$\mathcal{A}_f(\cdot) \in L^\infty_*([0,T], \mathcal{L}(X_{j+1}, X_j))\ \textit{for}\ 0 \ \leq \ j \ \leq m-1 \ .$$

(ii) Assume $d^{l_1}\mathcal{A}_f(t) \cdot S^{l_2}(t) \ = \ 0 \ \ \forall l_1, l_2 \geq 1.$
Conclusion:

$$S^{j+1}(t)\Phi \ = \ \Big(\sum_{k=0}^{j-1}(-1)^{k+1}\mathcal{A}_f^k(t) d^{j-k}\mathcal{A}_f(t) \ + \ (-1)^{j+1}\mathcal{A}_f^{j+1}(t)\Big)\Phi \tag{5}$$

$$\forall t \in [0,T], j \geq 1\ \forall \Phi \in D^{j+1}(t) \subseteq X_{j+1} \ .$$

REMARK 1.5

(i) Condition (ii) in theorem 1.4 is satisfied for certain wave equations, where $\mathcal{A}_f(t)$ and $S^j(t)$ are 2×2 matrices of operators.

(ii) For $\Phi \in X_{j+1} \backslash D^{j+1}(t)$ we define $S^{j+1}(t)\Phi$ by the expression above.

(iii) The evaluation of the recursive definition of $S^j(t)$ leading to (5) is justified by assumption 1.4 (i). If $\Phi \in D^j(t)$, Φ is in general not in the domain of $\mathcal{A}(t)$ or its powers.

2 INHOMOGENEOUS WAVE EQUATIONS

In this section we consider Cauchy problems for abstract wave equations of the following form:

$$\begin{cases} \ddot{x}(t) + Ax(t) = \theta(t) \ \forall t \in [0,T] \\ x(0) = x_0, \quad \dot{x}(0) = y_0 \\ x(t) \in D(A) \ \ \forall t \in [0,T] \end{cases} \tag{6}$$

where $x : [0,T] \to H$ is the unknown function with values in a Hilbert space H. We choose the following setting:

ASSUMPTIONS 2.1 Consider

(i) Hilbert spaces H, V with $V \hookrightarrow H$;

(ii) a selfadjoint operator $A : D(A) \to H$ with $D(A) \subseteq V$ and $D(A^{1/2}) = V$;

(iii) a scale of Hilbert spaces

$$H = H_0 \hookleftarrow H_1 \hookleftarrow \ldots \hookleftarrow H_{m+1}$$

such that $A : D(A) \cap H_{j+2} \to H_j$ is continuous for $j = 1, \ldots, m-1$.

(iv) the condition

$$x \in D(A); Ax \in H_j \Rightarrow$$

$$x \in H_{j+2} \text{ and } \mid x \mid_{H_{j+2}} \leq \gamma(\mid Ax \mid_{H_j} + \mid x \mid_{H_0})$$

with some $\gamma > 0$, $j = 1, \ldots, m-1$;

(v) a function $\theta : [0,T] \to H$ with $\theta \in L^\infty([0,T], H_j)$, $j = 1, \ldots, m+1$;

(vi) the existence of a continuous operator $A_f : H_2 \to H_0$ coinciding with A on $D(A) \cap H_2$, having continuous restrictions

$$A_f : H_{j+2} \to H_j \ , \ j = 1, \ldots, m-1 \ ;$$

(vii) initial conditions $x_0 \in H_{m+1}$; $y_0 \in H_m$.

Now we can establish the situation of theorem 1.4 making the following choices:

DEFINITION 2.2

(i) $\mathcal{X} := \begin{pmatrix} V \\ H \end{pmatrix}$; $\mathcal{X}_j := \begin{pmatrix} V \cap H_{j+1} \\ H_j \end{pmatrix}$, $j = 1, \ldots, m$;

$D(\mathcal{A}_{hom}) := \begin{pmatrix} D(A) \\ V \end{pmatrix}$; $\mathcal{Y} := D(\mathcal{A}_{hom}) \cap \mathcal{X}_1$; $\mathcal{Y}_j := \mathcal{Y} \cap \mathcal{X}_j$, $j = 1, \ldots, m$;

$\mathcal{A}_{hom} : D(\mathcal{A}_{hom}) \to \mathcal{X}$ with $\mathcal{A}_{hom} := \begin{pmatrix} 0 & -Id \\ A & 0 \end{pmatrix}$;

$\Theta(t) := \begin{pmatrix} 0 \\ \theta(t) \end{pmatrix}$, $t \in [0,T]$;

$\mathcal{A}_{hom,f} : \mathcal{X}_{j+1} \to \mathcal{X}_j$ with $\mathcal{A}_{hom,f} := \begin{pmatrix} 0 & -Id \\ A_f & 0 \end{pmatrix}$, $j = 1, \ldots, m-1$.

(ii) $X := \begin{pmatrix} \mathcal{X} \\ I\!R \end{pmatrix}$; $X_j := \begin{pmatrix} \mathcal{X}_j \\ I\!R \end{pmatrix}$, $j = 1, \ldots, m$; $D(\mathcal{A}) := \begin{pmatrix} D(\mathcal{A}_{hom}) \\ I\!R \end{pmatrix}$;

$Y_j := \begin{pmatrix} \mathcal{Y}_j \\ I\!R \end{pmatrix}$, $j = 1, \ldots, m$;

$\mathcal{A}(t) : D(\mathcal{A}) \to X$ with $\mathcal{A}(t) = \begin{pmatrix} \mathcal{A}_{hom} & -\Theta(t) \\ 0 & 0 \end{pmatrix}$, $t \in [0,T]$;

$\mathcal{A}_f(t) : X_{j+1} \to X_j$ with $\mathcal{A}_f(t) = \begin{pmatrix} \mathcal{A}_{hom,f} & -\Theta(t) \\ 0 & 0 \end{pmatrix}$, $t \in [0,T], j = 1, \ldots, m-1$.

LEMMA 2.3

(i) With

$$\underline{u}(t) := \begin{pmatrix} x(t) \\ \dot{x}(t) \end{pmatrix}, \quad u(t) := \begin{pmatrix} \underline{u}(t) \\ k(t) \end{pmatrix}$$

(where $k : I\!R \to I\!R$),

$$\underline{\phi} := \begin{pmatrix} x_0 \\ y_0 \end{pmatrix}, \quad \Phi := \begin{pmatrix} \underline{\phi} \\ \mathbf{1} \end{pmatrix}$$

(where $\mathbf{1}(t) := 1 \ \forall t \in I\!R$), (6) is equivalent to

$$\begin{cases} \dot{\underline{u}}(t) + \mathcal{A}_{hom}\,\underline{u}(t) = \Theta(t); \quad \dot{k}(t) = 0 \ \forall t \in [0,T] \\ \underline{u}(0) = \underline{\phi}, \ k(0) = 1 \\ \underline{u}(t) \in D(\mathcal{A}_{hom}) \ \forall t \in [0,T] \end{cases} \tag{7}$$

*and to (3) **together** with $u(t) \in D_j(t) \ \ \forall t \in [0,T]$, i.e. the situation in Kato's theory. The reduction of (7) to (3) is suggested in [11][3].*

(ii) The hypotheses of theorem 1.3 are satisfied. Therefore, if $\Phi \in D^j(0)$, we have existence and uniqueness of a solution of (3) with the present notations, with the regularity property (4).

REMARK 2.4 Hypothesis (P1) in theorem 1.3 follows from $\mathcal{A}_{hom} \in G(\mathcal{X}, 1, 0)$ (cf. [1] or [2]),
$\mathcal{A}(t) = \begin{pmatrix} \mathcal{A}_{hom} & 0 \\ 0 & 0 \end{pmatrix} + \begin{pmatrix} 0 & -\Theta(t) \\ 0 & 0 \end{pmatrix}$ and the perturbation result proposition 1.2 in [11].

Now we want to express the socalled compatibility condition $\Phi \in D^j(0) = D^j(0)$ in 2.3 (ii) in terms of x_0, y_0, A_f and θ.

[3]The subscript 'hom' corresponds to the fact, that $\dot{\underline{u}}(t) + \mathcal{A}_{hom}\,\underline{u}(t) = 0$ is equivalent to the homogeneous wave equation, whereas $\dot{u} + \mathcal{A}u = 0$ describes the inhomogeneous one.

The recursive definition of the $D^j(t)$ in definition 1.2 reads in our case for $t \in [0,T]$ and $j \in \mathbb{N}_0$:

$$\left(\begin{pmatrix} x \\ y \end{pmatrix} \atop \mathbf{1}\right) \in D^{j+1}(t) \Longleftrightarrow \left(\begin{pmatrix} x \\ y \end{pmatrix} \atop \mathbf{1}\right) \in D^j(t) \text{ and}$$

$$S^k(t)\left(\begin{pmatrix} x \\ y \end{pmatrix} \atop \mathbf{1}\right) \in Y_{j+1-k} \text{ for } 0 \le k \le j \ .$$

DEFINITION 2.5 For $x, y \in H_0$, $j \in \{0, \ldots, m\}$ and $t \in [0,T]$ we say: (x,y) satisfies condition $(C_{j,t}^{A,\theta}) :\Leftrightarrow$

$$\left(\begin{pmatrix} x \\ y \end{pmatrix} \atop \mathbf{1}\right) \in D^j(t)$$

Note that (x,y) satisfies condition $(C_{0,t}^{A,\theta}) \Leftrightarrow x \in V$ and $y \in H_0$. $(C_{j,t}^{A,\theta})$ is called j-th compatibility condition associated with A and θ at the time t and can be evaluated:

THEOREM 2.6 *For $j \in \{1, \ldots, m\}$, $x \in H_{j+1}$ and $y \in H_j$ holds:*
(x,y) satisfies $(C_{j,t}^{A,\theta}) \iff$
$x \in V$ and

$$(-A_f)^{(k-1)/2} y \ - \ \textstyle\sum_{l=0,\ l\ odd}^{k-2} (-A_f)^{(l-1)/2} d^{k-1-l} \theta(t)$$
$$\in \ D(A) \cap H_{j+1-k} \text{ and}$$
$$(-A_f)^{(k+1)/2} x - (-A_f)^{(k-1)/2} \theta(t) \ - \ \textstyle\sum_{l=0,\ l\ even}^{k-2} (-A_f)^{(l-1)/2} d^{k-1-l} \theta(t)$$
$$\in \ V \cap H_{j-k}$$

if k is odd and

$$(-A_f)^{k/2} x - (-A_f)^{(k-2)/2} \theta(t) \ - \ \textstyle\sum_{l=0,\ l\ odd}^{k-2} (-A_f)^{(l-1)/2} d^{k-1-l} \theta(t)$$
$$\in \ D(A) \cap H_{j+1-k} \text{ and}$$
$$(-A_f)^{k/2} y \ - \ \textstyle\sum_{l=0,\ l\ even}^{k-2} (-A_f)^{(l-1)/2} d^{k-1-l} \theta(t)$$
$$\in \ V \cap H_{j-k}$$

if k is even, for $1 \le k \le j-1$.

REMARK 2.7 In the recursion for the $D^j(t)$ in definition 1.2, the construction guaranties, that all elements are in the correct domains of the operators which are supposed to act on them. Evaluating this formula, we may loose this property, which is the reason for the requirement of the existence of the 'formal' extension A_f. In applications, this extension is often canonically given by the formal differential operator.

Theorem 1.4 implies thus the following statement:

THEOREM 2.8 *Let the data* (x_0, y_0) *from 2.1 satisfy condition* $(C^{A,\theta}_{i,0})$, *for some* $j \in \{1, \ldots, m\}$.
Conclusion:
We have existence and uniqueness of a solution $x(\cdot) \in C^2([0,T], H_0)$ *of (6). It satisfies*

$$x(\cdot) \in C^{j+1-i}([0,T], H_{j+1}) \ , \quad i = 0, \ldots, j+1$$

and

$$(x(t), y(t)) \text{ satisfies } (C^{A,\theta}_{j,t}) \text{ for } t \in [0,T] \ .$$

3 APPLICATION TO INTERACTION PROBLEMS

We recall the notion of interaction problems (cf. [1], [3]):

DATA 3.1

(i) Consider numbers $n \in I\!N$, $k_1, \ldots, k_n \in I\!N$ and connected open sets $\Omega_i \subseteq I\!R^{k_i}$, $i = 1, \ldots, n$, with C^∞–boundaries.

(ii) Consider real numbers $d_0, d_1 > 0$ and realvalued functions $a_i^{lj} \in C^\infty(\overline{\Omega}_i) \cap L^\infty(\Omega_i)$, $l, j = 1, \ldots, k_i$, $i = 1, \ldots, n$ with the following property (ellipticity):

$$\sum_{l,j=1}^{k_i} a_i^{lj}(x)\xi_l\xi_j \geq d_0 \sum_{j=1}^{k_i} \mid \xi_j \mid^2 \quad \forall (\xi_1, \ldots, \xi_{k_i}) \in I\!R^{k_i}, x \in \Omega_i, i = 1, \ldots, n$$

and symmetry

$$a_i^{lj}(x) = a_i^{jl}(x) \ \ \forall\, l, j = 1, \ldots, k_i; x \in X_i; i = 1, \ldots, n.$$

Consider further $c_i \in C^\infty(\overline{\Omega}_i)$ with $c_i \geq d_1$ for $i = 1, \ldots, n$.

(iii) Define $a_i : H^1(\Omega_i) \times H^1(\Omega_i) \longrightarrow \mathbb{C}$ by

$$a_i(u_i, v_i) := \int_{\Omega_i} [\sum_{l,j=1}^{k_i} a_i^{lj}(x)\partial_l u_i(x)\overline{\partial_j v_i(x)}]d^{k_i}x + \int_{\Omega_i} c_i(x)u_i(x)\overline{v_i(x)}d^{k_i}x \ ,$$

where $\partial_i := \frac{\partial}{\partial x_i}$. A version of Gårding's inequality (e.g. [16] chapter III) shows that a_i is a continuous, symmetric, coercive, positive sesquilinear form.

(iv) For $u_i \in H^2(\Omega_i)$ denote

$$A_i u_i(x) := -\sum_{l,j=1}^{k_i} \partial_j a_i^{lj}(x)\partial_l u_i(x) + c_i(x)u_i(x), \quad \forall x \in \Omega_i, \quad i = 1, \ldots, n.$$

ASSUMPTIONS 3.2 Define

$$\tilde{V} := \prod_{i=1}^n H^1(\Omega_i) \text{ and } \tilde{H} := \prod_{i=1}^n L^2(\Omega_i) \ .$$

Take the sum of the inner products of the factors as inner products of $\tilde{V}$ and $\tilde{H}$. Let V be a closed subspace of $\tilde{V}$ and H the closure of V in $\tilde{H}$. Therefore: $V \hookrightarrow H$. Endow V and H with the structures induced by $\tilde{V}$ and $\tilde{H}$ and identify H with H'. Thus

$$V \hookrightarrow H \ = \ H' \hookrightarrow V'.$$

DEFINITION 3.3 Define $a : V \times V \longrightarrow \mathbb{C}$ by $a((u_i),(v_i)) = \sum_{i=1}^n a_i(u_i,v_i)$. Clearly, a is also a continuos, symmetric, coercive, positive sesquilinear form.

Now we see by Friedrichs Extension :

THEOREM 3.4 *Define* $\tilde{A} : V \longrightarrow V'$ *by* $\tilde{A}u(v) := a(u,v) \quad \forall u,v \in V$ *and* $A : D(A) \subseteq H \longrightarrow H$ *by*

$$D(A) := \{u \in V : \tilde{A}u \in H' = H\}\ , \quad A := \tilde{A}\mid_{D(A)} .$$

Conclusion:

$(A, D(A))$ *is selfadjoint, positive and* $D(A^{1/2}) = V$.

DEFINITION 3.5 The space V, assumed in 3.2, is called interaction space. The operator $A : D(A) \longrightarrow H$ constructed in 3.4, using the data 3.1, is called interaction operator.

In this setting, assumption 2.1 (iv) is not automatically satisfied (e.g. the Dirichlet problem for the Laplace operator on a domain with corners does not have the H^2-regularity). So we conserve it and give examples at the end of the section.

ASSUMPTIONS 3.6 Assume 2.1 (iv).

Further we introduce the inhomogeneous part of the equation and initial conditions:

DATA 3.7 Consider

(i) $\theta_i \in C^\infty(\overline{[0,T] \times \Omega_i})$, $i = 1,\ldots,n$;

(ii) $x_0 \in \prod_{i=1}^n H^{m+1}(\Omega_i)$; $y_0 \in \prod_{i=1}^n H^m(\Omega_i)$;

With these data and assumptions we can establish the situation of section 2:

DEFINITION 3.8

(i) $H_j := \prod_{i=1}^n H^j(\Omega_i)$, $j = 1,\ldots,m-1$.

(ii) $A_f : H_2 \to H_0$ with $A_f(u_i) := (A_i u_i)_{j=1,\ldots,n}$.

COROLLARY 3.9 *With the assumptions and choices in this section we have the conclusions of theorem 2.8.*

So we have existence, uniqueness and regularity of solutions of (6), where A is now a concrete interaction operator describing wave equations with the spatial parts A_i

on Ω_i and interaction conditions. These conditions can be given in concrete terms using an abstract Green's formula (cf. [16], some special applications cf. [3]).

However, the assumption of H^2-regularity 3.6 is not satisfied in all possible cases (e.g. it is *not* satisfied for boundary value problems on domains with corners or with mixed Dirichlet - Neumann conditions).

CONCRETIZATIONS 3.10 Assumption 3.6 is satisfied in the following cases:

(i) Consider a Cauchy problem for wave equations on a locally elementary ramified space (cf. [3], definition 3.2; locally, this is a subset of a family of halfspaces with suitably identified boundaries ('book')). The H^2-regularity is proved in [3], theorem 3.3, based on a result of S. Nicaise for elementary ramified spaces (cf. [14]). Further results cf. [1], [4]. An important special case is the case of onedimensional networks (cf. also [5]).

(ii) Consider the interaction problem (1) but with (T_i) replaced by (I_i) as indicated in section 0. Assume that the Fourier integral operators satisfy the conditions of the theorem of Egoroff (cf. [17]). Then, by similarity, our system is equivalent to a wave equation with pseudodifferential spatial part and transmission conditions (T_i). The symbol of the spatial part is given by the pullback of the symbol of $c_i\Delta$ under the function whose graph is the canonical relation associated to F_i. The H^2-regularity 3.6 can be shown by an argument using an algebra of the Boutet de Monvel-type (to be published elsewhere). 3.6 is clear by [14], if the F_i are coordinate transformations, for the spatial part is then differential.

The compatibility conditions given by theorem 2.6 have an especially simple form in the following special case:

PROPOSITION 3.11 *Consider the assumptions of this section for* $m = \infty$.
Assume $\partial_t^j \theta_i(0,\cdot) \equiv 0 \quad \forall j \in \mathbb{N}$.
Let $\theta := (\theta_i)_{i=1,\ldots,n}$ *Conclusion:*
x_0, y_0 *satisfy* $(C_{j,0}^{A,\theta}) \ \forall\, j \in \mathbb{N} \Leftrightarrow \ x_0, y_0 \ \in \ D(A^j) \ \ \forall j \in \mathbb{N}$.
Both sides are especially true, if $x_0 = y_0 \equiv 0$.

REMARK 3.12 For $n = 1$ and $V := H_0^1(\Omega_1)$, where $\partial\Omega$ is smooth and $x_0 = y_0 \equiv 0$, 3.11 gives the regularity result of Lions-Magenes [12] Vol. 2, p. 131 (the H^2-regularity 3.6 is given by classical methods, cf. [16]).

REFERENCES

[1] ALI MEHMETI, F. Lokale und globale Lösungen linearer und nichtlinearer hyperbolischer Evolutionsgleichungen mit Transmission, Dissertation, Joh. Gutenberg-Universität, Mainz (1987).

[2] ALI MEHMETI, F. Global Existence of Solutions of Semilinear Evolution Equations with Interaction, in: B.-W. Schulze, H. Triebel, eds., *Symp. 'Part. Diff. Equ.', Holzhau 1988* (Teubner-Texte zur Mathematik 112, Leipzig 1989) 11-23.

[3] ALI MEHMETI, F. Regular Solutions of Transmission and Interaction Problems for Wave Equations, *Math. Meth. Appl. Sci.* **11** (1989) 665-685.

[4] ALI MEHMETI, F. Global Existence for Semilinear Evolution Equations with Applications to Interaction Problems, Preprint, Mainz (1989).

[5] BELOW, J.V. Classical Solvability of Linear Parabolic Equations on Networks, *J. Diff. Eq.* **72** (1988) 316-337.

[6] BREZIS, H. Monotonicity Methods in Hilbert Spaces and Some Applications to Nonlinear Partial Differential Equations, in: E. Zarantonello, ed., *Contributions to Nonlinear Functional Analysis* (1971) 101-156.

[7] DUISTERMAAT, J.J. *Fourier Integral Operators* (Courant Inst. of Math. Sci., New York University, 1973).

[8] GRAMSCH, B. Zum Einbettungssatz von Rellich bei Sobolevräumen, *Math. Zeitschr.* **106** (1968) 81-87.

[9] GRAMSCH, B. Asymptotik der Eigenwerte stark elliptischer Operatoren auf kompakten verzweigten Räumen, unpublished (1979).

[10] HANSEN, S. Singularities of Transmission Problems, *Math Ann.* **268** (1984) 233-253.

[11] KATO, T. Linear and Quasilinear Equations of Evolution of Hyperbolic Type, *C.I.M.E., II ciclo, Cortona* (1976) 125-191.

[12] LIONS, J.L., MAGENES, E. *problèmes aux limites nonhomogenes et applications,* vol. 2, travaux et recherches math. 18 (Dunod, Paris, 1968).

[13] LUMER,G. Espaces ramifiés et diffusions sur les réseaux topologiques, *C.R. Acad. Sc. Paris*, t. 291, Série A (1980) 627-630.

[14] NICAISE, S. Problèmes de Cauchy posés en norme uniforme sur les espaces ramifiés élémentaires, *C.R. Acad. Sc. Paris, t. 303*, Série I, no. 10 (1986) 443-446.

[15] NOSMAS, J.CL. Paramétrix du problème de transmission pour l'equation des ondes; *J. Math. Pures et Appl.* **56** (1977) 423-435.

[16] SHOWALTER, R. E. *Hilbert Space Methods for Partial Differential Equations* (Pitman, London, San Franzisco, Melbourne, 1977).

[17] TRÈVES, F. *Introduction to Pseudodifferential and Fourier Integral Operators* (Plenum Press, New York and London, 1980).

F. Ali Mehmeti
Fachbereich Mathematik
Johannes Gutenberg-Universität
Saarstraße 21
D-6500 Mainz
Fed. Rep. of Germany

Operator Theory:
Advances and Applications, Vol. 50

INTERPOLATION PROBLEMS, EXTENSIONS OF SYMMETRIC OPERATORS AND REPRODUCING KERNEL SPACES I

Daniel Alpay, Piet Bruinsma, Aad Dijksma, Henk de Snoo

The aim of this paper is to study interpolation problems for pairs of functions of the extended Nevanlinna class using two different approaches, namely the Kreĭn–Langer theory of extensions of symmetric operators and the de Branges theory of Hilbert spaces of analytic functions, and to make explicit various links between them. In this part we consider the Kreĭn–Langer theory. In the next part we shall treat these problems from the point of view of the de Branges theory.

0. Introduction

In this paper we study the following version of the well known interpolation problem: *Given*:

(0.1) points $w_1, w_2, \ldots, w_m \in \mathbb{C}\backslash\mathbb{R}$, not necessarily distinct,

(0.2) nonnegative integers $r_1, r_2, \ldots, r_m$, and

(0.3) two $m \times n$ matrix functions $\mathcal{K}(\ell)$, $\mathcal{L}(\ell)$, defined and locally holomorphic on $\mathbb{C}\backslash\mathbb{R}$, with rows denoted by $\mathcal{K}_i(\ell)$, $\mathcal{L}_i(\ell)$, respectively, such that

(0.4) $(D_\ell)^p \left(\mathcal{L}_i(\ell)\mathcal{K}_j(\bar{\ell})^* - \mathcal{K}_i(\ell)\mathcal{L}_j(\bar{\ell})^* \right) |_{\ell=w_i} = 0, \; 0 \le p \le \min(r_i, r_j)$, if $w_i = \bar{w}_j$,

where $(D_\ell)^p$ stands for the differential expression $(1/p!)(d/d\ell)^p$. *Find*: necessary and sufficient conditions on the data (0.1)–(0.3) which ensure the existence of a pair of $n \times n$ matrix functions $\mathcal{M}(\ell), \mathcal{N}(\ell)$, defined and locally holomorphic on $\mathbb{C}\backslash\mathbb{R}$, satisfying for all $\ell \in \mathbb{C}\backslash\mathbb{R}$

$$(0.5) \quad \begin{cases} \operatorname{rank}(\mathcal{M}(\ell)^* : \mathcal{N}(\ell)^*) = n & \text{(nondegeneracy)}, \\ \mathcal{M}(\bar{\ell})^*\mathcal{N}(\ell) - \mathcal{N}(\bar{\ell})^*\mathcal{M}(\ell) = 0 & \text{(symmetry)}, \\ \operatorname{Im}\ell . \operatorname{Im}\mathcal{M}(\ell)^*\mathcal{N}(\ell) \ge 0 & \text{(nonnegativity)}, \end{cases}$$

which solves the interpolation problem

$$\text{(IP)} \quad (\mathcal{K}_i(\ell)\mathcal{N}(\ell))^{(p)} |_{\ell=w_i} = (\mathcal{L}_i(\ell)\mathcal{M}(\ell))^{(p)} |_{\ell=w_i}, \quad 1 \le i \le m, \; 0 \le p \le r_i,$$

and *determine*: the set of all solutions, when these conditions are met. The points w_i in (0.1) are called the interpolation points, the matrix functions $\mathcal{K}(\ell)$, $\mathcal{L}(\ell)$ in (0.3) and their derivatives determine the directions in which the matrix functions $\mathcal{M}(\ell)$, $\mathcal{N}(\ell)$ are to be interpolated and the integers r_i in (0.2) fix the order of the derivatives. Obviously, in order to formulate the problem (IP) one only needs the values of the $1{\times}n$ vectors $\mathcal{K}_i^{(p)}(w_i)$ and $\mathcal{L}_i^{(p)}(w_i)$, $i=1,2,\dots,m$, $p=0,1,\dots,r_i$, rather than the complete matrix functions $\mathcal{K}(\ell)$, $\mathcal{L}(\ell)$ defined on $\mathbb{C}\backslash\mathbb{R}$, which interpolate them and which can always be constructed. The conditions on these values when $w_i=\bar{w}_j$ formulated in (0.4) are necessary and sufficient to ensure that the Lyapunov equation associated with the interpolation problem (IP) to be defined below has a solution. For a proof of this fact we refer to the Appendix at the end of this paper. Clearly, the functions $\mathcal{K}(\ell)$ and $\mathcal{L}(\ell)$ are not uniquely determined by the values $\mathcal{K}_i^{(p)}(w_i)$ and $\mathcal{L}_i^{(p)}(w_i)$, $0\le i\le m$, $0\le p\le r_i$, satisfying (0.4). Nevertheless we use these functions, because they give rise to a simple and convenient notation. The above formulation of the problem is partly inspired by the paper of Ball [B1].

The interpolation problem (IP) has a long history and originates with the work of Pick and Nevanlinna at the beginning of this century. Interpolation involving derivatives goes back to Hermite and Fejér. For a historical survey we refer to the book of Rosenblum and Rovnyak [RR], and for more recent references to the lecture notes of Dym [D1], the monograph of Katsnelson [K] and the survey paper by Ball [B2]. A new feature in our interpolation problem (IP) is that we look for solutions which belong to the extended Nevanlinna class $\mathbf{N}^n$, consisting of elements, called Nevanlinna pairs, which can be described by ordered pairs $(\mathcal{M}(\ell),\mathcal{N}(\ell))$ of locally holomorphic $n{\times}n$ matrix functions $\mathcal{M}(\ell)$, $\mathcal{N}(\ell)$ satisfying (0.5). In the classical formulation of the interpolation problem $\mathcal{M}(\ell)$ is set identically equal to I, the identity matrix, and then one looks for solutions $\mathcal{N}(\ell)$ of (IP) which are Nevanlinna functions, i.e., satisfy (0.5) with $\mathcal{M}(\ell)=I$. Note that if the pair $(\mathcal{M}(\ell),\mathcal{N}(\ell))$ is a solution of the general interpolation problem (IP) then the pair $(\mathcal{M}(\ell)\mathcal{G}(\ell),\mathcal{N}(\ell)\mathcal{G}(\ell))$ is also a solution of this problem for any locally holomorphic invertible $n{\times}n$ matrix function $\mathcal{G}(\ell)$. This projective property of the solution pairs is also reflected in the projective definition of the extended Nevanlinna class. This class contains besides the usual Nevanlinna functions, these are the ordered pairs $(\mathcal{M}(\ell),\mathcal{N}(\ell))$ in which the first entry is invertible on $\mathbb{C}\backslash\mathbb{R}$ (except for a set of isolated points) and which can be identified with the functions $\mathcal{N}(\ell)\mathcal{M}(\ell)^{-1}$, also the nonproper elements, the pairs in which the first entry is not invertible. In the scalar case the extended Nevanlinna class is obtained by adding the function ∞ to the class of all Nevanlinna functions. For more details we refer to Definition 1.4 in Section 1. In Section 3 of this paper we give a necessary and sufficient condition in terms of the data (0.1)–(0.3), to ensure that all solutions of the interpolation problem (IP) are classical Nevanlinna functions. Another feature of the interpolation problem (IP) is that it includes the two sided interpolation problems for Nevanlinna functions. To see this, assume for a moment that in the data, for some indices i and j, the interpolation points are complex conjugates of each other, $w_i = \bar{w}_j$, and that the Nevanlinna function $\mathcal{N}(\ell)$ is a solution of the interpolation problem (IP), that is, that the pair $(I,\mathcal{N}(\ell))$ is a solution. Then we have that $\mathcal{N}(\ell)$ satisfies the equality

$$(\mathcal{K}_i(\ell)\mathcal{N}(\ell))^{(p)}\,|_{\ell=w_i} = \mathcal{L}_i^{(p)}(w_i),\ \ p=0,1,\ldots,r_i$$

and by taking adjoints and using the relation $\mathcal{N}(\ell)^* = \mathcal{N}(\bar{\ell})$, we find that $\mathcal{N}(\ell)$ also satisfies the equality

$$(\mathcal{N}(\ell)\mathcal{K}_j(\bar{\ell})^*)^{(p)}\,|_{\ell=w_i} = \mathcal{L}_j^{(p)}(\bar{w}_i)^*,\ \ p=0,1,\ldots,r_j.$$

Here, the first equality implies that $\mathcal{N}(\ell)$ must meet some left sided interpolation requirements at w_i, by which we mean, that certain combinations of the rows of $\mathcal{N}(\ell)$ must have prescribed values and derivatives at w_i, whereas the second equality implies that $\mathcal{N}(\ell)$ must satisfy some right sided interpolation requirements, that is, that certain combinations of the columns of $\mathcal{N}(\ell)$ must assume prescribed values and derivatives at the same point w_i. Thus $\mathcal{N}(\ell)$ is subjected to two sided interpolation requirements.

Associated with the data (0.1)–(0.3) is the socalled Lyapunov equation alluded to above and defined in the following manner. Put $r = \Sigma_{i=1}^{m}(r_i+1)$, let

$$V = (V_{ip}) = (V_{10}\vdots\ldots\vdots V_{1r_1}\vdots\ldots\vdots V_{i0}\vdots\ldots\vdots V_{ir_i}\vdots\ldots\vdots V_{m0}\vdots\ldots\vdots V_{mr_m})$$

and

$$W = (W_{ip}) = (W_{10}\vdots\ldots\vdots W_{1r_1}\vdots\ldots\vdots W_{i0}\vdots\ldots\vdots W_{ir_i}\vdots\ldots\vdots W_{m0}\vdots\ldots\vdots W_{mr_m})$$

be the $n{\times}r$ matrices defined by

$$V_{ip} = (1/p!)\,\mathcal{K}_i^{(p)}(w_i)^*, \qquad W_{ip} = (1/p!)\,\mathcal{L}_i^{(p)}(w_i)^*$$

and let $Z = (Z_{ij}^{pq})$ be the Jordan matrix with $Z_{ij}^{pq} = 0$ if $i \neq j$ and

$$Z_{ii}^{pq} = \begin{cases} \bar{w}_i, & \text{if } p=q, \\ 1, & \text{if } q=p+1, \\ 0, & \text{otherwise.} \end{cases}$$

Here and elsewhere in the sequel when we write $X=(X_{ij}^{pq})$ we mean that X is an $m\times m$ block matrix $X=(X_{ij})$ whose i,j–th entry X_{ij} is an $(r_i+1)\times(r_j+1)$ matrix with entries X_{ij}^{pq}. Note that in these formulas here and below the lower indices i,j start with 1, whereas the upper indices p,q start with 0. The Lyapunov equation associated with V,W and Z is the matrix equation

$$\mathbb{P}Z - Z^*\mathbb{P} = V^*W - W^*V,$$

in which $\mathbb{P}$ is the unknown $r\times r$ matrix. If $\mathbb{P}$ is a solution then so are $\mathbb{P}^*$ and $\frac{1}{2}(\mathbb{P}+\mathbb{P}^*)$, and hence, if the Lyapunov equation has a solution, it also has a hermitian solution. The hermitian solutions of the Lyapunov equation will be called Pick matrices associated with the interpolation problem (IP). For statements concerning the existence and the analytic representation of solutions of the Lyapunov equation we refer to the Appendix. The Lyapunov equation and its Pick matrix solutions play important roles in the study of the interpolation problem as is clear from the next theorem, which implies that a necessary and sufficient condition for the existence of a solution of the interpolation problem (IP) is that the Lyapunov equation has a solution $\mathbb{P}\geq 0$. In one direction we can be more explicit and for this we need the following definition. If $(\mathcal{M}(\ell),\mathcal{N}(\ell))$ is a Nevanlinna pair we define the matrix $\mathbb{P}_{\mathcal{M},\mathcal{N}} = ((\mathbb{P}_{\mathcal{M},\mathcal{N}})_{ij}^{pq})$ by

$$(0.6)\quad (\mathbb{P}_{\mathcal{M},\mathcal{N}})_{ij}^{pq} = (D_\ell)^p(D_{\bar\lambda})^q(\ell\mathcal{K}_i(\ell)+\mathcal{L}_i(\ell)).$$

$$\cdot\frac{\hat{\mathcal{N}}(\bar\ell)^*\hat{\mathcal{M}}(\bar\lambda)-\hat{\mathcal{M}}(\bar\ell)^*\hat{\mathcal{N}}(\bar\lambda)}{\ell-\bar\lambda}(\lambda\mathcal{K}_j(\lambda)+\mathcal{L}_j(\lambda))^*\,|_{\ell=w_i,\lambda=w_j},$$

where the matrix functions $\hat{\mathcal{M}}(\ell)$ and $\hat{\mathcal{N}}(\ell)$ are given by

$$(0.7)\quad \hat{\mathcal{M}}(\ell)=\mathcal{M}(\ell)(\ell\mathcal{M}(\ell)+\mathcal{N}(\ell))^{-1},\qquad \hat{\mathcal{N}}(\ell)=\mathcal{N}(\ell)(\ell\mathcal{M}(\ell)+\mathcal{N}(\ell))^{-1}.$$

Because of the symmetry condition in (0.5) the kernel

$$\frac{\hat{\mathcal{N}}(\bar\ell)^*\hat{\mathcal{M}}(\bar\lambda)-\hat{\mathcal{M}}(\bar\ell)^*\hat{\mathcal{N}}(\bar\lambda)}{\ell-\bar\lambda}$$

has a sesquianalytic continuation, i.e., an analytic continuation simultaneously in ℓ and $\bar\lambda$, to the points ℓ and $\bar\lambda$ with $\ell=\bar\lambda$ and hence formula

(0.6) also makes sense for those indices i,j for which $w_i = \bar{w}_j$.

THEOREM 0.1. *(i) If the Nevanlinna pair* $(\mathcal{M}(\ell),\mathcal{N}(\ell))$ *is a solution of the interpolation problem* (IP), *then the matrix* $\mathbb{P}_{\mathcal{M},\mathcal{N}}$ *is a nonnegative solution of the Lyapunov equation. (ii) Conversely, if the Lyapunov equation has a nonnegative solution, then the interpolation problem* (IP) *has a solution also.*

Part (i) of Theorem 0.1 and the necessity of the condition $\mathbb{P} \geq 0$ for the existence of a solution of the interpolation problem follow easily from an operator representation of Nevanlinna pairs and this will be shown in Section 1. The sufficiency of the condition and the formulas for the parametrization of all solutions can be proved in various ways and a number of approaches to solve this problem as well as related problems have appeared in the recent literature. For detailed accounts we refer to [B1,2], [BGR], [BH], [BR], [D1,2], [F1,2,3] and [Ko]. Other sources of information will be the forthcoming books by Ball, Gohberg and Rodman and by Dym. In the present note and the one to follow we want to solve the above problem using two different methods and to discuss the relations between the two.

In this paper we concentrate on the method based on the extension theory of symmetric relations and of isometric operators developed by Kreĭn, Langer and others, see [Kr], [KL1–5] and also [LT1], [LS]. It can be summarized as follows. With each Pick matrix $\mathbb{P} \geq 0$ for the interpolation problem (IP), there can be associated in a natural way a nondensely defined symmetric relation S in a Hilbert space of dimension $\leq r+n$ and we show in Section 3 that the minimal selfadjoint relation extensions of S in possibly larger Hilbert spaces are in one to one correspondence with all solutions $(\mathcal{M}(\ell),\mathcal{N}(\ell))$ of the interpolation problem (IP) which satisfy $\mathbb{P}_{\mathcal{M},\mathcal{N}} = \mathbb{P}$. Since the Hilbert space is finite dimensional S has equal defect numbers. If $\mathbb{P} \geq 0$ and has a nontrivial kernel then the defect numbers may be zero, if $\mathbb{P} > 0$ they are strictly positive. In the case that the defect numbers are zero, S is selfadjoint and the interpolation problem (IP) has one solution with $\mathbb{P}_{\mathcal{M},\mathcal{N}} = \mathbb{P}$, that is, up to multiplication from the right by a matrix function. If the defect numbers are positive, S has infinitely many nonisomorphic minimal selfadjoint extensions and the interpolation problem has infinitely many solutions. The extension theory implies that the class of minimal selfadjoint extensions of S can be parametrized by means of the elements of the extended Nevanlinna class. This parametrization involves a linear fractional transformation induced by a socalled resolvent matrix associated with S. Each resolvent matrix together with the operator representation of Nevanlinna pairs mentioned above leads in a very simple way to a solution matrix through which all solutions of the interpolation problem (IP) with $\mathbb{P}_{\mathcal{M},\mathcal{N}}=\mathbb{P}$ can be parametrized. A solution matrix $U(\ell)$ associated with a nonsingular $\mathbb{P}$ is a 2×2 block matrix $U(\ell)=(U_{ij}(\ell))$ with entries $U_{ij}(\ell)$ which are rational $n\times n$ matrix functions having poles in the set $\{w_1,\bar{w}_1,w_2,\bar{w}_2,\ldots,w_m,\bar{w}_m\}$, such that the formula

$$(\mathcal{M}(\ell),\mathcal{N}(\ell))=(U_{21}(\ell)\mathcal{B}(\ell)+U_{22}(\ell)\mathcal{A}(\ell),U_{11}(\ell)\mathcal{B}(\ell)+U_{12}(\ell)\mathcal{A}(\ell))$$

establishes a one to one correspondence between all Nevanlinna pair solutions $(\mathcal{M}(\ell),\mathcal{N}(\ell))$ of the interpolation problem (IP) with $\mathbb{P}_{\mathcal{M},\mathcal{N}}=\mathbb{P}$ and all Nevanlinna pairs $(\mathcal{A}(\ell),\mathcal{B}(\ell))$ in the extended Nevanlinna class $\mathbf{N}^n$. Canonical and finite dimensional extensions of S give rise to rational solutions. The connection between their degrees and the dimension of the extending spaces remains to be worked out. We note that nonrational solutions may correspond to selfadjoint extensions of S which have an absolutely continuous spectrum. We show in Section 3 that if $\mathbb{P}>0$, the following matrix functions are solution matrices associated with $\mathbb{P}$

$$U_a(\ell)=\begin{pmatrix} I & 0\\ 0 & I\end{pmatrix}+(\ell-a)\begin{pmatrix} W\\ V\end{pmatrix}(Z-\ell)^{-1}\mathbb{P}^{-1}(Z-a)^{-*}(W^*:V^*)\begin{pmatrix} 0 & I\\ -I & 0\end{pmatrix},$$

$$U_\infty(\ell)=\begin{pmatrix} I & 0\\ 0 & I\end{pmatrix}+\begin{pmatrix} W\\ V\end{pmatrix}(Z-\ell)^{-1}\mathbb{P}^{-1}(W^*:V^*)\begin{pmatrix} 0 & I\\ -I & 0\end{pmatrix},$$

where V, W and Z are the matrices defined above and $a\in\mathbb{R}$. For any boundedly invertible operator T we denote by T^{-*} the operator $(T^{-1})^*$. Solution matrices of these forms have appeared elsewhere in the literature, see, e.g., [N1,2], [KP], [K], [Ko] and [D2]. The two solution matrices differ from each other by a $\left(\begin{smallmatrix} 0 & -I\\ I & 0\end{smallmatrix}\right)$-unitary factor from the right, but note that they have the additional property that $U_a(\ell)$ tends to $U_\infty(\ell)$ if $a\to\infty$. See Section 3 for more details. The method of extensions of symmetric relations or of isometric operators (through resolvent matrices) is very general and has been applied to a variety of problems in classical analysis. In some cases it gives rise to polynomial or entire solution matrices, see for instance [KL2–5].

Sz.-Nagy and Koranyi in [KSz] studied an interpolation problem with possibly infinitely many interpolation points (but simpler in that it involves interpolation of full matrix values and no conditions on derivatives) also by reducing the problem to finding selfadjoint extensions of a symmetric operator. For a treatment of interpolation problems similar to the one described in the previous paragraph we refer to [KHJ], which was brought to our attention by Professor Katsnelson after the manuscript was completed. Finally, we mention that like Dym in [D1,2] also Kimura in [Ki] emphasized the connection between the interpolation problem and the Lyapunov equation.

We outline the contents of the paper. In Section 1 we briefly present the necessary facts concerning linear relations in Hilbert spaces. We define the extended Nevanlinna class in a Hilbert space setting and discuss some of the properties of its elements. For the definition in the matrix case, see, e.g., [K]. In particular, we prove a representation of a Nevanlinna pair in terms of the compression of the resolvent of a minimal selfadjoint relation, uniquely determined up to isomorphisms, see Theorem 1.5. As stated above this representation leads to a proof of the necessity of the condition $\mathbb{P}\geq 0$ for the existence of a solution of the interpolation problem (IP). In the interpolation problem (IP) Theorem 1.5 plays the same role as the Riesz–Herglotz representation theorem (or its operator version) does in the classical interpolation problem. In Section 2 we review the Kreĭn–Langer extension theory based on the important formula due to M.G. Kreĭn which relates resolvents of selfadjoint extensions of a symmetric relation to elements of an extended Nevanlinna class via the Q–function. We recall the definitions of a Q–function, a module space $\mathfrak{L}$ and of an $\mathfrak{L}$–resolvent matrix associated with symmetric relations. In order to obtain the solution matrices $U_a(\ell)$ and $U_\infty(\ell)$, we construct some specific $\mathfrak{L}$–resolvent matrices $W_a(\ell)$ and $W_\infty(\ell)$ for the relation S associated with the interpolation problem (IP). In doing so we have simplified some of the arguments in the paper [LT2]. The relation between $W_a(\ell)$ and $W_\infty(\ell)$ becomes most apparent if we consider the corresponding coresolvent matrices for isometric operators and use the Cayley transformation. Therefore we also repeat the definitions of a Q–function, a module space $\mathfrak{L}$ and of an $\mathfrak{L}$–coresolvent matrix associated with isometric operators and show how they are related via the Cayley transformation to the corresponding notions associated with symmetric relations. This theory is also of independent interest. The main results of this paper are contained in Section 3 where we prove the parametrizations of the solutions of the interpolation problem (IP) mentioned above. It is possible to generalize some of these results to ones involving Nevanlinna pairs whose kernels have κ negative squares, cf. [Go]. Finally, in the Appendix we present some results pertaining to the Lyapunov equation.

Concerning the second part of this paper, which is in preparation we remark the following. The Q–function, the resolvent matrix and the solution matrix generate in a natural way positive functions of the type considered by de Branges in his theory of reproducing kernel Hilbert spaces of analytic functions, see [dB1–4], [AD1,2] and [ABDS]. Moreover, in [D2] an interpolation problem similar to (IP) is studied from the point of view of de Branges spaces. Thus the de Branges theory and the Kreĭn–Langer theory have many points of contact and these we shall explore in part II. In particular we study the links between the resolvent matrix in this paper and certain models in terms of reproducing kernel spaces.

We thank Professor Heinz Langer for his suggestion to treat the interpolation problem using the theory of resolvent matrices and his encouragement.

1. Nevanlinna pairs

A linear relation T in a Hilbert space $(\mathfrak{H}, [.,.])$ is a linear manifold T in $\mathfrak{H}^2=\mathfrak{H}\oplus\mathfrak{H}$, the direct sum Hilbert space consisting of all pairs $\{f,g\}$, $f,g\in\mathfrak{H}$. The graph of a linear operator T in $\mathfrak{H}$ is a linear relation and we shall often identify T with its graph. Note that T is (the graph of) an operator if and only if the multivalued part $T(0)=\{g\in\mathfrak{H}\,|\,\{0,g\}\in T\}$ of T is equal to $\{0\}$. In the sequel we shall only consider linear relations and linear operators that are closed without stating this explicitly each time. The definitions of the domain $\mathfrak{D}(T)$, the range $\mathfrak{R}(T)$, the null space $\nu(T)$, the inverse T^{-1}, the adjoint T^*, the Cayley transform $C_\mu(T)$ and its inverse $F_\mu(T)$, the resolvent set $\rho(T)$, the spectrum $\sigma(T)$ and its subdivisions $\sigma_c(T)$, $\sigma_p(T)$, etc., the product ST, the sum $T+S$, etc., can be extended in an obvious way from operators to relations T and S. For example,

$$T+S=\{\{f,g+k\}\,|\,\{f,g\}\in T,\{f,k\}\in S\}$$

(note the difference with the notation $\dot{+}$ for the sum of two linear relations in $\mathfrak{H}^2$: $T\dot{+}S=\{\{f+h,g+k\}\,|\,\{f,g\}\in T,\{h,k\}\in S\}$) and

$$\rho(T)=\{\ell\in\mathbb{C}\mid(T-\ell)^{-1}\in\mathbf{L}(\mathfrak{H})\},$$

where $\mathbf{L}(\mathfrak{H})$ designates the space of bounded linear operators on $\mathfrak{H}$, we add ∞ to $\rho(T)$ if $T\in\mathbf{L}(\mathfrak{H})$. The adjoint T^* of T is defined by

$$T^*=\{\{f,g\}\in\mathfrak{H}^2\mid[g,u]-[f,v]=0 \text{ for all } \{u,v\}\in T\}.$$

Finally, by $M_\ell(T)$, $\Omega_\ell(T)$ and $N_\ell(T)$, $\ell\in\mathbb{C}$ we denote the sets

$$M_\ell(T)=\{\{\varphi,\psi\}\in T\mid\psi=\ell\varphi\},\quad \Omega_\ell(T)=\{\{\varphi,\psi\}\in T\mid\psi-\ell\varphi\in\nu(T-\ell)\},$$

$$N_\ell(T)=\{\{\varphi,\psi\}\in T\mid\psi-\ell\varphi\in\nu(T-\ell),\ \varphi\perp\nu(T-\ell)\}$$

and we also introduce

$$M_\infty(T)=\{\{\varphi,\psi\}\in T\mid\varphi=0\},\quad \Omega_\infty(T)=\{\{\varphi,\psi\}\in T\mid\varphi\in T(0)\},$$

$$N_\infty(T)=\{\{\varphi,\psi\}\in T\mid\varphi\in T(0),\ \psi\perp T(0)\}.$$

Note that $\mathfrak{D}(M_\ell(T))=\nu(T-\ell)$, $\mathfrak{R}(\Omega_\ell(T)-\ell)\subset\nu(T-\ell)$, $\ell\in\mathbb{C}$, and $\mathfrak{R}(M_\infty(T))=T(0)$, $\mathfrak{D}(\Omega_\infty(T))\subset T(0)$. Furthermore, if T is closed,

$$\Omega_\ell(T)=M_\ell(T)\dot{+}N_\ell(T),\quad \ell\in\mathbb{C}\cup\{\infty\},\quad \text{direct sum in } \mathfrak{H}^2.$$

To prove this equality for $\ell\in\mathbb{C}$, consider $\{\varphi,\psi\}\in\Omega_\ell(T)$. Denoting by φ_1 the orthogonal projection of φ onto $\nu(T-\ell)$, we find that $\{\varphi_1,\ell\varphi_1\}\in M_\ell(T)$ and that

$$\{\varphi,\psi\}-\{\varphi_1,\ell\varphi_1\}=\{\varphi-\varphi_1,\psi-\ell\varphi_1\}\in N_\ell(T),$$

which implies that $\{\varphi,\psi\}\in M_\ell(T)\dot{+}N_\ell(T)$. Hence $\Omega_\ell(T)\subset M_\ell(T)\dot{+}N_\ell(T)$ and as the other inclusion is trivial, the equality follows. For $\ell=\infty$ the argument is similar. The following results are easy to prove and are left to the reader.

LEMMA 1.1. *(i) T is a closed linear relation in $\mathfrak{H}$ with $\mu\in\rho(T)$ if and only if there exist operators $A,B\in\mathbf{L}(\mathfrak{H})$ with $(B-\mu A)^{-1}\in\mathbf{L}(\mathfrak{H})$, or $A^{-1}\in\mathbf{L}(\mathfrak{H})$ if $\mu=\infty$, such that $T=\{\{Ah,Bh\}\mid h\in\mathfrak{H}\}$. In either case we have $(T-\mu)^{-1}=A(B-\mu A)^{-1}$ or $T=BA^{-1}$ if $\mu=\infty$. (ii) Let T, S be linear relations with $\mu\in\rho(T)$, $\bar{\mu}\in\rho(S)$ and repesentations $T=\{\{Ah,Bh\}\mid h\in\mathfrak{H}\}$, $S=\{\{Ch,Dh\}\mid h\in\mathfrak{H}\}$, where A,B,C and $D\in\mathbf{L}(\mathfrak{H})$ are such that $(B-\mu A)^{-1}$, $(D-\bar{\mu}C)^{-1}\in\mathbf{L}(\mathfrak{H})$, or A^{-1}, $C^{-1}\in\mathbf{L}(\mathfrak{H})$ if $\mu=\infty$. Then $T=S^*$ if and only if $D^*A=C^*B$.*

We call the linear relation T dissipative if $\operatorname{Im}[g,f]\geq 0$ for all $\{f,g\}\in T$, symmetric if $T\subset T^*$, selfadjoint if $T=T^*$, contractive if $[g,g]\leq[f,f]$ for all $\{f,g\}\in T$, isometric if $T^{-1}\subset T^*$ and unitary if $T^{-1}=T^*$. Note that in the last three cases T is automatically (the graph of) an operator. For $\mu\in\mathbb{C}\backslash\mathbb{R}$ we define the Cayley transform C_μ and the inverse Cayley transform F_μ by

$$C_\mu(T) = \{\{g-\mu f, g-\bar{\mu} f\} \mid \{f,g\} \in T\},$$

$$F_\mu(T) = \{\{g-f, \mu g-\bar{\mu} f\} \mid \{f,g\} \in T\}.$$

Clearly, $F_\mu(C_\mu(T)) = T$ and $C_\mu(F_\mu(T)) = T$ for all linear relations T. For $\mu \in \mathbb{C}\backslash\mathbb{R}$ the map $T \to C_\mu(T)$ is bijection between all selfadjoint (symmetric) relations and all unitary (isometric) operators, while for $\mu \in \mathbb{C}^-(\mathbb{C}^+)$ this map is a bijection between all dissipative relations and all contractive (expansive) operators. The linear relation T in a Hilbert space $\mathfrak{K}$ is called an extension of the linear relation S in $\mathfrak{H}$ if $\mathfrak{H} \subset \mathfrak{K}$, the inner products of $\mathfrak{K}$ and $\mathfrak{H}$ coincide on $\mathfrak{H}$ and $S \subset T$; it is called canonical if $\mathfrak{K} = \mathfrak{H}$. We denote by $P_{\mathfrak{H}}$ the orthogonal projection of $\mathfrak{K}$ onto $\mathfrak{H}$. In the sequel we shall consider selfadjoint extensions of a given symmetric relation S and unitary extensions of a given isometric operator V in $\mathfrak{H}$. We remark that such extensions always exist, but they need not necessarily be canonical. Canonical extensions exist if and only if the defect numbers of the given relations are equal. For the symmetric relation S this means that for some $\mu \in \mathbb{C}\backslash\mathbb{R}$ $\dim \nu(S^* - \mu) = \dim \nu(S^* - \bar{\mu})$, while for the isometric operator V this means that $\dim \mathfrak{D}(V)^\perp = \dim \mathfrak{R}(V)^\perp$. Of course, one statement follows from the other by using the Cayley transformation. If A in the Hilbert space $\mathfrak{K}$ is a selfadjoint extension of S, then A and (or) $\mathfrak{K}$ are (is) called minimal if

$$\mathfrak{K} = \text{c.l.s.}\{(I + (\ell - \mu)(A - \ell)^{-1})f \mid f \in \mathfrak{H},\ \ell \in \rho(A)\},$$

where c.l.s. stands for closed linear span. Applying the Cayley transformation, we obtain a corresponding definition for a unitary extension W in $\mathfrak{K}$ of V: W and (or) $\mathfrak{K}$ are (is) called minimal if

$$\mathfrak{K} = \text{c.l.s.}\{(I - zW)^{-1})f \mid f \in \mathfrak{H},\ z = 0, \text{ or } 1/z \in \rho(W)\}.$$

We refer to Section 2 for more details concerning this correspondence via the Cayley transformation. Here we continue with the case of symmetric relations.

Let S be a closed symmetric relation in the Hilbert space $\mathfrak{H}$, then the socalled Von Neuman's identity

$$S^* = S \dotplus M_\ell(S^*) \dotplus M_{\bar{\ell}}(S^*), \text{ direct sums in } \mathfrak{H}^2,\ \ell \in \mathbb{C}\backslash\mathbb{R},$$

is valid, cf. [DdS]. If $\mathfrak{R}(S-\lambda)$ is closed for some $\lambda \in \mathbb{R}$ then we have that

$$S^* = S \dotplus \Omega_\lambda(S^*), \quad S \cap \Omega_\lambda(S^*) = M_\lambda(S).$$

It is clear that $S \dotplus \Omega_\lambda(S^*) \subset S^*$. In order to show the reverse inclusion we let $\{f,g\} \in S^*$. Then $g-\lambda f = k-\lambda h+\alpha$ for some $\{h,k\} \in S$ and $\alpha \in \nu(S^*-\lambda)$. Hence

$$\{f,g\} - \{h,k\} = \{f-h, \lambda(f-h)+\alpha\} \in \Omega_\lambda(S^*),$$

which implies that $\{f,g\} \in S \dotplus \Omega_\lambda(S^*)$. Thus we have shown the desired equality. In particular, the above shows that when $\Re(S-\lambda) = \mathfrak{H}$ for some $\lambda \in \mathbb{R}$, then $M_\lambda(S^*) = \{\{0,0\}\} = \Omega_\lambda(S^*)$ and this implies that S is selfadjoint. If $\Re(S-\lambda)$ is closed and $\nu(S-\lambda) = \{0\}$, then $S \dotplus \Omega_\lambda(S^*)$ is a direct sum in $\mathfrak{H}$ and $\Re(\Omega_\lambda(S^*)-\lambda) = \nu(S^*-\lambda)$. Note that now we have the following version of Von Neumann's identity: if $\Re(S-\lambda)$ is closed and $\nu(S-\lambda) = \{0\}$ then

$$S^* = S \dotplus M_\lambda(S^*) \dotplus N_\lambda(S^*), \text{ direct sums in } \mathfrak{H}^2, \ \lambda \in \mathbb{R}.$$

If $\mathfrak{D}(S)$ is closed, then

$$S^* = S \dotplus \Omega_\infty(S^*), \ S \cap \Omega_\infty(S^*) = M_\infty(S).$$

If $\mathfrak{D}(S)$ is closed and $S(0) = \{0\}$, then $S^* = S \dotplus \Omega_\infty(S^*)$ is a direct sum in $\mathfrak{H}^2$, $\mathfrak{D}(\Omega_\infty(S^*)) = S^*(0)$. Thus we have obtained another version of Von Neumann's identity: if $\mathfrak{D}(S)$ is closed and $S(0) = \{0\}$, then

$$S^* = S \dotplus M_\infty(S^*) \dotplus N_\infty(S^*), \text{ direct sums in } \mathfrak{H}^2.$$

PROPOSITION 1.2. *Let S be a closed symmetric relation in $\mathfrak{H}$. (i) If, for some $\lambda \in \mathbb{R}$, $\Re(S-\lambda)$ is closed, then $A = S \dotplus M_\lambda(S^*)$ is a canonical selfadjoint extension of S with $\nu(A-\lambda) = \nu(S^*-\lambda)$. If $\mathfrak{D}(S)$ is closed, then $A = S \dotplus M_\infty(S^*)$ is a canonical selfadjoint extension of S with $A(0) = S^*(0)$. In particular, under the condition that $\Re(S-\lambda)$, for some $\lambda \in \mathbb{R}$, or $\mathfrak{D}(S)$ is closed, S has equal defect numbers. (ii) If, for some $\lambda \in \mathbb{R}$, $\Re(S-\lambda)$ is closed and $\nu(S-\lambda) = \{0\}$, then $A = S \dotplus N_\lambda(S^*)$ is a canonical extension of S with $\lambda \in \rho(A)$. If $\mathfrak{D}(S)$ is closed and $S(0) = \{0\}$, then $A = S \dotplus N_\infty(S^*)$ is a canonical selfadjoint extension of S with $\infty \in \rho(A)$.*

Proof. The first part follows the Lemmas preceding Theorems 2 and 6 in [CdS], which state that $S \dotplus M_\lambda(S^*)$ is selfadjoint if and only if

$$\Re(S-\lambda) = \Re(S^*-\lambda) \cap (\Re(S-\lambda))^c,$$

where the superscript c stands for the closure of the set in $\mathfrak{H}$, and that $S \dotplus M_\infty(S^*)$ is selfadjoint if and only if

$$\mathfrak{D}(S) = \mathfrak{D}(S^*) \cap (\mathfrak{D}(S))^c.$$

The remaining parts of (i) are easy to verify and left to the reader. To show (ii), we first observe that

$$\mathfrak{R}(\Omega_\lambda(S^*)-\lambda)=\mathfrak{R}(N_\lambda(S^*)-\lambda)=\nu(S^*-\lambda).$$

Therefore, $\mathfrak{R}(A-\lambda)=\mathfrak{R}(S-\lambda)+\nu(S^*-\lambda)=\mathfrak{H}$ and in order to prove that A is selfadjoint it suffices by the Hellinger–Toeplitz Theorem to show that A is symmetric. Let $\{f,g\}\in A$, then it can be written as

$$\{f,g\}=\{h,k\}+\{\alpha,\beta\},\ \{h,k\}\in S,\ \{\alpha,\beta\}\in N_\lambda(S^*),$$

and

$$[g,f]=[k,h]+[k,\alpha]+[\beta,h]+[\beta,\alpha].$$

Now, $\{h,k\}\in S$ and $\{\alpha,\beta\}\in S^*$, which implies that $[k,h]\in\mathbb{R}$ and $[k,\alpha]+[\beta,h]\in\mathbb{R}$, and $\{\alpha,\beta\}\in N_\lambda(S^*)$ implies that $[\beta,\alpha]=[\beta-\lambda\alpha,\alpha]+\lambda[\alpha,\alpha]=\lambda[\alpha,\alpha]\in\mathbb{R}$. We conclude that $[g,f]\in\mathbb{R}$ for all $\{f,g\}\in A$ and hence A is symmetric. In order to show (ii) when $\mathfrak{D}(S)$ is closed and $S(0)=\{0\}$, we observe that

$$\mathfrak{D}(\Omega_\infty(S^*))=\mathfrak{D}(N_\infty(S^*))=S^*(0).$$

Hence $\mathfrak{D}(A)=\mathfrak{H}$ and since, as is easy to see, A is symmetric, it is selfadjoint. This completes the proof of the proposition.

If A in $\mathfrak{K}$ is a selfadjoint extension of S in $\mathfrak{H}$ we denote by $R(\ell)$ the compression of its resolvent to $\mathfrak{H}$:

$$R(\ell)=P_{\mathfrak{H}}(A-\ell)^{-1}|_{\mathfrak{H}},\ \ell\in\rho(A). \tag{1.1}$$

The function $R(\ell)$ is locally holomorphic on $\mathbb{C}\backslash\mathbb{R}$ with values in $\mathbf{L}(\mathfrak{H})$ and satisfies

$R(\ell)(S-\ell)\subset I$, for all $\ell\in\mathbb{C}\backslash\mathbb{R}$,

$R(\ell)=R(\bar{\ell})^*$, for all $\ell\in\mathbb{C}\backslash\mathbb{R}$, and

the kernel $\mathsf{K}_R(\ell,\lambda)=\dfrac{R(\ell)-R(\lambda)^*}{\ell-\bar{\lambda}}-R(\lambda)^*R(\ell)$ is nonnegative on $\mathbb{C}\backslash\mathbb{R}$.

A function with these properties is called a generalized resolvent of S. The following theorem is well known, see for example [DLS2].

THEOREM 1.3. *(i) If $R(\ell)$ is a generalized resolvent of S then there exist a minimal selfadjoint extension A of S, uniquely determined up to isomorphisms, such that the equality* (1.1) *is valid. (ii) If $R(\ell)$ is defined and holomorphic on a neighborhood $\mathcal{O}$ in $\mathbb{C}^+$ of a point $\ell_0\in\mathbb{C}^+$ with values in*

$\mathbf{L}(\mathfrak{H})$ *and has the properties that* $R(\ell_0)(S-\ell_0)\subset I$ *and the selfadjoint operator* $\mathsf{K}_R(\ell,\ell)$ *is nonnegative for all* $\ell\in\mathcal{O}$, *then* $R(\ell)$ *has unique extension to a generalized resolvent of* S.

Hence the equality (1.1) establishes a one to one correspondence between all (equivalence classes of isomorphic copies of) minimal selfadjoint extensions A of S on the one hand and all generalized resolvents $R(\ell)$ on the other. Clearly, a minimal selfadjoint extension is canonical if and only if the kernel associated with the corresponding generalized resolvent is identically equal to zero, i.e., if and only if $R(\ell)$ satisfies the resolvent equation. In the following definition and also elsewhere in the sequel we speak about locally meromorphic functions defined on $\mathbb{C}\backslash\mathbb{R}$ and certain relations between them. Whenever we say that such a relation holds on a set we mean to say that this relation is valid for those values of the argument that are contained in the set and for which the expressions in the relation make sense. Recall that a kernel $\mathsf{K}(\ell,\lambda)$ which is defined for ℓ,λ in some set Ω and has values in $\mathbf{L}(\mathfrak{H})$ is called nonnegative if $\mathsf{K}(\ell,\lambda)^*=\mathsf{K}(\lambda,\ell)$ on Ω and if for each $n\in\mathbb{N}$ and all choices of $\ell_i\in\Omega$ and $h_i\in\mathfrak{H}$, $i=1,2,\ldots,n$, the hermitian nxn matrix $([\mathsf{K}(\ell_i,\ell_j)h_i,h_j])$ is nonnegative.

DEFINITION 1.4. *A family of closed linear relations* $\mathcal{T}(\ell)$ *in* $\mathfrak{H}$ *defined for* $\ell\in\mathbb{C}\backslash\mathbb{R}$ *is called a Nevanlinna pair if* $\mathcal{T}(\ell)$ *can be written in the form*

(1.2) $\quad \mathcal{T}(\ell)=\{\{\mathcal{A}(\ell)f,\mathcal{B}(\ell)f\}\,|\,f\in\mathfrak{H}\}$ *on* $\mathbb{C}\backslash\mathbb{R}$,

where $\mathcal{A}(\ell)$ *and* $\mathcal{B}(\ell)$ *are functions, defined and locally meromorphic on* $\mathbb{C}\backslash\mathbb{R}$ *with values in* $\mathbf{L}(\mathfrak{H})$, *which satisfy the following requirements:*

(*a*) $\quad (\mathcal{B}(\ell)+\ell\mathcal{A}(\ell))^{-1}\in\mathbf{L}(\mathfrak{H})$ *on* $\mathbb{C}\backslash\mathbb{R}$ (*nondegeneracy*),

(*b*) $\quad \mathcal{A}(\bar{\ell})^*\mathcal{B}(\ell)-\mathcal{B}(\bar{\ell})^*\mathcal{A}(\ell)=0$ *on* $\mathbb{C}\backslash\mathbb{R}$ (*symmetry*), *and*

(*c*) $\quad$ *the kernel* $\mathsf{K}_{\mathcal{A},\mathcal{B}}(\ell,\lambda)=\dfrac{\mathcal{A}(\lambda)^*\mathcal{B}(\ell)-\mathcal{B}(\lambda)^*\mathcal{A}(\ell)}{\ell-\bar{\lambda}}$ *is nonnegative on* $\mathbb{C}\backslash\mathbb{R}$.

If $\mathcal{T}(\ell)$ *is a Nevanlinna pair and is of the above form, we say that it is determined or generated by the ordered pair* $(\mathcal{A}(\ell),\mathcal{B}(\ell))$ *and instead of writing the equality* (1.2) *we frequently use the shorthand notation* $\mathcal{T}(\ell)=(\mathcal{A}(\ell),\mathcal{B}(\ell))$. *We denote the class of all Nevanlinna pairs by* $\mathbf{N}(\mathfrak{H})$ *and call it the extended Nevanlinna class.*

According to Lemma 1.1 (*i*) condition (*a*) above is equivalent to $-\ell\in\rho(\mathcal{T}(\ell))$ and, since by condition (*c*) $\mathcal{T}(\ell)$ is dissipative for $\operatorname{Im}\ell>0$, it follows that $\mathcal{T}(\ell)$ is maximal dissipative and hence $\mathbb{C}^{-}\subset\rho(\mathcal{T}(\ell))$, $\operatorname{Im}\ell>0$, see, e.g., [DdS]. Note also that on account of Lemma 1.1 (ii) condition (b) is equivalent to $\mathcal{T}(\ell)^{*}=\mathcal{T}(\bar{\ell})$ on $\mathbb{C}\backslash\mathbb{R}$. It is easy to see that two ordered pairs $(\mathcal{A}_1(\ell),\mathcal{B}_1(\ell))$ and $(\mathcal{A}_2(\ell),\mathcal{B}_2(\ell))$ determine the same Nevanlinna pair if and only if there exists a locally meromorphic function $\mathcal{G}(\ell)$ on $\mathbb{C}\backslash\mathbb{R}$ with values in $\mathbf{L}(\mathfrak{H})$ which has an inverse with the same properties, such that $\mathcal{A}_1(\ell)\mathcal{G}(\ell)=\mathcal{A}_2(\ell)$ and $\mathcal{B}_1(\ell)\mathcal{G}(\ell)=\mathcal{B}_2(\ell)$ on $\mathbb{C}\backslash\mathbb{R}$. Clearly, this condition defines an equivalence relation between ordered pairs $(\mathcal{A}(\ell),\mathcal{B}(\ell))$ of functions $\mathcal{A}(\ell)$ and $\mathcal{B}(\ell)$ satisfying (a)–(c) of Definition 1.4 and $\mathbf{N}(\mathfrak{H})$ can be seen as the set of all equivalence classes. In fact, the formula $\mathcal{T}(\ell)=(\mathcal{A}(\ell),\mathcal{B}(\ell))$ establishes a one to one correspondence between $\mathcal{T}(\ell)\in\mathbf{N}(\mathfrak{H})$ determined by $(\mathcal{A}(\ell),\mathcal{B}(\ell))$ and the equivalence class containing this ordered pair. In the sequel we make no distinction between the ordered pair, the equivalence class containing it and the corresponding family of linear relations, and refer to each of them as a Nevanlinna pair. If $\mathcal{T}(\ell)=(\mathcal{A}(\ell),\mathcal{B}(\ell))\in\mathbf{N}(\mathfrak{H})$, then $\mathcal{T}(\ell)$ is (the graph of) a meromorphic function on $\mathbb{C}\backslash\mathbb{R}$ if and only if $\mathcal{A}(\ell)$ has an inverse which is locally meromorphic on $\mathbb{C}\backslash\mathbb{R}$ and has values in $\mathbf{L}(\mathfrak{H})$. In this case $\mathcal{T}(\ell)=(I,\mathcal{B}(\ell)\mathcal{A}(\ell)^{-1})=\mathcal{B}(\ell)\mathcal{A}(\ell)^{-1}$, which is a Nevanlinna function in the usual sense.

THEOREM 1.5. *If for ℓ in an open set $\mathcal{O}\subset\mathbb{C}^{+}$ the relation $\mathcal{T}(\ell)$ is defined by $\mathcal{T}(\ell)=\{\{\mathcal{A}(\ell)h,\mathcal{B}(\ell)h\}\mid h\in\mathfrak{H}\}$, where $\mathcal{A}(\ell)$ and $\mathcal{B}(\ell)$ are functions, defined and holomorphic on $\mathcal{O}$ with values in $\mathbf{L}(\mathfrak{H})$, such that $(\mathcal{B}(\ell)+\ell\mathcal{A}(\ell))^{-1}\in\mathbf{L}(\mathfrak{H})$ and $\operatorname{Im}\mathcal{A}(\ell)^{*}\mathcal{B}(\ell)\geq 0$ for all $\ell\in\mathcal{O}$, then it can be uniquely extended to an element in $\mathbf{N}(\mathfrak{H})$. Furthermore, the following statements are equivalent.*

(i) $\mathcal{T}(\ell)\in\mathbf{N}(\mathfrak{H})$.

(ii) *There exist a Hilbert space $\mathfrak{K}$, a selfadjoint relation A in $\mathfrak{K}$ and a linear mapping $\Gamma:\mathfrak{H}\to\mathfrak{K}$ with $\Gamma^{*}\Gamma=I$ such that $\mathcal{T}(\ell)=(\mathcal{A}(\ell),\mathcal{B}(\ell))$ with*

$$\mathcal{A}(\ell)=-\Gamma^{*}(A-\ell)^{-1}\Gamma \text{ and } \mathcal{B}(\ell)=\Gamma^{*}(I+\ell(A-\ell)^{-1})\Gamma.$$

The selfadjoint relation A in (ii) can be chosen Γ minimal, i.e., such that for some $\mu\in\mathbb{C}\backslash\mathbb{R}$

$$\mathfrak{K}=\text{c.l.s.}\{(I+(\ell-\mu)(A-\ell)^{-1})\Gamma h\mid h\in\mathfrak{H},\ \ell\in\rho(A)\}$$

in which case it is uniquely determined by $\mathcal{T}(\ell)\in\mathbf{N}(\mathfrak{H})$ up to isomorphisms.

Proof. Put $R(\ell)=-\mathcal{A}(\ell)(\mathcal{B}(\ell)+\ell\mathcal{A}(\ell))^{-1}$. Then $R(\ell)$ satisfies the hypotheses of Theorem 1.3 (*ii*) and hence can be extended uniquely to a generalized resolvent of the trivial symmetric relation $S=\{\{0,0\}\}$. The formula

$$\mathcal{T}(\ell)=\{\{-R(\ell)f,(I+\ell R(\ell))f\}\mid f\in\mathfrak{H}\}$$

extends the given family of relations to all $\ell\in\mathbb{C}\backslash\mathbb{R}$ and defines a Nevanlinna pair. It is easy to see that by analytic continuation this extension is unique. This proves the first part of the theorem. To prove the implication $(i)\Rightarrow(ii)$ we assume (i) and use the same argument as above to conclude that $\mathcal{T}(\ell)=(-R(\ell),I+\ell R(\ell))$, where $R(\ell)$ is a generalized resolvent of $S=\{\{0,0\}\}$. Hence, on account of Theorem 1.3 (*i*), there exists a minimal selfadjoint extension A of S in some Hilbert space $\mathfrak{K}$ such that $R(\ell)$ admits a representation of the form (1.1). It follows that

$$\mathcal{T}(\ell)=(\,-\Gamma^*(A-\ell)^{-1}\Gamma,\,\Gamma^*(I+\ell(A-\ell)^{-1})\Gamma\,),$$

where Γ equals the restriction operator to $\mathfrak{H}$, so that $\Gamma^*=P_{\mathfrak{H}}$ and $\Gamma^*\Gamma=I$, the identity on $\mathfrak{H}$. This implies (*ii*). Moreover, the minimality of A implies the Γ-minimality. If also

$$\mathcal{T}(\ell)=(\,-\Gamma_1^*(A_1-\ell)^{-1}\Gamma_1,\Gamma_1^*(I+\ell(A_1-\ell)^{-1})\Gamma_1\,),$$

where A_1 is a selfadjoint relation in a Hilbert space $\mathfrak{K}_1$, $\Gamma_1:\mathfrak{H}\to\mathfrak{K}_1$ is a linear mapping with $\Gamma_1^*\Gamma_1=I$, such that A_1 is Γ_1-minimal, then

$$\Gamma^*(A-\ell)^{-1}\Gamma=\Gamma_1^*(A_1-\ell)^{-1}\Gamma_1$$

and via

$$W\Gamma h=\Gamma_1 h,\ W(A-\ell)^{-1}\Gamma h=(A_1-\ell)^{-1}\Gamma_1 h,\ \ell\in\mathbb{C}\backslash\mathbb{R},\ h\in\mathfrak{H},$$

one can construct a unitary mapping $W:\mathfrak{K}\to\mathfrak{K}_1$ such that $W\Gamma=\Gamma_1$ and

$$A_1=\{\{Wf,Wg\}\mid\{f,g\}\in A\}.$$

This proves the last statement in the theorem. Now assume (*ii*), then

$$\mathcal{B}(\ell)+\ell\mathcal{A}(\ell)=I,\ \text{for all}\ \ell\in\mathbb{C}\backslash\mathbb{R}$$

and using the resolvent equation, we obtain that

$$\mathsf{K}_{\mathcal{A},\mathcal{B}}(\ell,\lambda)=\Gamma^*(A-\bar{\lambda})^{-1}(I-\Gamma\Gamma^*)(A-\ell)^{-1}\Gamma,\ \text{for all}\ \ell,\lambda\in\mathbb{C}\backslash\mathbb{R}.$$

The first equality implies condition (*a*) and the latter conditions (*b*) and (*c*) of Definition 1.4, which shows that (*i*) is valid. This completes the proof.

Theorem 1.5 implies in particular that each Nevanlinna pair contains an ordered pair $(\mathcal{A}(\ell),\mathcal{B}(\ell))$ in which $\mathcal{A}(\ell)$ and $\mathcal{B}(\ell)$ are both locally holomorphic on all of $\mathbb{C}\backslash\mathbb{R}$, satisfy the conditions (a)–(c) of Definition 1.4 for all $\ell,\lambda\in\mathbb{C}\backslash\mathbb{R}$ and even have the property that $\mathcal{B}(\ell)+\ell\mathcal{A}(\ell)=I$ for all $\ell\in\mathbb{C}\backslash\mathbb{R}$. Such an ordered pair is uniquely determined by the Nevanlinna pair it generates and is given by

$$\hat{\mathcal{A}}(\ell)=\mathcal{A}(\ell)(\ell\mathcal{A}(\ell)+\mathcal{B}(\ell))^{-1},\quad \hat{\mathcal{B}}(\ell)=\mathcal{B}(\ell)(\ell\mathcal{A}(\ell)+\mathcal{B}(\ell))^{-1},$$

cf. (0.7). If $\mathfrak{H}=\mathbb{C}^n$ we write $\mathbf{N}^n$ instead of $\mathbf{N}(\mathbb{C}^n)$. In this case condition (a) in Definition 1.4 may be replaced by

$$\operatorname{rank}(\mathcal{A}(\ell)^*:\mathcal{B}(\ell)^*)=n \text{ on } \mathbb{C}\backslash\mathbb{R}.$$

It is now easy to see that each ordered pair $(\mathcal{M}(\ell),\mathcal{N}(\ell))$ satisfying the conditions (0.5) in the Introduction is a generating element for a Nevanlinna pair in the class $\mathbf{N}^n$ and that each Nevanlinna pair in $\mathbf{N}^n$ contains such an element uniquely determined up to multiplication from the right by a locally holomorphic invertible $n\times n$ matrix function.

COROLLARY 1.6. *The multivalued part $\mathcal{T}(\ell)(0)$ of $\mathcal{T}(\ell)\in\mathbf{N}(\mathfrak{H})$ is independent of $\ell\in\mathbb{C}\backslash\mathbb{R}$. In addition, if in the representation in (ii) of Theorem 1.5 we have that $\mathcal{A}(\ell)f=0$ for some $\ell\in\mathbb{C}\backslash\mathbb{R}$, then $\{\mathcal{A}(\ell)f,\mathcal{B}(\ell)f\}=\{0,f\}$ and $\{0,\Gamma f\}\in A$. Conversely, if $\{0,\Gamma f\}\in A$, then $\mathcal{A}(\ell)f=0$ for all $\ell\in\mathbb{C}\backslash\mathbb{R}$.*

Proof. If $\mathcal{A}(\ell)f=0$, then with $h=(A-\ell)^{-1}\Gamma f$ we have that $\Gamma^*h=0$ and $\{h,\Gamma f+\ell h\}\in A$. Hence

$$0=[\Gamma f+\ell h,h]-[h,\Gamma f+\ell h]=(\ell-\bar{\ell})[h,h],$$

which implies that $h=0$ and therefore $\{0,\Gamma f\}\in A$. Conversely, if $\{0,\Gamma f\}\in A$, then for all $\ell\in\mathbb{C}\backslash\mathbb{R}$ $\{0,\Gamma f\}\in A-\ell$ and $(A-\ell)^{-1}\Gamma f=0$, so that $\mathcal{A}(\ell)f=0$. These arguments show that $\mathcal{T}(\ell)(0)=\Gamma^*A(0)$, which proves the first statement in the corollary. The proofs of the remaining statements are left to the reader.

For another proof of the first statement in Corollary 1.6 we refer to [DdS]. The next proposition is based on Theorem 1.5 and contains part (i) of Theorem 0.1. For the proof of Theorem 0.1 (ii) we refer to Theorem 3.3 in Section 3. In the following we shall use without further explanation the formulas in the Introduction and the Appendix. In particular we refer the reader to formula (0.6) for the definition of the matrix $\mathbb{P}_{\mathcal{M},\mathcal{N}}=((\mathbb{P}_{\mathcal{M},\mathcal{N}})_{ij}^{pq})$.

PROPOSITION 1.7. *Let the Nevanlinna pair* $(\mathcal{M}(\ell),\mathcal{N}(\ell))\in\mathbf{N}^n$ *have the operator representation*

$$(\mathcal{M}(\ell),\mathcal{N}(\ell)) = (-\Gamma^*(A-\ell)^{-1}\Gamma, \Gamma^*(I+\ell(A-\ell)^{-1})\Gamma)$$

with A *and* Γ *as in Theorem* 1.5 *with* $\mathfrak{H}=\mathbb{C}^n$. *Then the entries of the corresponding matrix* $\mathbb{P}_{\mathcal{M},\mathcal{N}}$ *are given by*

$$(1.3)\quad (\mathbb{P}_{\mathcal{M},\mathcal{N}})_{ij}^{pq} = (D_\ell)^p(D_{\bar\lambda})^q(\ell\mathcal{K}_i(\ell)+\mathcal{L}_i(\ell))\Gamma^*(A-\ell)^{-1}(I-\Gamma\Gamma^*). \cdot(A-\lambda)^{-*}\Gamma(\lambda\mathcal{K}_j(\lambda)+\mathcal{L}_j(\lambda))^*|_{\ell=w_i,\lambda=w_j}$$

and hence $\mathbb{P}_{\mathcal{M},\mathcal{N}}\geq 0$. *If in addition* $(\mathcal{M}(\ell),\mathcal{N}(\ell))$ *is a solution of the interpolation problem* (IP), *then the matrix* $\mathbb{P}_{\mathcal{M},\mathcal{N}}$ *is a solution of the Lyapunov equation and its entries can be written as*

$$(1.4)\quad (\mathbb{P}_{\mathcal{M},\mathcal{N}})_{ij}^{pq} = \Big((D_\ell)^p\left(\mathcal{K}_i(\ell)\Gamma^*+(\ell\mathcal{K}_i(\ell)+\mathcal{L}_i(\ell))\Gamma^*(A-\ell)^{-1}\right)|_{\ell=w_i}\Big)\cdot \cdot\Big((D_\lambda)^q\left(\mathcal{K}_j(\lambda)\Gamma^*+(\lambda\mathcal{K}_j(\lambda)+\mathcal{L}_j(\lambda))\Gamma^*(A-\lambda)^{-1}\right)|_{\lambda=w_j}\Big)^*.$$

Proof. It is easy to see that $\hat{\mathcal{M}}(\ell)$, $\hat{\mathcal{N}}(\ell)$ defined by (0.7) can be written as

$$\hat{\mathcal{M}}(\ell) = -\Gamma^*(A-\ell)^{-1}\Gamma,\ \hat{\mathcal{N}}(\ell)=\Gamma^*(I+\ell(A-\ell)^{-1})\Gamma$$

and that, on account of the resolvent equation for A,

$$\frac{\hat{\mathcal{N}}(\bar\ell)^*\hat{\mathcal{M}}(\bar\lambda)-\hat{\mathcal{M}}(\bar\ell)^*\hat{\mathcal{N}}(\bar\lambda)}{\ell-\bar\lambda} = \Gamma^*(A-\ell)^{-1}(I-\Gamma\Gamma^*)(A-\bar\lambda)^{-1}\Gamma,$$

which implies (1.3). The nonnegativity of $\mathbb{P}_{\mathcal{M},\mathcal{N}}$ follows from (1.3), the fact that, since $\Gamma^*\Gamma=I$, $(I-\Gamma\Gamma^*)\geq 0$, and, for example, the Cauchy formula for derivatives of holomorphic operator functions. If the pair $\mathcal{M}(\ell)$, $\mathcal{N}(\ell)$ is a solution of the interpolation problem (IP) then so is the pair $\hat{\mathcal{M}}(\ell)$, $\hat{\mathcal{N}}(\ell)$ and hence, since $\ell\hat{\mathcal{M}}(\ell)+\hat{\mathcal{N}}(\ell)=I$,

$$(1.5)\quad (\mathcal{K}_i(\ell)-(\ell\mathcal{K}_i(\ell)+\mathcal{L}_i(\ell))\hat{\mathcal{M}}(\ell))^{(p)}|_{\ell=w_i}=0,\quad i=1,2,\ldots,m,\ p=0,1,\ldots r_i,$$

$$(1.6)\quad ((\ell\mathcal{K}_i(\ell)+\mathcal{L}_i(\ell))\hat{\mathcal{N}}(\ell)-\mathcal{L}_i(\ell))^{(p)}|_{\ell=w_i}=0,\quad i=1,2,\ldots,m,\ p=0,1,\ldots r_i.$$

The equalities (1.5) imply that the righthand side of (1.4) does not alter when we insert in between the two factors the expression $(I-\Gamma\Gamma^*)$. Again since $\Gamma^*\Gamma=I$, $\Gamma^*(I-\Gamma\Gamma^*)=0$, it follows that the righthand side of (1.4) coincides with the righthand side of (1.3) and hence the equality (1.4) is valid. From (1.5) and (1.6) it follows that $\tilde{\mathcal{K}}(\ell)$ and $\tilde{\mathcal{L}}(\ell)$ defined by (A.2) in the Appendix interpolate the data of the interpolation problem (IP) and consequently, by (A.3) and Proposition A.2, $\mathbb{P}_{\mathcal{M},\mathcal{N}}$ is a solution of the Lyapunov equation. This completes the proof.

We denote by $\mathbf{R}^2(\mathfrak{H})$ the set of all 2x2 matrices $V=(V_{ij})$ with entries $V_{ij}\in\mathbf{L}(\mathfrak{H})$. For example, the solution matrix U defined in the Introduction has values $U(\ell)\in\mathbf{R}^2(\mathbb{C}^n)$. Let $V\in\mathbf{R}^2(\mathfrak{H})$, then for any relation T in $\mathfrak{H}$ we define its V-transform $\mathfrak{t}_V(T)$ by

$$\mathfrak{t}_V(T)=\{\{V_{21}g+V_{22}f,\ V_{11}g+V_{12}f\}\mid\{f,g\}\in T\}.$$

If $T=\{\{Ah,Bh\}\mid h\in\mathfrak{H}\}$ with $A,B\in\mathbf{L}(\mathfrak{H})$, then

$$\mathfrak{t}_V(T)=\{\{(V_{21}B+V_{22}A)h,(V_{11}B+V_{12}A)h\}\mid h\in\mathfrak{H}\}$$

and if furthermore the operator $(V_{21}B+V_{22}A)^{-1}\in\mathbf{L}(\mathfrak{H})$, then

$$\mathfrak{t}_V(T)=(V_{11}B+V_{12}A)(V_{21}B+V_{22}A)^{-1}.$$

This transformation could be called the right V-transform of T (since the inverse is on the right) as opposed to a similar transform ${}_V\mathfrak{t}(T)$ in [LT2] which could be called the left transform. This transform is defined for those $V\in\mathbf{R}^2(\mathfrak{H})$ for which V_{21} and linear relations T in $\mathfrak{H}$ for which $T+V_{11}V_{21}^{-1}$ are boundedly invertible and satisfies, if also $(V_{11}V_{21}^{-1}V_{22}-V_{12})^{-1}\in\mathbf{L}(\mathfrak{H})$,

$${}_V\mathfrak{t}(T)=\mathfrak{t}_W(T)\ \text{ with } W=\begin{pmatrix}0&-I\\I&0\end{pmatrix}V^{-1}\begin{pmatrix}0&-I\\I&0\end{pmatrix}.$$

Note that the Cayley transform and its inverse are examples of V-transforms. It is easy to see that if $V,W\in\mathbf{R}^2(\mathfrak{H})$, then $\mathfrak{t}_{VW}=\mathfrak{t}_V\circ\mathfrak{t}_W$. In the sequel we shall need the following simple result.

LEMMA 1.8. *If $V\in\mathbf{R}^2(\mathfrak{H})$ and $V_{21}^{-1}\in\mathbf{L}(\mathfrak{H})$, then we have that*

$$\mathfrak{t}_V(T)=V_{11}V_{21}^{-1}-(V_{11}V_{21}^{-1}V_{22}-V_{12})(T+V_{21}^{-1}V_{22})^{-1}V_{21}^{-1}.$$

Proof. The following equalities are easily verified one after the other:

$$T+V_{21}^{-1}V_{22}=\{\{f,g+V_{21}^{-1}V_{22}f\}\mid\{f,g\}\in T\},$$

$$(T+V_{21}^{-1}V_{22})^{-1}=\{\{g+V_{21}^{-1}V_{22}f,f\}\mid\{f,g\}\in T\},$$

$$(T+V_{21}^{-1}V_{22})^{-1}V_{21}^{-1}=\{\{V_{21}g+V_{22}f,f\}\mid\{f,g\}\in T\},$$

$$(V_{11}V_{21}^{-1}V_{22}-V_{12})(T+V_{21}^{-1}V_{22})^{-1}V_{21}^{-1}=$$
$$\{\{V_{21}g+V_{22}f,V_{11}V_{21}^{-1}V_{22}f-V_{12}f\}\mid\{f,g\}\in T\},$$

$$V_{11}V_{21}^{-1}-(V_{11}V_{21}^{-1}V_{22}-V_{12})(T+V_{21}^{-1}V_{22})^{-1}V_{21}^{-1}=$$

$$\{\{V_{21}g+V_{22}f, V_{11}g+V_{11}V_{21}^{-1}V_{22}f-V_{11}V_{21}^{-1}V_{22}f+V_{12}f\} \mid \{f,g\}\in T\} = t_V(T).$$

This completes the proof.

2. Resolvent matrices

The theory of resolvent matrices originates from M.G. Kreĭn [Kr] and has been studied in the papers with Saakjan [KS1,2] and Šmuljan [KSm]. Extensions of the theory to a context of indefinite inner product spaces are given in [LS] and in [KL3]. Resolvent matrices for symmetric linear relations are studied in the paper [LT2]. As we shall show in the next section there is a simple solution matrix for the interpolation problem (IP) which can be obtained from a certain limit of resolvent matrices for a symmetric relation S. In order to show that this limit remains a resolvent matrix for S we include the theory of resolvent matrices associated with isometric operators. This case is also of interest of its own. In our treatment we simplify some of the arguments in [LT2].

We consider a closed symmetric relation S in a Hilbert space $\mathfrak{H}$ and assume that S has equal defect numbers so that it has canonical selfadjoint extensions. We fix one of these, A_0, say, and denote its resolvent by $R_0(\ell)$. For $\ell,\mu\in\rho(A_0)$ the operator $I+(\ell-\mu)R_0(\ell)$ is boundedly invertible from $\mathfrak{H}$ onto $\mathfrak{H}$ and is a bijection from $\nu(S^*-\mu)$ onto $\nu(S^*-\ell)$. Let $\mathfrak{G}$ be a Hilbert space and, for a fixed point $\mu\in\mathbb{C}\backslash\mathbb{R}$, let $\Gamma_{\mathfrak{G},\bar{\mu}}$ be a boundedly invertible mapping from $\mathfrak{G}$ onto $\nu(S^*-\bar{\mu})$. Then the operator $\Gamma_{\bar{\mu}}(\ell)\in\mathbf{L}(\mathfrak{G},\mathfrak{H})$ defined by

$$\Gamma_{\bar{\mu}}(\ell) = \big(I+(\ell-\bar{\mu})R_0(\ell)\big)\Gamma_{\mathfrak{G},\bar{\mu}}, \qquad \ell\in\rho(A_0),$$

is a boundedly invertible mapping with range $\mathfrak{R}(\Gamma_{\bar{\mu}}(\ell)) = \nu(S^*-\ell)$. Its adjoint maps $\nu(S^*-\ell)$ one to one onto $\mathfrak{G}$ and coincides with the zero operator on $\mathfrak{R}(S-\bar{\ell})$. Note that $\Gamma_{\bar{\mu}}(\bar{\mu}) = \Gamma_{\mathfrak{G},\bar{\mu}}$. With the function $\Gamma_{\bar{\mu}}(\ell)$ we associate a so-called Q-function $Q(\ell)$ of the closed symmetric linear relation S in $\mathfrak{H}$ in the following way: $Q(\ell)$ is defined on $\rho(A_0)$, with values in $\mathbf{L}(\mathfrak{G})$, such that it satisfies

$$\frac{Q(\ell)-Q(\lambda)^*}{\ell-\bar{\lambda}} = \Gamma_{\bar{\mu}}(\lambda)^*\Gamma_{\bar{\mu}}(\ell), \qquad \ell,\lambda\in\rho(A_0).$$

One can show (see, e.g., [LT1]) that $Q(\ell)$ is defined by this relation up to a constant selfadjoint operator $C_Q\in\mathbf{L}(\mathfrak{G})$ and for any $\tau\in\mathbb{C}\backslash\mathbb{R}$ it has the form

$$Q(\ell)=C_Q+\Gamma_{\bar{\mu}}(\tau)^*(\ell-\mathrm{Re}\tau+(\ell-\tau)(\ell-\bar{\tau})(A_0-\ell)^{-1})\Gamma_{\bar{\mu}}(\tau).$$

Hence $Q(\ell)$ is a locally holomorphic function on $\mathbb{C}\backslash\mathbb{R}$ with values in $\mathbf{L}(\mathfrak{G})$, $Q(\ell)^*=Q(\bar{\ell})$ and $\mathrm{Im}\,Q(\ell)$ is positive and boundedly invertible for $\ell\in\mathbb{C}^+$. These properties are characteristic for Q–functions, i.e., if $Q(\ell)$ has these properties, then it is the Q–function of a closed simple symmetric relation (and hence an) operator S in some Hilbert space $\mathfrak{H}$, and S and a canonical extension A_0 are up to isomorphisms uniquely determined. The Q–function plays an important role in the parametrization of all selfadjoint extensions of S. The formula, due to Kreĭn,

$$R(\ell)=R_0(\ell)-\Gamma_{\bar{\mu}}(\ell)(Q(\ell)+\mathcal{T}(\ell))^{-1}\Gamma_{\bar{\mu}}(\bar{\ell})^* \tag{2.1}$$

establishes a one to one correspondence between all generalized resolvents (and hence all minimal selfadjoint extensions) of S and the Nevanlinna pairs $\mathcal{T}(\ell)=(\mathcal{A}(\ell),\mathcal{B}(\ell))\in\mathbf{N}(\mathfrak{G})$. In this formula $(Q(\ell)+\mathcal{T}(\ell))^{-1}=\mathcal{A}(\ell)(Q(\ell)\mathcal{A}(\ell)+\mathcal{B}(\ell))^{-1}$ is a bounded operator on $\mathfrak{G}$ for $\ell\in\mathbb{C}\backslash\mathbb{R}$.

From now on we shall only consider symmetric relations S in $\mathfrak{H}$ with finite equal defect numbers n, say. Let $\mathfrak{L}\subset\mathfrak{H}$ be a subspace with $\dim\mathfrak{L}=n$. A point $\ell\in\mathbb{C}$ is an $\mathfrak{L}$–regular point for S if $\mathfrak{H}$ can be decomposed as

$$\mathfrak{H}=\mathfrak{R}(S-\ell)\dot{+}\mathfrak{L},\qquad \text{direct sum,}$$

and the point ∞ is called $\mathfrak{L}$–regular for S if

$$\mathfrak{H}=\mathfrak{D}(S)\dot{+}\mathfrak{L},\qquad \text{direct sum.}$$

By $\rho_S(\mathfrak{L})$ we denote the set of all points ℓ in the extended complex plane such that ℓ and $\bar{\ell}$ are $\mathfrak{L}$–regular points for S. Thus by definition $\rho_S(\mathfrak{L})$ is symmetric with respect to the real axis. It follows that $\rho_S(\mathfrak{L})\cap\rho(A_0)$ is open and if $\rho_S(\mathfrak{L})\cap\rho(A_0)\neq\emptyset$, then $\rho(A_0)\backslash\rho_S(\mathfrak{L})$ is empty or a discrete subset of $\rho(A_0)$. We shall call $\mathfrak{L}$ a module space for S if $\rho_S(\mathfrak{L})\cap\rho(A_0)\neq\emptyset$, or, equivalently, $\rho_S(\mathfrak{L})\cap\mathbb{C}\backslash\mathbb{R}\neq\emptyset$. In the following $\mathfrak{L}$ will be a fixed module space for S and G a bijection from $\mathfrak{G}$ onto $\mathfrak{L}$. It can be shown (cf. [LT2], Lemma 3.1) that

$$\rho_S(\mathfrak{L})\cap\rho(A_0)=\{\ell\in\rho(A_0)\mid\Gamma_{\bar{\mu}}(\ell)^*G \text{ and } \Gamma_{\bar{\mu}}(\bar{\ell})^*G:\mathfrak{G}\to\mathfrak{G} \text{ are bijective}\}.$$

Associated with $\mathfrak{L}$ are two operators $\mathcal{P}(\ell)$ and $\mathcal{Q}(\ell)$. By definition $\mathcal{P}(\ell)$ is the projection of $\mathfrak{H}$ onto $\mathfrak{L}$ parallel to $\mathfrak{R}(S-\ell)$ and the operator $\mathcal{Q}(\ell)$ is defined by

$$\mathcal{Q}(\ell)=R_0(\ell)(I-\mathcal{P}(\ell)).$$

Clearly, $\mathcal{P}(\ell)$ does not depend on the particular choice of the canonical extension A_0 and, since $I-\mathcal{P}(\ell)$ maps $\mathfrak{H}$ onto $\mathfrak{R}(S-\ell)$, neither does $\mathcal{Q}(\ell)$. Observe that for $\ell\in\rho_S(\mathfrak{L})\cap\rho(A_0)$

$$\mathcal{P}(\ell)=G(\Gamma_{\bar{\mu}}(\bar{\ell})^*G)^{-1}\Gamma_{\bar{\mu}}(\bar{\ell})^*.$$

If $\infty\in\rho_S(\mathfrak{L})\cap\rho(A_0)$ we define $\mathcal{P}(\infty)$ as the projection of $\mathfrak{H}$ onto $\mathfrak{L}$ parallel to $\mathfrak{D}(S)$ and

$$\mathcal{Q}(\infty)=(A_0-\mu)(I-\mathcal{P}(\infty))=(S-\mu)(I-\mathcal{P}(\infty)).$$

Recall that $\infty\in\rho(A_0)$ implies that A_0 is a bounded operator and for this to be possible S has to be a (not necessarily densely defined) bounded operator at the outset. If $\infty\in\rho_S(\mathfrak{L})\cap\rho(A_0)$, then

$$(2.2)\qquad \mathcal{P}(\ell)\to\mathcal{P}(\infty) \text{ and } \mathcal{Q}(\ell)\to 0, \quad \text{as } \ell\to\infty.$$

The first limit we shall prove below, see (2.18) further on, while the second is a simple consequence of the first. The mapping $\ell\to G^*R(\ell)G$, $\ell\in\rho_S(\mathfrak{L})$, where $R(\ell)$ is a generalized resolvent of S, is called an $\mathfrak{L}$–resolvent for S. Clearly, it has values in $\mathbf{L}(\mathfrak{G})$ and to each $\mathfrak{L}$–resolvent for S there corresponds precisely one generalized resolvent of S.

Definition 2.1. *A function $W(\ell)$ defined on $\rho_S(\mathfrak{L})$ with values in $\mathbf{R}^2(\mathfrak{G})$ is called an $\mathfrak{L}$–resolvent matrix for S if it has the following properties:*

(i) $W(\ell)$ is locally holomorphic and invertible on $\rho_S(\mathfrak{L})$.

(ii) $W_{22}(\ell)\mathcal{A}(\ell)+W_{21}(\ell)\mathcal{B}(\ell)$ is invertible for each $\ell\in\rho_S(\mathfrak{L})$ and each Nevanlinna pair $(\mathcal{A}(\ell),\mathcal{B}(\ell))\in\mathbf{N}(\mathfrak{G})$.

(iii) The formula

$$G^*R(\ell)G=t_{W(\ell)}(\mathcal{T}(\ell))=(W_{12}(\ell)\mathcal{A}(\ell)+W_{11}(\ell)\mathcal{B}(\ell))(W_{22}(\ell)\mathcal{A}(\ell)+W_{21}(\ell)\mathcal{B}(\ell))^{-1}$$

establishes a one to one correspondence between all $\mathfrak{L}$–resolvents of S and all Nevanlinna pairs $\mathcal{T}(\ell)=(\mathcal{A}(\ell),\mathcal{B}(\ell))\in\mathbf{N}(\mathfrak{G})$.

It is clear from this definition that if we multiply an $\mathfrak{L}$–resolvent matrix for S by any locally holomorphic function $\beta(\ell)\neq 0$ and from the right by any $\left(\begin{smallmatrix}0 & -I\\ I & 0\end{smallmatrix}\right)$–unitary matrix U, we again obtain an $\mathfrak{L}$–resolvent matrix for S. In fact, an $\mathfrak{L}$–resolvent matrix for S is uniquely determined modulo such multiplications, i.e., if $W(\ell)$ and $\widetilde{W}(\ell)$ are two $\mathfrak{L}$–resolvent matrices for S, then $W(\ell)=\beta(\ell)\widetilde{W}(\ell)U$ for some such $\beta(\ell)$ and U. This follows easily from the analogous result concerning $\mathfrak{L}$–coresolvent matrices for the isometric operators, see below. We are now ready to exhibit some specific $\mathfrak{L}$–resolvent matrices for S.

THEOREM 2.2. *Under the assumptions made in the preceding paragraphs and the notations given there, the function*

$$W(\ell)=\begin{pmatrix} G^*R_0(\ell)G(\Gamma_{\bar\mu}(\bar\ell)^*G)^{-1} & G^*R_0(\ell)G(\Gamma_{\bar\mu}(\bar\ell)^*G)^{-1}Q(\ell)-G^*\Gamma_{\bar\mu}(\ell) \\ (\Gamma_{\bar\mu}(\bar\ell)^*G)^{-1} & (\Gamma_{\bar\mu}(\bar\ell)^*G)^{-1}Q(\ell) \end{pmatrix}$$

is an $\mathfrak{L}$-resolvent matrix for S and satisfies

$$\text{(2.3)}\qquad \frac{W(\ell)\left(\begin{smallmatrix} 0 & -I \\ I & 0\end{smallmatrix}\right)W(\lambda)^*-\left(\begin{smallmatrix} 0 & -I \\ I & 0\end{smallmatrix}\right)}{\ell-\bar\lambda}=\begin{pmatrix} G^*Q(\ell) \\ -G^{-1}P(\ell)\end{pmatrix}\begin{pmatrix} G^*Q(\lambda) \\ -G^{-1}P(\lambda)\end{pmatrix}^*.$$

In particular, if $a\in\mathbb{R}\cap\rho(A_0)\cap\rho_S(\mathfrak{L})$, then $W(a)$ is $\left(\begin{smallmatrix} 0 & -I \\ I & 0\end{smallmatrix}\right)$-unitary and

$$\text{(2.4)}\qquad W(\ell)W(a)^{-1}=I+(\ell-a)\begin{pmatrix} G^*Q(\ell) \\ -G^{-1}P(\ell)\end{pmatrix}\begin{pmatrix} G^*Q(a) \\ -G^{-1}P(a)\end{pmatrix}^*\begin{pmatrix} O & I \\ -I & O\end{pmatrix}$$

is also an $\mathfrak{L}$-resolvent matrix for S and satisfies (2.3).

Note that the $\mathfrak{L}$-resolvent matrix $W(\ell)W(a)^{-1}$ in (2.4) is independent of the parameter μ and also that it does not tend to a limit when $a\to\infty$ and $\infty\in\rho(A_0)\cap\rho_S(\mathfrak{L})$. In the next theorem we give an $\mathfrak{L}$-resolvent matrix $W_a(\ell)$ parametrized by $a\in\mathbb{R}\cap\rho(A_0)\cap\rho_S(\mathfrak{L})$ for which the limit

$$\text{(2.5)}\qquad W_\infty(\ell)=\lim_{a\to\infty}W_a(\ell)$$

does exist, provided $\infty\in\rho(A_0)\cap\rho_S(\mathfrak{L})$, and is an $\mathfrak{L}$-resolvent matrix for S also. To that end we first define the operator $\mathcal{R}(\ell)$ by

$$\text{(2.6)}\qquad \mathcal{R}(\ell)=\frac{2}{\mu-\bar\mu}(\ell-\bar\mu)\big(I-\mathcal{P}(\ell)+(\ell-\mu)\mathcal{Q}(\ell)\big)=\frac{2}{\mu-\bar\mu}(\ell-\bar\mu)\big(I+(\ell-\mu)R_0(\ell)\big)\big(I-\mathcal{P}(\ell)\big).$$

If $\infty\in\rho(A_0)\cap\rho_S(\mathfrak{L})$, then $\mathcal{R}(\ell)$ can be written as

$$\mathcal{R}(\ell)=\frac{2}{\mu-\bar\mu}(\ell-\bar\mu)(A_0-\mu)R_0(\ell)(I-\mathcal{P}(\ell))$$

and consequently,

$$\mathcal{R}(\infty)=\lim_{\ell\to\infty}\mathcal{R}(\ell)=-\frac{2}{\mu-\bar\mu}(A_0-\mu)(I-\mathcal{P}(\infty)),$$

since $\lim_{\ell\to\infty}\ell R_0(\ell)=-I$, as $A_0\in\mathbf{L}(\mathfrak{H})$.

THEOREM 2.3. *If $a\in\mathbb{R}\cap\rho(A_0)\cap\rho_S(\mathfrak{L})$, then the matrix function $W_a(\ell)$*

defined by

$$W_a(\ell)=\begin{pmatrix}\frac{1}{2(a-\mu)} & \frac{\mu+\bar{\mu}-2a}{a-\mu}G^*G\\ O & 2(\ell-\bar{\mu})\end{pmatrix}+$$

$$-\frac{\ell-a}{a-\mu}\begin{pmatrix}G^*\mathcal{Q}(\ell)\\ -G^{-1}\mathcal{P}(\ell)\end{pmatrix}\begin{pmatrix}G^*\mathcal{R}(a)+G^*\mathcal{P}(a)\\ -G^{-1}\mathcal{P}(a)\end{pmatrix}^*\begin{pmatrix}O & \mu-\bar{\mu}\\ \frac{1}{2}I & O\end{pmatrix}$$

is an $\mathfrak{L}$*−resolvent matrix for* S*. If* $\infty\in\rho(A_0)\cap\rho_S(\mathfrak{L})$*, then also*

$$W_\infty(\ell)=\begin{pmatrix}O & -2G^*G\\ O & 2(\ell-\bar{\mu})\end{pmatrix}+\begin{pmatrix}G^*\mathcal{Q}(\ell)\\ -G^{-1}\mathcal{P}(\ell)\end{pmatrix}\begin{pmatrix}G^*\mathcal{R}(\infty)+G^*\mathcal{P}(\infty)\\ -G^{-1}\mathcal{P}(\infty)\end{pmatrix}^*\begin{pmatrix}O & \mu-\bar{\mu}\\ \frac{1}{2}I & O\end{pmatrix}$$

is an $\mathfrak{L}$*−resolvent matrix for* S *and* (2.5) *is valid. Moreover, both matrix functions satisfy* (2.3).

As to the proofs of Theorems 2.2 and 2.3 we remark that it follows immediately from Kreĭn's formula (2.1) and Lemma 1.8 that $W(\ell)$ is an $\mathfrak{L}$–resolvent matrix for S. The identity (2.3) can be verified directly and easily leads to (2.4). If we define the constant matrix E and the matrix function $H(\ell)$ by

$$E=\operatorname{diag}\left((4i\operatorname{Im}\mu)^{-1}I,I\right), \tag{2.7}$$

$$H(\ell)=\begin{pmatrix}\frac{\mu-\bar{\mu}}{2(\ell-\mu)(\ell-\bar{\mu})} & -\frac{\ell-\operatorname{Re}\mu}{(\ell-\mu)(\ell-\bar{\mu})}G^*G\\ O & I\end{pmatrix}, \tag{2.8}$$

we find after some straightforward calculations, using the obvious identities

$$G^{-1}\mathcal{P}(\ell)G=I,\ (I-\mathcal{P}(\ell))G=0,$$

that for $a\in\mathbb{R}\cap\rho(A_0)\cap\rho_S(\mathfrak{L})$

$$W_a(\ell)=W(\ell)W(a)^{-1}2(a-\bar{\mu})H(a)E$$

and that $2(a-\bar{\mu})H(a)E$ is a $\left(\begin{smallmatrix}0 & -I\\ I & 0\end{smallmatrix}\right)$–unitary matrix. Hence, $W_a(\ell)$ is an $\mathfrak{L}$–resolvent matrix for S. Obviously, the matrix function $W_\infty(\ell)$ defined as in the last part of Theorem 2.3 satisfies (2.5), but this does not immediately imply that $W_\infty(\ell)$ is also an $\mathfrak{L}$–resolvent matrix for S. To prove this we need to resort to analogous results for isometric operators. In the following paragraphs we discuss Q–functions and $\mathfrak{L}$–coresolvent matrices for isometric operators and give some explicit representations of such matrices, see

Theorem 2.5. Using the Cayley transform we then prove that $W_\infty(\ell)$ is an $\mathfrak{L}$–resolvent matrix for S and give second proofs of Theorem 2.2 and the first part of Theorem 2.3.

Let V be a closed isometric operator in $\mathfrak{H}$ and let W be a unitary extension in a Hilbert space $\mathfrak{K}$ of V, i.e., $\mathfrak{H}\subset\mathfrak{K}$, the inner products on $\mathfrak{H}$ and $\mathfrak{K}$ coincide on $\mathfrak{H}$ and $V\subset W$. The extension W is called canonical if $\mathfrak{K}=\mathfrak{H}$ and W and $\mathfrak{K}$ are called minimal if

$$\mathfrak{K}=\text{c.l.s.}\,\{\,(I-zW)^{-1}f\mid f\in\mathfrak{H},\ 1/z\in\rho(W)\,\}.$$

We denote by $C(z)$ the compression of the coresolvent $(I-zW)^{-1}(I+zW)$ of W to the space $\mathfrak{H}$:

$$(2.9)\qquad C(z)=P_{\mathfrak{H}}(I-zW)^{-1}(I+zW)|_{\mathfrak{H}},\quad z=0,\ \text{or}\ 1/z\in\rho(W).$$

The function $C(z)$ is a locally holomorphic function on $\mathbb{C}\backslash\partial\mathbb{D}$ with values in $\mathbf{L}(\mathfrak{H})$. It satisfies $C(z)^*=-C(1/\bar{z})$, $C(z)(I-zV)=I+zV$ and the kernel

$$\frac{C(z)+C(w)^*}{1-z\bar{w}}$$

is nonnegative. A function with these properties is called a generalized coresolvent of V. Each such $C(z)$ determines, uniquely up to isomorphisms, a minimal unitary extension W of V such that (2.9) is valid, see [DLS1].

In the following we assume that the defect numbers of V are equal, we fix a canonical extension W_0 of V and denote the corresponding generalized coresolvent by $C_0(z)$. It can be shown that $(I-zW_0^*)^{-1}$ maps $\mathfrak{D}(V)^\perp$ onto $\mathfrak{R}(I-\bar{z}V)^\perp$. Let $\mathfrak{G}$ be a Hilbert space and let $\Delta_{\mathfrak{G}}$ be a boundedly invertible mapping from $\mathfrak{G}$ onto $\mathfrak{D}(V)^\perp$. We define the mapping $\Delta(z)$ by

$$\Delta(z)=(I-zW_0^*)^{-1}\Delta_{\mathfrak{G}},\quad z\in\rho_i(w_0),$$

where $\rho_i(w_0)$ stands for the set $\rho_i(W_0)=\{z\in\mathbb{C}\,|\,z=0,\ \text{or}\ 1/z\in\rho(W_0)\}$. Then

$$(2.10)\qquad \Delta(\bar{w})-\Delta(1/z)=(1-z\bar{w})(I-zW_0)^{-1}\Delta(\bar{w})$$

and $\Delta(z)^*$ maps $\mathfrak{R}(I-\bar{z}V)^\perp$ one to one onto $\mathfrak{G}$. Associated with $\Delta(z)$ is a so–called Q–function $q(z)$ which satisfies

$$(2.11)\qquad \frac{q(z)+q(w)^*}{1-z\bar{w}}=\Delta(\bar{z})^*\Delta(\bar{w}).$$

It follows that $q(z)$ is defined by this relation up to a constant selfadjoint operator $C_q\in\mathbf{L}(\mathfrak{G})$ and has the form

$$q(z) = iC_q + \tfrac{1}{2}\Delta_{\mathfrak{G}}^*(I - zW_0)^{-1}(I + zW_0)\Delta_{\mathfrak{G}}.$$

It is a locally holomorphic function on $\mathbb{C}\backslash\partial\mathbb{D}$ with values in $\mathbf{L}(\mathfrak{G})$, $q(z)^* = -q(1/\bar{z})$ and $\operatorname{Re} q(z)$ is positive and boundedly invertible for $z \in \mathbb{D}$. Conversely, if $q(z)$ has these properties, then it is the Q-function of a closed simple isometric operator V in some Hilbert space $\mathfrak{H}$ and V together with a unitary extension W_0 are uniquely determined up to isomorphisms. The Q-function $q(z)$ plays an important role in the parametrization of all unitary extensions of V in possibly larger Hilbert spaces. According to a result of Kreĭn these unitary extensions W of V can be parametrized by

$$(2.12) \quad C(z) = C_0(z) + 2\Delta(1/z)(q(z) + t(z))^{-1}\Delta(\bar{z})^*,$$

where $C(z)$ is given by (2.9) and the parameter $t(z)$ runs through the Carathéodory class $\mathbf{C}(\mathfrak{G})$, see, e.g., [DLS3]. Here, the class $\mathbf{C}(\mathfrak{G})$ is defined by

$$\mathbf{C}(\mathfrak{G}) = \{\, t(z) \mid it(z(\ell)) \in \mathbf{N}(\mathfrak{G}) \,\} =$$

$$= \{\, (\mathcal{A}(z), \mathcal{B}(z)) \mid (\mathcal{A}(z(\ell)), i\mathcal{B}(z(\ell))) \in \mathbf{N}(\mathfrak{G}) \,\},$$

where $z(\ell) = (\ell - \mu)/(\ell - \bar{\mu})$ for some $\mu \in \mathbb{C}^+$. The pair $t(z) = (A(z), B(z)) \in \mathbf{C}(\mathfrak{G})$ will be called a Carathéodory pair. It is easy to see that in formula (2.12) $(q(z) + t(z))^{-1}$ is a bounded operator on $\mathfrak{G}$.

We assume that V has finite and equal defect numbers n, say. If $\mathfrak{L} \subset \mathfrak{H}$ is a subspace with $\dim \mathfrak{L} = n$ we call a point $z \in \mathbb{C}$ an $\mathfrak{L}$-regular point for V if $\mathfrak{H}$ can be decomposed as

$$\mathfrak{H} = \mathfrak{R}(I - zV) + \mathfrak{L}, \qquad \text{direct sum.}$$

By $\rho_V(\mathfrak{L})$ we denote the set of all points z such that z and $1/\bar{z}$ are $\mathfrak{L}$-regular points for V. The set $\rho_V(\mathfrak{L}) \cap \rho_i(W_0)$ is open and $\rho_i(W_0) \backslash \rho_V(\mathfrak{L})$ is empty or a discrete subset of $\rho_i(W_0)$. We shall call $\mathfrak{L}$ a module space for V if $\rho_V(\mathfrak{L}) \cap \rho_i(W_0) \neq \emptyset$. From now on $\mathfrak{L}$ will be a fixed module space for V and G a bijection from a Hilbert space $\mathfrak{G}$ onto $\mathfrak{L}$. Associated with $\mathfrak{L}$ are the operators $\mathsf{P}(z)$ and $\mathsf{Q}(z)$ defined for $z \in \rho_V(\mathcal{L})$ as follows. $\mathsf{P}(z)$ is the projection of $\mathfrak{H}$ onto $\mathfrak{L}$ parallel to $\mathfrak{R}(I - zV)$ and

$$\mathsf{Q}(z) = I + C_0(z)(I - \mathsf{P}(z)) = \mathsf{P}(z) + 2(I - zW_0)^{-1}(I - \mathsf{P}(z)).$$

Clearly, $\mathsf{P}(z)$ does not depend on the particular choice of the canonical extension W_0 and, since $I - \mathsf{P}(z)$ maps $\mathfrak{H}$ onto $\mathfrak{R}(I - zV)$, neither does $\mathsf{Q}(z)$. For

$z \in \rho_V(\mathfrak{L}) \cap \rho_i(W_0)$ we have that $\Delta(\bar{z})^* G$ is boundedly invertible on $\mathfrak{G}$ and

$$\mathsf{P}(z) = G(\Delta(\bar{z})^* G)^{-1} \Delta(\bar{z})^*.$$

The mapping $z \to G^* C(z) G$, $z \in \rho_V(\mathfrak{L})$, where $C(z)$ is a generalized coresolvent of V is called an $\mathfrak{L}$–coresolvent of V. It has values in $\mathbf{L}(\mathfrak{G})$ and to each $\mathfrak{L}$–coresolvent of V there corresponds precisely one generalized coresolvent of the operator V.

DEFINITION 2.4. *A function $V(z)$ defined on $\rho_V(\mathfrak{L})$ with values in $\mathbf{R}^2(\mathfrak{G})$ is called an $\mathfrak{L}$–coresolvent matrix for V if it has the following properties:*

(i) $V(z)$ is locally holomorphic and invertible on $\rho_V(\mathfrak{L})$.

(ii) $V_{22}(z)\mathcal{A}(z) + V_{21}(z)\mathcal{B}(z)$ is invertible for each $z \in \rho_V(\mathfrak{L})$ and each Carathéodory pair $(\mathcal{A}(z), \mathcal{B}(z)) \in \mathbf{C}(\mathfrak{G})$.

(iii) The formula

$$G^* C(z) G = \mathfrak{t}_{V(z)}(\mathfrak{t}(z)) = (V_{12}(z)\mathcal{A}(z) + V_{11}(z)\mathcal{B}(z))(V_{22}(z)\mathcal{A}(z) + V_{21}(z)\mathcal{B}(z))^{-1}$$

establishes a one to one correspondence between all $\mathfrak{L}$–coresolvents of V and all Carathéodory pairs $\mathfrak{t}(z) = (\mathcal{A}(z), \mathcal{B}(z)) \in \mathbf{C}(\mathfrak{G})$.

If we multiply an $\mathfrak{L}$–coresolvent matrix for V by any nonvanishing locally holomorphic function and from the right by any $\left(\begin{smallmatrix} 0 & I \\ I & 0 \end{smallmatrix}\right)$–unitary matrix, we obtain another $\mathfrak{L}$–coresolvent matrix for V and an $\mathfrak{L}$–coresolvent matrix for V is uniquely determined modulo such factors, see [KSm]. Kreĭn's formula (2.12) will be the starting point for the proof of the next theorem in which we display some explicit $\mathfrak{L}$–coresolvent matrices for V.

THEOREM 2.5. *Under the assumptions made in the preceding paragraphs and the notations given there, the function*

$$V(z) = \begin{pmatrix} G^* C_0(z) G (\Delta(\bar{z})^* G)^{-1} & \tfrac{1}{2} G^* C_0(z) G (\Delta(\bar{z})^* G)^{-1} \mathsf{q}(z) + G^* \Delta(1/z) \\ (\Delta(\bar{z})^* G)^{-1} & \tfrac{1}{2} (\Delta(\bar{z})^* G)^{-1} \mathsf{q}(z) \end{pmatrix} \tag{2.13}$$

is an $\mathfrak{L}$-coresolvent matrix for V and satisfies for $z, w \in \rho_V(\mathfrak{L}) \cap \rho_i(W_0)$

$$\frac{V(z)\left(\begin{smallmatrix} 0 & I \\ I & 0 \end{smallmatrix}\right)V(w)^* - \left(\begin{smallmatrix} 0 & I \\ I & 0 \end{smallmatrix}\right)}{1 - z\bar{w}} = \tfrac{1}{2} \begin{pmatrix} G^* \mathsf{Q}(z) \\ -G^{-1} \mathsf{P}(z) \end{pmatrix} \begin{pmatrix} G^* \mathsf{Q}(w) \\ -G^{-1} \mathsf{P}(w) \end{pmatrix}^*. \tag{2.14}$$

In particular, if $\zeta \in \partial\mathbb{D} \cap \rho_V(\mathfrak{L}) \cap \rho_i(W_0)$, then $V(\zeta)$ is $\left(\begin{smallmatrix} 0 & I \\ I & 0 \end{smallmatrix}\right)$-unitary and consequently,

$$V(z)V(\zeta)^{-1} = I + \tfrac{1}{2}(1-z\bar{\zeta}) \begin{pmatrix} G^*\mathsf{Q}(z) \\ -G^{-1}\mathsf{P}(z) \end{pmatrix} \begin{pmatrix} G^*\mathsf{Q}(\zeta) \\ -G^{-1}\mathsf{P}(\zeta) \end{pmatrix}^* \begin{pmatrix} O & I \\ I & O \end{pmatrix}$$

is also an $\mathfrak{L}$*-coresolvent matrix for* V *and satisfies* (2.14).

Proof. Kreĭn's formula (2.12) and Lemma 1.8 imply that the equality

$$G^*C(z)G = \mathfrak{t}_{V(z)}(\tfrac{1}{2}\mathfrak{t}(z))$$

with $V(z)$ defined by (2.13) is valid and that it establishes a one to one correspondence between the coresolvents $C(z)$ of V and $\mathfrak{t}(z) \in \mathbf{C}(\mathfrak{G})$. Hence $V(z)$ is an $\mathfrak{L}$–coresolvent matrix for V. Making use of the formulas (2.10), (2.11) and the coresolvent equation for the canonical extension W_0:

$$\frac{C_0(z)+C_0(w)^*}{1-z\bar{w}} = 2(I-wW_0)^{-*}(I-zW_0)^{-1},$$

one can check in a straightforward manner that $V(z)$ satisfies the stated identity. The special case follows from this identity and the relation

$$V(\zeta)^{-1} = \begin{pmatrix} 0 & I \\ I & 0 \end{pmatrix} V(\zeta)^* \begin{pmatrix} 0 & I \\ I & 0 \end{pmatrix}.$$

We now return to the beginning of this section and Theorems 2.2 and 2.3. Recall that we have assumed that S is a closed symmetric relation in a Hilbert space $\mathfrak{H}$ with finite and equal defect numbers n, $\mathfrak{L}$ is a module space for S and that we have fixed a canonical selfadjoint extension A_0 of S, a Hilbert space $\mathfrak{G}$ with $\dim\mathfrak{G} = n$ and two bijective mappings $\Gamma_{\mathfrak{G},\bar{\mu}}:\mathfrak{G}\rightarrow\nu(S^*-\bar{\mu})$ and $G:\mathfrak{G}\rightarrow\mathfrak{L}$. Here μ will be a fixed point in $\mathbb{C}^+$. Put $V = C_\mu(S)$, the Cayley transform of S, and

$$z(\ell) = (\ell-\mu)/(\ell-\bar{\mu}),\ \ z(\infty) = 1,\ \ z(\bar{\mu}) = \infty.$$

Then V is a closed isometric operator in $\mathfrak{H}$ and $\mathfrak{R}(S-\ell) = \mathfrak{R}(I-z(\ell)V)$. In particular, $\mathfrak{D}(V) = \mathfrak{R}(S-\mu)$ and $\mathfrak{R}(V) = \mathfrak{R}(S-\bar{\mu})$, so that the defect numbers of V are both equal to n. If A is a selfadjoint extension of S in a possibly larger Hilbert space and $W = C_\mu(A)$, then W is a unitary extension of V and it follows from the identity

$$(2.15)\quad (I-z(\ell)W)^{-1} = \frac{\ell-\bar{\mu}}{\mu-\bar{\mu}}\,(I+(\ell-\mu)(A-\ell)^{-1}),$$

or, equivalently, from

$$(I-z(\ell)W)^{-1}(I+z(\ell)W) = \frac{2(\ell-\mu)(\ell-\bar{\mu})}{\mu-\bar{\mu}}(A-\ell)^{-1} + 2\frac{\ell-\operatorname{Re}\mu}{\mu-\bar{\mu}}$$

that

$$(2.16)\quad R(\ell)=\frac{\mu-\bar{\mu}}{2(\ell-\mu)(\ell-\bar{\mu})}C(z(\ell))-\frac{\ell-\mathrm{Re}\mu}{(\ell-\mu)(\ell-\bar{\mu})}.$$

In particular, if $A=A_0$, the canonical extension of S, then $W_0=C_\mu(A_0)$ is a canonical unitary extension of V and (2.15) becomes

$$(I-z(\ell)W_0)^{-1}=\frac{\ell-\bar{\mu}}{\mu-\bar{\mu}}(I+(\ell-\mu)R_0(\ell)).$$

Clearly, $\mathfrak{L}$ is also a module space for V and

$$\mathsf{P}(z(\ell))=\mathcal{P}(\ell),\quad \mathsf{Q}(z(\ell))=\mathcal{R}(\ell)+\mathcal{P}(\ell),$$

where $\mathcal{R}(\ell)$ is given by (2.6). These equalities imply

$$(2.17)\quad H(\ell)\begin{pmatrix} G^*\mathsf{Q}(z(\ell)) \\ -G^{-1}\mathsf{P}(z(\ell)) \end{pmatrix}=\begin{pmatrix} G^*\mathcal{Q}(\ell) \\ -G^{-1}\mathcal{P}(\ell) \end{pmatrix}+\begin{pmatrix} \frac{1}{\ell-\mu}G^* \\ O \end{pmatrix},$$

where $H(\ell)$ is defined by (2.8). Furthermore, it is easy to see that $\infty\in\rho(A_0)\cap\rho_S(\mathfrak{L})$ if and only if $1\in\rho_V(\mathfrak{L})\cap\rho_i(W_0)$ and if one of these conditions holds then

$$(2.18)\quad \lim_{\ell\to\infty}\mathcal{P}(\ell)=\lim_{\ell\to\infty}\mathsf{P}(z(\ell))=\lim_{z\to 1}\mathsf{P}(z)=\mathsf{P}(1)=\mathcal{P}(\infty).$$

Hence (2.2) is valid. Note that $\mathfrak{R}(I-\mathsf{P}(1))=\mathfrak{R}(I-V)=\mathfrak{D}(S)=\mathfrak{R}(I-\mathcal{P}(\infty))$. Choosing $\Delta_{\mathfrak{G}}=\Gamma_{\mathfrak{G},\bar{\mu}}$, we obtain the following equalities

$$\Delta((1/z(\ell))=\frac{\ell-\mu}{\bar{\mu}-\mu}\Gamma_{\bar{\mu}}(\ell),\quad \Delta(\overline{z(\ell)})^*=\frac{\ell-\bar{\mu}}{\mu-\bar{\mu}}\Gamma_{\bar{\mu}}(\bar{\ell})^*$$

and

$$\mathcal{Q}(\ell)=(\mu-\bar{\mu})\mathfrak{q}(z(\ell)),$$

where in the formula for $\mathcal{Q}(\ell)$ we have set $C_{\mathcal{Q}}=i(\mu-\bar{\mu})C_q$ and $\tau=\bar{\mu}$. With these equalities the Kreĭn's formula (2.12) for the isometric case can be transformed into its counterpart (2.1) for the symmetric case if we put

$$(2.19)\quad \mathcal{T}(\ell)=(\mu-\bar{\mu})\mathfrak{t}(z(\ell)).$$

We are now in a position to give a second proof of Theorem 2.2.

Proof of Theorem 2.2 *via Theorem* 2.5. Straightforward calculations yield that

$$(2.20)\quad W(\ell)=2(\ell-\bar{\mu})H(\ell)V(z(\ell))E,$$

where E and $H(\ell)$ are given by (2.7) and (2.8), repectively, and $V(z)$ is the $\mathfrak{L}$–coresolvent matrix for V in (2.13). Using formulas (2.16), (2.19) and (2.20) we obtain

$$G^*R(\ell)G = \mathrm{t}_{H(\ell)}G^*C(z(\ell))G = \mathrm{t}_{H(\ell)}\mathrm{t}_{V(z(\ell))}(\tfrac{1}{2}t(z(\ell))) =$$

$$= \mathrm{t}_{H(\ell)}\mathrm{t}_{V(z(\ell))}\mathrm{t}_E(\mathcal{T}(\ell)) = \mathrm{t}_{W(\ell)}(\mathcal{T}(\ell)),$$

which implies the first part of Theorem 2.2. The equality (2.3) follows from (2.20), the corresponding formula in Theorem 2.5 and the relations

$$1 - z(\ell)\overline{z(\lambda)} = -\frac{(\ell-\bar{\lambda})(\mu-\bar{\mu})}{(\ell-\bar{\mu})(\bar{\lambda}-\mu)}$$

and

$$H(\ell)\begin{pmatrix} G^*\mathsf{Q}(z(\ell)) \\ -G^{-1}\mathsf{P}(z(\ell)) \end{pmatrix}\begin{pmatrix} G^*\mathsf{Q}(z(\lambda)) \\ -G^{-1}\mathsf{P}(z(\lambda)) \end{pmatrix}^* H(\lambda)^* =$$

$$= \begin{pmatrix} G^*\mathcal{Q}(\ell) \\ -G^{-1}\mathcal{P}(\ell) \end{pmatrix}\begin{pmatrix} G^*\mathcal{Q}(\lambda) \\ -G^{-1}\mathcal{P}(\lambda) \end{pmatrix}^* + \begin{pmatrix} ((\ell-\mu)(\bar{\lambda}-\bar{\mu}))^{-1}G^*G & -(\ell-\mu)^{-1} \\ -(\bar{\lambda}-\bar{\mu})^{-1} & 0 \end{pmatrix},$$

which can be derived from (2.17).

Proof of Theorem 2.3 *via Theorem* 2.5. Since by Theorem 2.5 $V(z)V(\zeta)^{-1}$ is an $\mathfrak{L}$–coresolvent matrix for V if $\zeta \in \partial\mathbb{D} \cap \rho_V(\mathfrak{L}) \cap \rho_i(W_0)$, the same arguments as in the beginning of the previous proof of Theorem 2.2 give that for $a \in (\mathbb{R} \cup \{\infty\}) \cap \rho_S(\mathfrak{L}) \cap \rho(A_0)$ the matrix function

$$2(\ell-\bar{\mu})H(\ell)V(z(\ell))V(z(a))^{-1}E$$

is an $\mathfrak{L}$–resolvent matrix for S. Straightforward calculations show that this matrix function coincides with the matrix function $W_a(\ell)$ of the theorem for the case that $a \in \mathbb{R}$ as well as for the case that $a = \infty$. The last statement in Theorem 2.3 follows from the corresponding statement in Theorem 2.5 by direct computation and the formula

$$H(\ell)\left(\begin{smallmatrix} 0 & I \\ I & 0 \end{smallmatrix}\right)H(\lambda)^* = -\tfrac{1}{2}(1 - z(\ell)\overline{z(\lambda)})\,\mathcal{D}(\ell,\lambda),$$

where the matrix function $\mathcal{D}(\ell,\lambda)$ has entries

$$\begin{pmatrix} ((\ell-\mu)(\bar{\lambda}-\bar{\mu}))^{-1}GG^* & ((\ell-\bar{\lambda})(\ell-\mu))^{-1}(\bar{\lambda}-\mu) \\ -((\ell-\bar{\lambda})(\bar{\lambda}-\bar{\mu}))^{-1}(\ell-\bar{\mu}) & O \end{pmatrix}.$$

This completes the proof.

3. Interpolation via resolvent matrices

We now turn to the interpolation problem

(IP) $\quad (\mathcal{K}_i(\ell)\mathcal{N}(\ell))^{(p)}|_{\ell=w_i} = (\mathcal{L}_i(\ell)\mathcal{M}(\ell))^{(p)}|_{\ell=w_i}, \quad 1\le i\le m,\ 0\le p\le r_i,$

with data described by (0.1)–(0.3) in the Introduction. We shall say that $\mathcal{T}(\ell)\in \mathbf{N}^n$ is a solution of this problem if $\mathcal{T}(\ell)=(\mathcal{M}(\ell),\mathcal{N}(\ell))$ for some pair of matrix functions $\mathcal{M}(\ell)$, $\mathcal{N}(\ell)$ satisfying condition (0.5) in the Introduction which solves the interpolation problem (IP). We often identify the equivalence class $\mathcal{T}(\ell)$ with this ordered pair which generates it, see also the remarks following Definition 1.4. Throughout this section we shall use the following notations: $\mathbb{C}^k$ stands for the usual k dimensional complex linear space provided with the usual inner product $(u,v)_k = v^*u$, $u,v\in\mathbb{C}^k$. We put $r=\Sigma_{i=1}^m(r_i+1)$ as before and fix a basis $\{\varepsilon_{ip}\,|\,i=1,2,\ldots,m,\ \ p=0,1,\ldots,r_i\}$ for $\mathbb{C}^r$, which is orthonormal with respect to this inner product, i.e., satisfies $(\varepsilon_{ip},\varepsilon_{jq})_r=\delta_{ij}\delta_{pq}$. Then every $u\in\mathbb{C}^r$ can be written as $u=\sum u_{ip}\varepsilon_{ip}$, where $u_{ip}\in\mathbb{C}$ and the sum is taken over all indices $i=1,2,\ldots,m$, $p=0,1,\ldots,r_i$. Define the linear operators $Z:\mathbb{C}^r\to\mathbb{C}^r$ and $V,W:\mathbb{C}^r\to\mathbb{C}^n$ by their action on the basis elements ε_{ip} of $\mathbb{C}^r$ as follows:

$$Z\varepsilon_{ip}=\begin{cases}\bar{w}_i\varepsilon_{ip} & p=0,\\ \bar{w}_i\varepsilon_{ip}+\varepsilon_{i(p-1)} & p\ge 1,\end{cases}$$

$$V\varepsilon_{ip}=(1/p!)\,\mathcal{K}_i^{(p)}(w_i)^* \quad\text{and}\quad W\varepsilon_{ip}=(1/p!)\,\mathcal{L}_i^{(p)}(w_i)^*.$$

Note that Z has a Jordan block structure and maps $\text{l.s.}\{\varepsilon_{ip}\,|\,p=0,1,\ldots,k\}$ into itself, $k=0,1,..,r_i$. We shall frequently identify these operators with their matrix representations which were considered in the Introduction. They form the building stones for the Lyapunov equation

(3.1) $\quad \mathbb{P}Z-Z^*\mathbb{P}=V^*W-W^*V$

associated with the interpolation problem (IP) in which $\mathbb{P}$ is the unknown $r\times r$ matrix. In accordance with the notation introduced here and in the Introduction, we have $\mathbb{P}=(\mathbb{P}_{ij}^{pq})$, with $\mathbb{P}_{ij}^{pq}=(\mathbb{P}\varepsilon_{jq},\varepsilon_{ip})_r$ and

(3.2) $\quad (\mathbb{P}u,v)_r=\sum \bar{v}_{ip}\mathbb{P}_{ij}^{pq}u_{jq}.$

A hermitian solution of the Lyapunov equation, if it exists, will be called a

Pick matrix associated with interpolation problem (IP). In the following paragraphs we shall associate with each nonnegative Pick matrix $\mathbb{P}$ a triplet consisting of a Hilbert space $\mathfrak{H}$, a symmetric linear relation S and a linear mapping $G:\mathbb{C}^n \to \mathfrak{H}$ which satisfies $G^*G=I$. We shall refer to this triplet $\mathfrak{H}$, S and G as the model associated with $\mathbb{P}$ and call S the $\mathbb{P}$ symmetric relation. The model plays an important role in the description of all solutions of the interpolation problem (IP).

We begin by considering an arbitrary hermitian nonnegative $r \times r$ matrix $\mathbb{P}=(\mathbb{P}_{ij}^{pq})$. We put $\mathfrak{H}=\mathfrak{H}_1 \oplus \mathfrak{H}_2$, where $\mathfrak{H}_1=\mathbb{C}^r$ equipped with the nonnegative inner product defined by (3.2), $\mathfrak{H}_2=\mathbb{C}^n$ with the usual inner product and $\oplus$ stands for the orthogonal sum. We write an element in $\mathfrak{H}$, x, say, as $x=\begin{pmatrix} x_1 \\ x_2 \end{pmatrix}$ with $x_j \in \mathfrak{H}_j$ and denote the inner product on $\mathfrak{H}$ by $[.,.]$, so that

$$[x,y]=(\mathbb{P}x_1,y_1)_r+(x_2,y_2)_n.$$

If $\mathbb{P}>0$, this inner product is positive definite. If $\mathbb{P}\geq 0$ and $\nu(\mathbb{P}) \neq \{0\}$, then the space $(\mathfrak{H},[.,.])$ is degenerate and in order to get a Hilbert space we adapt the definition of $\mathfrak{H}$ in the following manner. We replace $\mathfrak{H}_1$ by its quotient space $\hat{\mathfrak{H}}_1=\mathfrak{H}_1/N_0$, the elements of which are the usual equivalence classes $\hat{u}=u+N_0$, where $N_0=\{u\in\mathbb{C}^r \mid (\mathbb{P}u,u)_r=0\}$, and provide it with the usual quotient space inner product. The orthogonal sum $\hat{\mathfrak{H}}_1 \oplus \mathfrak{H}_2$ will still be denoted by $\mathfrak{H}$ and the inner product on $\mathfrak{H}$, which now is positive definite again, by $[.,.]$ as before. In $\mathfrak{H}$ we define the linear relation S by

$$S=\{\{\begin{pmatrix} \hat{u} \\ -Vu \end{pmatrix}, \begin{pmatrix} (Zu)^{\wedge} \\ Wu \end{pmatrix}\} \mid u\in\mathfrak{H}_1\}.$$

As simple examples show S may be multivalued. If $\mathbb{P}>0$, then S is an operator with $\dim \mathfrak{D}(S)^{\perp}=n$.

PROPOSITION 3.1. *The relation S is symmetric in the Hilbert space $\mathfrak{H}$ if and only if $\mathbb{P}$ is a nonnegative solution of the Lyapunov equation* (3.1), *and then S has equal defect numbers $\leq n$. If $\mathbb{P}$ is a positive Pick matrix, then the defect numbers are equal to n.*

Proof. The first statement follows from the equality

$$[\begin{pmatrix} (Zu)^{\wedge} \\ Wu \end{pmatrix}, \begin{pmatrix} \hat{v} \\ -Vv \end{pmatrix}]-[\begin{pmatrix} \hat{u} \\ -Vu \end{pmatrix}, \begin{pmatrix} (Zv)^{\wedge} \\ Wv \end{pmatrix}]=((\mathbb{P}Z-Z^*\mathbb{P}-V^*W+W^*V)u,v)_r,$$

which is valid for all u, $v\in\mathfrak{H}_1$. Straightforward calculations show that for all $\mu\in\mathbb{C}\backslash\{w_i \mid i=1,2,\ldots,m\}$

$$\nu(S^*-\mu)=\Re(S-\bar{\mu})^{\perp}=\{\begin{pmatrix}\hat{\varphi}\\ \psi\end{pmatrix} \mid \mathbb{P}\varphi=-(Z^*-\mu)^{-1}(W^*+\mu V^*)\psi,\ \varphi\in\mathfrak{H}_1,\ \psi\in\mathfrak{H}_2\},$$

which implies the last two statements.

To complete the construction of the model, we define $G:\mathbb{C}^n\to\hat{\mathfrak{H}}$ by $Gf=\begin{pmatrix}0\\ f\end{pmatrix}$. From the equalities

$$(G^*\hat{\varepsilon}_{ip},g)_n=[\hat{\varepsilon}_{ip},Gg]=\left[\begin{pmatrix}\varepsilon_{ip}\\ 0\end{pmatrix},\begin{pmatrix}0\\ g\end{pmatrix}\right]=0,$$

$$(G^*\begin{pmatrix}0\\ f\end{pmatrix},g)_n=\left[\begin{pmatrix}0\\ f\end{pmatrix},\begin{pmatrix}0\\ g\end{pmatrix}\right]=(f,g)_n$$

it follows that $G^*\hat{\varepsilon}_{ip}=0$ and $G^*\begin{pmatrix}\hat{x}_1\\ x_2\end{pmatrix}=x_2$ so that $G^*G=I$ on $\mathbb{C}^n$. It is easy to verify that for $\mu\in\mathbb{C}$ we have that $Gf\in\nu(S^*-\bar{\mu})\Leftrightarrow f\in\Re(W+\mu V)^{\perp}$ and that

$$(3.3)\qquad Gf\in S^*(0)\Leftrightarrow f\in\Re(V)^{\perp}.$$

Finally, we have $\Re(S-\ell)+\Re(G)$ is a direct sum in $\mathfrak{H}$ for some (and hence for all) $\ell\in\rho(Z)$ if and only if $\mathbb{P}>0$. If $\mathbb{P}>0$ and if $\mathbb{P}$ is a solution of the Lyapunov equation (3.1), then it is not difficult to see that $\mathfrak{L}=\Re(G)$ is a module space for the $\mathbb{P}$ symmetric relation S and

$$\rho_S(\mathfrak{L})=\{\infty\}\cup\mathbb{C}\backslash\{w_1,\bar{w}_1,w_2,\bar{w}_2,\ldots,w_m,\bar{w}_m\}.$$

We end the discussion about the model with the following result, for which it is relevant to observe first that of the elements $\mathfrak{H}$, S and G of the triplet only the mapping G is independent of $\mathbb{P}$.

Proposition 3.2. *Let $\mathbb{P}$ be a nonnegative solution of the Lyapunov equation* (3.1) *and let A be selfadjoint extension in a Hilbert space $\mathfrak{K}$ of the $\mathbb{P}$ symmetric relation S. Then A is a minimal extension of S, i.e.,*

$$\mathfrak{K}=\text{c.l.s.}\{(I+(\ell-\mu)(A-\ell)^{-1})x \mid x\in\mathfrak{H},\ \ell\in\rho(A)\},$$

if and only if A is G minimal, i.e.,

$$\mathfrak{K}=\text{c.l.s.}\{(I+(\ell-\nu)(A-\ell)^{-1})x \mid x\in\Re(G),\ \ell\in\rho(A)\}.$$

(Here μ, ν are some fixed points in $\mathbb{C}\backslash\mathbb{R}$.)

Proof. Clearly, if A is G minimal it is a minimal extension. We prove the other implication. For this, on account of the resolvent equation, it suffices to show that

$$\mathfrak{H}\subset\text{l.s.}\{(I+(\ell-\nu)(A-\ell)^{-1})x \mid x\in\Re(G),\ \ell\in\rho(A)\}.$$

From the definition of the $\mathbb{P}$ symmetric relation S and the fact that $S\subset A$, we easily obtain that for $i=1,2,\ldots,m$

$$\begin{pmatrix} \hat{\varepsilon}_{i0} \\ -V\varepsilon_{i0} \end{pmatrix} = (S-\bar{w}_i)^{-1}\begin{pmatrix} 0 \\ (W+\bar{w}_iV)\varepsilon_{i0} \end{pmatrix} \in (A-\bar{w}_i)^{-1}\Re(G),$$

$$\begin{pmatrix} \hat{\varepsilon}_{ip} \\ -V\varepsilon_{ip} \end{pmatrix} = (S-\bar{w}_i)^{-1}\begin{pmatrix} \hat{\varepsilon}_{i(p-1)} \\ (W+\bar{w}_iV)\varepsilon_{ip} \end{pmatrix} = (A-\bar{w}_i)^{-1}\begin{pmatrix} \hat{\varepsilon}_{i(p-1)} \\ (W+\bar{w}_iV)\varepsilon_{ip} \end{pmatrix}, \quad 1\le p\le r_i.$$

Clearly, we have that

$$\mathfrak{H}_2=\Re(G)\subset\{(I+(\ell-\nu)(A-\ell)^{-1})x\mid x\in\Re(G),\ \ell\in\rho(A)\}.$$

Using these relations we find by mathematical induction and repeated application of the resolvent equation that each $\hat{\varepsilon}_{ip}$ belongs to the linear span of the set $\{(I+(\ell-\nu)(A-\ell)^{-1})x\mid x\in\Re(G),\ \ell\in\rho(A)\}$. This implies the desired inclusion and completes the proof.

By Theorem 0.1 (i), proved in Section 1, if $(M(\ell),N(\ell))$ is a solution of (IP) then $\mathbb{P}_{M,N}$ defined by (0.6) is a nonnegative Pick matrix. Since every symmetric relation in a Hilbert space has at least one selfadjoint extension (possibly in a space larger than the given one), the following result immediately implies that a converse, namely part (ii) of Theorem 0.1, is also true.

THEOREM 3.3. *(i) Suppose that the Lyapunov equation* (3.1) *has a hermitian nonnegative solution $\mathbb{P}$. Then for each selfadjoint extension A of the $\mathbb{P}$ symmetric relation S in a possibly larger Hilbert space the pair $(M(\ell),N(\ell))$ defined by*

(3.4) $$M(\ell)=-G^*(A-\ell)^{-1}G,\quad N(\ell)=I+\ell G^*(A-\ell)^{-1}G$$

belongs to the class $\mathbf{N}^n$, is a solution of the interpolation problem (IP) *and satisfies $\mathbb{P}_{M,N}=\mathbb{P}$. (ii) Conversely, associated with each solution $(M(\ell),N(\ell))\in\mathbf{N}^n$ of the interpolation problem* (IP), *there exists a uniquely determined pair $\mathbb{P}$, A, consisting of a hermitian nonnegative solution $\mathbb{P}$ of the Lyapunov equation and a minimal selfadjoint Hilbert space extension A of the $\mathbb{P}$ symmetric relation S, which has the property that $(M(\ell),N(\ell))$ can be represented as*

(3.5) $$(M(\ell),N(\ell))=(-G^*(A-\ell)^{-1}G,\ I+\ell G^*(A-\ell)^{-1}G).$$

Here A is uniquely determined up to isomorphisms leaving the space $\mathfrak{H}$ of the *model associated with* $\mathbb{P}$ *invariant and* $\mathbb{P}=\mathbb{P}_{\mathcal{M},\mathcal{N}}$.

Proof. We begin by proving the converse assertion and assume that the pair $(\mathcal{M}(\ell),\mathcal{N}(\ell))\in\mathbf{N}^n$ is a solution of the interpolation problem (IP). Then $\mathbb{P}_{\mathcal{M},\mathcal{N}}$ is a nonnegative solution of the Lyapunov equation, see Theorem 0.1 (*i*), and $(\mathcal{M}(\ell),\mathcal{N}(\ell))$ admits a representation of the form

$$(3.6)\quad (\mathcal{M}(\ell),\mathcal{N}(\ell))=(-\Gamma^*(A-\ell)^{-1}\Gamma, I+\ell\Gamma^*(A-\ell)^{-1}\Gamma),$$

where A is a selfadjoint relation in a Hilbert space $\mathfrak{K}$ and Γ is a linear mapping from $\mathbb{C}^n$ to $\mathfrak{K}$ such that $\Gamma^*\Gamma=I$ and A is Γ–minimal, see Theorem 1.5. We first show that there exist a subspace $\mathfrak{K}_0$ of $\mathfrak{K}$ with $\Re(\Gamma)\subset\mathfrak{K}_0$ and a symmetric relation S_0 in $\mathfrak{K}_0$ such that A in $\mathfrak{K}$ is a minimal extension of S_0 in $\mathfrak{K}_0$ and such that the triplet $\mathfrak{K}_0$, S_0 and Γ is isomorphic to the model $\mathfrak{H}$, S and G, associated with $\mathbb{P}_{\mathcal{M},\mathcal{N}}$ which has been constructed above. The latter means that there exists an isomorphism $U:\mathfrak{H}\to\mathfrak{K}_0$ such that $UG=\Gamma$ and

$$(3.7)\quad S_0=\{\{Ux,Uy\}\mid\{x,y\}\in S\}.$$

We proceed as follows. For any two locally holomorphic $n\times 1$ vector functions $P(\ell)$, $Q(\ell)$ on $\rho(A)$ we define by mathematical induction the functions $f^p(P,Q;\ell)$, $\ell\in\rho(A)$, with values in $\mathfrak{K}$ by:

$$(3.8)\quad f^0(P,Q;\ell)=\big((A-\ell)^{-1}\Gamma(Q(\ell)+\ell P(\ell))+\Gamma P(\ell)\big),$$

$$(3.9)\quad f^p(P,Q;\ell)=(A-\ell)^{-1}f^{p-1}(P,Q;\ell)\ +\ f^0(D^pP,D^pQ;\ell),\ \ p\geq 1.$$

Here we denote by D^p the p–th derivative $(D_\ell)^p=(1/p!)(d/d\ell)^p$. It is easy to see that $f^p(P,Q;\ell)$ is locally holomorphic on $\rho(A)$ for each $p\geq 0$ and using mathematical induction one can show that

$$(3.10)\quad f^p(P,Q;\ell)=\sum_{k=0}^{p}(A-\ell)^{-k}f^0(D^{p-k}P,D^{p-k}Q;\ell)=D^pf^0(P,Q;\ell).$$

In the following we take for $P(\ell)$, $Q(\ell)$ the vector functions $\mathcal{K}_i^\sharp(\ell)=\mathcal{K}_i(\bar{\ell})^*$, $\mathcal{L}_i^\sharp(\ell)=\mathcal{L}_i(\bar{\ell})^*$, respectively, and for ℓ the values $\bar{\varpi}_i$. We define the elements $e_{ip}\in\mathfrak{K}$ by

$$(3.11)\quad e_{ip}=(D_\ell)^p((A-\ell)^{-1}\Gamma(\mathcal{L}_i^\sharp(\ell)+\ell\mathcal{K}_i^\sharp(\ell))+\Gamma\mathcal{K}_i^\sharp(\ell))\,|_{\ell=\bar{\varpi}_i}=f^p(\mathcal{K}_i^\sharp,\mathcal{L}_i^\sharp;\bar{\varpi}_i),$$

denote the linear span of these elements by $\mathfrak{K}_1$ and put $\mathfrak{K}_0=\mathfrak{K}_1+\Re(\Gamma)$. Straightforward calculations yield the formulas

$$(3.12)\quad (\mathbb{P}_{\mathcal{M},\mathcal{N}})_{ij}^{pq}=[e_{jq},e_{ip}],\qquad 1\leq i,j\leq m,\ \ 0\leq p\leq r_i \text{ and } 0\leq q\leq r_j,$$

$$[\Gamma f,e_{ip}]=0,\qquad 0\leq i\leq m,\ \ 0\leq p\leq r_i.$$

The first one follows from formula (1.4) and the definition of e_{ip} and the second one from the fact that $(\mathcal{M}(\ell),\mathcal{N}(\ell))$ is a solution of the interpolation problem (IP) and the representation (3.6). They imply that the linear mapping $U:\mathfrak{H}\to\mathfrak{K}_0$ determined by

$$U\binom{\hat{\varepsilon}_{ip}}{0}=e_{ip},\ U\binom{0}{f}=\Gamma f$$

is an isomorphism between $\mathfrak{H}$ and $\mathfrak{K}_0$. Now, define the linear relation S_0 in $\mathfrak{K}_0$ by (3.7). Then S_0 is symmetric and $S_0\subset A$. To prove the inclusion we substitute in (3.9) for $f^0(D^pP,D^pQ;\ell)$ the righthand side of (3.8) with P, Q replaced by D^pP, D^pQ, respectively, and obtain

(3.13) $\{f^p(P,Q;\ell)-\Gamma(D_\ell)^pP(\ell),\, f^{p-1}(P,Q;\ell)+\ell f^p(P,Q;\ell)+\Gamma(D_\ell)^pQ(\ell)\}\in A,$

where we have set $f^{-1}(P,Q;\ell)=0$. Making the same replacements for P, Q and ℓ as we did before we find that

$$\{U\binom{\hat{\varepsilon}_{ip}}{-V\varepsilon_{ip}},U\binom{(Z\varepsilon_{ip})^{\wedge}}{W\varepsilon_{ip}}\}=\{e_{ip}-\Gamma V\varepsilon_{ip},\, e_{i,p-1}+\bar{w}_ie_{ip}+\Gamma W\varepsilon_{ip}\}\in A$$

and by taking linear combinations of these elements we get the desired inclusion. Clearly, the Γ-minimality of A implies that A is a minimal extension of S. The proof of the uniqueness will be postponed until the end of the proof of part (i). This completes the proof of part (ii) of the theorem.

To prove the first part, we assume that $\mathbb{P}$ is a nonnegative solution of the Lyapunov equation and that $\mathfrak{H}$, S and G constitute the model associated with $\mathbb{P}$ which has been constructed just before the theorem. Now, suppose that A in a Hilbert space $\mathfrak{K}$ is a minimal selfadjoint extension of the $\mathbb{P}$ symmetric relation S and define $\mathcal{M}(\ell)$, $\mathcal{N}(\ell)$ by (3.4). Clearly, $(\mathcal{M}(\ell),\mathcal{N}(\ell))\in\mathbf{N}^n$ and we have to show that it is a solution of the problem (IP) and that $\mathbb{P}_{\mathcal{M},\mathcal{N}}=\mathbb{P}$. To prove that it is a solution we define $f^p(P,Q;\ell)$ by (3.8) and (3.9) with Γ replaced by G and note that (3.10) and (3.13) remain valid. We denote the righthand side of (3.11) with $\Gamma=G$ by e_{ip} and claim that

(3.14) $e_{ip}=\hat{\varepsilon}_{ip},\quad i=1,2,\ldots,m,\ p=0,1,\ldots,r_i.$

For $p=0$ we get from the definition of S and (3.13) with P, Q and ℓ replaced by $\mathcal{K}_i^\sharp$, $\mathcal{L}_i^\sharp$ and $\bar{w}_i$, respectively, that

$$\begin{pmatrix} \hat{\varepsilon}_{i0} \\ -V\varepsilon_{i0} \end{pmatrix} = (S-\bar{w}_i)^{-1}\begin{pmatrix} 0 \\ (W+\bar{w}_i V)\varepsilon_{i0} \end{pmatrix} = (A-\bar{w}_i)^{-1}G(W+\bar{w}_i V)\varepsilon_{i0} = e_{i0} - GV\varepsilon_{i0},$$

which implies that $e_{i0} = \hat{\varepsilon}_{i0}$. We continue with mathematical induction and assume that the claim is true for $p \geq 1$. Then we get in the same way as before that

$$\begin{pmatrix} \hat{\varepsilon}_{ip} \\ -V\varepsilon_{ip} \end{pmatrix} = (S-\bar{w}_i)^{-1}\begin{pmatrix} \hat{\varepsilon}_{i(p-1)} \\ (W+\bar{w}_i V)\varepsilon_{ip} \end{pmatrix} = (A-\bar{w}_i)^{-1}(\hat{\varepsilon}_{i(p-1)} + G(W+\bar{w}_i V)\varepsilon_{ip}) =$$

$$= (A-\bar{w}_i)^{-1}(e_{i(p-1)} + G(W+\bar{w}_i V)\varepsilon_{ip}) = e_{ip} - GV\varepsilon_{ip}$$

and this proves the claim. It follows that

$$((D_\ell)^p(\mathcal{K}_i(\ell)\mathcal{N}(\ell) - \mathcal{L}_i(\ell)\mathcal{M}(\ell)))^*|_{\ell=w_i} =$$

$$= (D_\ell)^p\left(\mathcal{K}_i^\sharp(\ell) + G^*(A-\ell)^{-1}G(\ell\mathcal{K}_i^\sharp(\ell) + \mathcal{L}_i^\sharp(\ell))\right)|_{\ell=\bar{w}_i} =$$

$$= G^*(D_\ell)^p\left(G\mathcal{K}_i^\sharp(\ell) + (A-\ell)^{-1}G(\ell\mathcal{K}_i^\sharp(\ell) + \mathcal{L}_i^\sharp(\ell))\right)|_{\ell=\bar{w}_i} =$$

$$= G^* e_{ip} = G^*\hat{\varepsilon}_{ip} = 0,$$

which implies that $(\mathcal{M}(\ell), \mathcal{N}(\ell))$ is a solution of the interpolation problem (IP), which is what we wanted to prove. Now that we have established this, (3.12) is valid, see Proposition 1.7, and hence on account of (3.14)

$$(\mathbb{P}_{\mathcal{M},\mathcal{N}})_{ij}^{pq} = [e_{jq}, e_{ip}] = [\hat{\varepsilon}_{jq}, \hat{\varepsilon}_{ip}] = [\varepsilon_{jq}, \varepsilon_{ip}] = \mathbb{P}_{ij}^{pq},$$

which shows that $\mathbb{P}_{\mathcal{M},\mathcal{N}} = \mathbb{P}$. This proves part (i) of the theorem and the last series of equalities establishes the uniqueness of the nonnegative Pick matrix in part (ii). Assume that A and A' are minimal selfadjoint extensions of the $\mathbb{P}$ symmetric relation S for which (3.5) is valid. Then by Proposition 3.2 they are G minimal and by Theorem 1.5 they are unitarily equivalent. From the way the isomorphism is constructed in the proof of Theorem 1.5 it can be seen that when the isomorphism is restricted to $\mathfrak{H}$ it coincides with the identity on $\mathfrak{H}$. This completes the proof of the theorem.

As an immediate consequence of Theorem 3.3 we mention the following result.

COROLLARY 3.4. *Let $\mathbb{P}$ be a nonnegative Pick matrix associated with the interpolation problem* (IP). *Then formula* (3.5) *establishes a one to one correspondence between all (equivalence classes of isomorphic copies of)* *minimal selfadjoint Hilbert space extensions A of the $\mathbb{P}$ symmetric relation S and all solutions $(\mathcal{M}(\ell),\mathcal{N}(\ell))\in\mathbf{N}^n$ of the interpolation problem* (IP) *which satisfy the additional restraint $\mathbb{P}_{\mathcal{M},\mathcal{N}}=\mathbb{P}$.*

The restraint $\mathbb{P}_{\mathcal{M},\mathcal{N}}=\mathbb{P}$ in the corollary amounts to imposing additional interpolation conditions on the derivatives in those interpolation points w_i, for which $\bar{w}_i$ is also an interpolation point. Corollary 3.4 implies that if $\mathbb{P}$ is a nonnegative solution of the Lyapunov equation the interpolation problem (IP) together with the additional restraint $\mathbb{P}_{\mathcal{M},\mathcal{N}}=\mathbb{P}$ has either precisely one solution or infinitely many ones. The former case occurs if and only if the Lyapunov equation has a unique nonnegative solution $\mathbb{P}$ and the $\mathbb{P}$ symmetric relation S is selfadjoint and the latter case occurs in all other situations: (i) the Lyapunov equation has a unique solution $\mathbb{P}\geq 0$ and the $\mathbb{P}$ symmetric relation S has defect numbers larger than zero, or (ii) the Lyapunov equation has infinitely many nonnegative solutions.

In the interpolation problem (IP) we look for solutions $(\mathcal{M}(\ell),\mathcal{N}(\ell))$, which belong to the extended Nevanlinna class $\mathbf{N}^n$. On the other hand in the classical interpolation problem one looks for solutions, which are functions, i.e., of the form $(I,\mathcal{N}(\ell))$, or which are inverses of functions, i.e., of the form $(\mathcal{M}(\ell),I)$. The next corollary of Theorem 3.3 shows when all solutions of the interpolation problem (IP) are functions or inverses of functions.

COROLLARY 3.5. *Suppose that the Lyapunov equation* (3.1) *has a nonnegative solution.* (i) *A necessary and sufficient condition for all solutions of the interpolation problem* (IP) *to be functions is that the mapping V is surjective, i.e., that*

$$\text{(3.15)}\quad \text{l.s.}\{\,\mathcal{K}_i^{(p)}(w_i)^*\mid i=1,2,\ldots,m,\ \ p=0,1,\ldots,r_i\}=\mathbb{C}^n.$$

(ii) *A necessary and sufficient condition for all solutions of the interpolation problem* (IP) *to be inverses of functions is that the mapping W is surjective, i.e., that*

$$\text{l.s.}\{\,\mathcal{L}_i^{(p)}(w_i)^*\mid i=1,2,\ldots,m,\ \ p=0,1,\ldots,r_i\}=\mathbb{C}^n.$$

Proof. Assume that (3.15) is satisfied and let $(\mathcal{M}(\ell),\mathcal{N}(\ell))$ be a solution of the problem (IP). Without loss of generality we may suppose that the functions $\mathcal{M}(\ell)$ and $\mathcal{N}(\ell)$ have the representation (3.4). Assume that $\mathcal{M}(\lambda)f=0$ for some $\lambda\in\mathbb{C}\backslash\mathbb{R}$ and some $f\in\mathbb{C}^n$. It follows from Corollary 1.6 that

$\{0,Gf\}\in A$, and since $Gf\in\mathfrak{H}$, $\{0,Gf\}\in S^*$. By (3.3) this implies $f\in\Re(V)^{\perp}=\{0\}$. Hence $\mathcal{M}(\ell)$ is invertible for all $\ell\in\mathbb{C}\setminus\mathbb{R}$ and $(\mathcal{M}(\ell),\mathcal{N}(\ell))$ is a function.

Conversely, assume that (3.15) is not satisfied, i.e., assume that $\Re(V)\neq\mathbb{C}^n$ and let $0\neq f\in\mathbb{C}^n\ominus\Re(V)$. Consider the model $\mathfrak{H}$, S and G associated with a nonnegative solution of the Lyapunov equation, constructed in the beginning of the section. Then $Gf\neq 0$ and by (3.3) we have that $Gf\in S^*(0)$. By Proposition 1.2 (i) $A=S\dot{+}M_\infty(S^*)$ is a canonical selfadjoint extension of S with $S^*(0)=A(0)$ and hence $Gf\in A(0)$. Theorem 3.3 (i) implies that the pair $(\mathcal{M}(\ell),\mathcal{N}(\ell))$ with $\mathcal{M}(\ell)$, $\mathcal{N}(\ell)$ defined by (3.4) is a solution of the problem (IP). It has the additional property $\mathcal{M}(\ell)f=0$ for all $\ell\in\mathbb{C}\setminus\mathbb{R}$, which implies that it is not a function. This completes the proof of (i).

The statement (ii) follows from (i): simply rewrite the interpolation problem (IP) as

$$(-\mathcal{L}_i(\ell)(-\mathcal{M}(\ell)))^{(p)}\,|_{\ell=w_i}=(\mathcal{K}_i(\ell)\mathcal{N}(\ell))^{(p)}\,|_{\ell=w_i},\quad 1\leq i\leq m,\ 0\leq p\leq r_i,$$

and observe that $(\mathcal{M}(\ell),\mathcal{N}(\ell))\in\mathbf{N}^n$ if and only if $(\mathcal{N}(\ell),-\mathcal{M}(\ell))\in\mathbf{N}^n$. This completes the proof.

We now come to the construction of some solution matrices and assume for the remainder of this section that the Lyapunov equation (3.1) has a hermitian solution $\mathbb{P}>0$. To apply the results of Section 2 we consider the model $\mathfrak{H}$, S and G associated with $\mathbb{P}$ constructed in the beginning of the section. Then, as observed before $\mathfrak{L}=\Re(G)=\mathbb{C}^n$ is a module space for the relation S. In Section 2 we considered an arbitrary linear space $\mathfrak{G}$ and a bijection from $\mathfrak{G}$ onto $\mathfrak{L}$. In this section we put $\mathfrak{G}=\mathbb{C}^n$ and take for the bijection the mapping G of the model. If $\mathcal{M}(\ell),\mathcal{N}(\ell)$ has the representation (3.4) then we may write

$$\begin{pmatrix}\mathcal{N}(\ell)\\ \mathcal{M}(\ell)\end{pmatrix}=\begin{pmatrix}I+\ell G^*(A-\ell)^{-1}G\\ -G^*(A-\ell)^{-1}G\end{pmatrix}=\begin{pmatrix}\ell & I\\ -I & 0\end{pmatrix}\begin{pmatrix}G^*(A-\ell)^{-1}G\\ I\end{pmatrix}$$

and hence $(\mathcal{M}(\ell),\mathcal{N}(\ell))=\mathfrak{t}_{K(\ell)W(\ell)}(\mathcal{T}(\ell))$ for some $\mathcal{T}(\ell)\in\mathbf{N}^n$ corresponding to the generalized resolvent $R(\ell)=P_{\mathfrak{H}}(A-\ell)^{-1}|_{\mathfrak{H}}$ of S, where $W(\ell)$ is a right $\mathfrak{L}$-resolvent matrix for S, see Definition 2.1, and $K(\ell)=\left(\begin{smallmatrix}\ell & I\\ -I & 0\end{smallmatrix}\right)$. Note that $K(\ell)$ is $\left(\begin{smallmatrix}0 & -I\\ I & 0\end{smallmatrix}\right)$ unitary for $\ell\in\mathbb{R}$. Thus Theorem 3.3 leads to the following result.

COROLLARY 3.6. *Let $\mathbb{P}>0$ be a solution of the Lyapunov equation* (3.1) *and put $\mathfrak{L}=\Re(G)$. Then the mapping $W(\ell)\to U(\ell)=\left(\begin{smallmatrix}\ell & I\\ -I & 0\end{smallmatrix}\right)W(\ell)$ gives a one to one* *correspondence between all right $\mathfrak{L}$–resolvent matrices $W(\ell)$ of the $\mathbb{P}$ symmetric relation S and all solution matrices $U(\ell)$ for the interpolation problem* (IP) *associated with* $\mathbb{P}$.

Using Theorem 2.3 we can exhibit explicitly some solution matrices for the interpolation problem (IP).

THEOREM 3.7. *Assume that the Lyapunov equation* (3.1) *has a positive solution* $\mathbb{P}$. *Then the following two matrix functions* $U_a(\ell)$, $a\in\mathbb{R}$, *and* $U_\infty(\ell)$ *are solution matrices of the interpolation problem* (IP) *associated with* $\mathbb{P}$:

$$U_a(\ell)=\begin{pmatrix} I & 0\\ 0 & I\end{pmatrix}+(\ell-a)\begin{pmatrix} W\\ V\end{pmatrix}(Z-\ell)^{-1}\mathbb{P}^{-1}(Z-a)^{-*}(W^*:V^*)\begin{pmatrix} 0 & I\\ -I & 0\end{pmatrix},$$

$$U_\infty(\ell)=\begin{pmatrix} I & 0\\ 0 & I\end{pmatrix}+\begin{pmatrix} W\\ V\end{pmatrix}(Z-\ell)^{-1}\mathbb{P}^{-1}(W^*:V^*)\begin{pmatrix} 0 & I\\ -I & 0\end{pmatrix},$$

where the $n\times r$ *matrices* W, V *and the* $r\times r$ *matrix* Z *are defined at the beginning of this section. For* $\ell\in\mathbb{R}$ *both* $U_a(\ell)$, $a\in\mathbb{R}$, *and* $U_\infty(\ell)$ *are* $\left(\begin{smallmatrix} 0 & -I\\ I & 0\end{smallmatrix}\right)$*–unitary.*

Proof. We consider the model and the corresponding relations given at the beginning of this section. Proposition 1.2 (*ii*) implies that for each $a\in\mathbb{R}\cup\{\infty\}$ there exists a canonical selfadjoint extension A_0 of the symmetric operator S in the space $\mathfrak{H}$ of the model, such that $a\in\rho(A_0)$. Hence $W_a(\ell)$, $a\in\mathbb{R}$, and $W_\infty(\ell)$ in Theorem 2.3 are right $\mathfrak{L}$–resolvent matrices for S.

First we express the operators $G^*\mathcal{Q}(\ell)$, $-G^{-1}\mathcal{P}(\ell)$ in terms of the data. We refer to Section 2 for the definition of the operators $\mathcal{P}(\ell)$, $\mathcal{Q}(\ell)$, etc. Note that $G^{-1}=G^*$ on $\mathfrak{L}=\mathfrak{H}_2$. For $\ell\in\rho_S(\mathfrak{L})$ and $x\in\mathfrak{H}$, we have that $\mathcal{P}(\ell)x\in\mathfrak{L}$, $(I-\mathcal{P}(\ell))x\in\mathfrak{R}(S-\ell)$ and hence there exist $y_j\in\mathfrak{H}_j$, $j=1,2$, such that

$$\mathcal{P}(\ell)\begin{pmatrix} x_1\\ x_2\end{pmatrix}=\begin{pmatrix} 0\\ y_2\end{pmatrix},\quad \begin{pmatrix} x_1\\ x_2-y_2\end{pmatrix}=(I-\mathcal{P}(\ell))\begin{pmatrix} x_1\\ x_2\end{pmatrix}=\begin{pmatrix} (Z-\ell)y_1\\ (W+\ell V)y_1\end{pmatrix}.$$

It follows that $y_1=(Z-\ell)^{-1}x_1$, $y_2=x_2-(W+\ell V)(Z-\ell)^{-1}x_1$ and

$$\mathcal{P}(\ell)x=\begin{pmatrix} 0 & 0\\ -(W+\ell V)(Z-\ell)^{-1} & I\end{pmatrix}\begin{pmatrix} x_1\\ x_2\end{pmatrix},$$

$$G^*\mathcal{P}(\ell)x=\left(-(W+\ell V)(Z-\ell)^{-1}:I\right)\begin{pmatrix} x_1\\ x_2\end{pmatrix}.$$

Since $S\subset A_0$, we find, using the same notation, that

$$\mathcal{Q}(\ell)x=R_0(\ell)(I-\mathcal{P}(\ell))x=(A_0-\ell)^{-1}\begin{pmatrix} (Z-\ell)y_1\\ (W+\ell V)y_1\end{pmatrix}=$$

$$= \begin{pmatrix} y_1 \\ -Vy_1 \end{pmatrix} = \begin{pmatrix} (Z-\ell)^{-1} & 0 \\ -V(Z-\ell)^{-1} & 0 \end{pmatrix} \begin{pmatrix} x_1 \\ x_2 \end{pmatrix},$$

which implies that

$$G^*\mathcal{Q}(\ell)x = \left(-V(Z-\ell)^{-1} : 0\right) \begin{pmatrix} x_1 \\ x_2 \end{pmatrix}.$$

It follows that

$$\begin{pmatrix} \ell & I \\ -I & 0 \end{pmatrix} \begin{pmatrix} G^*\mathcal{Q}(\ell) \\ -G^{-1}\mathcal{P}(\ell) \end{pmatrix} = \begin{pmatrix} W(Z-\ell)^{-1} & -I \\ V(Z-\ell)^{-1} & 0 \end{pmatrix}$$

and

$$\begin{pmatrix} I & 0 \\ \lambda & I \end{pmatrix} \begin{pmatrix} G^*\mathcal{Q}(\lambda) \\ -G^{-1}\mathcal{P}(\lambda) \end{pmatrix} = \begin{pmatrix} -V(Z-\lambda)^{-1} & 0 \\ W(Z-\lambda)^{-1} & -I \end{pmatrix}.$$

These equalities are between operators from the direct sum $\mathfrak{H}=\mathfrak{H}_1\oplus\mathfrak{H}_2$ to $\mathbb{C}^{2n}$. So, when we calculate the adjoints of these operators we must take into account the $\mathbb{P}$ inner product on $\mathfrak{H}$. We obtain

$$(3.16) \quad \begin{pmatrix} \ell & I \\ -I & 0 \end{pmatrix} \begin{pmatrix} G^*\mathcal{Q}(\ell) \\ -G^{-1}\mathcal{P}(\ell) \end{pmatrix} \begin{pmatrix} G^*\mathcal{Q}(\lambda) \\ -G^{-1}\mathcal{P}(\lambda) \end{pmatrix}^* \begin{pmatrix} I & \bar{\lambda} \\ 0 & I \end{pmatrix} =$$

$$= \begin{pmatrix} W(Z-\ell)^{-1} & -I \\ V(Z-\ell)^{-1} & 0 \end{pmatrix} \begin{pmatrix} \mathbb{P}^{-1} & 0 \\ 0 & I \end{pmatrix} \begin{pmatrix} -(Z-\lambda)^{-*}V^* & (Z-\lambda)^{-*}W^* \\ 0 & -I \end{pmatrix} =$$

$$= \begin{pmatrix} W \\ V \end{pmatrix} (Z-\ell)^{-1}\mathbb{P}^{-1}(Z-\lambda)^{-*}(W^* : V^*) \begin{pmatrix} 0 & I \\ -I & 0 \end{pmatrix} + \begin{pmatrix} 0 & I \\ 0 & 0 \end{pmatrix}.$$

Now, let $W(\ell)$ be any holomorphic matrix function satisfying (2.3). Then it follows from $W(\bar{\lambda})\begin{pmatrix} 0 & -I \\ I & 0 \end{pmatrix}W(\lambda)^* = \begin{pmatrix} 0 & -I \\ I & 0 \end{pmatrix}$ that

$$\begin{pmatrix} \ell & I \\ -I & 0 \end{pmatrix} W(\ell) = \begin{pmatrix} \ell & I \\ -I & 0 \end{pmatrix} \mathcal{X}(\ell,\lambda) \begin{pmatrix} I & \bar{\lambda} \\ 0 & I \end{pmatrix} \begin{pmatrix} \bar{\lambda} & I \\ -I & 0 \end{pmatrix} W(\bar{\lambda}),$$

where the matrix function $\mathcal{X}(\ell,\lambda)$ is defined by

$$\mathcal{X}(\ell,\lambda) = \begin{pmatrix} 0 & -I \\ I & 0 \end{pmatrix} + (\ell-\bar{\lambda}) \begin{pmatrix} G^*\mathcal{Q}(\ell) \\ -G^{-1}\mathcal{P}(\ell) \end{pmatrix} \begin{pmatrix} G^*\mathcal{Q}(\lambda) \\ -G^{-1}\mathcal{P}(\lambda) \end{pmatrix}^*.$$

Using (3.16) we obtain

$$(3.17) \quad \begin{pmatrix} \ell & I \\ -I & 0 \end{pmatrix} W(\ell) = \left[\begin{pmatrix} I & 0 \\ 0 & I \end{pmatrix} + (\ell-\bar{\lambda}) \begin{pmatrix} W \\ V \end{pmatrix} (Z-\ell)^{-1}\mathbb{P}^{-1}(Z-\lambda)^{-*}(-V^* : W^*) \right].$$

$$\cdot \begin{pmatrix} \bar{\lambda} & I \\ -I & 0 \end{pmatrix} W(\bar{\lambda}).$$

We shall apply this relation with $W(\ell)=W_a(\ell)$, $a\in\mathbb{R}$, and $\lambda=a$. It follows from Theorem 2.3 that

$$\begin{pmatrix} a & I \\ -I & 0 \end{pmatrix} W_a(a)=\frac{1}{a-\mu}\begin{pmatrix} \frac{1}{2}a & 2(|\mu|^2-a\mathrm{Re}\mu) \\ -\frac{1}{2} & 2(a-\mathrm{Re}\mu) \end{pmatrix}.$$

Since the two factors on the lefthand side are $\left(\begin{smallmatrix} 0 & -I \\ I & 0 \end{smallmatrix}\right)$ unitary, so are the product and its inverse D_a, which is given by

$$D_a=\frac{1}{a-\bar{\mu}}\begin{pmatrix} 2(a-\mathrm{Re}\mu) & 2(a\mathrm{Re}\mu-|\mu|^2) \\ \frac{1}{2} & \frac{1}{2}a \end{pmatrix}.$$

Hence $W_a(\ell)D_a$ is a resolvent matrix and by Corollary 3.6 the matrix function $U_a(\ell)$ defined by

$$(3.18)\quad U_a(\ell)=\left(\begin{smallmatrix} \ell & I \\ -I & 0 \end{smallmatrix}\right)W_a(\ell)D_a=$$

$$=\begin{pmatrix} I & 0 \\ 0 & I \end{pmatrix}+(\ell-a)\begin{pmatrix} W \\ V \end{pmatrix}(Z-\ell)^{-1}\mathbb{P}^{-1}(Z-a)^{-*}(W^*:V^*)\begin{pmatrix} 0 & I \\ -I & 0 \end{pmatrix}$$

is a solution matrix. Clearly the matrix $D_\infty=\lim_{a\to\infty}D_a$ is $\left(\begin{smallmatrix} 0 & -I \\ I & 0 \end{smallmatrix}\right)$ unitary and given by

$$D_\infty=\begin{pmatrix} 2 & \mu+\bar{\mu} \\ 0 & \frac{1}{2} \end{pmatrix}.$$

The limit $W_\infty(\ell)=\lim_{a\to\infty}W_a(\ell)$ is a resolvent matrix (see Theorem 2.3) and so is $W_\infty(\ell)D_\infty$. Therefore the matrix function $U_\infty(\ell)=\left(\begin{smallmatrix} \ell & I \\ -I & 0 \end{smallmatrix}\right)W_\infty(\ell)D_\infty$ is a solution matrix and $U_\infty(\ell)=\lim_{a\to\infty}U_a(\ell)$. It follows from (3.18) that $U_\infty(\ell)$ coincides with the one in the theorem. (The formula for $U_\infty(\ell)$ can also be obtained in the following way. Similar to the calculations above one can show that

$$G^*(R(\infty)+P(\infty))x=\left(V-2(\mu-\bar{\mu})^{-1}(W+\mu V):I\right)\begin{pmatrix} x_1 \\ x_2 \end{pmatrix}.$$

and

$$G^*P(\infty)x=\left(V:I\right)\begin{pmatrix} x_1 \\ x_2 \end{pmatrix}.$$

Via Theorem 2.3 this leads to $\lim_{\lambda\to\infty}\left(\begin{smallmatrix} \lambda & I \\ -I & 0 \end{smallmatrix}\right)W_\infty(\lambda)=D_\infty^{-1}$. Applying (3.17) to $W(\ell)=W_\infty(\ell)$ and taking the limit $\lambda\to\infty$ we obtain the desired formula). This completes the proof.

Since $U_a(\ell)$, $a\in\mathbb{R}$, and $U_\infty(\ell)$ are solution matrices they are equal modulo multiplication by a nonvanishing holomorphic function and by multiplication from the right by a $\left(\begin{smallmatrix} 0 & -I \\ I & 0 \end{smallmatrix}\right)$–unitary matrix. As we have seen $U_a(\ell)$, $a\in\mathbb{R}$, is $\left(\begin{smallmatrix} 0 & -I \\ I & 0 \end{smallmatrix}\right)$–unitary for $\ell\in\mathbb{R}$; in particular it follows from this that

$$U_a(\infty)^{-1} = \begin{pmatrix} I & 0 \\ 0 & I \end{pmatrix} + \begin{pmatrix} W \\ V \end{pmatrix}(Z-a)^{-1}\mathbb{P}^{-1}(W^*:V^*)\begin{pmatrix} 0 & I \\ -I & 0 \end{pmatrix} = U_\infty(a).$$

Straightforward calculations show that

$$U_a(\ell)U_a(\infty)^{-1} = U_\infty(\ell), \qquad U_\infty(\ell)U_\infty(a)^{-1} = U_a(\ell),$$

$$\frac{U_\infty(\ell)\left(\begin{smallmatrix} 0 & -I \\ I & 0 \end{smallmatrix}\right)U_\infty(\lambda)^* - \left(\begin{smallmatrix} 0 & -I \\ I & 0 \end{smallmatrix}\right)}{\ell-\bar\lambda} = \begin{pmatrix} W \\ V \end{pmatrix}(Z-\ell)^{-1}\mathbb{P}^{-1}(Z-\lambda)^{-*}(W^*:V^*)$$

and

$$\frac{U_\infty(\lambda)^*\left(\begin{smallmatrix} 0 & -I \\ I & 0 \end{smallmatrix}\right)U_\infty(\ell) - \left(\begin{smallmatrix} 0 & -I \\ I & 0 \end{smallmatrix}\right)}{\ell-\bar\lambda} =$$

$$= \begin{pmatrix} 0 & -I \\ I & 0 \end{pmatrix}\begin{pmatrix} W \\ V \end{pmatrix}\mathbb{P}^{-1}(Z-\lambda)^{-*}\mathbb{P}(Z-\ell)^{-1}\mathbb{P}^{-1}(W^*:V^*)\begin{pmatrix} 0 & I \\ -I & 0 \end{pmatrix}.$$

Formulas, similar to the last two, can be given for the function $U_a(\ell)$. We omit the details. The representation in Theorem 3.7 may be viewed as a realization of the matrix function $U_\infty(\ell)$. The fact that $U_\infty(\ell)$ is $\left(\begin{smallmatrix} 0 & -I \\ I & 0 \end{smallmatrix}\right)$–unitary for $\ell\in\mathbb{R}$ is intimately connected with the fact that the Pick matrix $\mathbb{P}$ is a solution of the Lyapunov equation (3.1), cf. [AG]. We intend to return to these questions in the second part of this paper.

Appendix: The Lyapunov equation

If $\tilde{\mathcal{K}}(\ell)$, $\tilde{\mathcal{L}}(\ell)$ are two locally holomorphic $m\times n$ matrix functions on $\mathbb{C}\backslash\mathbb{R}$ with rows $\tilde{\mathcal{K}}_i(\ell)$, $\tilde{\mathcal{L}}_i(\ell)$, respectively, we denote by $\mathbb{R}_{\tilde{\mathcal{K}},\tilde{\mathcal{L}}}$ the $n\times n$ matrix $\mathbb{R}_{\tilde{\mathcal{K}},\tilde{\mathcal{L}}} = ((\mathbb{R}_{\tilde{\mathcal{K}},\tilde{\mathcal{L}}})_{ij}^{pq})$ with

$$(\mathbb{R}_{\tilde{\mathcal{K}},\tilde{\mathcal{L}}})_{ij}^{pq} = (D_\ell)^p(D_{\bar\lambda})^q\frac{\tilde{\mathcal{L}}_i(\ell)\tilde{\mathcal{K}}_j(\lambda)^* - \tilde{\mathcal{K}}_i(\ell)\tilde{\mathcal{L}}_j(\lambda)^*}{\ell-\bar\lambda}\Big|_{\ell=w_i,\lambda=w_j},$$

for those indices i,j for which $w_i\neq\bar w_j$, and for those with $w_i=\bar w_j$

$$(\mathbb{R}_{\tilde{\mathcal{K}},\tilde{\mathcal{L}}})_{ij}^{pq} = \sum_{n=0}^{p}\Big((D_\ell)^{p-n}\tilde{\mathcal{K}}_i(\ell)((D_\lambda)^{q+1+n}\tilde{\mathcal{L}}_j(\lambda))^* +$$

$$-(D_\ell)^{p-n}\tilde{\mathcal{L}}_i(\ell)((D_\lambda)^{q+1+n}\tilde{\mathcal{K}}_j(\lambda))^*\Big)\Big|_{\ell=w_i,\lambda=w_j}.$$

If for all indices i,j we have that $w_i \neq \bar{w}_j$ then obviously $\mathbb{R}_{\tilde{\mathcal{K}},\tilde{\mathcal{L}}}$ is hermitian. In the general case, it is not difficult to verify that $\mathbb{R}_{\tilde{\mathcal{K}},\tilde{\mathcal{L}}}$ is hermitian if and only if for the indices i,j with $w_i = \bar{w}_j$ it holds that

$$\text{(A.1)} \quad (D_\ell)^p \left(\tilde{\mathcal{L}}_i(\ell)\tilde{\mathcal{K}}_j(\bar{\ell})^* - \tilde{\mathcal{K}}_i(\ell)\tilde{\mathcal{L}}_j(\bar{\ell})^* \right) |_{\ell=w_i,\lambda=w_j} = 0, \quad 0 \leq p \leq r_i + r_j + 1.$$

In this appendix we want to study the solutions of the Lyapunov equation

$$\mathbb{P}Z - Z^*\mathbb{P} = V^*W - W^*V$$

associated with the interpolation problem (IP), which can be written in the form $\mathbb{P} = \mathbb{R}_{\tilde{\mathcal{K}},\tilde{\mathcal{L}}}$. Note that $\mathbb{P}_{\mathcal{M},\mathcal{N}}$ with $(\mathcal{M}(\ell),\mathcal{N}(\ell)) \in \mathbf{N}^n$ defined by (0.7) can be written as $\mathbb{P}_{\mathcal{M},\mathcal{N}} = \mathbb{R}_{\tilde{\mathcal{K}},\tilde{\mathcal{L}}}$ with

$$\text{(A.2)} \quad \tilde{\mathcal{K}}(\ell) = (\ell\mathcal{K}(\ell) + \mathcal{L}(\ell))\hat{\mathcal{M}}(\bar{\ell})^*, \quad \tilde{\mathcal{L}}(\ell) = (\ell\mathcal{K}(\ell) + \mathcal{L}(\ell))\hat{\mathcal{N}}(\bar{\ell})^*,$$

where $\hat{\mathcal{M}}(\ell)$ and $\hat{\mathcal{N}}(\ell)$ are given by (0.8). These functions satisfy

$$\text{(A.3)} \quad \left(\tilde{\mathcal{L}}(\ell)\tilde{\mathcal{K}}(\bar{\ell})^* - \tilde{\mathcal{K}}(\ell)\tilde{\mathcal{L}}(\bar{\ell})^* \right) = 0, \quad \ell \in \mathbb{C}\backslash\mathbb{R},$$

and hence also (A.1). For the following result we also refer to [G].

PROPOSITION A.1. *Let X be an $r{\times}r$ matrix. Then the equation*

$$\text{(A.4)} \quad \mathbb{P}Z - Z^*\mathbb{P} = X$$

has a solution if and only if X satisfies the relation

$$\text{(A.5)} \quad \sum_{k=0}^{h} X_{ij}^{k(h-k)} = 0, \quad 0 \leq h \leq \min\{r_i, r_j\}$$

for those i,j for which $w_i = \bar{w}_j$. If the interpolation points w_i are such that $w_i \neq \bar{w}_j$ for all indices i,j then (A.4) *has exactly one solution. If there are indices i,j with $w_i = \bar{w}_j$ and* (A.5) *is valid, then* (A.4) *has infinitely many solutions.*

Proof. Componentwise the Lyapunov equation reduces to the system of equations

$$(\bar{w}_j - w_i)\mathbb{P}_{ij}^{pq} + \mathbb{P}_{ij}^{p(q-1)} - \mathbb{P}_{ij}^{(p-1)q} = X_{ij}^{pq}, \ 1 \leq i,j \leq m, \quad 0 \leq p \leq r_i, \ 0 \leq q \leq r_j,$$

where we have set $\mathbb{P}_{ij}^{pq} = 0$ when one of the indices p, q is outside its range indicated above. Fix lower indices i and j. If $w_i \neq \bar{w}_j$, then the remaining system of equations has a unique solution $(\mathbb{P}_{ij}^{pq})$, for any choice of X_{ij}^{pq} on the righthand side, namely

$$\mathbb{P}_{ij}^{pq} = \sum_{k=0}^{p} \sum_{h=0}^{q} \frac{(-1)^{p-k+1}}{(w_i-\bar{w}_j)^{p-k+q-h+1}} \binom{p-k+q-h}{p-k} X_{ij}^{kh}.$$

For the case that $w_i=\bar{w}_j$ we simplify the notation and write $\mathbb{P}(p,q)$ for $\mathbb{P}_{ij}^{pq}$ and use a similar notation for X. Then if the equation (A.4) has a solution $\mathbb{P}$, we find that for $0 \le h \le \min\{r_i,r_j\}$

$$\sum_{k=0}^{h} X_{ij}^{k(h-k)} = \sum_{k=0}^{h} \mathbb{P}(k,h-1-k) - \mathbb{P}(k-1,h-k) = \mathbb{P}(h,-1) - \mathbb{P}(-1,h) = 0,$$

which implies (A.5). Conversely, assume that (A.5) is valid and that $r_i \le r_j$, and put

$$\text{(A.6)} \quad \mathbb{P}(p,q) = \sum_{k=0}^{p} X(k,p+q+1-k),$$

where the $X(k,h)$'s are chosen arbitrarily for $h \ge r_j+1$. By considering the cases $(p=0,\ q=0)$, $(p=0,\ q\ge 1)$, $(p\ge 1,\ q\ge 1)$ and $(p\ge 1,\ q=0)$ separately one can easily check that $\mathbb{P}$ satisfies

$$\text{(A.7)} \quad \mathbb{P}(p,q-1) - \mathbb{P}(p-1,q) = X(p,q)$$

and this is what we had to prove. For instance, in the last case $(p\ge 1,\ q=0)$ the equation becomes $\mathbb{P}(p-1,0) = -X(p,q)$, which is equivalent to

$$\sum_{k=0}^{p} X(k,p-k) = 0, \quad 1 \le p \le r_i.$$

The proof for the case in which $r_j \le r_i$ is similar and omitted. The arbitrariness of X in the above argument implies that the equation has infinitely many solutions.

The equation (A.4) has a hermitian solution if and only if X satisfies $X^* = -X$ and X satisfies (A.5). The Lyapunov equation associated with the interpolation problem is of the form (A.4) with $X = V^*W - W^*V$ which is skew hermitian and (0.5) coincides with (A.5), since

$$V_{ip} = (1/p!)\,\mathcal{K}_i^{(p)}(w_i)^*, \quad W_{ip} = (1/p!)\,\mathcal{L}_i^{(p)}(w_i)^*.$$

Hence Proposition A.1 substantiates some of the remarks made at the beginning of the Introduction. We shall say that two $m\times n$ matrix functions $\tilde{\mathcal{K}}(\ell)$, $\tilde{\mathcal{L}}(\ell)$ defined and locally holomorphic on $\mathbb{C}\backslash\mathbb{R}$ interpolate the data of the interpolation problem (IP) if

$$(1/p!)\,\widetilde{\mathcal{K}}_i^{(p)}(w_i)^* = V_{ip}, \quad (1/p!)\,\widetilde{\mathcal{L}}_i^{(p)}(w_i)^* = W_{ip}.$$

Thus the functions $\mathcal{K}(\ell)$ and $\mathcal{L}(\ell)$ used in the formulation of the interpolation problem interpolate the data.

PROPOSITION A.2. *Suppose that* $\widetilde{\mathcal{K}}(\ell)$ *and* $\widetilde{\mathcal{L}}(\ell)$ *interpolate the data. Then* $\mathbb{P} = \mathbb{R}_{\widetilde{\mathcal{K}},\widetilde{\mathcal{L}}}$ *is a solution of the Lyapunov equation if and only if for all indices* i,j *with* $w_i = \bar{w}_j$

$$(A.8) \quad (D_\ell)^p \Big(\widetilde{\mathcal{L}}_i(\ell)\widetilde{\mathcal{K}}_j(\bar{\ell})^* - \widetilde{\mathcal{K}}_i(\ell)\widetilde{\mathcal{L}}_j(\bar{\ell})^* \Big) \,|_{\ell=w_i,\lambda=w_j} = 0, \quad 0 \le p \le \max\{r_i, r_j\}.$$

Proof. The proof of the proposition follows from arguments similar to those used in the part of the proof of Proposition A.1 around the formulas (A.6) and (A.7) with

$$X(p,q) = \Big((D_\ell)^p \widetilde{\mathcal{K}}_i(\ell)((D_\lambda)^q \widetilde{\mathcal{L}}_j(\lambda))^* - (D_\ell)^p \widetilde{\mathcal{L}}_i(\ell)((D_\lambda)^q \widetilde{\mathcal{K}}_j(\lambda))^* \Big) \,|_{\ell=w_i,\lambda=w_j}.$$

It interesting to observe that the sets of conditions (A.3), (A.1), (A.8) and (0.5), corresponding to (A.5), form a decreasing sequence. Let w be one of the interpolation points in (0.1) of the problem (IP) and put

$$\mathbb{N}(w) = \{\, i \in \mathbb{N} \mid w_i = w \,\}.$$

Denote by $n(w)$ the number of elements in $\mathbb{N}(w)$ and by D_w the $n(w) \times 2n$ matrix whose i–th row is given by $\big(\mathcal{K}_i(w) : \mathcal{L}_i(w)\big)$. The next proposition is simple and its proof is left to the reader.

PROPOSITION A.3. *Assume that* $n(w) \le n$ *and* $\operatorname{rank} D_w = n(w)$ *for those interpolation points* w *for which* $\bar{w}$ *is also an interpolation point. If* $\mathbb{P}$ *is a solution of the Lyapunov equation, then* $\mathbb{P} = \mathbb{R}_{\widetilde{\mathcal{K}},\widetilde{\mathcal{L}}}$ *for some* $m \times n$ *matrix functions* $\widetilde{\mathcal{K}}(\ell)$ *and* $\widetilde{\mathcal{L}}(\ell)$ *which interpolate the data and satisfy* (A.8) *for those* i,j *with* $w_i = \bar{w}_j$.

REFERENCES

[ABDS] D. Alpay, P. Bruinsma, A. Dijksma, H.S.V. de Snoo, "A Hilbert space associated with a Nevanlinna function", to appear.

[AD1] D. Alpay, H. Dym, "Hilbert spaces of analytic functions, inverse scattering and operator models I", Integral Equations Operator Theory, 7 (1984), 589–641.

[AD2] D. Alpay, H. Dym, "Hilbert spaces of analytic functions, inverse scattering and operator models II", Integral Equations Operator Theory, 8 (1985), 145–180.

[AG] D. Alpay, I. Gohberg, "Unitary rational matrix functions", Operator Theory: Adv. Appl., 33 (1988), 175–222.

[B1] J.A. Ball, "Interpolation problems of Pick–Nevanlinna and Loewner types for meromorphic matrix functions", Integral Equations Operator Theory, 6 (1983), 804–840.

[B2] J.A. Ball, "Nevanlinna–Pick interpolation: Generalizations and applications", in Surveys of some recent results in operator theory (J.B. Conway, B.B. Morrel, eds.), Pitman research notes in mathematics 171, Longman, 1989, 51–94.

[BGR] J.A. Ball, I. Gohberg, L. Rodman, "Realization and interpolation of rational matrix functions", Operator Theory: Adv. Appl., 33 (1988), 1–71.

[BH] J.A. Ball, J.W. Helton, "Interpolation problems of Pick–Nevanlinna and Loewner types for meromorphic matrix functions: parametrization of the set of all solutions", Integral Equations Operator Theory, 9 (1986), 155–203.

[BR] J.A. Ball, A. C. M. Ran, "Local inverse spectral problems for rational matrix functions", Integral Equations Operator Theory, 10 (1987), 349–415.

[dB1] L. de Branges, "Some Hilbert spaces of analytic functions I", Trans. Amer. Math. Soc., 106 (1963), 445–468.

[dB2] L. de Branges, "Some Hilbert spaces of analytic functions II", Trans. Amer. Math. Soc., 11 (1965), 44–72.

[dB3] L. de Branges, "Some Hilbert spaces of analytic functions III", Trans. Amer. Math. Soc., 12 (1965), 149–186.

[dB4] L. de Branges, *Hilbert spaces of entire functions*, Prentice Hall, Englewood Cliffs, N.J., 1968 (French translation: *Espaces Hilbertiens de fonctions entières*, Masson et Cie, Paris 1972).

[CdS] E.A. Coddington, H.S.V. de Snoo, "Positive selfadjoint extensions of positive symmetric subspaces", Math. Z., 159 (1978), 203–214.

[D1] H. Dym, *J contractive matrix functions, reproducing kernel Hilbert spaces and interpolation*, Regional conference series in mathematics, 71, Amer. Math. Soc., Providence, R.I., 1989.

[D2] H. Dym, "On reproducing kernel spaces, J unitary matrix functions, interpolation and displacement rank", Operator Theory: Adv. Appl., 41 (1989), 173–239.

[DdS] A. Dijksma, H.S.V. de Snoo, "Self–adjoint extensions of symmetric subspaces", Pacific J. Math., 54 (1974), 71–100.

[DLS1] A. Dijksma, H. Langer, H.S.V. de Snoo, "Unitary colligations in Kreĭn spaces and their role in the extension theory of isometries and symmetric linear relations in Hilbert spaces", Functional Analysis II, Proceedings Dubrovnik 1985, Lecture Notes in Mathematics, 1242 (1986), 1–42.

[DLS2] A. Dijksma, H. Langer, H.S.V. de Snoo, "Hamiltonian systems with eigenvalue depending boundary conditions", Operator Theory: Adv. Appl., 35 (1988), 37–83.

[DLS3] A. Dijksma, H. Langer, H.S.V. de Snoo, "Generalized coresolvents of standard isometric operators and generalized resolvents of standard symmetric relations in Kreĭn spaces", to appear.

[F1] I.P. Fedchina, "A criterion for the solvability of the Nevanlinna-Pick tangential problem", Mat. Issled, 7 (1972), 4 (26), 213–227 (Russian).

[F2] I.P. Fedchina, "Descriptions of solutions of the tangential Nevanlinna-Pick problem", Doklady Akad. Nauk. Arm. SSR., 60 (1975), 37–42 (Russian).

[F3] I.P. Fedchina, "Tangential Nevanlinna-Pick problem with multiple points", Doklady Akad. Nauk Arm. SSR., 61 (1975), 214–218 (Russian).

[G] F.R. Gantmacher, *Matrizentheorie*, (2nd edition), Nauka, Moscow 1966 (Russian) (German translation: VEB Deutscher Verlag der Wissenschaften, Berlin, 1986).

[Go] L.B. Golinskii, "On a generalization of the matrix Nevanlinna-Pick problem", Proc. Arm. Acad. Sci., 18 (1983), 187–205 (Russian).

[K] V.E. Katsnelson, *Methods of J-theory in continuous interpolation problems of analysis*, Part I, VINITI, Kharkhov, 1982 (Russian) (Private translation by T. Ando, Sapporo, 1985).

[KHJ] V.E. Katsnelson, A. Ja. Heifetz, P.M. Juditskij, "Abstract interpolation problems and extension theory of isometric operators", in Operators in spaces of functions and problems in function theory: Collected scientific papers, Kiev, Naukova Dumka, 1987, 83–96 (Russian).

[Ki] H. Kimura, "Conjugation, interpolation and model matching in H^∞", Int. J. Control, 49 (1989), 269–307.

[Ko] I.V. Kovalishina, "Analytic theory of a class of interpolation problems" Math. USSR Izv., 22 (1984) 419–463 (Russian).

[KP] I.V. Kovalishina, V.P. Potapov, *Integral representations of hermitian positive functions*, VINITI, Kharkhov, 1982 (Russian) (Private translation by T. Ando, Sapporo, 1982).

[Kr] M.G. Kreĭn, "Fundamental aspects of the representation theory of hermitian operators with deficiency index (m,m)", Ukrain. Math. Zh., 1 (1949), 3–66 (Russian) (English translation: Amer. Math. Soc. Transl., (2) 97 (1970), 75–143).

[KL1] M.G. Kreĭn, H. Langer, "Über die Q-Funktion eines π-hermiteschen Operators im Raume Π_κ", Acta Sci. Math. (Szeged), 34 (1973), 191–230.

[KL2] M.G. Kreĭn, H. Langer, "Über einige Fortsetzungsprobleme, die eng mit der Theorie hermitescher Operatoren im Raume Π_κ zusammenhängen. I: Einige Funktionenklassen und ihre Darstellungen". Math. Nachr., 77 (1977), 187–236.

[KL3] M.G. Kreĭn, H. Langer, "Über einige Fortsetzungsprobleme, die eng mit der Theorie hermitescher Operatoren im Raume Π_κ zusammenhängen. II: Verallgemeinerte Resolventen, u-Resolventen und ganze Operatoren", J. Funct. Anal., 30 (1978), 390–447.

[KL4] M.G. Kreĭn, H. Langer, "On some extension problems which are closely connected with the theory of hermitian operators in a space Π_κ. III: Indefinite analogues of the Hamburger and Stieltjes moment problems", Part (I), Beiträge Anal., 14 (1979), 25–40; Part (II), Beiträge Anal., 15 (1981), 27–45.

[KL5] M.G. Kreĭn, H. Langer, "On some continuation problems which are closely related to the theory of operators in spaces Π_κ. IV: Continuous analogues of orthogonal polynomials on the unit circle with respect to an indefinite weight and related continuation problems for some classes of functions", J. Operator Theory, 13 (1985), 299–417.

[KS1] M.G. Kreĭn, Š.N. Saakjan, "Some new results in the theory of resolvents of Hermitian operators", Dokl. Akad. Nauk SSSR, 169 (1966),1269–1272 (Russian) (English translation: Sov. Math. Dokl., 7 (1966), 1086–1089).

[KS2] M.G. Kreĭn, Š.N. Saakjan, "The resolvent matrix of a Hermitian operator and the characteristic functions related to it", Funktsional. Anal. i Prilozhen, 4 (1970), 103–104 (Russian) (English translation: Functional Anal. Appl., 4 (1970), 258–259).

[KSm] M.G. Kreĭn, Ju. L. Šmuljan, "On fractional linear mappings with operator coefficients", Mat. Issled. 2 (3) (1967), 64–96 (Russian) (English translation: Amer. Math. Soc. Transl., (2) 103 (1974), 125–152).

[KSz] A. Koranyi, B. Sz.-Nagy, "Operatortheoretische Behandlung und Verallgemeinerung eines Problemkreises in der komplexen Funktionentheorie", Acta Math., 100 (1958), 171–202.

[LS] H. Langer, P. Sorjonen, "Verallgemeinerte Resolventen hermitescher und isometrischer Operatoren im Pontryaginraum", Ann. Acad. Sci. Fenn. Ser. A I Mat., 561 (1974), 1–45.

[LT1] H. Langer, B. Textorius, "On generalized resolvents and Q-functions of symmetric linear relations (subspaces) in Hilbert space", Pacific J. Math., 72 (1977), 135–165.

[LT2] H. Langer, B. Textorius, "L-resolvent matrices of symmetric linear relations with equal defect numbers; applications to canonical differential relations", Integral Equations Operator Theory, 5 (1982), 208–243.

[N1] A.A. Nudelman, "On a new problem of moment problem type", Dokl. Akad. Nauk SSSR, 233 (1977), 792–795 (Russian) (English translation: Sov. Math. Dokl., 18 (1977), 507–510).

[N2] A.A. Nudelman, "On a generalization of classical interpolation problems", Dokl. Akad. Nauk SSSR 256 (1981), 790–793 (Russian) (English translation: Sov. Math. Dokl., 23 (1981), 125–128)).

[RR] M. Rosenblum, J. Rovnyak, *Hardy classes and operator theory*, Oxford University Press, 1985.

D. Alpay
Department of Mathematics
The weizmann Institute of Science
Rehovot 76100
Israel

P. Bruinsma, A. Dijksma, H.S.V. de Snoo
Department of Mathematics
University of Groningen
Postbox 800
9700 AV Groningen
Nederland

Operator Theory:
Advances and Applications, Vol. 50

SOME EXTREMAL PROBLEMS RELATED TO MAJORIZATION

T. Ando and Y. Nakamura

Dedicated to Professor S. Koshi on his sixtieth birthday

Let $\mathcal{N}$ be the dual cone of the class of monotone decreasing functions on $[0,1]$, and $(X, \|\cdot\|_X)$ any rearrangement invariant Banach function space. We show that the distance from a function $f \in X$ to the cone $\mathcal{N} \cap X$ is attained at one and the same element, irrespectively of the choice of a norm $\|\cdot\|_X$. A similar result holds for the distance from f to the convex set of functions which are majorized by a given function $g \in X$ in the sense of Hardy, Littlewood and Pólya.

1. INTRODUCTION. Recently several authors have been interested in determining the distance from a selfadjoint operator (or matrix) to the convex hull of the unitary similarity orbit of another selfadjoint operator (or matrix).

The trace norm case is closely related to the mixing distance of state space (see Alberti and Uhlman [1]). Hiai and Nakamura [4] established distance formulas for trace norm and operator norm while Li and Tsing [5] proved that in the space of matrices the distance is attained by one and the same element of the convex hull, irrespectively of the choice of a unitarily invariant norm.

As far as unitarily invariant norms are concerned, the basic part of the problem is reduced to the commutative case, that is, the space of functions (or finite sequences).

In the present paper we treat first the problem of finding, for a given function $f(x)$ on $[0,1]$, the function $h(x)$ in the dual cone of the class of monotone decreasing functions that is in the minimum distance to $f(x)$, with respect to a rearrangement invariant norm. A main result is that $h(x)$ is independent of the choice of a rearrangement invariant norm. As a consequence, the distance from $f(x)$ to the convex set of functions, majorized by another given function $g(x)$, is attained at one and the same function, irrespectively of the choice of a rearrangement invariant norm. This result yields a continuous version of the result of Li and Tsing.

In the finial section we apply our method to derive the distance formulas of Hiai and Nakamura [4].

2. PRELIMINARIES. We consider real-valued measurable functions on the interval $[0,1]$ with Lebesgue measure $m(\cdot)$. $\mathcal{M}$ denotes the class of right-continuous decreasing functions while $\mathcal{N}$ denotes the class of integrable functions $h(x)$ such that

$$\int_0^x h(t)\,dt \leq 0 \qquad \text{for } x \in [0,1] \tag{1}$$

and

$$\int_0^1 h(t)\,dt = 0. \tag{2}$$

Then it is immediately seen that $h \in L^1$ is in $\mathcal{N}$ if and only if

$$\langle g|h\rangle \equiv \int_0^1 g(t)h(t)\,dt \leq 0 \qquad \text{for } g \in L^\infty \cap \mathcal{M}, \tag{3}$$

and that a right-continuous $g \in L^1$ is in $\mathcal{M}$ if and only if

$$\langle g|h\rangle \leq 0 \qquad \text{for } h \in L^\infty \cap \mathcal{N}. \tag{4}$$

For $f \in L^1$ its decreasing rearrangement is denoted by f^*, that is, $f^* \in \mathcal{M}$ and f^* is *equi-measurable* with f in the sense that

$$m(\{x : f^*(x) > \lambda\}) = m(\{x : f(x) > \lambda\}) \qquad \text{for } \lambda \in \mathbf{R}.$$

It is known (see [3], Chap. 2, §2) that f^* satisfies the following property

$$\int_0^x f^*(t)\,dt = \max_{m(\Delta)=x} \int_\Delta f(t)\,dt \qquad \text{for } x \in [0,1], \tag{5}$$

where Δ is a Borel subset of $[0,1]$.

For $f, g \in L^1$ we say that f *majorizes* g, in symbol $g \prec f$, in the sense of Hardy, Littlewood and Pólya if $g^* - f^* \in \mathcal{N}$, that is,

$$\int_0^x g^*(t)\,dt \leq \int_0^x f^*(t)\,dt \qquad \text{for } x \in [0,1] \tag{6}$$

and

$$\int_0^1 g(t)\,dt = \int_0^1 f(t)\,dt. \tag{7}$$

Majorization "$\prec$" induces a quasi-order in L^1. $f \prec g$ and $g \prec f$ simultaneously mean that f is equi-measurable with g.

Recall (see [3], Chap. 1, §1) that a $[0,\infty]$-valued convex, positively homogeneous functional ρ on L^1 is called a *Banach function norm* if

(i) $\rho(f) = 0$ if and only if $f(x) = 0$ a.e.,

(ii) $\rho(g) \le \rho(f)$ if $|g(x)| \le |f(x)|$ a.e.,

(iii) $\rho(f_n) \uparrow \rho(f)$ if $0 \le f_n(x) \uparrow f(x)$ a.e.,

(iv) $\rho(f) < \infty$ for $f \in L^\infty$.

It is known that the collection $X = X(\rho)$ of all f with $\rho(f) < \infty$ becomes a Banach space with norm $||f||_X = \rho(f)$.

Axiom (iii) is called the *Fatou property*. Sometimes a weaker condition is used instead for the definition of a Banach function space. (See [3], Chap. 1, Excercise 12.)

A Banach function norm ρ is said to be *rearrangement invariant* if $\rho(f) = \rho(g)$ whenever f and g are equi-measurable. In this case, the space $X = X(\rho)$ and its norm $||f||_X$ are said to be rearrangement invariant. L^p-norm ($1 \le p \le \infty$) and Orlicz space norms are rearrangement invariant.

It is known (see [3], Chap. 2, Theorem 4.6) that for any rearrangement invariant Banach space $(X, ||\cdot||_X)$

$$f \in X,\ g \prec f \quad \text{implies} \quad g \in X \text{ and } ||g||_X \le ||f||_X. \tag{8}$$

We remark that decreasing rearrangement and majorization are usually considered only for non-negative functions, even in [3]. Most of the results related to these notions, however, can be extended to the case of general real functions.

3. DISTANCE TO THE CONE $\mathcal{N}$. For $f \in L^1$, let

$$F(x) = \int_0^x f(t)\,dt \qquad \text{for } x \in [0,1] \tag{9}$$

and

$$\widetilde{F}(x) = \int_0^x f^*(t)\,dt \qquad \text{for } x \in [0,1]. \tag{10}$$

Then by (5) $F(x)$ and $\widetilde{F}(x)$ are continuous functions such that

$$F(x) \le \widetilde{F}(x) \qquad \text{for } x \in [0,1] \tag{11}$$

and

$$F(0) = \widetilde{F}(0) = 0, \quad F(1) = \widetilde{F}(1) = \int_0^1 f(t)\,dt. \tag{12}$$

Since f^* is decreasing, $\widetilde{F}(x)$ is a concave function. Since the class of concave functions is inf-stable, there is the (pointwise) minimum of all concave functions $G(x)$ such that $F(x) \le G(x)$ for $x \in [0,1]$. The minimum will be denoted by $\widehat{F}(x)$. Then we have by (11) and (12)

$$F(x) \le \widehat{F}(x) \le \widetilde{F}(x) \qquad \text{for } x \in [0,1] \tag{13}$$

and

$$\widehat{F}(0) = 0, \quad \widehat{F}(1) = \int_0^1 f(t)\,dt. \tag{14}$$

Since $\widehat{F}(x)$ is concave, its (right) derivative, denoted by $\boldsymbol{M}f$, is in $\mathcal{M}$;

$$\widehat{F}(x) = \int_0^x \boldsymbol{M}f(t)\,dt \qquad \text{for } x \in [0,1]. \tag{15}$$

Then it follows from (13), (14) and (15) that

$$\boldsymbol{M}f \prec f \quad \text{and} \quad f - \boldsymbol{M}f \in \mathcal{N}. \tag{16}$$

THEOREM 1. *Let $f \in L^1$. Then*

$$\boldsymbol{M}f = f - (f - \boldsymbol{M}f) \qquad \textit{with} \quad f - \boldsymbol{M}f \in \mathcal{N}$$

and

$$\boldsymbol{M}f \prec f - h \qquad \textit{for } h \in \mathcal{N}. \tag{17}$$

Therefore in any rearrangement invariant Banach space $(X, ||\cdot||_X)$

$$||\boldsymbol{M}f||_X = \min\{||f-h||_X : h \in X \cap \mathcal{N}\} \qquad \textit{for } f \in X. \tag{18}$$

PROOF. Take $h \in \mathcal{N}$, and let

$$H(x) = \int_0^x h(t)\,dt \qquad \text{for } x \in [0,1].$$

Then by (1) and (2) $H(x) \leq 0$ and $H(0) = H(1) = 0$, hence by definition

$$\widehat{F}(x) \leq (\widehat{F-H})(x) \quad \text{and} \quad \widehat{F}(1) = (\widehat{F-H})(1).$$

This means by (15) and (16) that

$$\boldsymbol{M}f \prec \boldsymbol{M}(f-h) \prec f - h.$$

Finally (18) follows from (16) and (17) via (8). ∎

4. ANOTHER CHARACTERIZATION OF $\boldsymbol{M}f$. Let f, F and $\widehat{F}$ be as in the preceding section. Let us start with a lemma, which is an immediate consequence of the definition of concavity.

LEMMA 2. *If $G(x)$ is a concave function on $[0,1]$ and $K(x)$ is a concave function on an interval $[\alpha, \beta] \subset [0,1]$ such that*

$$K(\alpha) \geq G(\alpha) \quad \textit{and} \quad K(\beta) \geq G(\beta),$$

then the function

$$L(x) = \begin{cases} \min\{G(x), K(x)\} & \textit{for } x \in (\alpha, \beta), \\ G(x) & \textit{for } x \notin (\alpha, \beta), \end{cases}$$

is concave.

LEMMA 3. *Represent the open set* $\{x : \widehat{F}(x) > F(x)\}$ *as the union of a finite or infinite number of disjoint open intervals* (α_i, β_i), $i = 1, 2, \cdots$. *Then*

$$\widehat{F}(x) = \begin{cases} \dfrac{x-\alpha_i}{\beta_i-\alpha_i}F(\beta_i) + \dfrac{\beta_i - x}{\beta_i-\alpha_i}F(\alpha_i) & \text{for } x \in (\alpha_i, \beta_i), \\ F(x) & \text{for } x \notin \bigcup_i(\alpha_i, \beta_i). \end{cases}$$

PROOF. Fix i. Since $F(0) = \widehat{F}(0)$ and $F(1) = \widehat{F}(1)$, we have by definition of (α_i, β_i)

$$F(\alpha_i) = \widehat{F}(\alpha_i) \quad \text{and} \quad F(\beta_i) = \widehat{F}(\beta_i). \tag{19}$$

Let

$$H(x) = \frac{x-\alpha_i}{\beta_i-\alpha_i}F(\beta_i) + \frac{\beta_i - x}{\beta_i-\alpha_i}F(\alpha_i). \tag{20}$$

Then since $\widehat{F}(x)$ is concave, $\widehat{F}(x) \geq H(x)$ on (α_i, β_i). Let

$$\gamma = \max_{\alpha_i \leq x \leq \beta_i} \{F(x) - H(x)\}. \tag{21}$$

Then we have $\gamma \geq 0$. Since by (19) and (20)

$$H(\alpha_i) + \gamma \geq \widehat{F}(\alpha_i) \quad \text{and} \quad H(\beta_i) + \gamma \geq \widehat{F}(\beta_i),$$

it follows from Lemma 2 that the function

$$L(x) = \begin{cases} \min\{\widehat{F}(x), H(x) + \gamma\} & \text{for } x \in (\alpha_i, \beta_i), \\ \widehat{F}(x) & \text{for } x \notin (\alpha_i, \beta_i), \end{cases} \tag{22}$$

is concave. Since by (21) and (22) $H(x) + \gamma \geq F(x)$, the minimality of $\widehat{F}(x)$ implies $L(x) \geq \widehat{F}(x)$. Take x_0 in $[\alpha_i, \beta_i]$, at which the maximum in (21) is attained. Then it follows that

$$H(x_0) + \gamma \geq L(x_0) \geq \widehat{F}(x_0) \geq F(x_0) = H(x_0) + \gamma,$$

hence $\widehat{F}(x_0) = F(x_0)$, which is possible only when $x_0 = \alpha_i$ or $= \beta_i$. Then we have $\gamma = 0$, and $F(x) \leq H(x)$ on (α_i, β_i). Therefore $\widehat{F}(x) = H(x)$ on (α_i, β_i). ∎

THEOREM 4. *When* $f \in L^2$, *the decreasing function* $\boldsymbol{M}f$ *is characterized as a solution of the extremal problem*

$$\|f - \boldsymbol{M}f\|_2 = \min\{\|f - g\|_2 : g \in L^2 \cap \mathcal{M}\}.$$

PROOF. It suffices to prove that

$$\|f - g\|_2 \geq \|f - \boldsymbol{M}f\|_2 \qquad \text{for } g \in L^2 \cap \mathcal{M}. \tag{23}$$

Since $f - \boldsymbol{M}f \in \mathcal{N}$ by (16), as in (3) we have

$$\langle g | f - \boldsymbol{M}f \rangle \leq 0.$$

Therefore

$$\begin{aligned} \|f - g\|_2^2 &= \|f - \boldsymbol{M}f\|_2^2 + 2\langle \boldsymbol{M}f - g | f - \boldsymbol{M}f \rangle + \|\boldsymbol{M}f - g\|_2^2 \\ &\geq \|f - \boldsymbol{M}f\|_2^2 + 2\langle \boldsymbol{M}f | f - \boldsymbol{M}f \rangle. \end{aligned} \tag{24}$$

Since $\boldsymbol{M}f$ is a decreasing function, and by definition (9) and (15)

$$\int_0^x (f - \boldsymbol{M}f)(t)\, dt = F(x) - \widehat{F}(x),$$

and

$$F(0) - \widehat{F}(0) = F(1) - \widehat{F}(1) = 0,$$

integration by parts leads to

$$\langle \boldsymbol{M}f | f - \boldsymbol{M}f \rangle = -\int_0^1 \{F(x) - \widehat{F}(x)\}\, d\boldsymbol{M}f(x). \tag{25}$$

In view of Lemma 3 we have

$$F(x) - \widehat{F}(x) = 0 \qquad \text{for } x \notin \bigcup_i (\alpha_i, \beta_i)$$

and

$$\boldsymbol{M}f(x) = \frac{F(\beta_i) - F(\alpha_i)}{\beta_i - \alpha_i} \qquad \text{for } x \in (\alpha_i, \beta_i),$$

hence

$$\int_0^1 \{F(x) - \widehat{F}(x)\}\, d\boldsymbol{M}f(x) = 0.$$

Now inequality (23) follows from (24) and (25). ∎

The extremal problem is observed in [2] in connection with isotonic regressions.

5. DISTANCE TO THE ORBIT UNDER DOUBLY STOCHASTIC OPERATORS. Recall (see [3], Chap. 2, Theorem 7.5) that for each $h \in L^1$ there is a measurable map φ from $[0,1]$ to itself such that

$$h^* \circ \varphi = h \qquad \text{and} \qquad m(\varphi^{-1}(\Delta)) = m(\Delta) \quad \text{for all Borel } \Delta. \tag{26}$$

It is known (see [3], Chap. 3, §2) that for $g \in L^1$

$$\{k \in L^1 : k \prec g\} = \{Tg : T \text{ is doubly stochastic}\}.$$

Here a linear operator T on L^1 is called *doubly stochastic* if it is positivity preserving, $T1 = 1$ and

$$\int_0^1 (Tf)(t)\, dt = \int_0^1 f(t)\, dt \qquad \text{for } f \in L^1.$$

Thus we may call the set $\{k \in L^1 : k \prec g\}$ the orbit of g under doubly stochastic operators.

LEMMA 5. *Let $f \in L^1$ and $F(x) = \int_0^x f(t)\,dt$. If $G(x)$ is a concave function on $[0,1]$ such that $F(x) + G(x)$ is concave, then the function $F(x) + G(x) - \widehat{F}(x)$ is concave.*

PROOF. With the notations of Lemma 3

$$F(x) + G(x) - \widehat{F}(x)$$
$$= \begin{cases} F(x) + G(x) - \dfrac{x - \alpha_i}{\beta_i - \alpha_i} F(\beta_i) - \dfrac{\beta_i - x}{\beta_i - \alpha_i} F(\alpha_i) & \text{for } x \in (\alpha_i, \beta_i), \\ G(x) & \text{for } x \notin \bigcup_i (\alpha_i, \beta_i). \end{cases}$$

Therefore by assumption and $F(x) \leq \widehat{F}(x)$, on (α_i, β_i) the function $F(x) + G(x) - \widehat{F}(x)$ is concave and

$$F(x) + G(x) - \widehat{F}(x) \leq G(x)$$

with equality at both end points. Then by Lemma 2 $F(x) + G(x) - \widehat{F}(x)$ is concave on $[0,1]$. ∎

THEOREM 6. *Let $h, g \in L^1$, and let φ be a measurable map satisfying (26). Then*

$$h - \boldsymbol{M}(h^* - g^*) \circ \varphi \prec g$$

and

$$\boldsymbol{M}(h^* - g^*) \circ \varphi \prec h - k \qquad \text{for } k \prec g.$$

Therefore in any rearrangement invariant Banach space $(X, \|\cdot\|_X)$ for any $h, g \in X$ the distance from h to the orbit of g under doubly stochastic operators is attained at $h - \boldsymbol{M}(h^* - g^*) \circ \varphi$, and

$$\|\boldsymbol{M}(h^* - g^*)\|_X = \min\{\|h - k\|_X : k \prec g\}. \tag{27}$$

PROOF. Let $f = h^* - g^*$ and $F(x) = \int_0^x f(t)\,dt$ and $G(x) = \int_0^x g^*(t)\,dt$. According to Lemma 5 we have $F(x) + G(x) - \widehat{F}(x)$ is concave, which means that $h^* - \boldsymbol{M}(h^* - g^*)$ belongs to $\mathcal{M}$. On the other hand, $\{h^* - \boldsymbol{M}(h^* - g^*)\} - g^* = (h^* - g^*) - \boldsymbol{M}(h^* - g^*)$ belongs to $\mathcal{N}$, hence by definition

$$h^* - \boldsymbol{M}(h^* - g^*) \prec g^*.$$

Then it follows from (26) that

$$h - \boldsymbol{M}(h^* - g^*) \circ \varphi \prec g^* \circ \varphi \prec g.$$

For any $k \prec g$, we have by definition $k^* - g^* \in \mathcal{N}$, hence by Theorem 1

$$\boldsymbol{M}(h^* - g^*) \prec (h^* - g^*) - (k^* - g^*) = h^* - k^*.$$

Finally it is known (see [3], Chap. 3, Theorem 7.4) that

$$h^* - k^* \prec h - k,$$

hence we have

$$\boldsymbol{M}(h^* - g^*) \circ \varphi \prec h - k.$$

Now (27) follows from that $\boldsymbol{M}(h^* - g^*) \circ \varphi$ and $\boldsymbol{M}(h^* - g^*)$ are equi-measurable. ∎

6. NORM OF $\boldsymbol{M}f$ IN SPECIAL CASES. On the basis of Lemma 3 we can effectively determine the L^∞-norm and L^1-norm of $\boldsymbol{M}f$.

THEOREM 7. *(1) When* $f \in L^\infty$,

$$||\boldsymbol{M}f||_\infty = \max\left\{\sup_{0<x<1} \frac{\int_0^x f(t)\,dt}{x},\ -\inf_{0<x<1} \frac{\int_x^1 f(t)\,dt}{1-x}\right\}. \tag{28}$$

(2) When $f \in L^1$,

$$||\boldsymbol{M}f||_1 = \sup_{0<x<1}\left\{\int_0^x f(t)\,dt - \int_x^1 f(t)\,dt\right\}. \tag{29}$$

PROOF. Let us use the notations in Lemma 3, and write

$$\{x : \widehat{F}(x) > F(x)\} = \bigcup_i (\alpha_i, \beta_i).$$

(1) Since $\boldsymbol{M}f$ is a right-continuous decreasing function,

$$||\boldsymbol{M}f||_\infty = \max\{\boldsymbol{M}f(0+), -\boldsymbol{M}f(1-)\}. \tag{30}$$

Further $\frac{\widehat{F}(x)}{x}$ and $\frac{\widehat{F}(1)-\widehat{F}(x)}{1-x}$ are decreasing functions on $(0,1)$, and

$$\boldsymbol{M}f(0+) = \sup_{0<x<1} \frac{\widehat{F}(x)}{x}, \qquad \boldsymbol{M}f(1-) = \inf_{0<x<1} \frac{\widehat{F}(1)-\widehat{F}(x)}{1-x}. \tag{31}$$

Since by (13) and (14)

$$\int_0^x f(t)\,dt = F(x) \leq \widehat{F}(x)$$

and

$$\int_x^1 f(t)\,dt = F(1) - F(x) \geq \widehat{F}(1) - \widehat{F}(x),$$

it follows from (30) and (31) that the right hand side of (28) is not greater than the left hand side.

Next we claim that there is a decreasing sequence $\{x_j\}$, converging to 0, such that $F(x_j) = \widehat{F}(x_j)$ for $j = 1, 2, \cdots$. In fact, if 0 is a limiting point of $\{\alpha_i\}$, we can take as $\{x_j\}$ a suitable subsequence of $\{\alpha_i\}$, because $F(\alpha_i) = \widehat{F}(\alpha_i)$ by definition. If 0 is not a limiting point of $\{\alpha_i\}$, $\widehat{F}(x)$ concides with $F(x)$ near 0. This will show that

$$\sup_{0<x<1} \frac{\widehat{F}(x)}{x} \leq \sup_{0<x<1} \frac{F(x)}{x}.$$

In a similar way we have

$$\inf_{0<x<1} \frac{\widehat{F}(1)-\widehat{F}(x)}{1-x} \geq \inf_{0<x<1} \frac{F(1)-F(x)}{x}.$$

Hence the right hand side of (28) is not smaller than the left hand side.

(2) By (13) and (14) we have, for any x,

$$\begin{aligned}\int_0^x f(t)\,dt - \int_x^1 f(t)\,dt &= 2F(x) - F(1)\\ &\le 2\widehat{F}(x) - \widehat{F}(1)\\ &= \int_0^x \boldsymbol{M}f(t)\,dt - \int_x^1 \boldsymbol{M}f(t)\,dt\\ &\le \int_0^1 |\boldsymbol{M}f(t)|\,dt = ||\boldsymbol{M}f||_1.\end{aligned}$$

Thus the right hand side of (29) is not greater than the left hand side.

Finally let

$$x = \begin{cases} 0 & \text{if } \boldsymbol{M}f \le 0,\\ \sup\{t : \boldsymbol{M}f(t) > 0\} & \text{if } \boldsymbol{M}f \not\le 0.\end{cases}$$

Since by Lemma 3 $\boldsymbol{M}f(t)$ is constant on each (α_i, β_i), this x can not belong to $\bigcup_i(\alpha_i, \beta_i)$, hence $F(x) = \widehat{F}(x)$ by definition. Then since

$$\boldsymbol{M}f(t)\begin{cases} \ge 0 & \text{for } t < x,\\ \le 0 & \text{for } t \ge x,\end{cases}$$

we have

$$\begin{aligned}||\boldsymbol{M}f||_1 &= \int_0^x \boldsymbol{M}f(t)\,dt - \int_x^1 \boldsymbol{M}f(t)\,dt\\ &= 2\widehat{F}(x) - \widehat{F}(1)\\ &= 2F(x) - F(1)\\ &= \int_0^x f(t)\,dt - \int_x^1 f(x)\,dt.\end{aligned}$$

Thus the right hand side of (29) is not smaller than the left hand side. ∎

REFERENCES

[1] Alberti, P. W. and Uhlmann, A., Stochasticity and partial order: Doubly stochastic maps and unitary mixing, D. Reidel, Dordrecht, 1982.

[2] Barlow, R. E., Bartholomew, D. J., Bremner, J. M., and Brunk, H. D., Statistical inference under order restrictions: The theory and applications of isotonic regression, John Wiley & Sons, London - New York, 1972.

[3] Bennett, C. and Sharpley, R., Interpolation of operators, Academic Press, New York, 1988.

[4] Hiai, F. and Nakamura, Y., Closed convex hulls of unitary orbits in von Neumann algebras, Trans. Amer. Math. Soc. 319 (1990), to appear.

[5] Li, C.-K. and Tsing, N.-K., Distance to the convex hull of unitary orbits, Linear Multilinear Alg. 25 (1989), 93–103.

Division of Applied Mathematics
Research Institute of Applied Electricity
Hokkaido University
Sapporo, 060 Japan

Operator Theory:
Advances and Applications, Vol. 50

DE BRANGES-ROVNYAK OPERATOR MODELS AND SYSTEMS THEORY: A SURVEY

Joseph A. Ball[1] and Nir Cohen

We arrive at the de Branges-Rovnyak space $\mathcal{D}(W)$ from the point of view of model theory, i.e., as the space associated with a canonical model for a general completely nonunitary contraction operator on a Hilbert space. Connections with systems theory make contact with the approach of Livsic and Brodskii, in particular, the role of the characteristic function as the transfer function of a unitary system. The connection between invariant subspaces and regular factorizations leads to the introduction of the overlapping spaces of de Branges-Rovnyak. This analysis applied to the particular factorization $0 = 0 \cdot W \cdot 0$ leads to the derivation of the various geometric decompositions of the unitary dilation space for a contraction, the cornerstone of the model theory of Sz.-Nagy-Foias.

Introduction.

Operator model theory originates with the work of Livsic and his associates in the Soviet Union in the 1950's (see Brodskii [1971]), and with the work of Sz.-Nagy and Foias in Eastern Europe (see Sz.-Nagy-Foias [1970]) and of de Branges and Rovnyak in the United States (see de Branges-Rovnyak [1966a, 1966b]) in the 1960's. From the pure operator theory point of view, this can be viewed as an attempt to develop a spectral theory for operators on a Hilbert space close to being unitary (or self-adjoint), akin to the highly successful theory that had been achieved for normal operators. In all three model theories, there was attached to the nonunitary operator to be studied a matrix- or operator-valued analytic function, called the characteristic function. The game then was to reduce the study of operators to problems concerning analytic operator-valued functions. If the values of the operator function are just as complicated as the original operator, nothing is gained; however, the scheme does have merit when the problems associated with general infinite dimensional spaces do not appear in individual values of the operator valued function (e.g., if these values act on a finite dimensional space or are compact or trace-class perturbations of some fixed value). In all three schemes, it was observed that function theoretic properties of the characteristic function (e.g. factorization, location of zeros) correspond to operator theoretic

[1]The first author was partially supported by National Science Foundation Grant DMS-8701625.

properties of the original operator (e.g. invariant subspaces, location of spectrum), and the goal usually was to use function theory to do operator theory. In the Livisc-Brodskii approach, the square of the Hilbert space norm is thought of an energy function and the nonunitary property of the operator under study means that it generates a system which does not conserve energy; the operator is then coupled to other input and output operators in a configuration called an operator node or colligation so that the enlarged closed system is energy conserving. The characteristic function of the node is an operator function whose values are operators from the input space to the output space which somehow measures the extent to which energy fails to be conserved in the original system. This more physical, applied point of view included an awareness of, and search for, potential applications in engineering and physics (see Livsic [1973] and Livshits-Yantsevich [1979]). The approach of Sz.-Nagy and Foias arrives at the characteristic function by analyzing the geometric structure of the Sz.-Nagy strong isometric dilation for the contraction operator T and using function models for isometric and unitary operators. In the de Branges-Rovnyak approach, one starts with the characteristic operator function $W(z) : \mathcal{F} \to \mathcal{E}$ and arrives at the model space as a generalized orthogonal complement (in the sense of de Branges' theory of minimal decompositions) of $W H^2_{\mathcal{F}}$ in $H^2_{\mathcal{E}}$ (where $H^2_{\mathcal{G}}$ is the Hilbert space of norm-square-summable power series with coefficient in the Hilbert space $\mathcal{G}$).

In the meantime, beginning in the early 1960's, Kalman and his associates (see Kalman-Falb-Arbib [1969]) were developing the so-called state space approach to systems theory. This turns out in many aspects to be simply an affine version of the Livsic-Brodskii theory, i.e. one has an open system (now called a state space or black box) which interacts with the outside world (now thought of as the observer or controller) through inputs and outputs, but in the Kalman theory one does not keep track of energy. Thus the Kalman theory is invariant under similarities rather than under unitary transformations. As further elaborated on by Rosenbrock [1970], Wonham [1979] and most recently Bart-Gohberg-Kaashoek [1979], the direction of the flow between function theory and operator theory is now reversed, i.e. the goal is to reduce the understanding of properties of the transfer function of the system to linear algebra properties of the linear transformations associated with the system. There are also connections between operator model theory and the physical theories of scattering and circuits, as has been pointed out by Helton [1972, 1974].

After several years of neglect, the de Branges-Rovnyak model theory has recently received renewed attention (see Sarason [1986a, 1986b, 1989], Nikolskii-Vasyunin [1986, 1989] and Ball-Kriete [1987]), spurred on partly by de Branges' recent solution of the Bieberbach conjecture (de Branges [1985]). The aim of this paper is to give a systematic, expository development of the original model spaces and model operators associated with de Branges and Rovnyak from the point of view of operator model theory and systems theory.

Besides discussing basic properties of the model, we focus on the connection between factorization and invariant subspaces; this is done in the three settings of abstract systems, abstract unitary systems, and de Branges-Rovnyak models for unitary systems. Particular attention is given to the construction due to de Branges and Rovnyak of overlapping spaces associated with nonregular factorizations. As a new application, we show how a special case of the construction encodes the geometry of the unitary dilation of a given contraction operator as worked out in Chapter II of Sz.-Nagy-Foias [1970].

Many of the results discussed here are well-known; we hope that the organization and the systematic exposition of the connections between model theory and systems theory ideas is new. In order not to interfere with the flow of the ideas in the main text, we postpone discussions of history of and detailed attributions for various results to a separate section entitled Notes at the end of the paper.

Sections 1 and 2 deal with the construction of a de Branges-Rovnyak model for a given completely nonunitary contraction operator. Section 3 is a survey of the ideas from systems theory which we shall need. Sections 4 and 5 redevelop these ideas in the context of abstract unitary systems and concrete de Branges-Rovnyak models respectively. Section 6 gives the connection between factorization and invariant subspaces. Finally Section 7 introduces the overlapping space model associated with nonregular factorizations.

1. Models for completely nonunitary contractions.

In this section we collect a succession of models for various classes of nonunitary contraction operators of increasing sophistication which ultimately lead to the most general de Branges-Rovnyak model for a completely nonunitary contraction operator on a Hilbert space.

All these models involve the standard vector-valued Hardy space over the unit disk $\mathbf{D}$ (see e.g. Rosenblum-Rovnyak [1986]). If $\mathcal{E}$ is a Hilbert space, we denote by $H^2_{\mathcal{E}} = H^2 \otimes \mathcal{E}$ the tensor product of H^2 with $\mathcal{E}$; this can be identified with functions $f(z)$ having a power series representation on the unit disk $\mathbf{D}$

$$f(z) = \sum_{j=0}^{\infty} f_j z^j$$

with coefficients $f_j \in \mathcal{E}$ which are square summable in norm ($\sum \|f_j\|^2 < \infty$). If $\mathcal{E} = \mathbb{C}$ is 1-dimensional, we write more simply H^2. The shift operator $S_{\mathcal{E}} : f(z) \to zf(z)$ on $H^2_{\mathcal{E}}$ as adjoint $S^*_{\mathcal{E}}$ given by

$$S^*_{\mathcal{E}} : f(z) \to [f(z) - f(0)]/z.$$

The shift operator became of particular interest to operator theorists since the classification of all its invariant subspaces in terms of matrix and operator inner functions via the Beurling-Lax-Halmos theorem (see Helson [1964]). From the point of model theory, the invariant subspaces themselves do not lead to anything new, since any restriction $S_{\mathcal{E}}|\mathcal{M}$ of $S_{\mathcal{E}}$ to an invariant subspace

$\mathcal{M}$ is unitarily equivalent to the shift $S_{\mathcal{F}}$ for some Hilbert space $\mathcal{F}$ with $dim\mathcal{F} \leq dim\mathcal{E}$. However, the restrictions of the adjoint $S_{\mathcal{E}}^*$ to various $S_{\mathcal{E}}^*$-invariant subspaces $\mathcal{M}^{\perp}$ leads to a rich class of operators as shown by the following first model theoretic result of Rota.

THEOREM 1.1. *(see Rota [1960]). Let T be a bounded linear operator on a separable Hilbert space $\mathcal{H}$ such that the spectrum $\sigma(T)$ of T is contained in the open unit disk $\mathcal{D}$. Then there exists an invariant subspace $\mathcal{X}_T$ for $S_{\mathcal{H}}^*$ on $H_{\mathcal{H}}^2$ such that T and $S_{\mathcal{H}}^*$ are similar*

$$T \underset{s}{\cong} S_{\mathcal{H}}^*|\mathcal{X}_T.$$

Specifically, if $\tilde{X}_T$ is defined by

$$\tilde{X}_T : x \to (I - zT)^{-1}x$$

for $x \in H$, then $\tilde{X}_T$ maps $\mathcal{H}$ into $H_{\mathcal{H}}^2$, $\mathcal{X}_T := Im\tilde{X}_T$ is a closed subspace of $H_{\mathcal{H}}^2$ invariant for $S_{\mathcal{H}}^*$, $\tilde{X}_T : \mathcal{H} \to \mathcal{X}_T$ is bounded with bounded inverse and

$$(S_{\mathcal{H}}^*|\mathcal{H}_T)\tilde{X}_T = \tilde{X}_T T.$$

PROOF: The proof is a simple matter of checking that the mapping $\tilde{X}_T$ has all the desired properties. Since $\sigma(T) \subset \mathbf{D}$, $(I - zT)^{-1}x$ is analytic on a neighborhood of $\mathbf{D}$ for each $x \in \mathcal{H}$, and hence is in $H_{\mathcal{H}}^2$. In fact, from the Neumann series expansion

$$(I - zT)^{-1} = \sum_{j=0}^{\infty} T^j z^j$$

we have

$$\|x\|^2 \leq \|(I - zT)^{-1}x\|_{H_{\mathcal{H}}^2}^2 = \sum_{j=0}^{\infty} \|T^j x\|^2 \leq \left(\sum_{j=0}^{\infty} \|T^j\|^2\right) \|x\|^2$$

where the series on the right converges by the spectral radius formula since $\sigma(T) \subset \mathbf{D}$. Hence the map $\tilde{X}_T$ is bounded above and below and is a homeomorphic bijection to its image $\mathcal{X}_T$. Finally

$$\begin{aligned} S_{\mathcal{H}}^*\tilde{X}_T x &= S_{\mathcal{H}}^* \left(\sum_{j=0}^{\infty} T^j z^j\right) x \\ &= \sum_{j=0}^{\infty} T^{j+1} x z^j \\ &= \tilde{X}_T(Tx). \quad \square \end{aligned}$$

The Rota model has several disadvantages. First, it is only a similarity model and secondly, the model space consists of functions with values in an infinite dimensional space if the original operator was on an infinite dimensional space, and hence is no simpler to study than the original abstract operator. The simplest model in the same spirit which yields unitary equivalence is the following refinement of the Rota model due to de Branges-Rovnyak.

THEOREM 1.2. *Let T be a contraction operator on a Hilbert space $\mathcal{H}$ for which*

$$\lim_{n\to\infty} T^n x = 0 \text{ for each } x \in \mathcal{H}.$$

Define

$$D_T = (I - T^*T)^{1/2}$$

and let $\mathcal{D}_T$ denote the closure of $Im D_T$ in $\mathcal{H}$. Then the map

$$X_T : x \to D_T(I - zT)^{-1}x, \quad x \in \mathcal{H}$$

maps $\mathcal{H}$ isometrically onto a closed subspace $\mathbf{H}(T)$ of $H^2_{\mathcal{D}_T}$ which is invariant for $S^*_{\mathcal{D}_T}$ and moreover

$$(S^*_{\mathcal{D}_T}|\mathbf{H}(T))X_T = X_T T.$$

Thus T is unitarily equivalent to $S^*_{\mathcal{D}_T}|\mathbf{H}_T$.

PROOF: Note that $X_T(x)$ has the power series representation

$$X_T(x) = \sum_{j=0}^{\infty} D_T T^j x z^j$$

and hence

$$\begin{aligned}\|X_T(x)\|^2_{\mathcal{H}_{\mathcal{D}_T}} &= \sum_{j=0}^{\infty} \|D_T T^j x\|^2 = \sum_{j=0}^{\infty} < T^{*j}(I - T^*T)T^j x,\ x > \\ &= \lim_{N\to\infty} \{\|x\|^2 - \|T^{N+1}x\|^2\} = \|x\|^2\end{aligned}$$

and hence X_T is an isometry onto its image $\mathbf{H}(T)$. Moreover, as in the proof of Theorem 1.1,

$$S^*_{\mathcal{D}_T} X_T x = S^*_{\mathcal{D}_T}\left(\sum_{j=0}^{\infty} D_T T^j x\right) = \sum_{j=0}^{\infty} D_T T^{j+1} x = X_T(Tx). \quad \square$$

The requirement that powers of T tend strongly to zero in Theorem 1.2 can be relaxed but at the cost of making the norm in the model space more difficult to compute. This will be done presently in Theorem 1.3. To prepare for this, we say that an operator T on a Hilbert space $\mathcal{H}$ is *completely nonisometric* (c.n.i.) if there is no nonzero invariant subspace for T on which T is isometric; if $\|T\| < 1$, this is equivalent to

$$\|T^n x\| = \|x\| \text{ for } n = 0, 1, 2, \ldots, x \in \mathcal{H} \Rightarrow x = 0.$$

Note that the map X_T is still defined and maps $\mathcal{H}$ into $H^2_{\mathcal{D}_T}$ if only $\|T\| \leq 1$. Indeed, since $\|T\| \leq 1$, certainly $\{\|T^{N+1}x\|\}_{N\geq 0}$ is a nonincreasing sequence bounded below by zero and hence

$\lim_{N\to\infty} \|T^{N+1}x\| \geq 0$ exists for each $x \in \mathcal{H}$. Then the computation in the proof of Theorem 1.2 gives

$$\|X_T(x)\|^2_{H^2_{\mathcal{D}_T}} = \|x\|^2_{\mathcal{H}} - \lim_{N\to\infty} \|T^{N+1}x\|^2_{\mathcal{H}} \leq \|x\|^2_{\mathcal{H}}$$

and hence X_T maps $\mathcal{H}$ contractively into $H^2_{\mathcal{D}_T}$. If $x \in \mathcal{H}$ is in $kerX_T$, than $0 = X_Tx(z) = \sum_{j=0}^{\infty}(D_TT^jx)z^j$ forces $D_TT^jx = 0$ for $j = 0, 1, 2, \ldots$, and hence

$$\begin{aligned} 0 = \|D_TT^jx\|^2 &=< T^j(I - T^*T)T^jx,\ x.> \\ &= \|T^jx\|^2 - \|T^{j+1}x\|^2 \end{aligned}$$

for all $j \geq 0$. Hence by induction $\|T^jx\|^2 = \|x\|^2$ for all j, so $KerX_T = \{0\}$ precisely when T is c.n.i. Moreover the intertwining property $S^*_{\mathcal{D}_T}X_T = X_TT$ still holds as before. If we endow $\mathbf{H}(T) = ImX_T$ with a new norm so as to make X_T isometric by definition

$$\|X_T(x)\|_{\mathbf{H}(T)} = \|x\|_{\mathcal{H}},$$

and we define R_T to be the action of $S^*_{\mathcal{D}_T}$ on the Hilbert space $\mathbf{H}(T)$, then X_T establishes a unitary equivalence between T and R_T. This establishes the following model for c.n.i. contractions.

THEOREM 1.3. *Let T be a c.n.i. contraction operator on the Hilbert space $\mathcal{H}$. Let $X_T : \mathcal{H} \to H^2_{\mathcal{D}_T}$ be defined by $X_T : h \to D_T(I - zT)^{-1}h$; and let $\mathbf{H}_T := ImX_T$ with norm defined by*

$$\|X_T(x)\|_{\mathbf{H}(T)} = \|x\|_{\mathcal{H}}.$$

Let R_T be the restriction of $S^*_{\mathcal{D}_T}$ to $\mathbf{H}(T)$, considered as an operator on the Hilbert space $\mathbf{H}(T)$. Then T and R_T are unitarily equivalent via the unitray transformation $X_T : \mathcal{H} \to \mathbf{H}(T)$:

$$R_TX_T = X_TT.$$

The space $\mathbf{H}(T)$ in Theorem 1.3 is a (in general not closed) submanifold of $H^2_{\mathcal{D}_T}$ which is invariant for the backward shift $S^*_{\mathcal{D}_T}$; moreover, it is a Hilbert space in its own norm with respect to which the inclusion map $i : \mathbf{H}(T) \to H^2_{\mathcal{D}_T}$ is contractive. To make this model space useful we need a more intrinsic way to compute norms which bypasses pulling back to the abstract space $\mathcal{H}$. A step in this direction is the identification of $\mathbf{H}(T)$ as a reproducing kernel Hilbert space. This will be done in the next section.

At this point it will be useful to obtain a formula for the adjoint $(R_T)^*$ of R_T on $\mathbf{H}(T)$. The formula involves a map ψ_T from the Hilbert space $\mathbf{H}(T)$ to $\mathcal{D}_{T^*}(= Im(I - TT^*)^{1/2})$. The adjoint $\psi^*_T : D_{T^*} \to \mathbf{H}(T)$ of this map is given by

$$\psi^*_T : x \to X_TD_{T^*}x = D_T(I - zT)^{-1}D_{T^*}x \tag{1.1}$$

Thus the map ψ_T from $\mathbf{H}(T)$ into the Hilbert space $\mathcal{D}_{T^*}$ is given by

$$\psi_T = D_{T^*}X_T^*$$

PROPOSITION 1.4. *Let T be a c.n.i. contraction and let R_T on $\mathbf{H}(T)$ be as in Theorem 1.3. Then $(R_T)^*$ is given by*

$$(R_T)^* : f(z) \to zf(z) - W_T(z)\psi_T(f)$$

where ψ_T is defined by (1.1) and $W_T(z) : \mathcal{D}_{T^*} \to \mathcal{D}_T$ is the operator function defined by

$$W_T(z) = [-T^* + zD_T(I - zT)^{-1}D_{T^*}] \,|\, \mathcal{D}_{T^*}. \tag{1.2}$$

The function $W_T(z)$ defined by (1.2) plays a key role also in the Sz.-Nagy-Foias model theory; indeed $W_T(z)$ is the characteristic operator function $\Theta_{T^*}(z)$ for T^* in the Sz.-Nagy-Foias theory. It will appear again in the next section when we identify the model spaces here as reproducing kernel spaces.

PROOF: By construction $R_TX_T = X_TT$ and $X_T : \mathcal{H} \to \mathbf{H}(T)$ is unitary. Thus

$$R_T^*X_T = (X_TX_T^*)R_T^*X_t = X_T(T^*X_T^*)X_T = X_TT^*.$$

Hence for $x \in \mathcal{H}$,

$$\begin{aligned}(R_T^*X_Tx)(z) &= (X_TT^*x)(z) = D_T(I - zT)^{-1}T^*x \\ &= zD_T(I - zT)^{-1}x - D_T(I - zT)^{-1}(zI - T^*)x \\ &= z(X_Tx)(z) - D_T(I - zT)^{-1}(zI - T^*)x.\end{aligned} \tag{1.3}$$

On the other hand, a basic identity concerning characteristic operator functions is

$$W_T(z)D_{T^*}x = D_T(I - zT)^{-1}(zI - T^*)x \tag{1.4}$$

(see identity (1.2) in Chapter VI of Sz.-Nagy-Foias [1970]). Combining (1.4) with (1.3) gives

$$\begin{aligned}(R_T)^*X_Tx &= z(X_Tx)(z) - W_T(z)D_{T^*}x \\ &= z(X_Tx)(z) - W_T(z)\psi_T(X_Tx).\end{aligned}$$

As $f(z) = (X_Tx)(z)$ is an arbitrary element of $\mathbf{H}(T)$ the formula in Proposition 1.4 follows. □

The above procedure guarantees a model operator for T, namely R_T, only if T is c.n.i. If, on the other hand, T^* is c.n.i. one can still use $(R_{T^*})^*$ as a model operator for T. We shall now define a larger class of operators, namely completely nonunitary (c.n.u.) operators, which contains both c.n.i. operators and their adjoints. We shall construct a model for this class which is essentially a combination of R_T and R_{T^*}. Specifically, if T is c.n.u., then there is no nonzero subspace invariant for both T and T^* on which T is unitary, i.e., on which both T and T^* are isometric. If $\|T\| \leq 1$, this is equivalent to

$$\|T^n x\| = \|x\| = \|T^{*n}x\| \; for \; n = 0,1,2,\ldots, \; x \in \mathcal{H} \Rightarrow x = 0.$$

This in turn is equivalent to the map $Z_T : \mathcal{H} \to H^2_{\mathcal{D}_T} \oplus H^2_{\mathcal{D}_{T^*}}$ defined by

$$Z_T : x \to \begin{bmatrix} X_T x \\ X_{T^*} x \end{bmatrix} = \begin{bmatrix} D_T(I - zT)^{-1}x \\ D_{T^*}(I - zT^*)^{-1}x \end{bmatrix} \tag{1.5}$$

having trivial kernel. Indeed, for $x \in \mathcal{H}$, $Z_T x = 0$ means that both $D_T(I-zT)^{-1}x = \sum_{j=0}^{\infty}(D_T T^j x)z^j$ and $D_{T^*}(I - zT)^{-1}x = \sum_{j=0}^{\infty}(D_{T^*}T^{*j}x)z^j$ are 0, and hence all coefficients $D_T T^j x$, $D_{T^*}T^{*j}$($j = 0,1,2,\ldots$) are zero. From $\|D_T T^j x\|^2 = \|T^j x\|^2 - \|T^{j+1}x\|^2 = 0$ and $\|D_{T^*}T^{*j}x\|^2 = \|T^{*j}x\|^2 - \|T^{*j+1}x\|^2 = 0$ for all j we get that $\|T^j x\|^2 = \|x\|^2 = \|T^{*j}x\|^2$ for all j and hence x is in a reducing subspace for T on which T is unitary.

We get a model space $\mathbf{D}(T) \subset H^2_{\mathcal{D}_T} \oplus H^2_{\mathcal{D}_{T^*}}$ for a c.n.u. operator T as the image of Z_T. The result is as follows.

THEOREM 1.5. *Let T be a c.n.u. contraction on $\mathcal{H}$. Define a map $Z_T : \mathcal{H} \to H^2_{\mathcal{D}_T} \oplus H^2_{\mathcal{D}_{T^*}}$ by (1.5) and let $\mathbf{D}(T)$ be the image of Z_T with the pullback norm*

$$\|Z_T x\|_{\mathbf{D}(T)} = \|x\|_{\mathcal{H}}$$

and define $W_T(z) : \mathcal{D}_{T^*} \to \mathcal{D}_T$ and $W_{T^*}(z) : \mathcal{D}_T \to \mathcal{D}_{T^*}$ as in (1.2).

Then the map $\hat{R}_T$ defined by

$$\hat{R}_T : \begin{bmatrix} f(z) \\ g(z) \end{bmatrix} \to \begin{bmatrix} [f(z) - f(0)]/2 \\ zg(z) - W_{T^*}(z)f(0) \end{bmatrix} \tag{1.6}$$

is a well-defined linear transformation on $\mathbf{D}(T)$ with adjoint given by

$$(\hat{R}_T)^* : \begin{bmatrix} f(z) \\ g(z) \end{bmatrix} \to \begin{bmatrix} zf(z) - W_T(z)g(0) \\ [g(z) - g(0)]/z \end{bmatrix} \tag{1.7}$$

and T is unitarily equivalent to $\hat{R}_T$ via the unitary transformation $Z_T : \mathcal{H} \to \mathbf{D}(T)$:

$$Z_T T = \hat{R}_T Z_T. \tag{1.8}$$

PROOF: We have already observed that Z_T is one-to-one if T is completely nonunitary, so the norm on $\mathbf{D}(T)$ is well-defined. Since T is contractive, then as in the proof of Theorem 1.3 we see that $\mathbf{D}(T) \subset H^2_{\mathcal{D}_T} \oplus H^2_{\mathcal{D}_{T^*}}$. Since $Z_T : \mathcal{H} \to \mathbf{D}(T)$ is then unitary by the definition of the norm on $\mathbf{D}(T)$, it remains only to verify the intertwining relation (1.8) and the formula (1.7) for the adjoint. For a given x in $\mathcal{H}$ the $H^2_{\mathcal{D}_T}$ or first component of $Z_T T x$ is equal to

$$X_T T x = D_T(I - zT)^{-1}Tx = S^*_{\mathcal{D}_T}[D_T(I - zT)^{-1}x]$$

which is identical to the first component of $\hat{R}_T Z_T x$. The second component of $Z_T T x$ is equal to $X_{T^*} T x$. By a computation as in Proposition 1.4,

$$\begin{aligned}(X_{T^*} T x)(z) &= z(X_{T^*} x)(z) - W_{T^*}(z) D_T x \\ &= z(X_{T^*} x)(z) - W_{T^*}(z) \cdot (X_T x)(0)\end{aligned}$$

which is identical to the second component of $\hat{R}_T Z_T x$. To establish the adjoint formula (1.7), we can use the symmetry between T and T^* built in this model. Namely, since Z_T is unitary,

$$(\hat{R}_T)^* Z_T = Z_T T^*. \tag{1.9}$$

If $J : H^2_{\mathcal{D}_T} \oplus H^2_{\mathcal{D}_{T^*}} \to H^2_{\mathcal{D}_{T^*}} \oplus H^2_{\mathcal{D}_T}$ is the flip-flop map

$$J : \begin{bmatrix} f(z) \\ g(z) \end{bmatrix} \to \begin{bmatrix} g(z) \\ f(z) \end{bmatrix},$$

then J is a unitary transformation from $\mathbf{D}(T)$ onto $\mathbf{D}(T^*)$ such that

$$Z_{T^*} = J Z_T.$$

By (1.9) applied to T^* in place of T we know

$$Z_{T^*} T^* = \hat{R}_{T^*} Z_{T^*}$$

Hence

$$(J \hat{R}_{T^*} J) Z_T = J \hat{R}_{T^*} Z_{T^*} = J Z_{T^*} T^* = Z_T T^*. \tag{1.10}$$

Comparing (1.9) and (1.10) we condlude that

$$(\hat{R}_T)^* = J \hat{R}_{T^*} J.$$

But the formula for $(\hat{R}_T)^*$ in (1.7) is easily identified as the formula for $J\hat{R}_{T^*} J$, if $\hat{R}_T$ in general is given by (1.6). □

Remark. If T is a contraction which is not c.n.u., the construction in Theorem 1.5 still applies. In this case, the catch is that the map $Z_T : \mathcal{H} \to \mathbf{D}(T)$ has a kernel

$$Ker Z_T = \mathcal{H}_u = [\bigcap_{j \geq 0} Ker D_T T^j] \cap [\bigcap_{j \geq 0} Ker D_{T^*} T^{*j}]$$

equal to the largest reducing subspace for T on which T is unitary. It is still the case that the completely nonunitary part $T_{cnu} = T|\mathcal{H}_u^\perp$ of T is unitarily equivalent to $\hat{R}_T$ on $\mathbf{D}(T)$ via the unitary isomorphism $Z_T|(\mathcal{H}_u)^\perp : (\mathcal{H}_u)^\perp \to \mathbf{D}(T)$. A similar remark applies for Theorem 1.3 and the completely nonisometric part of T.

2. Models as reproducing kernel Hilbert spaces.

In general, suppose $\mathcal{E}$ is a Hilbert space (sometimes called the coefficient space) and $\mathcal{K}$ is a Hilbert space of $\mathcal{E}$-valued functions defined on some set S. Then $\mathcal{K}$ is called a *reproducing kernel Hilbert space* if the point evaluations $e(s) : \mathcal{K} \to \mathcal{E}$ defined by $e(s) : f \to f(s)$ are continuous for each s in S. In this case there is a function K, called the *reproducing kernel* for $\mathcal{K}$, defined on $S \times S$ with values in the space of operators on $\mathcal{E}$ for which

(i) $K(\cdot, s)\ x \in \mathcal{K}$ for each $s \in S$ and $x \in \mathcal{E}$

and

(ii) $< f(s), x >_{\mathcal{E}} = < f, K(\cdot, s)x >_{\mathcal{K}}$ for each $f \in \mathcal{K},\ x \in \mathcal{E},\ s \in S$.

If S is a domain in the complex plane and functions in $\mathcal{K}$ are analytic on S, then $K(t, s)$ on S has the additional property of being analytic in the first variable and conjugate analytic in the second.

If the function $K : S \times S \to \mathcal{E}$ is the reproducing kernel function for a reproducing kernel Hilbert space $\mathcal{K}$, then necessarily

$$0 \leq \| \sum_{j=1}^{N} \bar{c}_j K(\cdot, s_j) x_j \|_{\mathcal{K}}^2 = \sum_{i=1}^{N} \sum_{j=1}^{N} < K(s_i, s_j) x_j,\ x_i >_{\mathcal{E}} c_i \bar{c}_j$$

for all choices of scalars $c_1, \ldots, c_n \in \mathbb{C}$, so the $N \times N$ matrix

$$[< K(s_i, s_j) x_j,\ x_i >_{\mathcal{E}}]_{1 \leq i,\ j \leq N}$$

is positive semidefinite for all choices of N points $s_1, \ldots, s_n$ in S and N vectors $x_1, \ldots, x_N$ in $\mathcal{E}$. Such a matrix function $K : S \times S \to \mathcal{E}$ is said to be *positive definite* on S. Conversely, any positive definite function K on S is the reproducing kernel function for a reproducing kernel Hilbert space over S, a result going back at least to Aronszajn in [1950].

Examples of reproducing kernel Hilbert spaces important for the purposes of this paper are the space $H^2_{\mathcal{E}}$ with reproducing kernel $K(z, w) = (1 - z\bar{w})^{-1} I_{\mathcal{E}}$ $(z, w \in \mathcal{D})$ and the orthogonal complement $\mathcal{X}$ of an invariant subspace $\mathcal{M}$ of $H^2_{\mathcal{E}}$. Indeed, by the Beurling-Lax-Halmos theorem, any such invariant subspace $\mathcal{M}$ has the form $\mathcal{M} = \Theta H^2_{\mathcal{F}}$ where $\{\Theta, \mathcal{F}, \mathcal{E}\}$ is an inner function, i.e. $\Theta \in H^\infty_{\mathcal{L}(\mathcal{F}, \mathcal{E})}$ (the space of bounded analytic functions on $\mathbf{D}$ with values in the space of operators $\mathcal{L}(\mathcal{F}, \mathcal{E})$ from the Hilbert space $\mathcal{F}$ into $\mathcal{E}$) with the almost everywhere defined boundary values $\Theta(e^{it})$ on the unit circle isometric operators. Then $\mathcal{X} = H^2_{\mathcal{E}} \ominus \Theta H^2_{\mathcal{F}}$ is a reproducing kernel Hilbert space with reproducing kernel equal to

$$K_\Theta(z, w) = (I - z\bar{w})^{-1} [I_{\mathcal{E}} - \Theta(z)\Theta(w)^*].$$

The following result identifies $\mathbf{H}(T)$ as a reproducing kernel Hilbert space and gives an explicit formula for its reproducing kernel.

THEOREM 2.1. *Let T be a c.n.i. operator on $\mathcal{H}$ and let $\mathbf{H}(T)$ be the model space as in Theorem 1.3. Then $\mathbf{H}(T)$ is a reproducing kernel Hilbert space over the unit disk $\mathbf{D}$ with reproducing kernel given by*

$$K_T(z,w) = D_T(I - zT)^{-1}(I - \bar{w}T^*)^{-1}D_T|\mathcal{D}_T \tag{2.1a}$$

$$= (1 - z\bar{w})^{-1}[I - W_T(z)W_T(w)^*], \tag{2.1b}$$

where $W_T(z) : \mathcal{D}_{T^*} \to \mathcal{D}_T$ is defined by (1.2)

$$W_T(z) = [-T^* + zD_T(I - zT)^{-1}D_{T^*}]\,|\mathcal{D}_{T^*}.$$

PROOF: Let $x \in \mathcal{H}$ and consider $f(z) = D_T(I - zT)^{-1}x \in \mathbf{H}(T)$. Then for any $c \in \mathcal{D}_T$ and $w \in \mathcal{D}$,

$$\begin{aligned} < f(w), c >_{\mathcal{D}_T} &=< D_T(I - wT)^{-1}x, c >_{\mathcal{D}_T} \\ &=< x, (I - \bar{w}T^*)^{-1}D_Tc >_{\mathcal{H}} \\ &=< X_{T^x}, X_T(I - \bar{w}T^*)^{-1}D_Tc >_{\mathbf{H}(T)} \\ &=< f\, X_T(I - \bar{w}T^*)^{-1}D_Tc >_{\mathbf{H}(T)} \end{aligned}$$

where, by the definition of X_T,

$$\begin{aligned} &[X_T(I - \bar{w}T^*)^{-1}D_Tc](z) = \\ &\qquad D_T(I - zT)^{-1}(I - \bar{w}T^*)^{-1}D_Tc. \end{aligned}$$

This verifies the first formula (2.1a) for $K_T(z,w)$. The second formula (2.1b) is then a straightforward computation using the definition (1.2) of $W_T(z)$ (see Sz.-Nagy-Foias [1970], identity (1.4) in Chapter VI). □

The reproducing kernel function for the space $\mathbf{D}(T)$ given in Theorem 1.5 can be found in a similar way.

THEOREM 2.2. *Let T be a c.n.u. contraction operator on $\mathcal{H}$ and let $\mathbf{D}(T)$ be the model space as in Theorem 1.5. Then $\mathbf{D}(T)$ is a reproducing kernel Hilbert space over $\mathbf{D}$ with reproducing kernel function $\hat{K}_T(z,w) : \mathcal{D}_T \oplus \mathcal{D}_{T^*} \to \mathcal{D}_T \oplus \mathcal{D}_{T^*}$ given by*

$$\begin{aligned} \hat{K}_T(z,w) &= \begin{bmatrix} D_T(I - zT)^{-1} \\ D_{T^*}(I - zT^*)^{-1} \end{bmatrix} [(I - \bar{w}T^*)^{-1}D_T,\ (I - \bar{w}T)^{-1}D_{T^*}] \\ &= \begin{bmatrix} D_T(I - zT)^{-1}(I - \bar{w}T^*)^{-1}D_T & D_T(I - zT)^{-1}(I - \bar{w}T)^{-1}D_{T^*} \\ D_{T^*}(I - zT^*)^{-1}(I - \bar{w}T^*)^{-1}D_T & D_{T^*}(I - zT)^{-1}(I - \bar{w}T)^{-1}D_{T^*} \end{bmatrix} \end{aligned} \tag{2.2a}$$

or equivalently by

$$\hat{K}_T(z,w) = \begin{bmatrix} (1 - z\bar{w})^{-1}[I - W_T(z)W_T(w)^*] & (z - \bar{w})^{-1}[W_T(z) - W_T(\bar{w})] \\ (z - \bar{w})^{-1}[W_{T^*}(z) - W_{T^*}(w)] & (1 - z\bar{w})^{-1}[I - W_{T^*}(z)W_{T^*}(w)^*] \end{bmatrix} \tag{2.2b}$$

PROOF: We compute for $x \in \mathcal{H}, w \in \mathcal{D}, c \in \mathcal{D}_T$ and $d \in \mathcal{D}_{T^*}$,

$$
\begin{aligned}
&< (Z_T x)(w), c \oplus d >_{\mathcal{D}_T \oplus \mathcal{D}_{T^*}} \\
&=< \begin{bmatrix} D_T(I - wT)^{-1}x \\ D_{T^*}(I - WT^*)^{-1}x \end{bmatrix}, \begin{bmatrix} c \\ d \end{bmatrix} >_{\mathcal{D}_T \oplus \mathcal{D}_{T^*}} \\
&=< x, [(I - \bar{w}T^*)^{-1}D_T, (I - \bar{w}T)^{-1}D_{T^*}] \begin{bmatrix} c \\ d \end{bmatrix} >_{\mathcal{H}} \\
&=< Z_T x, Z_T[(I - \bar{w}T^*)^{-1}D_T, (I - \bar{w}T)^{-1}D_{T^*}] \begin{bmatrix} c \\ d \end{bmatrix} >_{\mathcal{D}(T)}
\end{aligned}
$$

By definition of Z_T,

$$
\begin{aligned}
&Z_T[(I - \bar{w}T^*)^{-1}D_T, (I - \bar{w}T)^{-1}D_{T^*}] \begin{bmatrix} c \\ d \end{bmatrix} (z) \\
&\begin{bmatrix} D_T(I - zT)^{-1} \\ D_{T^*}(I - zT^*)^{-1} \end{bmatrix} [(I - \bar{w}T^*)^{-1}D_T, (I - \bar{w}T)^{-1}D_{T^*}] \begin{bmatrix} c \\ d \end{bmatrix}.
\end{aligned}
$$

This establishes the first formula (2.2a) for $\hat{K}_T(z, w)$. The second one (2.2b) then follows from various identities for characteristic operator functions. □

The second formula (2.2b) for $K_T(z, w)$ and the fact that K_T, as a reproducing kernel function, is positive definite shows that necessarily the characteristic function $W_T(z)$ is analytic with contractive values on the unit disk.

3. Discrete-time linear systems.

Let us recall the formula (1.2) for the characteristic operator function $W_T(z)$:

$$W_T(z) = [-T^* + zD_T(I - zT)^{-1}D_{T^*}]|\mathcal{D}_{T^*}.$$

We now think of $W_T(z)$ as defining a multiplication operator (or input-output (IO) map)

$$\hat{u}(z) \in H^2_{\mathcal{D}_{T^*}} \to \hat{y}(z) = W_T(z)\hat{u}(z) \in H^2_{\mathcal{D}_{T^*}} \tag{3.1}$$

from $H^2_{\mathcal{D}_{T^*}}$ into $H^2_{\mathcal{D}_T}$. The formula for $W_T(z)$ is highly nonlinear with respect to the variable z. To linearize it, we introduce a new variable $\hat{x}(z)$ by

$$\hat{x}(z) = z(I - zT)^{-1}D_{T^*}\hat{u}(z).$$

Then the IO map (3.1) can be expressed as

$$
\begin{aligned}
z^{-1}\hat{x}(z) &= T\hat{x}(z) + D_{T^*}\hat{u}(z), \quad \hat{x}(0) = 0 \\
\hat{y}(z) &= D_T\hat{x}(z) - T^*\hat{u}(z).
\end{aligned}
\tag{3.2}
$$

In general, if $\hat{f}(z)$ has power series representation $\hat{f}(z) = \sum_{n=0}^{\infty} f(n)z^n$, we define the inverse Z-transform of f to be the sequence of Taylor coefficients $\{f(n)\}_{n \geq 0}$. If we apply the inverse Z-transform to the system (3.2) we get the system of difference equation

$$
\begin{aligned}
x(n + 1) &= Tx(n) + D_{T^*}u(n), \qquad x(0) = 0 \\
y(n) &= D_T x(n) - T^*u(n)
\end{aligned}
\tag{3.3}
$$

which induces an IO map $\{u(n)\}_{n\geq 0} \to \{y(n)\}_{n\geq 0}$ from $l^{2+}_{\mathcal{D}_{T^*}}$ into $l^{2+}_{\mathcal{D}_T}$. In general $l^{2+}_{\mathcal{E}}$ is the space of norm-square-summable $\mathcal{E}$-valued sequences indexed by the nonnegative integers; more generally $l^{+}_{\mathcal{E}}$ will denote the space of all $\mathcal{E}$-valued sequences (not necessarily norm square summable).

In general, if $\mathcal{F} : l^{+}_{\mathcal{U}} \to l^{+}_{\mathcal{Y}}$ is any IO map (where $\mathcal{U}$ and $\mathcal{Y}$ are Hilbert spaces) we say that it has a *state space realization* θ if there exists a Hilbert space $\mathcal{X}$ and bounded linear transformations

$$A : \mathcal{X} \to \mathcal{X},\ B : \mathcal{U} \to \mathcal{X},\ C : \mathcal{X} \to \mathcal{Y},\ d : \mathcal{U} \to \mathcal{Y}$$

for which $\mathcal{F}(\{u(n)\}_{n\geq 0}) = \{y(n)\}_{n\geq 0}$ exactly when

$$y(n) = Cx(n) + Du(n) \tag{3.4}$$

where the sequence of states $\{x(n)\}_{n\geq 0}$ is determined by

$$x(n+1) = Ax(n) + Bu(n), \quad x(0) = 0. \tag{3.5}$$

Upon taking Z-transforms $\{f(n)\}_{n\geq 0} \to \sum_{n=0}^{\infty} f(n)z^n$ in (3.4), (3.5) we see that the Z-transform $\hat{y}(z)$ of the output $\{y_n\} = \mathcal{F}(\{u(n)\})$ is determined by the Z-transform $\hat{u}(z)$ of the input sequence $\{u_n\}_{n\geq 0}$ by

$$\hat{y}(z) = W_\theta(z)\hat{u}(z)$$

where

$$W_\theta(z) = D + zC(I - zA)^{-1}B \tag{3.6}$$

is called the *transfer function* of the linear system θ associated with the equations (3.4) and (3.5). This linear system we may also write in matrix form as

$$\theta : \begin{bmatrix} A & B \\ C & D \end{bmatrix} : \mathcal{X} \oplus \mathcal{U} \to \mathcal{X} \oplus \mathcal{Y}. \tag{3.7}$$

The matrix $\begin{bmatrix} A & B \\ C & D \end{bmatrix}$ we sometimes refer to as the *system matrix* associated with θ. We shall refer to the system θ as *unitary, coisometric*, or *contractive* if the system matrix $\begin{bmatrix} A & B \\ C & D \end{bmatrix}$ has the associated property. When we wish to make the state space $\mathcal{X}$, the input space $\mathcal{U}$ and the output space $\mathcal{Y}$ explicit, besides (3.7) we sometimes write $\theta = (A, B, C, D, ; \mathcal{X}, \mathcal{U}, \mathcal{Y})$, or more simply $\theta = (A, B, C, D)$ when $\mathcal{X}, \mathcal{U}, \mathcal{Y}$ are understood. Thus the characteristic function $W_T(z)$ is the transfer function for the particular system

$$\theta_T : \begin{bmatrix} T & D_{T^*} \\ D_T & -T^* \end{bmatrix} : \mathcal{H} \oplus \mathcal{D}_{T^*} \to \mathcal{H} \oplus \mathcal{D}_T. \tag{3.8}$$

Note that the system matrix $\begin{bmatrix} T & D_{T^*} \\ D_T & -T^* \end{bmatrix}$ for θ_T in (3.8) is unitary as a map from $\mathcal{H} \oplus \mathcal{D}_{T^*} \to \mathcal{H} \oplus \mathcal{D}_T$ and hence θ_T is an example of a unitary system. We shall return to unitary systems in the next section.

If $W(z) : \mathcal{U} \to \mathcal{Y}$ is an operator function in the form (3.6) of a transfer function of a linear system θ, we say that $\theta = (A, B, C, D)$ is a *realization* of $W(z)$. If $W(z)$ has a realization (A, B, C, D) it is clear from the formula that $W(z)$ is analytic in a neighborhood of the origin. The converse question, namely, given an operator function $W(z)$ which is analytic on a neighborhood of the origin, to find a system θ so that $W(z) = W_\theta(z)$, is known as the realization problem in systems theory and has a positive solution (see Fuhrmann [1981]). We shall see instances of this later.

It is natural to ask how two systems θ_1 and θ_2 are related if they have the same transfer function $W_{\theta_1}(z) = W_{\theta_2}(z)$. The following facts are well known (see Helton [1974], Fuhrmann [1981]). If (A, B, C, D) is one realization for $W(z)$, then so is $(\tilde{A}, \tilde{B}, \tilde{C}, \tilde{D})$ where

$$\tilde{A} = \begin{bmatrix} A_{11} & A_{12} & A_{13} \\ 0 & A & A_{23} \\ 0 & 0 & A_{33} \end{bmatrix}, \qquad \tilde{B} = \begin{bmatrix} B_1 \\ B \\ 0 \end{bmatrix},$$
$$\tilde{C} = [\, 0 \quad C \quad C_1 \,], \qquad \tilde{D} = D.$$

This statement is verified immediately by observing that

$$(zI - \tilde{A})^{-1} = \begin{bmatrix} (zI - A_{11})^{-1} & * & * \\ 0 & (zI - A)^{-1} & * \\ 0 & 0 & (zI - A_{33})^{-1} \end{bmatrix}.$$

We say that $(\tilde{A}, \tilde{B}, \tilde{C}, \tilde{D})$ is a *dilation* of (A, B, C, D) or that (A, B, C, D) is a *compression* of $(\tilde{A}, \tilde{B}, \tilde{C}, \tilde{D})$. Another realization for the same $W(z)$ is $(SAS^{-1}, SB, CS^{-1}, D)$ where S is any invertible linear transformation on the state space $\mathcal{X}$ of the system $\theta = (A, B, C, D)$. In this case we say that $(SAS^{-1}, SB, CS^{-1}, D)$ is *similar* to (A, B, C, D). In a rough sense these are the only two kinds of ways two different systems can realize the same transfer function as the following discussion will make precise.

We say that the pair (A, B) is *approximately controllable* if span $\{ImA^jB : j = 0, 1, 2, \dots\}$ is dense in $\mathcal{X}$, and that the pair (C, A) is *observable* if

$$\bigcap_{j \geq 0} KerCA^j = \{0\}.$$

A *system* (A, B, C, D) is said to be *approximately controllable* if the pair (A, B) is approximately controllable, and to be *observable* if the pair (C, A) is observable; a system which is both *approximately controllable* and *observable* is often said to be *minimal*. The following is a well-known decomposition for linear systems (see Fuhrmann [1981], Helton [1974]).

THEOREM 3.1. *Let $\theta = (A, B, C, D)$ be a linear system such that the state space $\mathcal{X}$, the input space $\mathcal{U}$ and the output space $\mathcal{Y}$ are all Hilbert spaces. Then there exists an orthogonal decomposition of the state space*

$$\mathcal{X} = \mathcal{X}_{c\bar{o}} \oplus \mathcal{X}_{co} \oplus \mathcal{X}_{\bar{c}\bar{o}} \oplus \mathcal{X}_{\bar{c}o}$$

so that with respect to this decomposition the system has the form

$$\left(\begin{bmatrix} A_{c\bar{o}} & A_{12} & A_{13} & A_{14} \\ 0 & A_{co} & A_{23} & A_{24} \\ 0 & 0 & A_{\bar{c}\bar{o}} & A_{34} \\ 0 & 0 & 0 & A_{\bar{c}o} \end{bmatrix}, \begin{bmatrix} B_{c\bar{o}} \\ B_{co} \\ 0 \\ 0 \end{bmatrix}, [0, C_{co}, 0, C_{\bar{c}o}], D\right)$$

where the system

$$\theta_c = \left(\begin{bmatrix} A_{c\bar{o}} & A_{12} \\ 0 & A_{co} \end{bmatrix}, \begin{bmatrix} B_{c\bar{o}} \\ B_{co} \end{bmatrix}, [0, C_{co}], D\right)$$

is approximately controllable, the system

$$\theta_0 = \left(\begin{bmatrix} A_{co} & A_{24} \\ 0 & A_{\bar{c}o} \end{bmatrix}, \begin{bmatrix} B_{co} \\ 0 \end{bmatrix}, [C_{co}, C_{\bar{c}o}], D\right)$$

is observable, and the system

$$\theta_{co} = (A_{co}, B_{co}, C_{co}, D)$$

is both approximately controllable and observable. Moreover the transfer functions of $\theta, \theta_c, \theta_0$ *and* θ_{co} *are all the same:*

$$W_\theta(z) = W_{\theta_c}(z) = W_{\theta_0}(z) = W_{\theta_{co}}(z).$$

The import of Theorem 3.1 is that any realization (A, B, C, D) for a given operator function can be brought to an approximately controllable and observable one (A_0, B_0, C_0, D_0) while maintaining the same transfer function by performing a compression. Once the realization is brought to approximately controllable and observable form, it is essentially uniquely determined by its transfer function, as the following result shows. We include the proof since it will motivate a result concerning "closely connected systems" to come later in this section.

THEOREM 3.2. *(see Helton [1974], Theorem 3b.1) Suppose* $\theta_1 = (A_1, B_1, C_1, D_1)$ *and* $\theta_2 = (A_2, B_2, C_2, D_2)$ *are two approximately controllable and observable realizations for the same operator function* $W(z)$. *Then* $D_1 = D_2$ *and there is a densely defined (possibly unbounded) linear operator* $S : \mathcal{X}_1 \to \mathcal{X}_2$, *one-to-one with dense range, such that the identities*

$$SA_1 = A_2 S, \ SB_1 = B_2, \ C_1 = C_2 S$$

hold on a dense set.

PROOF: By looking at the Taylor coefficients in the power series expansion at the origin, we see that $W_{\theta_1}(z) = W_{\theta_2}(z)$ is equivalent to

$$C_1 A_1^n B_1 = C_2 A_2^n B_2 \quad \text{for } n = 0, 1, 2, \dots . \tag{3.9}$$

Since θ_1 is approximately controllable, the set $\mathcal{S}_1$ consisting of vectors x_1 in $\mathcal{X}_1$ of the form

$$x_1 = \sum_{j=0}^{N} A_1^j B_1 u_j \tag{3.10}$$

for some $N \geq 0$ and vectors $u_1, \ldots, u_N \in \mathcal{U}$ is dense in $\mathcal{X}_1$. For $x_1 \in \mathcal{S}_1$ of the form (3.8), define

$$Sx_1 = \sum_{j=0}^{N} A_2^j B_2 u_j \in \mathcal{X}_2. \tag{3.11}$$

Then S is a linear transformation on S_1 provided that it is well defined. To check

$$0 = \sum_{j=0}^{N} A_1^j B_1 u_j \Rightarrow 0 = \sum_{j=0}^{N} A_2^j B_2 u_j. \tag{3.12}$$

Since by assumption (C_2, A_2) is observable, to show that $0 = \sum_{j=0}^{N} A_2^j B_2 u_j$ is the same as to show that

$$C_2 A_2^k \left(\sum_{j=0}^{N} A_2^j B_2 u_j \right) = 0 \quad \text{for } k = 0, 1, 2, \ldots .$$

But we calculate

$$\begin{aligned} C_2 A_2^k \left(\sum_{j=0}^{N} A_2^j B_2 u_j \right) &= \sum_{j=0}^{N} C_2 A_2^{k+j} B_2 u_j \\ &= \sum_{j=0}^{N} C_1 A_1^{k+j} B_1 u_j \\ &= C_1 A_1^k \left(\sum_{j=0}^{N} A_1^j B_1 u_j \right) = C_1 A_1^k 0 = 0 \end{aligned}$$

where we used the assumption in (3.9). Thus (3.11) establishes that S is well-defined on $\mathcal{X}_1$. One checks that S is one-to-one by using the observability of (C_1, A_1). That S has dense range follows from the approximate controllability of (A_2, B_2). The intertwining property $SA_1 = A_2 S$ is trivial to check on S_1 and $SB_1 = B_2$ holds by definition since in fact $Im B_1 \subset S_1$. Finally, for $x_1 = \sum_{j=0}^{N} A_1^j B_1 u_j \in S_1$, note

$$\begin{aligned} C_2 S x_1 = C_2 S \left(\sum_{j=0}^{N} A_1^j B_1 u_j \right) &= C_2 \left(\sum_{j=0}^{N} A_2^j B_2 u_j \right) = \sum_{j=0}^{N} C_2 A_2^j B_2 u_j \\ = \sum_{j=0}^{N} C_1 A_1^j B_1 u_j &= C_1 \left(\sum_{j=0}^{N} A_1^j B_1 u_j \right) = C_1 x_1 \end{aligned}$$

so the intertwining relation $C_2 S = C_1$ holds on the same set $\mathcal{S}_1$. □

For finite dimensional systems the map S in Theorem 3.2 is bounded and boundedly invertible, but this no longer holds true in general for infinite dimensional systems. Two systems (A_i, B_i, C_i, D_i) $(i = 1, 2)$ related as in Theorem 3.2 we shall say are *quasisimilar*. It is easy to check that the converse of Theorem 3.2 is also true, namely: *any two quasi-similar systems have the same transfer function.*

A notion weaker than controllability or observability for a linear system and which has become a standard notion in the theory of unitary systems is that of *close connectedness.* We say that the linear system (A, B, C, D) is *closely connected* if

$$span\{p(A, A^*)B\mathcal{U},\ q(A, A^*)C^*\mathcal{Y} : p, q, \text{ words in two noncommuting symbols}\} \text{ is dense in } \mathcal{X}.$$

Equivalently, the smallest reducing subspace for A containing both ImB and ImC^* is all of $\mathcal{X}$. The approximate controllability condition is that $\mathcal{X}$ is the smallest invariant subspace for A containing ImB; hence approximate controllability alone already implies close connectedness. By taking adjoints in the definition of observability, we see that observability is equivalent to $\mathcal{X}$ being the smallest invariant subspace for A^* containing ImC^*; thus similarly, observability alone implies close connectedness. Note that the notions of approximately controllable and observable are independent of the choice of inner product on the spaces $\mathcal{X}, \mathcal{U}, \mathcal{Y}$; but the validity of the closely connected condition depends crucially on the choice of inner products.

The following is an easy decomposition for linear systems related to the property of close connectedness analogous to the decomposition in Theorem 3.1.

THEOREM 3.3. *If $\theta = (A, B, C, D)$ is a linear system, then the state space has an orthogonal decomposition*

$$\mathcal{X} = \mathcal{X}_0 \oplus \mathcal{X}_1$$

with respect to which A, B, C assume the block matrix forms

$$A = \begin{bmatrix} A_0 & 0 \\ 0 & A_1 \end{bmatrix}, \quad B = \begin{bmatrix} B_0 \\ 0 \end{bmatrix}, \quad C = [C_0, 0]$$

where $\theta_0(A_0, B_0, C_0, D)$ is closely connected. Moreover,

$$W_\theta(z) = W_{\theta_0}(z).$$

PROOF: Let $\mathcal{X}_0$ be the smallest reducing subspace for A containing both ImB and ImC^* and let $\mathcal{X}_1 = \mathcal{X}_0^\perp$, and check that the pair $(\mathcal{X}_0, \mathcal{X}_1)$ has all the required properties. □

We observed above that two systems $\theta_i = (A_i, B_i, C_i, D_i)$ have the same transfer function $W_{\theta_1}(z) = W_{\theta_2}(z)$ if and only if they have the same *moments* (sometimes also called *Markov parameters*)

$$C_1 A_1^j B_1 = C_2 A_2^j B_2 \quad \text{for } j = 0, 1, 2, \ldots. \tag{3.13}$$

If θ_1 and θ_2 are both approximately controllable and observable, we saw in Theorem 3.2 that this is enough to force θ_1 and θ_2 to be quasi-similar. However it is possible for two closely connected systems θ_1 and θ_2 to have the same moments (3.13) (and hence the same transfer functions) but not to be quasi-similar. To classify closely connected systems up to quasisimilarity requires additional invariants. Such additional invariants are provided by the set of "generalized moments" defined as follows. If $\theta = (A, B, C, D)$ is a linear system, then define the associated set of *generalized moments* to consist of the operator

$$D : \mathcal{U} \rightarrow \mathcal{Y}$$

together with the three mappings

$$\tau_1 : q \rightarrow B^* q(A, A^*) B$$
$$\tau_2 : q \rightarrow C q(A, A^*) B$$
$$\tau_3 : q \rightarrow C q(A, A^*) C^*$$

from words q in two noncommuting symbols to operators from $\mathcal{U}$ into $\mathcal{U}$, $\mathcal{U}$ into $\mathcal{Y}$ and $\mathcal{Y}$ into $\mathcal{Y}$ respectively. The usual collection of moments $\{CA^j B\}_{j \geq 0}$ can be identified with the map τ_2 restricted to words q involving only the first symbol A. The following result shows that the set of generalized moments determines a closely connected system even up to unitary equivalence.

THEOREM 3.4. *Let $\theta_j = (A_j, B_j, C_j, D_j)$ $(j = 1, 2)$ be two closley connected systems with the same set of generalized moments. Then there is a unitary transformation $U : \mathcal{X}_1 \rightarrow \mathcal{X}_2$ such that*

$$UA_1 = A_2 U, \quad UB_1 = B_2, \quad C_1 = C_2 U.$$

PROOF: The proof parallels that of Theorem 3.2. Since θ_1 is closely connected, a dense set $\mathcal{S}_1$ is $\mathcal{X}_1$ is formed by elements of the form

$$x_1 = \sum_{j=1}^{N} [p_j(A_1, A_1^*) B_1 u_j + q_j(A_1, A_1^*) C_1^* y_j] \tag{3.14}$$

where N is a positive integer, $p_1, \ldots, p_N$, $q_1, \ldots, q_N$ are words in two noncommuting symbols, $u_1, \ldots, u_N$ are elements of $\mathcal{U}_1$ and $y_1, \ldots, y_N$ are elements of $\mathcal{Y}_1$. For x_1 of the form (3.14), define

$$Ux_1 = \sum_{j=1}^{N} [p_j(A_2, A_2^*) B_2 u_j + q_j(A_2, A_2^*) C_2^* y_j]$$

We check that U is isometric; this also then will verify that U is well-defined. Compute

$$\begin{aligned} \|Ux_1\|^2 = &\sum_{j=1}^{N} \sum_{j=1}^{N} \{< p_j(A_2, A_2^*) B_2 u_j,\ p_k(A_2, A_2^*) B_2 u_k > \\ &+ 2Re < p_j(A_2, A_2^*) B_2 u_j,\ q_k(A_2, A_2^*) C_2^* y_k > \\ &+ < q_j(A_2, A_2^*) C_2^* y_j,\ q_k(A_2, A_2^*) C_2^* y_k >\} \\ = &\sum_{j=1}^{N} \sum_{k=1}^{N} \{< B_2^* p_k(A_2, A_2^*)^* p_j(A_2, A_2^*) B_2 u_j,\ u_k > \\ &+ 2Re < C_2 q_k(A_2, A_2^*)^* p_j(A_2, A_2^*) B_2 u_j,\ y_k > \\ &+ < C_2 q_k(A_2, A_2^*)^* q_j(A_2, A_2^*) C_2^* y_j, y_k >\}. \end{aligned} \tag{3.15}$$

The hypothesis that θ_1 and θ_2 have the same set of generalized moments implies that

$$B_2^* p_k(A_2, A_2^*)^* p_j(A_2, A_2^*) B_2 = B_1^* p_k(A_1, A_1^*)^* p_j(A_1, A_1^*) B_1$$
$$C_2 q_k(A_2, A_2^*)^* p_j(A_2, A_2^*) B_2 = C_1 q_k(A_1, A_1^*)^* p_j(A_1, A_1^*) B_1$$

and

$$C_2 q_k(A_2, A_2^*)^* q_j(A_2, A_2^*) C_2^* = C_1 q_k(A_1, A_1^*)^* q_1(A_1, A_1^*) C_1^*$$

for all words p_k, p_j, q_k, q_j. Plugging these identities into (3.15) and unravelling them gives $\|Ux_1\|^2 = \|x_1\|^2$ as desired. That $U\mathcal{S}_1$ is dense in $\mathcal{X}_2$ follows from the assumption that θ_2 is closely connected. Thus U is densely defined with dense image and is isometric on its domain, and hence extends uniquely by continuity to a unitary map from $\mathcal{X}_1$ onto $\mathcal{X}_2$. The intertwining relations

$$UA_1 = A_2U, \quad UB_1 = B_2, \quad C_1 = C_2U$$

are easily checked from the definition of U on the dense set $\mathcal{S}_1$ just as in the proof of Theorem 3.2. □

4. Contractive and Unitary Systems.

In this section we specialize the discussion in the preceding section to unitary systems, i.e. systems $\theta = (A, B, C, D)$ for which the system matrix

$$\begin{bmatrix} A & B \\ C & D \end{bmatrix} : \mathcal{X} \oplus \mathcal{U} \to \mathcal{X} \oplus \mathcal{Y}$$

is unitary. We have already pointed out that an example is the system $\theta_T = (T, D_T, D_{T^*}, -T^*)$ with $\mathcal{X} = \mathcal{H}$, $\mathcal{U} = \mathcal{D}_{T^*}$, $Y = \mathcal{D}_T$ associated with any contraction operator T on an abstract Hilbert space $\mathcal{H}$. We saw in Section 2 that the transfer function $W_{\theta_T}(z)$ for the system (i.e. the characteristic operator function $W_T(z)$) is analytic with contractive values on the open unit disk $\mathbf{D}$. That this is a general phenomenon for unitary (and more gneerally for contractive) systems can be seen directly as follows.

THEOREM 4.1. *Suppose $\theta = (A, B, C, D)$ is a contractive linear system, i.e.*

$$\|\tilde{x}\|^2 + \|y\|^2 \leq \|x\|^2 + \|u\|^2$$

whenever

$$\begin{bmatrix} A & B \\ C & D \end{bmatrix} \begin{bmatrix} x \\ u \end{bmatrix} = \begin{bmatrix} \tilde{x} \\ y \end{bmatrix}$$

for $x, \ \tilde{x} \in X, \ u \in U, \ y \in Y$. Then the transfer function $W_\theta(z) = D + zC(I - zA)^{-1}B$ is analytic with contractive values on the unit disk $\mathbf{D}$.

PROOF: To show that $W_\theta(z)$ is an analytic contractive operator function on $\mathbf{D}$ amounts to showing that the multiplication operator $\hat{u}(z) \to W_\theta(z)\hat{u}(z) = \hat{y}(z)$ is a contraction from $H^2_{\mathcal{U}}$ into $H^2_{\mathcal{Y}}$. By the general discussion in Section 3, one sees after an application of the inverse Z-transform that this is the same as verifying that

$$\|\{y(n)\}_{n\geq 0}\|_{l_y^{2+}} \leq \|\{u(n)\}_{n\geq 0}\|^2_{l_u^{2+}}$$

whenever $\{y(n)\}_{n\geq 0}$ is generated from $\{u(n)\}_{n\geq 0}$ via the system of recurrence relations

$$\begin{aligned} x(n+1) &= Ax(n) + Bu(n), \qquad x(0) = 0 \\ y(n) &= Cx(n) + Du(n). \end{aligned}$$

By the assumption that $\begin{bmatrix} A & B \\ C & D \end{bmatrix}$ is contractive, we have

$$\|x(n+1)\|^2 + \|y(n)\|^2 \leq \|x(n)\|^2 + \|u(n)\|^2$$

for $n = 0, 1, 2, \ldots$. An easy inductive argument gives

$$\|x(n+1)\|^2 \leq \sum_{j=0}^{n} \|u(j)\|^2 - \sum_{j=0}^{n} \|y(j)\|^2$$

Taking limits gives that $\{y(n)\}_{n\geq 0} \in l_y^{2+}$ whenever $\{u(n)\}_{n\geq 0} \in l_{\mathcal{U}}^{2+}$ and that $\|\{y(n)\}_{n\geq 0}\|^2_{l_y^{2+}} \leq \|\{u(n)\}_{n\geq 0}\|^2_{l_u^{2+}}$ as required. □

The notions of approximately controllable, observable and closely connected apply in particular to unitary and contractive systems. The following result shows how for unitary systems the structure of the main operator A alone determines these system theoretic characteristics.

THEOREM 4.2. *Suppose $\theta = (A, B, C, D)$ is a unitary system. Then:*

(1) θ is approximately controllable if and only if A^* is completely nonisometric.

(2) θ is observable if and only if A is completely nonisometric.

(3) θ is closely connected if and only if A is completely nonunitary.

PROOF: Controllability of θ by definition means that span $\{A^j B\mathcal{U} : j = 0, 1, 2, \ldots\}$ is dense in $\mathcal{H}$. By duality this is equivalent to

$$\bigcap_{j\geq 0} Ker(B^* A^{*j}) = \{0\}.$$

A vector x being in $\bigcap_{j\geq 0} Ker(B^* A^{*j})$ means that

$$\begin{aligned} 0 &= \|B^* A^{*j} x\|^2 = < BB^* A^{*j} x, A^{*j} x > \\ &= < (I - AA^*) A^{*j} x, A^{*j} x > \\ &= \|A^{*j} x\|^2 - \|A^{*j+1} x\|^2 \end{aligned}$$

for $j = 0, 1, 2, \ldots$ (Here we used that $AA^* + BB^* = I$, a consequence of $\begin{bmatrix} A & B \\ C & D \end{bmatrix}$ being unitary). By induction this is equivalent to

$$\|x\|^2 = \|A^{*j} x\|^2$$

for $j = 0, 1, 2, \ldots$, i.e., to x being in the isometric invariant subspace for A^*. This proves (1). Property (2) can be proved directly by a similar argument, or by applying (1) and a duality argument, since (A, B, C, D) is observable if and only if (A^*, C^*, B^*, D^*) is approximately controllable. For (3), again apply a duality argument. By definition θ is closely connected if and only if span $\{p(A, A^*)B\mathcal{U} + q(A, A^*)C^*\mathcal{Y} : p, q, \in \mathcal{W}\}$ is dense in $\mathcal{X}$, where $\mathcal{W}$ is the set of polynomials in two noncommuting variables. By computing the orthogonal complement, we see that this is the same as

$$[\bigcap_{p\in\mathcal{W}} Ker B^* p(A, A^*)] \cap [\bigcap_{q\in\mathcal{W}} Ker C q(A, A^*)] = \{0\}. \tag{4.1}$$

Using the relations

$$AA^* = I - BB^*, \qquad A^*A = I - C^*C$$

and induction, we see that (4.1) is equivalent to

$$[\bigcap_{j\geq 0} Ker B^* A^{*j}] \cap [\bigcap_{j\geq 0} Ker C A^j] = \{0\} \tag{4.2}$$

Note next that an element $x \in \mathcal{X}$ is in the space on the left in (4.2) if and only if

$$\begin{aligned} 0 &= \|B^* A^{*j} x\|^2 = < BB^* A^{*j} x, A^{*j} x > \\ &= \|A^{*j} x\|^2 - \|A^{*j+1} x\|^2 \end{aligned}$$

and

$$\begin{aligned} 0 &= \|CA^j x\|^2 = < C^* C A^j x, A^j x > \\ &= \|A^j x\|^2 - \|A^{j+1} x\|^2 \end{aligned}$$

for all $j = 0, 1, 2, \ldots$, or equivalently, if and only if

$$\|x\|^2 = \|A^j x\|^2 = \|A^{*j} x\|^2$$

for $j = 0, 1, 2, \ldots$, i.e., if and only if x is in the largest reducing subspace $\mathcal{X}_u$ for A on which A is unitary. □

The following result is the analogue of the decomposition in Theorem 3.1 for contractive systems. The proof is straightforward and so will be omitted.

THEOREM 4.3. *Let $\theta = (A, B, C, D)$ be a contractive linear system. Then θ has a decomposition as in Theorem 3.1 where $\theta_c, \theta_0, \theta_{co}$ in addition are also contractive systems.*

We point out that in general if the original system θ is a unitary system it may not be possible to arrange that the controllable compression θ_c, the observable compression θ_0 or the minimal compression θ_{co} also be a unitary system. On the other hand, it is always possible to compress a unitary system to a smaller closely connected system which is also unitary and which has the same transfer function.

THEOREM 4.4. *Suppose $\theta = (A, B, C, D)$ is a unitary linear system. Then θ has a decomposition as in Theorem 3.3 with the additional property that the closely connected subsystem θ_0 is also unitary.*

As a final result in this section, we show that, unlike the case for general affine closely connected systems (see Theorem 3.4), for unitary closely connected systems θ the transfer function $W_\theta(z)$ alone is a complete unitary invariant.

THEOREM 4.5. *Suppose $\theta_j(A_j, B_j, C_j, D_j)$ $(j = 1,2)$ are two unitary closely connected systems with the same transfer functions*

$$W_{\theta_1}(z) = W_{\theta_2}(z) \ for \ z \in \mathcal{D}.$$

Then θ_1 and θ_2 are unitarily equivalent, i.e. $D_1 = D_2$ and there exists a unitary transformation $U : \mathcal{X}_1 \to \mathcal{X}_2$ such that

$$UA_1 = A_2U, \quad UB_1 = B_2, \quad C_1 = C_2U.$$

PROOF: By Theorem 3.4 we need only show that two unitary systems θ_1 and θ_2 have the same set of generalized moments if they have the same transfer functions, i.e., if they have the same set of moments

$$D_1 = D_2 \ and \ C_1A_1^jB_1 = C_2A_2^jB_2, \quad j = 0, 1, 2, \ldots .$$

Since $\begin{bmatrix} A_j & B_j \\ C_j & D_j \end{bmatrix}$ is unitary, we have the relations

$$\begin{aligned} A_j^*A_j + C_j^*C_j = I, \quad & A_jA_j^* + B_jB_j^* = I \\ A_j^*B_j + C_j^*D_j = 0, \quad & A_jC_j^* + B_jD_j^* = 0 \\ B_j^*B_j + D_j^*D_j = I, \quad & C_jC_j^* + D_jD_j^* = I \end{aligned}$$

holding for $j = 1,2$. For q a word in two noncommuting symbols, we must consider expressions of the three types $B_j^*q(A_j, A_j^*)B_j$, $C_jq(A_j, A_j^*)B_j$ and $C_jq(A_j, A_j^*)C_j^*$. If q is empty, we have

$$\begin{aligned} &B_1^*B_1 = I - D_1^*D_1 = I - D_2^*D_2 = B_2^*B_2 \\ &C_1B_1 = z^{-1}[W_{\theta_1}(z) - W_{\theta_1}(0)] = z^{-1}[W_{\theta_2}(z) - W_{\theta_2}(0)] = C_2B_2 \\ &C_1C_1^* = I - D_1D_1^* = I - D_2D_2^* = C_2C_2^*. \end{aligned}$$

If q is a monomial in A we have

$$\begin{aligned} &B_1^*A_1^jB_1 = -D_1^*C_1A_1^{j-1}B_1 = -D_2^*C_2A_2^{j-1}B_2 = B_2^*A_2^jB_2, \\ &C_1A_1^jB_1 = C_2A_2^jB_2, \\ &C_1A_1^jC_1^* = -C_1A_1^{j-1}B_1D_1^* = -C_2A_2^{j-1}B_2D_2^* = C_2A_2^jC_2^*. \end{aligned}$$

If q is a monomial in A^*, the identity of the corresponding moments follows upon taking adjoints in the above. If q involves powers of both A and A^*, use the relations

$$A_j^* A_j = I - C_j^* C_j, \quad A_j A_j^* = I - B_j B_j^*$$

and induction to reduce the identity for moments corresponding to new words q to the identity of moments corresponding to words already known. In the end one gets that θ_1 and θ_2 have the same set of generalized moments as required. □

5. de Branges-Rovnyak model unitary systems.

In Section 1 we started with an abstract contraction operator T on a Hilbert space $\mathcal{H}$ and produced a model space $\mathbf{D}(T)$ and associated model operator $\hat{R}_T$, where the reproducing kernel function for $\mathbf{D}(T)$ (and hence $\mathbf{D}(T)$ itself) was completely determined by the characteristic operator function $W_T(z)$ for T. In this section, we reverse the process. Starting with a contractive operator function $W(z)$, we may define a reproducing kernel Hilbert space $\mathcal{D}(W)$ and a model operator $\hat{R}_W$, and, under certain hypotheses and identifications, recover W as the characteristic operator function of $\hat{R}_W$, or somewhat more generally, as the transfer function of a unitary system.

An operator valued function $W(z)$ analytic on $\mathbf{D}$ with values equal to contraction operators from the Hilbert space $\mathcal{F}$ into the Hilbert space $\mathcal{E}$ we denote in various ways, namely, $W(z) : \mathcal{F} \to \mathcal{E}$ or $\{W(z), \mathcal{F}, \mathcal{E}\}$ when we wish to make explicit the initial space $\mathcal{F}$ and the final space $\mathcal{E}$.

To begin this business, we need the following result. It can be proved by using the Riesz-Herglotz representation for a function analytic on the unit disk with positive real part there; we refer to Ball [1975] where it is proved more generally in the context of models for noncontractions.

PROPOSITION 5.1. *Suppose $\{W(z), \mathcal{F}, \mathcal{E}\}$ is a contractive analytic operator function on the unit disk $\mathbf{D}$. Then the kernel function*

$$\hat{K}_W(z,w) = \begin{bmatrix} (1-z\bar{w})^{-1}[I - W(z)W(w)^*] & (z-\bar{w})^{-1}[W(z) - W(\bar{w})] \\ (z-\bar{w})^{-1}[\tilde{W}(z) - \tilde{W}(w)] & (1-z\bar{w})^{-1}[I - \tilde{W}(z)\tilde{W}(w)^*] \end{bmatrix} \tag{5.1}$$

where

$$\tilde{W}(z) := W(\bar{z})^*$$

is a positive definite kernel function over $\mathbf{D}$.

Once we have Proposition 5.1, then we can define a Hilbert space $\mathcal{D}(W)$ to be the completion of the span of $(\mathcal{E} \oplus \mathcal{F})$-valued analytic functions of the form

$$\hat{K}_W(\cdot, w) \begin{bmatrix} u \\ v \end{bmatrix},$$

where $w \in \mathbf{D}$, $u \in \mathcal{E}$, $v \in \mathcal{F}$, with inner product given by

$$< \sum_j \hat{K}_W(\cdot, w_j) \begin{bmatrix} u_j \\ v_j \end{bmatrix}, \sum_i \hat{K}_W(\cdot, z_i) \begin{bmatrix} x_i \\ y_i \end{bmatrix} >_{\mathcal{D}(W)}$$

$$\sum_i \sum_i < \hat{K}_W(z_i, w_j) \begin{bmatrix} u_j \\ v_j \end{bmatrix}, \begin{bmatrix} x_i \\ y_i \end{bmatrix} > .$$

This is just the construction of the reproducing kernel Hilbert space $\mathcal{D}(W)$ having $\hat{K}_W$ as its reproducing kernel function. By direct computation one can verify that the mapping on kernel functions given by

$$\begin{aligned} \hat{R}_W : \hat{K}_W(\cdot, w) \begin{bmatrix} x \\ y \end{bmatrix} &\to \bar{w}\hat{K}_W(\cdot, w) \begin{bmatrix} x \\ 0 \end{bmatrix} - \hat{K}_W(\cdot, 0) \begin{bmatrix} 0 \\ \tilde{W}(\bar{w})x \end{bmatrix} \\ &+ \bar{w}^{-1}\hat{K}_W(\cdot, w) \begin{bmatrix} 0 \\ y \end{bmatrix} - \bar{w}^{-1}\hat{K}_W(\cdot, 0) \begin{bmatrix} 0 \\ y \end{bmatrix} \end{aligned} \tag{5.2}$$

can be expressed more explicitly as

$$\hat{R}_W : \begin{bmatrix} f(z) \\ g(z) \end{bmatrix} \to \begin{bmatrix} [f(z) - f(0)]/z \\ zg(z) - \tilde{W}(z)f(0) \end{bmatrix}. \tag{5.3}$$

Its adjoint, applied to kernel functions, is given by

$$\begin{aligned} (\hat{R}_W)^* : \hat{K}_W(\cdot, w) \begin{bmatrix} x \\ y \end{bmatrix} &\to \bar{w}^{-1}\hat{K}_W(\cdot, w) \begin{bmatrix} x \\ 0 \end{bmatrix} - \bar{w}^{-1}\hat{K}_W(\cdot, 0) \begin{bmatrix} x \\ 0 \end{bmatrix} \\ &+ \bar{w}\hat{K}_W(\cdot, w) \begin{bmatrix} 0 \\ y \end{bmatrix} - \hat{K}_W(\cdot, 0) \begin{bmatrix} W(\bar{w})y \\ 0 \end{bmatrix}, \end{aligned} \tag{5.4}$$

or more explicitly as

$$(\hat{R}_W)^* : \begin{bmatrix} f(z) \\ g(z) \end{bmatrix} \to \begin{bmatrix} zf(z) - W(z)g(0) \\ [g(z) - g(0)]/z \end{bmatrix}. \tag{5.5}$$

Motivation comes from the case examined in Sections 1 and 2 where the space is $\mathbf{D}(T)$ and W is a characteristic function W_T. Since the span of the kernel functions is invariant under the transformation given in (5.3), by an approximation argument, we conclude that $\hat{R}_W$ defines a bounded linear operator on all of $\mathcal{D}(W)$. Similarly, the formula (5.5) defines a bounded linear operator on all of $\mathcal{D}(W)$ which turns out to be the adjoint $(\hat{R}_W)^*$ of the operator $\hat{R}_W$ defined by (5.3).

Let us recall the setting of Section 1 for the case where $W(z) = W_T(z)$ is a characteristic operator function. In this case, $W(z) = W_T(z)$ is the transfer function of the unitary system

$$\theta_T : \begin{bmatrix} T & D_{T^*} \\ D_T & -T^* \end{bmatrix} : \begin{bmatrix} \mathcal{H} \\ \mathcal{D}_{T^*} \end{bmatrix} \to \begin{bmatrix} \mathcal{H} \\ \mathcal{D}_T \end{bmatrix} \tag{5.6}$$

and if T is c.n.u. then $Z_T : \mathcal{H} \to \mathcal{D}(W_T)$ is a unitary transformation. If we use Z_T to identify $\mathcal{H}$ with $\mathcal{D}(W_T)$, we get a new unitary system

$$\theta : \begin{bmatrix} Z_T T Z_T^* & Z_T D_{T^*} \\ D_T Z_T^* & -T^* \end{bmatrix} : \begin{bmatrix} \mathcal{D}(W_T) \\ \mathcal{D}_{T^*} \end{bmatrix} \to \begin{bmatrix} \mathcal{D}(W_T) \\ \mathcal{D}_T \end{bmatrix} \tag{5.7}$$

which also has $W_T(z)$ as its transfer function. From the computation in Section 1, we know

$$Z_T T Z_T^* = \hat{R}_T = \hat{R}_{W_T}, \tag{5.8}$$

$$\begin{aligned} Z_T D_{T^*} : u \to & \begin{bmatrix} D_T(I - zT)^{-1} D_{T^*} \\ D_{T^*}(I - zT^*)^{-1} D_{T^*} u \end{bmatrix} \\ = & \begin{bmatrix} z^{-1}(W_T(z) - W_T(0))x \\ [I - W_{T^*}(z) W_{T^*}(0)^*]x \end{bmatrix} \\ = & \hat{K}_{W_T}(\cdot, 0) \begin{bmatrix} 0 \\ x \end{bmatrix} \end{aligned} \tag{5.9}$$

$$D_T Z_T^* \begin{bmatrix} f \\ g \end{bmatrix} = D_T u = f(0) \tag{5.10}$$

if $\begin{bmatrix} f \\ g \end{bmatrix} = Z_T u$ for a $u \in \mathcal{H}$. In general, if $\{W(z), \mathcal{F}, \mathcal{E}\}$ is a contractive operator function, let $\hat{\varphi}_W : \mathcal{D}(W) \to \mathcal{E}$ and $\hat{\psi}_W : \mathcal{D}(W) \to \mathcal{F}$ be the evaluation maps at 0 in the first and second component respectively:

$$\hat{\varphi}_W : \begin{bmatrix} f \\ g \end{bmatrix} \to f(0) \tag{5.11}$$

$$\hat{\psi}_W : \begin{bmatrix} f \\ g \end{bmatrix} \to g(0) \tag{5.12}$$

Thus the adjoint maps $\hat{\varphi}_W^* : \mathcal{E} \to \mathcal{D}(W)$ and $\hat{\psi}_W^* : \mathcal{F} \to \mathcal{D}(W)$ are given by

$$\hat{\varphi}_W^* : x \to \hat{K}_W(\cdot, 0) \begin{bmatrix} x \\ 0 \end{bmatrix} = \begin{bmatrix} [I - W(z)W(0)^*]x \\ z^{-1}(\tilde{W}(z) - \tilde{W}(0))x \end{bmatrix}$$

and

$$\hat{\psi}_W^* : y \to \hat{K}_W(\cdot, 0) \begin{bmatrix} 0 \\ y \end{bmatrix} = \begin{bmatrix} z^{-1}(W(z) - W(0))x \\ [I - \tilde{W}(z)\tilde{W}(0)^*]x \end{bmatrix}$$

using that $W_{T^*}(z) = \tilde{W}_T(z)$ and the notation (5.11), (5.12), we see from (5.8), (5.9) and (5.10) that the system θ in (5.7) can be written as

$$\theta : \begin{bmatrix} \hat{R}_{W_T} & \psi_{W_T}^* \\ \varphi_{W_T} & W_T(0) \end{bmatrix} : \begin{bmatrix} \mathcal{D}(W_T) \\ \mathcal{D}_{T^*} \end{bmatrix} \to \begin{bmatrix} \mathcal{D}(W_T) \\ \mathcal{D}_T \end{bmatrix} \tag{5.13}$$

We have verified all of the following result for the case that W is known to be the characteristic operator function W_T for a c.n.u. contraction operator T. The result can be checked directly for any contractive operator function $\{W(z), \mathcal{F}, \mathcal{E}\}$ but we do not include a proof here.

THEOREM 5.2 *Let $\{W(z), \mathcal{F}, \mathcal{E}\}$ be a contractive operator function. Then*

$$\theta : \begin{bmatrix} \hat{R}_W & \hat{\psi}_W^* \\ \hat{\varphi}_W & W(0) \end{bmatrix} : \begin{bmatrix} \mathcal{D}(W) \\ \mathcal{F} \end{bmatrix} \to \begin{bmatrix} \mathcal{D}(W) \\ \mathcal{E} \end{bmatrix}$$

is a unitary system with transfer function equal to $W(z)$:

$$W_\theta(z) = W(z)$$

A system of the form $\theta = (\hat{R}_W, \psi_W^*, \varphi_W, W(0))$ with state space $\mathcal{X} = \mathcal{D}(W)$, input space $\mathcal{F}$ and output space $\mathcal{E}$ associated with a contractive operator function $\{W(z), \mathcal{F}, \mathcal{E}\}$ we shall call a (*de Branges-Rovnyak*) *model unitary system.*

An analogous result holds in connection with the model space $\mathcal{H}(W)$, the reproducing kernel space associated with the reproducing kernel function

$$K_W(z,w) = \frac{I - W(z)W(w)^*}{I - z\bar{w}}$$

where $\{W(z), \mathcal{F}, \mathcal{E}\}$ is a contractive operator function. Other approaches are possible, but since we have already introduced $\mathcal{D}(W)$, we here simply identify $\mathcal{H}(W)$ as the projection of $\mathcal{D}(W)$ onto the first component, with

$$\|f(z)\|^2_{\mathcal{H}(W)} = \min\{\| \begin{bmatrix} f(z) \\ g(z) \end{bmatrix} \|^2_{\mathcal{D}(w)} : g \in H^2_{\mathcal{F}} \text{ such that } \begin{bmatrix} f(z) \\ g(z) \end{bmatrix} \in \mathcal{D}(W)\}.$$

Then we see that

$$R_W : f(z) \to [f(z) - f(0)]/z$$

is a contraction operator on $\mathcal{H}(W)$. For the case that $W = W_T$ for a c.n.u. contraction operator T, we know that $W_T(z)$ is the transfer function for the unitary system:

$$\theta_T : \begin{bmatrix} T & D_{T^*} \\ D_T & -T^* \end{bmatrix} : \begin{bmatrix} \mathcal{H} \\ \mathcal{D}_{T^*} \end{bmatrix} \to \begin{bmatrix} \mathcal{H} \\ \mathcal{D}_T \end{bmatrix}$$

If we let P_{cni} be the orthogonal projection of $\mathcal{H}$ onto the completely nonisometric subspace $\mathcal{H}_{cni}$ for T

$$\mathcal{H}_{cni} = \{\bigcap_{j \geq 0} Ker D_T T^J\}^\perp,$$

then since $D_T = D_T P_{cni}$ and $P_{cni}T = P_{cni}TP_{cni}$ we see that $W_T(z)$ is also the transfer function for the coisometric system

$$\tilde{\theta}_T : \begin{bmatrix} P_{cni}T & P_{cni}D_{T^*} \\ D_T & -T^* \end{bmatrix} : \begin{bmatrix} \mathcal{H}_{cni} \\ \mathcal{D}_{T^*} \end{bmatrix} \to \begin{bmatrix} \mathcal{H}_{cni} \\ \mathcal{D}_T \end{bmatrix}$$

Recall that the map

$$X_T : x \to D_T(I - zT)^{-1}x$$

is a unitary map from $\mathcal{H}_{cni}$ onto $\mathcal{H}(W_T)$. Hence $W_T(z)$ is also the transfer function for the coisometric system

$$\theta : \begin{bmatrix} X_T(P_{cni}T)X_T^* & X_T D_{T^*} \\ D_T X_T^* & -T^* \end{bmatrix} : \begin{bmatrix} \mathcal{H}(W) \\ \mathcal{D}_{T^*} \end{bmatrix} \to \begin{bmatrix} \mathcal{H}(W) \\ \mathcal{D}_T \end{bmatrix} \tag{5.14}$$

Next it works out that

$$X_T(P_{cni}T)X_T^* = R_T = R_{W_T} \tag{5.15}$$

$$X_T D_{T^*} : x \to D_T(I - zT)^{-1} D_{T^*} x = \frac{W_T(z) - W_T(0)}{z} x \tag{5.16}$$

and

$$D_T X_T^* : f(z) \to f(0). \tag{5.17}$$

In general, for a contractive operator function $W(z)$ define maps $\varphi_W : \mathcal{H}(W) \to \mathcal{E}$ and $\psi_W : \mathcal{H}_W \to \mathcal{F}$ by

$$\varphi_W : f(z) \to f(0) \tag{5.18}$$

and

$$\psi_W^* : x \to \frac{W(z) - W(0)}{z} x. \tag{5.19}$$

Finally, using the notation (5.18) and (5.19) and the identifications (5.15), (5.16) and (5.17) we see that the system θ in (5.14) may be written as

$$\theta : \begin{bmatrix} R_{W_T} & \psi_{W_T}^* \\ \varphi_{W_T} & W_T(0) \end{bmatrix} : \begin{bmatrix} \mathcal{H}(W_T) \\ \mathcal{D}_{T^*} \end{bmatrix} \to \begin{bmatrix} \mathcal{H}(W_T) \\ \mathcal{D}_T \end{bmatrix}.$$

This establishes the following result for the case that the contractive operator function $W(z)$ has the form $W_T(z)$ of a characteristic operator function. The general case can be established by direct computations but we do not present the details here.

THEOREM 5.3. *Let $\{W(z), \mathcal{F}, \mathcal{E}\}$ be a contractive operator function, and define maps $\varphi_W : \mathcal{H}(W) \to \mathcal{E}$ and $\psi_W : \mathcal{H}(W) \to \mathcal{F}$ by (5.18) and (5.19). Then the system*

$$\theta_W : \begin{bmatrix} R_W & \psi_W^* \\ \varphi_W & W(0) \end{bmatrix} : \begin{bmatrix} \mathcal{H}(W) \\ \mathcal{F} \end{bmatrix} \to \begin{bmatrix} \mathcal{H}(W) \\ \mathcal{E} \end{bmatrix}$$

is coisometric and has transfer function equal to $W(z)$:

$$W_\theta(z) = W(z).$$

If $\theta = (A, B, C, D)$ is an abstract unitary system with transfer function $W_\theta(z) = D + zC(I - zA)^{-1}B$, it is sometimes useful to model A with $\hat{R}_{W_\theta}$ on $\mathcal{D}(W_\theta)$ rather than with $\hat{R}_A$ on $\mathcal{D}(W_A)$; in fact the characteristic operator function W_A for A is essentially the same (after some normalizations and identifications) as the transfer function W_θ for the system θ, but in applications it may be clumsy to carry around the needed identifications. This goal is handled by the following result.

THEOREM 5.4. *Suppose that*

$$\theta : \begin{bmatrix} A & B \\ C & B \end{bmatrix} : \begin{bmatrix} \mathcal{X} \\ \mathcal{U} \end{bmatrix} \to \begin{bmatrix} \mathcal{X} \\ \mathcal{Y} \end{bmatrix}$$

is a unitary system with transfer function W_θ. Then the map

$$Z_\theta : x \to \begin{bmatrix} C(I - zA)^{-1} \\ B^*(I - zA)^{-1} \end{bmatrix} x$$

is a partial isometry mapping the main space $\mathcal{X}$ for θ onto the model space $\mathcal{D}(W_\theta)$, with kernel equal to the unitary subspace $\mathcal{X}_u$ for A:

$$KerZ_\theta = \mathcal{X}_u := [\bigcap_{j\geq 0} KerD_A A^k] \cap [\bigcap_{j\geq 0} KerD_{A^*} A^{*j}].$$

Moreover, the transformed system

$$\tilde{\theta} : \begin{bmatrix} Z_\theta A Z_\theta^* & Z_\theta B \\ C Z_\theta^* & D \end{bmatrix} : \begin{bmatrix} \mathcal{D}(W_\theta) \\ \mathcal{U} \end{bmatrix} \to \begin{bmatrix} \mathcal{D}(W_\theta) \\ \mathcal{Y} \end{bmatrix}$$

coincides with the model unitary system θ_{W_θ} associated with the contractive operator function W_θ as in Theorem 5.2.

PROOF: Define a space $\tilde{\mathcal{D}}$ to be the image of Z_θ with the pull-back norm:

$$\|Z_\theta x\|^2_{\tilde{\mathcal{D}}} = \|P_{KerZ_\theta^\perp} x\|^2_{\mathcal{X}}.$$

Using the general identities for a transfer function

$$W_\theta(z) = D + zC(I - zA)^{-1}B$$
$$\frac{I - W_\theta(z)W_\theta(w)^*}{1 - z\bar{w}} = C(I - zA)^{-1}(I - \bar{w}A^*)^{-1}C^*$$
$$\frac{W_\theta(z) - W_\theta(w)}{z - \bar{w}} = C(I - zA)^{-1}(I - \bar{w}A)^{-1}B$$
$$\frac{I - \tilde{W}_\theta(z)\tilde{W}_\theta(\bar{w})^*}{1 - z\bar{w}} = B^*(I - zA^*)^{-1}(I - \bar{w}A)^{-1}B$$

and

$$\frac{\tilde{W}_\theta(z) - \tilde{W}_\theta(w)}{z - \bar{w}} = B^*(I - zA^*)^{-1}(I - \bar{w}A^*)^{-1}C^*,$$

one can show as in the proof of Theorem 2.1 that $\tilde{\mathcal{D}}$ is a reproducing kernel Hilbert space with reproducing kernel equal to $\hat{K}_{W_\theta}(z,w)$. By noting the Taylor expansions

$$C(I - zA)^{-1}x = \sum_{j=0}^{\infty}(CA^j x)z^j$$

and

$$B^*(I - zA^*)^{-1}x = \sum_{j=0}^{\infty}(B^*A^{*j}x)z^j$$

we see that

$$KerZ_\theta = [\bigcap_{j\geq 0} KerB^*A^{*j}] \cap [\bigcap_{j\geq 0} KerCA^J]$$

Note that $x \in KerB^*A^{*j}$ for $j = 0, 1, 2, \ldots$ is the same as

$$0 = \|B^*A^{*j}x\|^2 =< BB^*A^{*j}x, A^{*j}x >$$
$$=< (I - AA^*)A^{*j}x,\ A^{*j}x >$$
$$= \|A^{*j}x\|^2 - \|A^{*j+1}x\|^2$$

for $j = 0, 1, 2, \ldots$, or equivalently

$$\|x\| = \|A^{*j}x\| \; for \; j \geq 0.$$

Similarly, using that $C^*C = I - A^*A$, we see that $x \in KerCA^j$ for $j = 0, 1, 2, \ldots$ is the same as

$$\|x\| = \|A^j x\| \; for \; j \geq 0.$$

Hence $KerZ_\theta$ is precisely the unitary subspace X_u for the main operator A.

It remains only to check the intertwining relations

$$Z_\theta A = \hat{R}_{W_\theta} Z_\theta \tag{5.20}$$

$$Z_\theta B = \hat{\psi}^*_W \tag{5.21}$$

$$C Z^*_\theta = \hat{\varphi} W \tag{5.22}$$

$$D = W_\theta(0). \tag{5.23}$$

Relation (5.23) is true by definition of W_θ. If $\begin{bmatrix} f(z) \\ g(z) \end{bmatrix} = Z_\theta x$ then $CZ^*_\theta \begin{bmatrix} f \\ g \end{bmatrix} = C_x = \hat{\varphi}_W \left(\begin{bmatrix} f \\ g \end{bmatrix} \right)$ and (5.22) follows. Similarly,

$$\begin{aligned} (Z_\theta B_u) &= C(I - \cdot A)^{-1} Bu \\ &= \hat{K}_{W_\theta}(\cdot, 0) \begin{bmatrix} 0 \\ u \end{bmatrix} \\ &= \hat{\psi}^*_{W_\theta} u \end{aligned}$$

and (5.21) follows. The relation (5.20) follows in the same way as the proof of (1.8) in Theorem 1.5. □

In Theorem 5.4, note that Z_θ gives a unitary equivalence between the completely nonunitary part A_{cnu} of A

$$A_{cnu} = A \,|\, \{[\bigcap_{j \geq 0} KerD_A A^j] \cap [\bigcap_{j \geq 0} KerD_{A^*} A^{*j}]\}^\perp$$

and $\hat{R}_{W_\theta}$. It follows in particular that $\hat{R}_{W_\theta}$ is completely nonunitary. From Theorem 4.1 we conclude that the model unitary system in Theorem 5.2 is closely connected if $W = W_\theta$ is the transfer function of a unitary system. But Theorem 5.2 can also be viewed as a realization theorem; exhibited there is a unitary system θ for which a given contractive operator function $W(z)$ is the transfer function $W_\theta(z)$. Hence we pick up the following corollary to Theorem 5.4.

COROLLARY 5.5. *If $W(z)$ is any contractive analytic operator function, then W has a closely connected realization. Specifically the model unitary system in Theorem 5.2 is closely connected with transfer function equal to W, and the model operator $\hat{R}_W$ on $\mathcal{D}(W)$ is a completely nonunitary contraction.*

6. Factorization and invariant subspaces.

Suppose we are given two systems $\theta_i = (A_i, B_i, C_i, D_i)$ with state space $\mathcal{X}_i$, input space $\mathcal{U}_i$ and output space $\mathcal{Y}_i$, $i = 1, 2$, such that the output space $\mathcal{Y}_1$ for the first system is the same as the input space $\mathcal{U}_2$ for the second. We may then generate a new system θ by feeding in the output of the first system as the input for the second (see Figure 6.1).

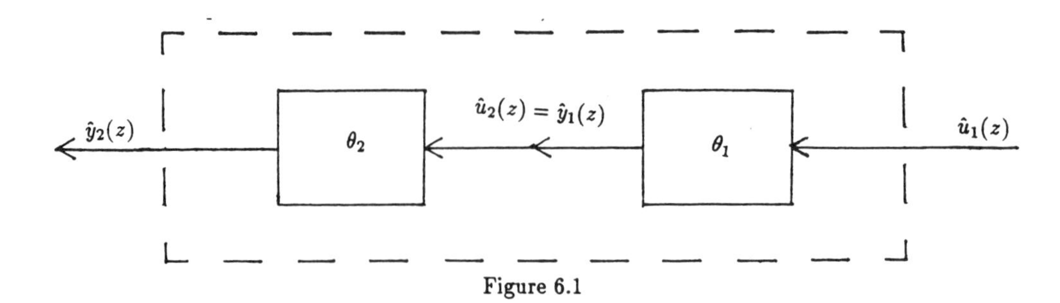

Figure 6.1

Then since $W_\theta(z) : \hat{u}_1(z) \to \hat{y}_2(z)$ where $W_{\theta_1}(z) : \hat{u}_1(z) \to \hat{y}_1(z)$, $W_{\theta_2}(z) : \hat{u}_2(z) \to \hat{y}_2(z)$ and $\hat{u}_2(z) = \hat{y}_1(z)$, we see that on the one hand

$$\hat{y}_2(z) = W_\theta(z)\hat{u}_1(z)$$

while on the other

$$\hat{y}_2(z) = W_{\theta_2}(z)\hat{u}_2(z) = W_{\theta_2}(z)\hat{y}_1(z) = W_{\theta_2}(z)W_{\theta_1}(z)\hat{u}_1(z).$$

We conclude that the transfer function $W_\theta(z)$ for the composite system θ (called the *cascade connection* $\theta_1\theta_2$ of θ_1 and θ_2) is simply the product of the transfer functions of the component systems θ_1 and θ_2:

$$W_{\theta_2\theta_1}(z) = W_{\theta_2}(z)W_{\theta_1}(z).$$

At the level of the system equations we have

$$\begin{aligned} x_2(n+1) &= A_2x_2(n) + B_2u_2(n) \quad & x_1(n+1) &= A_1x_1(n) + B_1u_1(n) \\ y_2(n) &= C_2x_2(n) + D_2u_2(n) \quad & y_1(n) &= C_1x_1(n) + D_1u_1(n) \end{aligned}$$
$$u_2(n) = y_1(n).$$

Eliminating $u_2(n) = y_1(n)$ and simplifying gives

$$\begin{bmatrix} x_2(n+1) \\ x_1(n+1) \end{bmatrix} = \begin{bmatrix} A_2 & B_2C_1 \\ 0 & A_1 \end{bmatrix} \begin{bmatrix} x_2(n) \\ x_1(n) \end{bmatrix} + \begin{bmatrix} B_2D_1 \\ B_1 \end{bmatrix} u_1(n)$$
$$y_2(n) = [C_2, \ D_2C_1] \begin{bmatrix} x_2(n) \\ x_1(n) \end{bmatrix} + D_2D_1u_1(n).$$

Thus the composite system $\theta = \theta_2\theta_1$ has system matrix

$$\begin{aligned}\theta = \theta_2\theta_1 : &\begin{bmatrix} A_2 & B_2C_1 & B_2D_1 \\ 0 & A_1 & B_1 \\ C_2 & D_2C_1 & D_2D_1 \end{bmatrix} \\ &= \begin{bmatrix} A_2 & 0 & B_2 \\ 0 & I & 0 \\ C_2 & 0 & D_2 \end{bmatrix} \begin{bmatrix} I & 0 & 0 \\ 0 & A_1 & B_1 \\ 0 & C_1 & D_1 \end{bmatrix} : \begin{bmatrix} \mathcal{X}_2 \\ \mathcal{X}_1 \\ \mathcal{U}_1 \end{bmatrix} \to \begin{bmatrix} \mathcal{X}_2 \\ \mathcal{X}_1 \\ \mathcal{Y}_2 \end{bmatrix},\end{aligned} \tag{6.1}$$

Conversely, it is of interest to understand how a given system $\theta = (A, B, C, D)$ can be represented as a cascade composition of two component subsystems θ_1 and θ_2. Formula (6.1) for $\theta_2\theta_1$ gives the tipoff; if there is a decomposition $\mathcal{X} = \mathcal{X}_2 \oplus \mathcal{X}_1$ of the state space for $\theta = (A, B, C, D)$ with respect to which θ assumes the form (6.1) of a cascade composition of two smaller systems θ_1 and θ_2, then $\mathcal{X}_2 \oplus \{0\}$ is an invariant subspace for A. For general affine systems this interplay between invariant subspaces for the main operator A of θ and factorization of the transfer function $W_\theta(z)$ is developed systematically in Bart-Gohberg-Kaashoek [1979] for the case where D is invertible, and is extended to the case where D may be singular in Cohen [1983]. Here we concentrate on unitary systems.

From the factored form of the system matrix for $\theta_2\theta_1$ in (6.1), it is clear that $\theta_2\theta_1$ is again a unitary system if each of θ_1 and θ_2 is a unitary system. However, it can happen that $\theta_2\theta_1$ is not closely connected even when θ_1 and θ_2 are closely connected. On the other hand if a closely connected system θ has a cascade decomposition as $\theta = \theta_2\theta_1$, then each of the component systems θ_1 and θ_2 must be also closely connected. If $\{W(z), \mathcal{E}, \mathcal{F}\}$ is a contractive operator function with a representation as a product

$$W(z) = W_2(z) \cdot W_1(z) \tag{6.2}$$

of two contractive operator function $\{W_2(z), \mathcal{G}, \mathcal{E}\}$ and $\{W_1(z), \mathcal{F}, \mathcal{E}\}$, we say that the factorization (6.2) is a *regular factorization* if the system $\theta = \theta_2\theta_1$ is closely connected whenever θ_2 and θ_1 are closely connected systems for which $W_2(z) = W_{\theta_2}(z)$ and $W_1(z) = W_{\theta_1}(z)$ (note that then $W_\theta(z) = W_{\theta_2\theta_1}(z) = W_{\theta_2}(z)W_{\theta_1}(z) = W_2(z)W_1(z)$). The connection between factorization and invariant subspaces for contractive operator functions and unitary systems is laid out in the next theorem.

THEOREM 6.1. *Let $\theta = (A, B, C, D; \mathcal{X}, \mathcal{U}, \mathcal{Y})$ be a unitary system, and suppose $\mathcal{M} \subset \mathcal{X}$ is an invariant subspace for A. Let the decomposition of $\mathcal{X}$*

$$\mathcal{X} = \mathcal{M} \oplus \mathcal{M}^\perp$$

induce block matrix decompositions for A, B, C:

$$A = \begin{bmatrix} A_{\beta\beta} & A_{\beta\alpha} \\ 0 & A_{\alpha\alpha} \end{bmatrix}, \quad B = \begin{bmatrix} B_\beta \\ B_\alpha \end{bmatrix}, \quad C = [C_\beta C_\alpha]$$

Let $\mathcal{F}$ be a Hilbert space with

$$dim\mathcal{F} = dim\{\mathcal{M}^{\perp} \ominus Im[A_{\alpha\alpha}B_{\alpha}]\}.$$

Choose operators $C_1 : \mathcal{M}^{\perp} \to \mathcal{F}$ and $D_1 : \mathcal{U} \to \mathcal{F}$ so that

$$\begin{bmatrix} A_{\alpha\alpha} & B_{\alpha} \\ C_1 & D_1 \end{bmatrix} : \begin{bmatrix} \mathcal{M}^{\perp} \\ \mathcal{U} \end{bmatrix} \to \begin{bmatrix} \mathcal{M}^{\perp} \\ \mathcal{F} \end{bmatrix}$$

is unitary. Define additional opertors $A_1, B_1, A_2, B_2, C_2, D_2$ by

$$\begin{aligned} A_1 &= A_{\alpha\alpha}, & B_1 &= B_2 \\ A_2 &= A_{\beta\beta}, & B_2 &= A_{\beta\alpha}C_1^* + B_{\beta}D_1^* \\ C_2 &= C_{\beta}, & D_2 &= C_{\alpha}C_1^* + DD_1^* \end{aligned}$$

and let θ_1 and θ_2 be the associated systems

$$\theta_1 = (A_1, B_1, C_1, D_1; \mathcal{M}^{\perp}, \mathcal{U}, \mathcal{F})$$
$$\theta_2 = (A_2, B_2, C_2, D_2; \mathcal{M}, \mathcal{F}, \mathcal{Y}).$$

Then θ_1 and θ_2 are unitary systems such that $\theta = \theta_2\theta_1$, and every cascade decomposition of θ arises in this way. Moreover, if θ is closely connected, then

$$W_{\theta}(z) = W_{\theta_2}(z)W_{\theta_1}(z)$$

is a regular factorization of $W_{\theta}(z)$ and every regular factorization of $W_{\theta}(z)$ arises in this way.

PROOF: First suppose $\mathcal{M}$ is an invariant subspace for A. Since θ is a unitary system we have the identity

$$\begin{bmatrix} A_{\beta\beta} & A_{\beta\alpha} & B_{\alpha} \\ 0 & A_{\alpha\alpha} & B_{\alpha} \\ C_{\beta} & C_{\alpha} & D \end{bmatrix} \begin{bmatrix} A_{\beta\beta}^* & 0 & C^* \\ A_{\beta\beta}^* & A_{\alpha\alpha}^* & C_{\alpha}^* \\ B_{\alpha}^* & B_{\alpha}^* & D^* \end{bmatrix} = \begin{bmatrix} I & 0 & 0 \\ 0 & I & 0 \\ 0 & 0 & I \end{bmatrix}. \tag{6.3}$$

Hence in particular $[\,A_{\alpha\alpha} \quad B_{\alpha}\,]$ is coisometric. Then there exists $[\,C_1 \quad D_1\,]$ for which $\begin{bmatrix} A_{\alpha\alpha} & B_{\alpha} \\ C_1 & D_1 \end{bmatrix}$ is unitary. In particular we then have the identity

$$\begin{bmatrix} A_{\alpha\alpha} & B_{\alpha} \\ C_1 & D_1 \end{bmatrix} \begin{bmatrix} A_{\alpha\alpha}^* & C_1^* \\ B_{\alpha}^* & D_1^* \end{bmatrix} = \begin{bmatrix} I & 0 \\ 0 & I \end{bmatrix} \tag{6.4}$$

and $\theta_1 = (A_{\alpha\alpha}, B_{\alpha}, C_1, D_1)$ is a unitary system. From (6.1) we see that we should define $\theta_2 = (A_2, B_2, C_2, D_2)$ so that the factorization holds:

$$\begin{bmatrix} A_{\beta\beta} & A_{\beta\alpha} & B_{\alpha} \\ 0 & A_{\alpha\alpha} & B_{\alpha} \\ C_{\beta} & C_{\alpha} & D \end{bmatrix} = \begin{bmatrix} A_2 & 0 & B_2 \\ 0 & I & 0 \\ C_2 & 0 & D_2 \end{bmatrix} \begin{bmatrix} I & 0 & 0 \\ 0 & A_{\alpha\alpha} & B_{\alpha} \\ 0 & C_1 & D_1 \end{bmatrix}.$$

This suggests defining a 3×3 block operator matrix R by

$$R = [R_{ij}]_{1 \le i,j \le 3} = \begin{bmatrix} A_{\beta\beta} & A_{\beta\alpha} & B_\beta \\ 0 & A_{\alpha\alpha} & B_\alpha \\ C_\beta & C_\alpha & D \end{bmatrix} \begin{bmatrix} I & 0 & 0 \\ 0 & A_{\alpha\alpha}^* & C_1^* \\ 0 & B_\alpha^* & D_1^* \end{bmatrix}.$$

Multiplying out gives immediately that

$$R_{21} = 0.$$

From (6.3) we read off that

$$R_{12} = 0, R_{32} = 0, R_{22} = I.$$

From (6.4) we read off that

$$R_{13} = 0.$$

Hence R has the form

$$R = \begin{bmatrix} R_{11} & 0 & R_{13} \\ 0 & I & 0 \\ R_{31} & 0 & R_{33} \end{bmatrix}.$$

We now define operators A_2, B_2, C_2, D_2 by

$$A_2 = R_{11}, B_2 = R_{13}, C_2 = R_{31}, D_2 = R_{33}.$$

This agrees with the definition of A_1, B_1, C_1, D_1 in the statement of the theorem. Since R is the product of unitary operators, R itself is unitary, and hence $\theta_2 = (A_2, B_2, C_2, D_2)$ is a unitary system. From the definition of R, we see that the system matrix for θ has the factorization displayed in (6.1), and hence $\theta = \theta_2\theta_1$.

Conversely, if θ has a cascade decomposition $\theta = \theta_2\theta_1$ then the main operator A for θ has the upper triangular form

$$A = \begin{bmatrix} A_2 & B_2C_1 \\ 0 & A_1 \end{bmatrix}$$

so $\mathcal{X}_2 \oplus \{0\}$ is an invariant subspace for A; it is then straightforward to verify that one recovers the component systems θ_2, θ_1 via the recipe in the theorem based on this invariant subspace.

If $\theta = \theta_2\theta_1$ and θ is closely connected, then (as was remarked above), θ_2 and θ_1 are closely connected and $W_\theta(z) = W_{\theta_2}(z)W_{\theta_1}(z)$. Since $\theta = \theta_2\theta_1$ is closely connected, by definition this is a regular factorization of $W_\theta(z)$. Conversely, if $W_\theta(z) = W_2(z) \cdot W_1(z)$ is a regular factorization of $W_\theta(z)$, we may build closely connected unitary systems θ_2 and θ_1 which realize W_2 and W_1 as the respective transfer functions: $W_2(z) = W_{\theta_2}(z)$, $W_1(z) = W_{\theta_1}(z)$. Since the factorization $W_2(z) \cdot W_1(z)$ is assumed to be regular, the cascade composition system $\theta_2\theta_1$ is closely connected and has transfer function

$$W_{\theta_2\theta_1}(z) = W_{\theta_2}(z)W_{\theta_1}(z)$$

equal to the transfer function $W_\theta(z)$ for the closely connected unitary system θ. Theorem 4.4 now implies that θ and $\theta_2\theta_1$ are unitarily equivalent. This now enables us to produce an invariant subspace $\mathcal{M}$ for the main operator A of θ from which the recipe in the theorem produces the factorization $W_\theta(z) = W_{\theta_2}(z) \cdot W_{\theta_1}(z)$. □

Historically, operator theorists were interested in a result such as Theorem 6.1 as a tool for reducing the invariant subspace problem for a c.n.u. contraction operator T on a Hilbert space to the problem of factoring the characteristic operator function $W_T(z)$. If $W_T(z)$ is a matrix function (i.e., if $I - T^*T$ and $I - TT^*$ are finite rank) or if $W_T(z) - W_T(0)$ has trace class values for all z (i.e. if $I - T^*T$ is trace class), this approach led to many successes (see e.g. Sz.–Nagy–Foias [1970],Brodskii [1971]). It turns out that any contractive operator function $W(z)$ can be factored nontrivially as a product of two other contractive operator functions (see Sz.–Nagy–Foias [1970]); the difficulty to produce invariant subspaces is to produce nontrivial *regular* factorizations of a given $W(z)$. A number of more intrinsic function theoretic characterizations of when a given factorization $W(z) = W_2(z)W_1(z)$ is regular are known (see Sz.–Nagy–Foias [1964,1970,1974], Kriete [1976], Ball [1978]).

If W is the transfer function for a closely connected unitary system $\theta = (A, B, C, D)$ and the factorization $W = W_2 \cdot W_1$ is not regular, we may still form closely connected unitary systems θ_2 and θ_1 for which $W_2(z) = W_{\theta_2}(z)$ and $W_1(z) = W_{\theta_1}(z)$ and then form the cascade connection to obtain a system $\hat{\theta} = \theta_2\theta_1$ which has the same transfer function as θ:

$$W_{\hat{\theta}}(z) = W_{\theta_2}(z)W_{\theta_1}(z) = W(z) = W_\theta(z).$$

The difficulty is that the unitary system $\hat{\theta}$ is not closely connected, so Theorem 4.5 does not apply. Nevertheless, from Theorem 5.4 we can deduce that the main operator $\hat{A}$ for $\hat{\theta}$ and the main operator A for θ have the same completely nonunitary part; hence

$$\hat{A} \underset{u}{\cong} A \oplus U$$

where U is a unitary operator (and $\underset{u}{\cong}$ stands for unitary equivalence). Thus, in general, *if T is a c.n.u. contraction operator with characteristic operator function W_T, then (not necessarily regular) factorizations $W_T = W_2 \cdot W_1$ of W_T correspond to invariant subspaces (if not for T, at least) for $T' = T \oplus U$ where U is some unitary operator.*

7. Factorization, invariant subspaces and de Branges–Rovnyak models.

In this section we illustrate the ideas of the previous section for model unitary systems. Let $\{W_2(z), \mathcal{E}, \mathcal{G}\}$ and $\{W_1(z), \mathcal{G}, \mathcal{F}\}$ be two contractive operator functions for which the product $\{W(z) = W_2(z)W_1(z), \mathcal{E}, \mathcal{F}\}$ is defined. Then we may form the associated model unitary systems

$$\theta_{W_2} : \begin{bmatrix} \hat{R}_{W_2} & (\hat{\psi}_{W_2})^* \\ \varphi_{W_2} & W_2(0) \end{bmatrix} : \begin{bmatrix} \mathcal{D}(W_2) \\ \mathcal{G} \end{bmatrix} \to \begin{bmatrix} \mathcal{D}(W_2) \\ \mathcal{E} \end{bmatrix}$$

and

$$\theta_{W_1} : \begin{bmatrix} \hat{R}_{W_1} & (\hat{\psi}_{W_1})^* \\ \varphi_{W_1} & W_1(0) \end{bmatrix} : \begin{bmatrix} \mathcal{D}(W_1) \\ \mathcal{F} \end{bmatrix} \to \begin{bmatrix} \mathcal{D}(W_1) \\ \mathcal{G} \end{bmatrix}.$$

Then the cascade connection $\theta_{W_2}\theta_{W_1} = (A, B, C, D; \mathcal{D}(W_2) \oplus \mathcal{D}(W_1), \mathcal{F}, \mathcal{E})$ is given by

$$\begin{bmatrix} A & B \\ C & D \end{bmatrix} = \begin{bmatrix} \hat{R}_{W_2} & (\hat{\psi}_{W_2})^*\varphi_{W_1} & (\hat{\psi}_{W_2})^* W_1(0) \\ 0 & \hat{R}_{W_1} & (\hat{\psi}_{W_1})^* \\ \varphi_{W_2} & W_2(0)\hat{\psi}_{W_1} & W_2(0)W_1(0) \end{bmatrix}.$$

More explicitly

$$A : \begin{bmatrix} f_2(z) \\ g_2(z) \end{bmatrix} \oplus \begin{bmatrix} f_1(z) \\ g_1(z) \end{bmatrix} \to \begin{bmatrix} [f_2(z) - f_2(0)]/z + z^{-1}[W_2(z) - W_2(0)]f_1(0) \\ zg_2(z) - \tilde{W}_2(z)f_2(0) + [I - \tilde{W}_2(z)\tilde{W}_2(0)^*]f_1(0) \end{bmatrix} \oplus \begin{bmatrix} [f_1(z) - f_1(0)]/z \\ zg_1(z) - \tilde{W}_1(z)f_1(0) \end{bmatrix} \tag{7.1}$$

$$B : u \to \begin{bmatrix} z^{-1}[W_2(z) - W_2(0)]W_1(0)u \\ [I - \tilde{W}_2(z)\tilde{W}_2(0)^*]W_1(0)u \end{bmatrix} \oplus \begin{bmatrix} z^{-1}[W_1(z) - W_1(0)]u \\ [I - \tilde{W}_1(z)\tilde{W}_1(0)^*]u \end{bmatrix} \tag{7.2}$$

$$C : \begin{bmatrix} f_2(z) \\ g_2(z) \end{bmatrix} \oplus \begin{bmatrix} f_1(z) \\ g_1(z) \end{bmatrix} \to f_2(0) + W_2(0)f_1(0) \tag{7.3}$$

and

$$D : u \to W_2(0)W_1(0)u. \tag{7.4}$$

Then $\theta_{W_2}\theta_{W_1}$ has transfer function equal to the product $W = W_2W_1$, and hence the c.n.u. part of A is unitarily equivalent to the model operator $\hat{R}_W$ on $\mathcal{D}(W)$. The partial isometry

$$Z_{\theta_{W_2}\theta_{W_1}} : \mathcal{D}(W_2) \oplus \mathcal{D}(W_1) \to \mathcal{D}(W)$$

which implements this unitary equivalence is given by Theorem 5.4:

$$Z := Z_{\theta_{W_2}\theta_{W_1}} : x \to \begin{bmatrix} C(I - zA)^{-1} \\ B^*(I - zA^*)^{-1} \end{bmatrix} x$$

(where A, B, C are given by (7.1), (7.2) and (7.3)). This can be shown to be given explicitly as

$$Z : \begin{bmatrix} f_2(z) \\ g_2(z) \end{bmatrix} \oplus \begin{bmatrix} f_1(z) \\ g_1(z) \end{bmatrix} \to \begin{bmatrix} f_2(z) + W_2(z)f_1(z) \\ \tilde{W}_1(z)g_2(z) + g_1(z) \end{bmatrix}. \tag{7.5}$$

If $W = W_2 \cdot W_1$ is not a regular factorization then A has a nontrivial unitary part A_u and the unitary subspace for A is equal to $Ker\, Z$. One can go on to define an overlapping space $\mathcal{E}(W_2 \cdot W_1)$ associated with the factorization $W_2 \cdot W_1$ which models A_u (see de Branges [1970] and Ball [1978]). The definition is

$$\mathcal{E}(W_2 \cdot W_1) = \left\{ \begin{bmatrix} f \\ g \end{bmatrix} \in H^2_{\mathcal{G}} \oplus H^2_{\mathcal{G}} \middle| \begin{bmatrix} W_2 f \\ -g \end{bmatrix} \in \mathcal{D}(W_2) \text{ and } \begin{bmatrix} -f \\ \tilde{W}_1 g \end{bmatrix} \in \mathcal{D}(W_1) \right\} \tag{7.6}$$

with

$$\left\| \begin{bmatrix} f \\ g \end{bmatrix} \right\|^2_{\mathcal{E}(W_2 \cdot W_1)} = \left\| \begin{bmatrix} W_2 f \\ -g \end{bmatrix} \right\|^2_{\mathcal{D}(W-2)} + \left\| \begin{bmatrix} -f \\ \tilde{W}_1 g \end{bmatrix} \right\|^2_{\mathcal{D}(W_1)}. \tag{7.7}$$

Then it can be shown that the map

$$\mathcal{X} : \begin{bmatrix} f \\ g \end{bmatrix} \to \begin{bmatrix} W_2 f \\ -g \end{bmatrix} \oplus \begin{bmatrix} -f \\ \tilde{W}_1 g \end{bmatrix} \tag{7.8}$$

is a unitary map from $\mathcal{E}(W_2 \cdot W_1)$ onto $Ker\, Z \subset \mathcal{D}(W_2) \oplus \mathcal{D}(W_1)$. Moreover the map

$$U_{W_2 \cdot W_1} : \begin{bmatrix} f(z) \\ g(z) \end{bmatrix} \to \begin{bmatrix} [f(x) - f(0)]/z \\ zg(z) + f(0) \end{bmatrix} \tag{7.9}$$

is a unitary transformation on $\mathcal{E}(W_2 \cdot W_1)$ and satisfies the interwining relation

$$A\mathcal{X} = \mathcal{X}U. \tag{7.10}$$

This U is unitarily equivalent to the unitary part of A via $\mathcal{X}$ identified more explicitly as the reproducing kernel Hilbert space with reproducing kernel of the form

$$\hat{K}_{W_2 \cdot W_1}(z, w) = \begin{bmatrix} \frac{1}{2}(1 - z\overline{w})^{-1}(\varphi(z) + \varphi(w^*)) & \frac{1}{2}(z - \overline{w})^{-1}(\varphi(z) - \varphi(w)) \\ \frac{1}{2}(z - \overline{w})^{-1}(\tilde{\varphi}(z) - \tilde{\varphi}(\overline{w})) & \frac{1}{2}(1 - z\overline{w})^{-1}(\tilde{\varphi}(z) + \tilde{\varphi}(w))^* \end{bmatrix}$$

where $\varphi(z)$ is an anlytic operator function on $\mathbf{D}$ with positive real part (see de Branges [1970]). In Ball [1978], $\varphi(z)$ was identified more explicitly as

$$\varphi(z) = \frac{1}{s\pi} \int_0^{2\pi} \frac{1 + ze^{-it}}{1 - ze^{-it}} \Omega(t)\, dt$$

where $\Omega(t)$ is the parallel sum of $\Omega_2(t) := [I - W_2(e^{it})^* W_2(e^{it})]^{1/2}$ and $\Omega_1(t) := [I - W_1(e^{it}) W_1(e^{it})^*]^{1/2}$; when $\Omega_2(t)$ and $\Omega_1(t)$ are both invertible the parallel sum $\Omega(t) = \Omega_2(t) || \Omega_1(t)$ is determined by

$$\Omega(t)^{-1} = \Omega_2(t)^{-1} - \Omega_1(t)^{-1} - I. \tag{7.11}$$

In general, it is the case that

$$Im\, \Omega(t)^{1/2} = Im\, \Omega_2(t)^{1/2} \cap Im\, \Omega(t)^{1/2}. \tag{7.12}$$

From this one can read off one of the several known characterizations of when a factorization $W = W_2 \cdot W_1$ is regular: *$W = W_2 \cdot W_1$ is regular if and only if*

$$Im\, [I - W_2(e^{it})^* W_2(e^{it})]^{1/2} \cap Im\, [I - W_1(e^{it}) W_1(e^{it})^*]^{1/2} = \{0\}$$

for a.e. t, $0 \le t \le 2\pi$. The map $\hat{Z} := \begin{bmatrix} Z \\ \mathcal{X}^* \end{bmatrix} : \mathcal{D}(W_1) \oplus \mathcal{D}(W_2) \to \mathcal{D}(W) \oplus \mathcal{E}(W_2 \cdot W_1)$ then is unitary and satisfies the intertwining equation

$$\hat{Z}A = (\hat{R}_W \oplus U_{W_2 \cdot W_1})\hat{Z}.$$

Let us consider the special case where $\mathcal{F} = \{0\}$ and $W_1(z) = 0 : \{0\} \to \mathcal{G}$. Then $W(z) = W_2(z) \cdot 0 = 0 : \{0\} \to \mathcal{E}$. Then $\hat{R}_W$ can be computed explicitly and seen to be the adjoint of a shift. Since A (as in (7.1)) has $\hat{R}_W$ as its completely nonunitary part, it follows that A is coisometric. Note that A has the form $\begin{bmatrix} \hat{R}_W & * \\ 0 & * \end{bmatrix}$ on $\mathcal{D}(W_2) \oplus \mathcal{D}(0)$, and hence is a coisometric extension of the c.n.u. contraction $\hat{R}_{W_2}$. We also have the two equivalent ways of writing A, as in (7.1) on $\mathcal{D}(W_2) \oplus \mathcal{D}(0)$ or, after transforming by the unitary map $\hat{Z}$, as $\hat{R}_{0'} \oplus U_{W_2 \cdot 0}$ on $\mathcal{D}(0') \oplus \mathcal{E}(W_2 \cdot 0)$. (Here $0'$ stands for the zero function form $\{0\}$ to $\mathcal{E}$ while 0 stands for the zero function from $\{0\}$ to $\mathcal{G}$.) These two presentations for the coisometric extension of $\hat{R}_{W_2}$ correspond to the two different decompositions $\mathcal{H} \oplus M_+(\mathcal{L})$ and $M_+(\mathcal{L}_*) \oplus \mathcal{R}$ in the Sz.–Nagy–Foias analysis of the geometry of the isometric dilation of $(\hat{R}_{W_2})^*$ (see Sz.–Nagy–Foias [1970]).

One can also consider three–fold products $W = W_3 \cdot W_2 \cdot W_1$ and generate an array of model spaces and mappings between them, corresponding to different associations $W = W_3(W_2W_1) = (W_3W_2)W_1$. More precisely the following spaces can all be identified with each other by canonical unitary maps which intertwine canonical model operators on the respective spaces:

$$\mathcal{D}(W_3) \oplus \mathcal{D}(W_1) \oplus \mathcal{D}(W_1) \tag{7.13i}$$

$$\mathcal{D}(W_3) \oplus \mathcal{D}(W_2W_1) \oplus \mathcal{E}(W_2 \cdot W_1) \tag{7.13ii}$$

$$\mathcal{D}(W_3W_2W_1) \oplus \mathcal{E}(W_3 \cdot W_2W_1) \oplus \mathcal{E}(W_2 \cdot W_1) \tag{7.13iii}$$

$$\mathcal{D}(W_3W_2) \oplus \mathcal{D}(W_1) \oplus \mathcal{E}(W_3 \cdot W_2) \tag{7.13iv}$$

$$\mathcal{D}(W_3W_2W_1) \oplus \mathcal{E}(W_3W_2 \cdot W_1) \oplus \mathcal{E}(W_3 \cdot W_2) \tag{7.13v}$$

In de Branges [1970], some of this model calculus is worked out in detail.

We now illustrate the theory for three–fold factorizations for the special case when W_3 is the zero function with zero final space, W_1 is the zero function with zero initial space, and W_2 is a general contractive operator function. Then the model operator A on $\mathcal{D}(W_3) \oplus \mathcal{D}(W_2) \oplus \mathcal{D}(W_1)$ (i.e., the main operator for the unitary systems $\theta_{W_3}\theta_{W_2}\theta_{W_1}$) is unitary (since the transfer function for $\theta_{W_3}\theta_{W_2}\theta_{W_1}$) is $0 : \{0\} \to \{0\}$), and in fact can be identified as the minimal unitary dilation for $\hat{R}_{W_2}$. Then the various presentations (7.13i)–(7.13v) for the unitary dilation space of $\hat{R}_{W_2}$ correspond to the following geometric decompositions of the unitary dilation space of $(\hat{R}_{W_2})^*$ as presented in Sz.–Nagy–Foias [1970] (Chapter II):

$$M_-(\mathcal{L}_*) \oplus \mathcal{H} \oplus M_+(\mathcal{L}) \tag{7.14i}$$

$$M_-(\mathcal{L}_*) \oplus M_+(\mathcal{L}_*) \oplus \mathcal{R} \tag{7.14ii}$$

$$\{0\} \oplus M(\mathcal{L}_*) \oplus \mathcal{R} \tag{7.14iii}$$

$$M_-(\mathcal{L}) \oplus M_+(\mathcal{L}) \oplus \mathcal{R}_* \tag{7.14iv}$$

$$\{0\} \oplus M(\mathcal{L}) \oplus \mathcal{R}_*. \tag{7.14v}$$

This establishes another point of contact between the de Branges–Rovnyak and Sz.–Nagy–Foias model theories. The above discussion shows that the theory of three–fold factorization $W = W_3 \cdot W_2 \cdot W_1$ is rich and interesting for a trivial special case $W_3(z) : \mathcal{E} \to \{0\}$, $W_1(z) : \{0\} \to \mathcal{F}$. The general case should therefore also lead to other interesting applications, but up to this point has been ignored.

8. Notes.

Section 1. Generalizations of Rota's theorem (Rota [1960]) to other domains (including domains in higher dimensions) other than the unit disk **D** were obtained by Ball [1977], Herrero [1976] and Curto–Herrero [1985]. Theorem 1.2 is due to de Branges–Rovnyak [1966b]; a powerful generalization to models based on operators other than the shift was obtained by Agler [1982, 1988].

The unitary equivalence of abstract operators to parts of weighted shifts was studied in Müller [1988]; related models were given by Foias and Pearcy [1974] and Yadev and Bansal [1983]. The explicit map of a contraction operator T on a Hilbert space $\mathcal{H}$ to its model R_T on $\mathbf{H}(T)$ or $\hat{R}_T$ on $\mathbf{D}(T)$ in Theorems 1.3 and 1.5 and the introduction of the characteristic function $W_T(z)$ via the formula for R_T^* as in Proposition 1.4 comes from Ball [1973]. The developments in this section give an approach to the de Branges–Rovnyak model theory much in the same spirit as the presentation of the Sz.–Nagy–Foias model theory in Sz.–Nagy–Foias [1970].

Section 2. This derivation of the reproducing kernel for the space $\mathcal{D}(T)$ and $\mathcal{H}(T)$ comes from Ball [1973]. Reproducing kernel Hilbert (and more generally, Krein) spaces of the type $\mathcal{H}(W)$ and of other types studied by de Branges have recently appeared in the context of applications to the lossless inverse scattering problem and matricial Nevanlinna–Pick interpolation problems in Alpay–Dym [1984,1985] and Dym [1989].

Section 3. A general reference for linear systems is Kailath [1980]. The concept of close–connectedness is standard in the context of unitary systems but not usually introduced for general systems (see Curgus–Dijksma–Langer–de Snoo [1989]). Theorem 3.4 probably appears explicitly here for the first time. This result can also be seen as a particular illustrative example of a general theory of minimality and uniqueness for more general types of systems worked out in Gohberg–Kaashoek–Lerer [1988].

Section 4. Unitary systems in the more general framework of Krein spaces were part of the Livsic–Brodskii model theory almost from the beginning (see Brodskii [1971]); in this theory Theorem 4.4 is well–known. The contribution here is to separate the roles of "closely connected" and "unitary system" in the proof. Unitary systems (also called unitary colligations or unitary nodes) have recently been studied systematically by Dijksma–Langer–de Snoo [1986] (where applications to the study of self–adjoint extensions of a symmetric linear operator S are included) and Curgus–Dijksma–Langer–de Snoo [1989].

Section 5. This approach to the construction of $\mathcal{D}(W)$ and $\hat{R}_W$ starting from a contractive operator function W follows Ball [1975,1978]. The identification of the maps $\hat{R}_W$, $\hat{\psi}^*$, $\hat{\varphi}_W$, $W(0)$ to form a model unitary system along with the more flexible version in Theorem 5.4 comes from Ball [1978]. The fact that unitary systems characterize the equivalent Sz.–Nagy–Foias model theory has been pointed out by Brodskii [1978]. That the system associated with $\mathcal{H}(W)$ as in Theorem 5.3 is coisometric rather than unitary has recently been emphasized by de Branges [1986].

Theorem 5.2 can be viewed as a realization theorem for contractive operator functions; namely, given the function $W(z)$, Theorem 5.2 constructs a unitary system θ for which $W(z)$ is the transfer function. If $W(z)$ satisfies an additional hypothesis (namely, $\|W(0)u\| = \|u\| \Rightarrow u = 0$), then one can go on to show that $W(z)$ is essentially the characteristic operator function $W_T(z)$ for the c.n.u. contraction $T = \hat{R}_W$; this is a well–known result in the Sz.–Nagy–Foias framework (see Sz.–Nagy–Foias [1970]). The approach here was used in Ball [1975] to study the realization problem for characteristic functions of noncontractions; closely related work on the same problem using somewhat different techniques is Brodskii–Gohberg–Krein [1971], Clark [1974] and McEnnis [1979]. A more recent systematic and thorough account of the realization problem for characteristic operator functions and transfer functions of general J–unitary systems (or colligations) is Curgus–Dijksma–Langer–de Snoo [1989].

In the original work of de Branges and Rovnyak, $\mathcal{H}(W)$ was defined first as a generalized orthogonal complement of $B \cdot H^2_{\mathcal{F}}$ using the de Branges theory of minimal decompositions and then $\mathcal{D}(W)$ was defined via an extension process (see de Branges–Rovnyak [1966a,1966b]). The Sz.–Nagy–Foias model space is embedded in a direct sum of L^2–spaces and hence norms on model space functions are more easily computed; on an abstract level the Sz.–Nagy–Foias model arises from analyzing the geometry of the unitary dilation space for a given c.n.u. contraction (see Sz.–Nagy–Foias [1970] and Bercovici [1988] for a more recent treatment). Nikolskii–Vasyunin [1986] have recently presented a coordinate–free form of the Sz.–Nagy–Foias model; special choices of coordinates then lead to the usual Sz.–Nagy–Foias model, the de Branges–Rovnyak model and another form of the model due to Pavlov. For other connections between the Sz.–Nagy–Foias and de Branges–Rovnyak models, see Ball–Kriete [1987]. For a recent exploration of some of the function and operator theory associated with the de Branges–Rovnyak space $\mathcal{H}(W)$, see Sarason [1986a,1986b,1989]. Nikolskii–Khrushchev [1988] gives an overview of the Sz.–Nagy–Foias model and its applications to the theory of generalized interpolation and spectral expansions.

Section 6. The proof of Theorem 6.1 via the factorization of a block operator unitary matrix with unitary factors appears in Brodskii [1978]. Extension of the same idea to J–unitary systems was done in Genin–van Dooren–Kailath–Delosme–Morf [1983]. The same idea of factorization of the block operator matrix in the proof of Theorem 6.1 is the basis for Cohen's handling of the factorization problem for general affine systems (A, B, C, D) with D not invertible (see Cohen [1983]). The fact that nonregular factorizations $W = W_2 \cdot W_1$ lead to invariant subspaces for $A \oplus U$ rather than for A (for some unitary U) was discovered by Sz.–Nagy and Foias in the context of their model theory in Sz.–Nagy–Foias [1964].

Section 7. The results of this section are taken chiefly from de Branges [1970] and Ball [1978]. The connections of the special cases $0 = W \cdot 0$ and $0 = 0 \cdot W \cdot 0$ with the theory of isometric and unitary dilations appear here for the first time.

REFERENCES

J. Agler [1982], The Arveson extension theorem and coanalytic models, Integral Equations and Operator Theory 5, 608–631.

J Agler [1988], An abstract approach to model theory, in Surveys of Some Recent Results in Operator Theory Vol. II (ed. J. B. Conway & B. B. Morrel), Longman Scientific & Technical, pp. 1–23.

D. Alpay and H. Dym [1984], Hilbert spaces of analytic functions, inverse scattering and operator models I, Integral Equations and Operator Theory 7, 589–641.

D. Alpay and H. Dym [1985], Hilbert spaces of analytic functions, inverse scattering and operator models II, Integral Equations and Operator Theory 8, 145–180.

N. Aronszajn [1950], Theory of reproducing kernels, Trans. Amer. Math. Soc. 68, 337–404.

J. A. Ball [1973], Unitary Perturbations of Contractions, University of Virginia dissertation.

J. A. Ball [1975], Models for noncontractions, J. Math. Anal. Appl. 52, 235–254.

J. A. Ball [1977], Rota's theorem for general functional Hilbert spaces, Proc. Amer. Math. Soc. 64, 55–61.

J. A. Ball [1978], Factorization and model theory for contraction operators with unitary part, Memoirs Amer. Math. Soc. No. 198.

J. A. Ball and T. L. Kriete [1978], Operator-valued Nevanlinna–Pick kernels and the functional models for contraction operators, Integral Equations and Operator Theory 10, 17–61.

H. Bart, I. Gohberg and M. A. Kaashoek [1979], Minimal Factorization of Matrix and Operator Functions, Birkhäuser–Verlag OT1 (Basel).

H. Bercovici [1988], Operator Theory and Arithmetic in H^∞, Math. Surveys and Monographs No. 26, Amer. Math. Soc. (Providence).

L. de Branges [1970], Factorization and invariant subspaces, J. Math. Anal. Appl. 29, 163–200.

L. de Branges [1985], A proof of the Bieberbach conjecture, Acta Math. 154, 137–152.

L. de Branges [1986], Unitary linear systems whose transfer functions are Riemann mapping functions, in Operator Theory and Systems (ed. H. Bart, I. Gohberg, M. A. Kaashoek), Birkhäuser–Verlag OT19 (Basel), pp. 105–124.

L. de Branges and J. Rovnyak [1966a], Square Summable Power Series, Holt, Rinehart and Winston, New York.

L. de Branges and J. Rovnyak [1966b], Appendix on square summable power series, Canonical models in quantum scattering theory, in Perturbation Theory and its Applications in Quantum Mechanics (ed. C. H. Wilcox), Wiley (New York).

M. S. Brodskii [1971], Triangular and Jordan Representations of Linear Operators, Transl. Math. Monographs vol. 32, Amer. Math. Soc. (Providence).

M. S. Brodskii [1978], Unitary operator colligations and their characteristic functions, Russian Math. Surveys 22, 159–191.

V. M. Brodskii, I. C. Gohberg and M. G. Krein [1971], On characteristic functions of invertible operators, Acta Sci. Math. (Szeged) 32, 141–164.

D. N. Clark [1974], On models for noncontractions, Acta Sci. Math. (Szeged) 36, 5–16.

N. Cohen [1983], On minimal factorizations of rational matrix functions, Integral Equations and Operator Theory 6, 647–671.

B. Curgus, A. Dijksma, H. Langer and H. S. V. de Snoo [1989], Characteristic functions of unitary colligations and of bounded operators in Krein spaces, in The Gohberg Anniversary Collections Vol II (ed. H. Dym, S. Goldberg, M. A. Kaashoek, P. Lancaster), Birhäuser–Verlag OT41 (Basel), pp. 125–152.

R. Curto and D. Herrero [1985], On the closures of joint similarity orbits, Integral Equations and Operator Theory 8, 489–556.

A. Dijksma. H. Langer, H. S. V. de Snoo [1986], Characteristic function of unitary operator colligations in Π_κ-spaces, in Operator Theory and Systems (ed. H. Bart, I. Gohberg, M. A. Kaashoek), Birhäuser-Verlag OT19 (Basel), pp. 125–194.

H. Dym [1989], *J* Contractive Matrix Functions, Reproducing Kernel Hilbert Spaces, and

Interpolation, CBMS No. 71, Amer. Math. Soc. (Providence).

C. Foias and C. Pearcy [1974], A model for quasinilpotent operators, Mich. Math. J. 21, 399–404.

P. A. Fuhrmann [1981], Linear Systems and Operators in Hilbert Space, McGraw–Hill (New York).

Y. Genin, P. Van Dooren, T. Kailath, J. M. Delosme and M. Morf [1983], On Σ-lossless transfer functions and related questions, Linear Alg. Appl. 50, 251–275.

I. Gohberg, M. A. Kaashoek and L. Lerer [1988], Nodes and realizations of rational matrix functions: Minimality theory and applications, in Topics in Operator Theory and Interpolation (ed. I. Gohberg) Birkhäuser-Verlag OT29 (Basel), pp. 181–232.

H. Helson [1964], Invariant Subspaces, Academic Press (New York).

J. W. Helton [1972], The characteristic functions of operator theory and electrical network realization, Indiana Univ. Math. J. 22, 403–414.

J. W. Helton [1974], Discrete time systems, operator models, and scattering theory, J. Functional Anal. 16, 15–38.

D. Herrero [1976], A Rota universal model for operators with multiply connected spectrum, Rev. Roum. Math. Pures Appl. 21, 15–23.

T. Kailath [1980], Linear Systems, Prentice–Hall (Englewood Cliffs).

T. L. Kriete [1976], Canonical models and the self-adjoint parts of dissipative operators, J. Functional Anal. 23, 39–94.

M. S. Livsic [1973], Operators, Oscillations, Waves, Open Systems, Transl. Math. Monographs Vol 34, Amer. Math. Soc. (Providence).

M. S. Livshits and A. A. Yantsevich [1979], Operator Colligations in Hilbert Spaces, V. H. Winston & Sons (Washington, DC).

R. E. Kalman, P. L. Falb and M. Arbib [1969], Topics in Mathematical Systems Theory, McGraw-Hill (New York).

B. W. McEnnis [1979], Purely contractive analytic functions and characteristic functions of noncontractions, Acta Sci. Math. (Szeged) 41, 161–172.

V. Müller [1988], Models for operators using weighted shifts, J. Operator Theory 20, 3–20.

B. Sz.-Nagy and C. Foias [1964], Sur les contractions de léspace de Hilbert. IX. Factorisations de la function caractéristique. Sous-espaces invariants, Acta Sci. Math. (Szeged) 25, 283–316.

B. Sz.-Nagy and C. Foias [1970], Harmonic Analysis of Operators on Hilbert Space, Amer. Elsevier (New York).

B. Sz.-Nagy and C. Foias [1970], Regular factorizations of contractions. Proc. Amer. Math. Soc. 43, 91–93.

N. K. Nikolskii and S. V. Khrushchev [1988], A functional model and some problems in the spectral theory of functions, Proc. Steklov Institute of Mathematics, 3.

N. K. Nikolskii and V. I. Vasyunin [1986], Notes on two function models, in the Bierberbach Conjecture: Proceedings of the Symposium on the Occasion of the Proof, Math. Surveys, Amer. Math. Soc. (Providence), pp. 113–141.

N. K. Nikolskii and V. I. Vasyunin [1989], A unified approach to function models and the transcription problem, in the Gohberg Anniversary Collection Vol. II (ed. H. Dym, S. Goldberg, M. A. Kaashoek, P. Lancaster), Birhäuser-Verlag OT41 (Basel), pp. 405–434.

M. Rosenblum and J. Rovnyak [1985], Hardy Classes and Operator Theory, Oxford Univ. Press.

G. C. Rota [1960], On models for linear operators, Comm. Pure Apppl. Math. 13, 469–472.

H. H. Rosenbrock [1970], State–Space and Multivariable Theory, Wiley (New York).

D. Sarason [1986a], Shift–invariant spaces from the Brangesian point of view, in The Bieberbach Conjecture: Proceedings of the Symposium on the Occasion of the Proof, Math. Surveys, Amer. Math. Soc. (Providence).

D. Sarason [1986b], Doubly shift–invariant spaces in H^2, J. Operator Theory 16, 75–97.

D. Sarason [1989], Exposed points in H^1, I, in the Gohberg Anniversary Collection Vol II (ed. H. Dym, S. Goldberg, M. A. Kaashoek, P. Lancaster), Birhäuser–Verlag OT41 (Basel), pp. 485–496.

W. M. Wonham [1979], Linear Multivariable Control: A Geometric Approach, Springer–Verlag (Berlin).

B. S. Yadev and R. Bansel [1983], On Rota's models for linear operators, Rocky Mountain J. Math. 13, 553–556.

Department of Mathematics
Virginia Tech
Blacksburg, VA 24061
USA

Department of Theoretical Mathematics
The Weizmann Institute of Science
Rehovot 76100
ISRAEL

Operator Theory:
Advances and Applications, Vol. 50

BAND MATRICES AND DICHOTOMY

Asher Ben-Artzi* and Israel Gohberg

The invertibility and Fredholm properties of block band matrices are studied herein. The results are stated in terms of dichotomy of some associated companion systems.

1. INTRODUCTION

To state the main results of this paper, we need to define the notion of dichotomy for general systems. Let a system

$$B_{n+1}x_{n+1} = A_n x_n \qquad (n = 0, \pm 1, \ldots) \tag{1.1}$$

be given, where $(A_n)_{n=-\infty}^{\infty}$ and $(B_n)_{n=-\infty}^{\infty}$ are sequences of complex $r \times r$ matrices, and $x_n \in \mathbb{C}^r$. We will consider bounded sequences $(P_n)_{n=-\infty}^{\infty}$ of projections in $\mathbb{C}^r$, satisfying the following conditions:

$$\operatorname{Rank} P_n \qquad (n = 0, \pm 1, \ldots) \text{ is constant,} \tag{1.2}$$

$$\operatorname{Im}(A_n P_n) \subset \operatorname{Im}(B_{n+1}P_{n+1}) \quad \text{and} \quad \operatorname{Im}\big(B_{n+1}(I_r - P_{n+1})\big) \subset \operatorname{Im}\big(A_n(I_r - P_n)\big), \tag{1.3}$$

for $n = 0, \pm 1, \ldots$, and

$$\operatorname{Im}(B_{n+1}P_{n+1}) + \operatorname{Im}\big(A_n(I_r - P_n)\big) = \mathbb{C}^r \qquad (n = 0, \pm 1, \ldots), \tag{1.4}$$

where $I_r = (\delta_{ij})_{ij=1}^{r}$. Since $\operatorname{Rank} P_n$ $(n = 0, \pm 1, \ldots)$ is constant, equality (1.4) shows that the restricted mappings $A_n|_{Ker\, P_n} : \operatorname{Ker} P_n \to \operatorname{Im}\big(A_n(I_r - P_n)\big)$, and $B_{n+1}|_{Im\, P_{n+1}}$ $: \operatorname{Im} P_{n+1} \to \operatorname{Im}(B_{n+1}P_{n+1})$ $(n = 0, \pm 1, \ldots)$ are invertible. We are now in a position to define the dichotomy of the system (1.1). A bounded sequence of projections $(P_n)_{n=-\infty}^{\infty}$ with properties (1.2)–(1.4), is called a dichotomy for the system (1.1), if there exist positive numbers $a < 1$ and M, such that the following inequalities hold

$$\|(B_{n+j}|_{Im\, P_{n+j}})^{-1} A_{n+j-1} \cdots (B_{n+1}|_{Im\, P_{n+1}})^{-1} A_n P_n\| \le M a^j, \tag{1.5}$$

and

$$\|(A_{n-j}|_{Ker\, P_{n-j}})^{-1} B_{n-j+1} \cdots (A_{n-1}|_{Ker\, P_{n-1}})^{-1} B_n (I_r - P_n)\| \le M a^j, \tag{1.6}$$

* Supported by a Dr. Chaim Weizmann fellowship for scientific research.

for $n = 0, \pm 1, \ldots$; $j = 1, 2, \ldots$. Note that the matrix products appearing in these inequalities are well defined in view of (1.3). We call the integer $p = \operatorname{Rank} P_n$ ($n = 0, \pm 1, \ldots$) the rank of the dichotomy. This definition extends the definition given in [1; Section 2], and [6], which treat the case $B_n = I_r$ ($n = 0, \pm 1, \ldots$). See also [3], [7] and [8]. We now introduce the block mappings $(B_{n+1}|_{Im\,P_{n+1}}, A_n|_{Ker\,P_n})$: $\operatorname{Im} P_{n+1} \oplus \operatorname{Ker} P_n \to \mathbb{C}^r$, ($n = 0, \pm 1, \ldots$), where $\operatorname{Im} P_{n+1} \oplus \operatorname{Ker} P_n$ is the abstract direct sum. Note that (1.2) and (1.4) imply that the mappings $(B_{n+1}|_{Im\,P_{n+1}}, A_n|_{Ker\,P_n})$ are invertible ($n = 0, \pm 1, \ldots$).

In this paper we prove the following result.

THEOREM 1.1. *Let $L = (a_{ij})_{ij=-\infty}^{\infty}$ be an infinite matrix, where the entries a_{ij} are $r \times r$ complex matrices with $\sup_{ij} \|a_{ij}\| \leq \infty$, and such that $a_{ij} = 0$ whenever $i > j + m_1$ or $i < j - m_2$ (m_1 and m_2 are two nonnegative integers, not both zero). Then L defines an invertible operator in $\ell_r^2(\mathbb{Z})$ if and only if the system*

$$B_{n+1} x_{n+1} = A_n x_n \qquad (n = 0, \pm 1, \ldots), \tag{1.7}$$

where A_n and B_n ($n = 0, \pm 1, \ldots$) are $m_1 + m_2$ block matrices given by

(1.8)
$$A_n = \begin{pmatrix} I_r & 0 & \cdots & 0 & 0 \\ 0 & I_r & \cdots & 0 & 0 \\ \cdot & \cdot & \cdot & \cdot & \cdot \\ 0 & 0 & \cdots & I_r & 0 \\ 0 & 0 & \cdots & 0 & a_{n+m_1,n} \end{pmatrix}; \qquad B_n = \begin{pmatrix} 0 & \cdots & 0 & a_{n-m_2,n} \\ -I_r & \cdots & 0 & a_{n-m_2+1,n} \\ \cdot & \cdot & \cdot & \cdot \\ 0 & \cdots & 0 & a_{n+m_1-2,n} \\ 0 & \cdots & -I_r & a_{n+m_1-1,n} \end{pmatrix},$$

admits a dichotomy $(P_n)_{n=-\infty}^{\infty}$ with the following additional property

$$\sup_n \|(B_{n+1}|_{Im\,P_{n+1}}, A_n|_{Ker\,P_n})^{-1}\| < \infty. \tag{1.9}$$

Moreover, if L is invertible, then $G = (a_{ij})_{ij=0}^{\infty}$ defines a Fredholm operator in ℓ_r^2, and

$$\operatorname{index} G = p - m_1 r, \tag{1.10}$$

where p is the rank of the dichotomy.

The system (1.7)–(1.8) above, is called a companion system associated with L. We will also prove that Theorem 1.1 above, continues to hold if the sequences of matrices $(A_n)_{n=-\infty}^{\infty}$, and $(B_n)_{n=-\infty}^{\infty}$, are replaced by the following

(1.11)
$$\overline{A}_n = \begin{pmatrix} 0 & -I_r & \cdots & 0 \\ 0 & 0 & \cdots & 0 \\ \cdot & \cdot & \cdot & \cdot \\ 0 & 0 & \cdots & -I_r \\ a_{n,n-m_1} & a_{n,n-m_1+1} & \cdots & a_{n,n+m_2-1} \end{pmatrix};$$
$$\overline{B}_{n+1} = \begin{pmatrix} I_r & 0 & \cdots & 0 & 0 \\ 0 & I_r & \cdots & 0 & 0 \\ \cdot & \cdot & \cdot & \cdot & \cdot \\ 0 & 0 & \cdots & I_r & 0 \\ 0 & 0 & \cdots & 0 & a_{n,n+m_2} \end{pmatrix}.$$

The following terminology will be used throughout the paper. The blocks in a block matrix are $r \times r$ complex matrices, where r is a fixed positive integer. A block matrix (a_{ij}) is called an (m_1, m_2) *banded* matrix, if $a_{ij} = 0$ whenever $i < j - m_2$ or $i > j + m_1$. Here, m_1 and m_2 are two nonnegative integers, not both zero. An (m_1, m_2) banded block matrix (a_{ij}) is called *regular* if a_{ij} is invertible for $i = j - m_2$ and $i = j + m_1$. We will identify bounded operators in ℓ_r^2 or $\ell_r^2(\mathbf{Z})$, with their block matrices $(a_{ij})_{ij=0}^{\infty}$ or $(a_{ij})_{ij=-\infty}^{\infty}$. If $A = (a_{ij})$ is a block matrix, then we denote the block transpose of A by A^T, thus $A^T = (a_{ji})$. Finally, $I_n = (\delta_{ij})_{ij=1}^{n}$ $(n = 1, 2, \ldots)$.

This paper is divided into six sections. The first is the introduction. In the next section, we study in details the special case of two diagonal operators. The results of that section connect the dichotomy with the Riesz projection of operator pencils. Concerning the latter, we make use of a Theorem 2.1 of [5]. Theorem 1.1 is proved in Section 3, as an application of the results about two diagonal operators, and a nonstationary linearization. The uniqueness of the dichotomy is proved in Section 4. In Section 5 we apply our results to regular band matrices. Here we treat systems of the form $x_{n+1} = A_n x_n$ $(n = 0, \pm 1, \ldots)$ and show that in this case, the previous definition of dichotomy coincides with the classical definition. The last section contains applications to Fredholm band operators. These are band operators which admit invertible band extensions.

2. TWO DIAGONAL OPERATORS

In this section we prove the following two results.

THEOREM 2.1. *Let* $A = (\delta_{ij} A_j)_{ij=-\infty}^{\infty}$, *and* $B = (\delta_{i+1,j} B_j)_{ij=-\infty}^{\infty}$ *be two bounded operators, where* A_n *and* B_n $(n = 0, \pm 1, \ldots)$ *are* $r \times r$ *complex matrices. Then, the operator* $A + \lambda B$ *is invertible in* $\ell_r^2(\mathbf{Z})$ *for at least one* $\lambda \in \mathbf{T}$ $(\mathbf{T} = \{z \in \mathbf{C} \colon |z| = 1\})$, *if and only if, the system*

$$B_{n+1} x_{n+1} = A_n x_n \qquad (n = 0, \pm 1, \ldots), \tag{2.1}$$

admits a dichotomy $(P_n)_{n=-\infty}^{\infty}$ *satisfying the following additional condition*

$$\sup_n \|(B_{n+1}|_{Im\, P_{n+1}}, A_n|_{Ker\, P_n})^{-1}\| < \infty. \tag{2.2}$$

Moreover, assume that the system (2.1) *admits a dichotomy* $(P_n)_{n=-\infty}^{\infty}$ *satisfying* (2.2). *Then,* $A + \lambda B$ *is invertible for every* $\lambda \in \mathbf{T}$, *and a formula for the dichotomy is given by*

$$(\delta_{ij} P_j)_{ij=-\infty}^{\infty} = \frac{1}{2\pi i} \int_{\mathbf{T}} (\lambda B + A)^{-1} B d\lambda. \tag{2.3}$$

A formula for the projectors U_n $(n = 0, \pm 1, \ldots)$ *onto* $\mathrm{Im}(B_{n+1} P_{n+1})$ *and along* $\mathrm{Im}(A_n(I_r - P_n))$, *is the following*

$$(\delta_{ij} U_j)_{ij=-\infty}^{\infty} = \frac{1}{2\pi i} \int_{\mathbf{T}} B(\lambda B + A)^{-1} d\lambda. \tag{2.4}$$

THEOREM 2.2. *Let the operators A and B be as in Theorem 2.1. If $A+B$ is invertible, then $G = (\delta_{ij}A_j + \delta_{i+1,j}B_j)_{ij=0}^{\infty}$ is a Fredholm operator in ℓ_r^2, and the index of G is equal to the rank of the dichotomy of the system* (2.1).

The case of regular band matrices was considered in [2], and the case of two diagonal operators, with $B_n = I_r$, and A_n invertible, was treated earlier in [6].

These two theorems are immediate consequences of Propositions 2.3 and 2.4 below. Throughout this section we let A, B, and G to be the operators defined in Theorems 2.1 and 2.2.

PROPOSITION 2.3. *Assume that the system* (2.1) *admits a dichotomy $(P_n)_{n=-\infty}^{\infty}$ satisfying condition* (2.2). *Then, the operator $T = A + B$ is invertible in $\ell_r^2(\mathbf{Z})$, G is a Fredholm operator in ℓ_r^2, and the index of G is equal to the rank of the dichotomy.*

PROPOSITION 2.4. *Assume that the operator $A + \lambda B$ is invertible, for some $\lambda \in \mathbf{T}$. Then $A + \lambda B$ is invertible for every $\lambda \in \mathbf{T}$, and the system* (2.1) *admits a dichotomy satisfying conditions* (2.2)–(2.4).

It is clear that Propositions 2.3 and 2.4, and the uniqueness of the dichotomy (see Corollary 4.2), imply Theorems 2.1 and 2.2.

PROOF OF PROPOSITION 2.3. It follows from (2.2) that

$$\sup_n \|(B_{n+1}|_{Im\,P_{n+1}})^{-1}\| < \infty; \qquad \sup_n \|(A_n|_{Ker\,P_n})^{-1}\| < \infty. \tag{2.5}$$

For every integer n, let S_n be a matrix whose first p (respectively last $r-p$) columns form an orthonormal basis of $\operatorname{Im} P_n$ (respectively $\operatorname{Ker} P_n$), and let E_n be a matrix whose first p (respectively last $r-p$) columns form an orthonormal basis for $\operatorname{Im}(B_{n+1}P_{n+1})$ (respectively $\operatorname{Im}(A_n(I_r - P_n))$. It is clear that $\|S_n\| \le 2$, $\|E_n\| \le 2$ $(n = 0, \pm 1, \ldots)$. In addition, the boundedness of $(P_n)_{n=-\infty}^{\infty}$ leads to $\sup_n \|S_n^{-1}\| < \infty$.

Moreover, inequality (2.2), and the boundedness of $(A_n)_{n=-\infty}^{\infty}$ and $(B_n)_{n=-\infty}^{\infty}$, imply that $\inf_n\{\operatorname{dist}(u, \operatorname{Im}(A_n(I_r - P_n))) : u \in \operatorname{Im}(B_{n+1}P_{n+1}),\ \|u\| = 1\} > 0$. Therefore, $\sup_n \|E_n^{-1}\| < \infty$. Hence,

$$\sup_n\{\|S_n\|, \|S_n^{-1}\|, \|E_n\|, \|E_n^{-1}\|\} < \infty. \tag{2.6}$$

The relations (1.3) show that the following block decompositions, of order $(p, r-p) \times (p, r-p)$, hold

$$E_n^{-1}A_nS_n = \begin{pmatrix} R_n & 0 \\ 0 & L_n \end{pmatrix}; \quad E_{n-1}^{-1}B_nS_n = \begin{pmatrix} M_n & 0 \\ 0 & N_n \end{pmatrix} \qquad (n = 0, \pm 1, \ldots). \tag{2.7}$$

It is clear that

$$\sup_n\{\|R_n\|, \|L_n\|, \|M_n\|, \|N_n\|\} \le \sup_n\{\|A_n\|, \|B_n\|\} < \infty. \tag{2.8}$$

The boundedness condition (2.5) shows that L_n and M_{n+1} $(n = 0, \pm 1, \ldots)$ are invertible, with

$$\sup_n \{\|L_n^{-1}\|, \|M_n^{-1}\|\} < \infty. \tag{2.9}$$

Moreover, the dichotomy inequalities (1.5)–(1.6) imply

$$\|M_{n+j}^{-1} R_{n+j-1} \cdots M_{n+2}^{-1} R_{n+1} M_{n+1}^{-1} R_n\| \leq M a^j, \tag{2.10}$$

and

$$\|L_{n-j}^{-1} N_{n-j+1} \cdots L_{n-2}^{-1} N_{n-1} L_{n-1}^{-1} N_n\| \leq M a^j, \tag{2.11}$$

for $n = 0, \pm 1, \ldots;\ j = 1, 2, \ldots$.

We now prove that T is invertible. Define two operators in $\ell_r^2(\mathbf{Z})$ by $S = (\delta_{ij} S_i)_{ij=-\infty}^{\infty}$ and $D = \left(\delta_{ij} \begin{pmatrix} M_{i+1}^{-1} & 0 \\ 0 & L_i^{-1} \end{pmatrix} E_i^{-1}\right)_{ij=-\infty}^{\infty}$. By (2.6), (2.8), and (2.9), S and D are bounded and invertible operators in $\ell_r^2(\mathbf{Z})$. The operator DTS has the following block representation

$$DTS = \begin{pmatrix} \cdot & \cdot & \cdot & \cdot & \cdot \\ \cdots & \begin{pmatrix} M_0^{-1} R_{-1} & 0 \\ 0 & I_{r-p} \end{pmatrix} & \begin{pmatrix} I_p & 0 \\ 0 & L_{-1}^{-1} N_0 \end{pmatrix} & 0 & \cdots \\ \cdots & 0 & \begin{pmatrix} M_1^{-1} R_0 & 0 \\ 0 & I_{r-p} \end{pmatrix} & \begin{pmatrix} I_p & 0 \\ 0 & L_0^{-1} N_1 \end{pmatrix} & \cdots \\ \cdots & 0 & 0 & \begin{pmatrix} M_2^{-1} R_1 & 0 \\ 0 & I_{r-p} \end{pmatrix} & \cdots \\ \cdot & \cdot & \cdot & \cdot & \cdot \end{pmatrix}$$

Here and in the sequel, the central block in a double infinite matrix has indices $i = j = 0$. Therefore, DTS is unitarily equivalent to the direct sum of the following two operators, which act in $\ell_p^2(\mathbf{Z})$ and $\ell_{r-p}^2(\mathbf{Z})$ respectively:

$$T_1 = \begin{pmatrix} \cdot & \cdot & \cdot & \cdot & \cdot \\ \cdots & M_0^{-1} R_{-1} & I_p & 0 & \cdots \\ \cdots & 0 & M_1^{-1} R_0 & I_p & \cdots \\ \cdots & 0 & 0 & M_2^{-1} R_1 & \cdots \\ \cdot & \cdot & \cdot & \cdot & \cdot \end{pmatrix},$$

and

$$T_2 = \begin{pmatrix} \cdot & \cdot & \cdot & \cdot & \cdot \\ \cdots & I_{r-p} & L_{-1}^{-1} N_0 & 0 & \cdots \\ \cdots & 0 & I_{r-p} & L_0^{-1} N_1 & \cdots \\ \cdots & 0 & 0 & I_{r-p} & \cdots \\ \cdot & \cdot & \cdot & \cdot & \cdot \end{pmatrix}.$$

However, it is easily seen using (2.10) and (2.11) that T_1 and T_2 are invertible. Therefore DTS is invertible; and so is T, because D and S are invertible.

Similarly, define two bounded and invertible operators in ℓ_r^2 via $S^0 = (\delta_{ij}S_i)_{ij=0}^{\infty}$ and $D^0 = \left(\delta_{ij}\begin{pmatrix} M_{i+1}^{-1} & 0 \\ 0 & L_i^{-1} \end{pmatrix} E_i^{-1}\right)_{ij=0}^{\infty}$. As for T, it is easily seen that D^0GS^0 is unitarily equivalent to the direct sum of

$$G_1 = \begin{pmatrix} M_1^{-1}R_0 & I_p & 0 & \cdots \\ 0 & M_2^{-1}R_1 & I_p & \cdots \\ 0 & 0 & M_3^{-1}R_2 & \cdots \\ \cdot & \cdot & \cdot & \cdot \end{pmatrix},$$

and

$$G_2 = \begin{pmatrix} I_{r-p} & L_0^{-1}N_1 & 0 & \cdots \\ 0 & I_{r-p} & L_1^{-1}N_2 & \cdots \\ 0 & 0 & I_{r-p} & \cdots \\ \cdot & \cdot & \cdot & \cdot \end{pmatrix},$$

which act in ℓ_p^2 and ℓ_{r-p}^2 respectively. The inequalities (2.11) show that G_2 is equal to the sum of the identity and an operator with spectral radius less than 1. Thus G_2 is invertible. Similarly, the inequalities (2.10) imply that the operator

$$\begin{pmatrix} I_p & 0 & 0 & \cdots \\ M_2^{-1}R_1 & I_p & 0 & \cdots \\ 0 & M_3^{-1}R_2 & I_p & \cdots \\ \cdot & \cdot & \cdot & \cdot \end{pmatrix}$$

is invertible. Therefore, the operator

$$\begin{pmatrix} 0 & I_p & 0 & \cdots \\ 0 & M_2^{-1}R_1 & I_p & \cdots \\ 0 & 0 & M_3^{-1}R_2 & \cdots \\ \cdot & \cdot & \cdot & \cdot \end{pmatrix}$$

is Fredholm, and has index p. Hence, G_1, D^0GS^0, and finally G are Fredholm operators with index equal to p. □

PROOF OF PROPOSITION 2.4. Denote $T = A + B$. For every $\lambda \in \mathbb{T}$, the operator $U(\lambda) = (\delta_{ij}\lambda^j I_r)_{ij=-\infty}^{\infty}$ is a unitary operator in $\ell_r^2(\mathbb{Z})$, and the following inequality holds

$$A + \lambda B = U(\lambda)^* T U(\lambda) \qquad (\lambda \in \mathbb{T}). \tag{2.12}$$

Thus, $A + \lambda B$ is unitarily equivalent with T ($\lambda \in \mathbb{T}$). Since we assume that $A + \lambda B$ is invertible for some $\lambda \in \mathbb{T}$, then T is invertible. Hence, $A + \lambda B$ is invertible for every $\lambda \in \mathbb{T}$.

We now apply Theorem 2.1 of [5]. It follows, first that the two operators

$$(2.13)\qquad Q' = \frac{1}{2\pi i}\int_{\mathbf{T}}(\lambda B + A)^{-1}B\,d\lambda; \qquad Q'' = \frac{1}{2\pi i}\int_{\mathbf{T}} B(\lambda B + A)^{-1}d\lambda$$

are projections in $\ell_r^2(\mathbf{Z})$. In addition, the operators A and B have the following block decompositions

$$(2.14)\qquad B = \begin{pmatrix} B' & 0 \\ 0 & B'' \end{pmatrix} : \operatorname{Ker} Q' \oplus \operatorname{Im} Q' \to \operatorname{Ker} Q'' \oplus \operatorname{Im} Q'',$$

and

$$(2.15)\qquad A = \begin{pmatrix} A' & 0 \\ 0 & A'' \end{pmatrix} : \operatorname{Ker} Q' \oplus \operatorname{Im} Q' \to \operatorname{Ker} Q'' \oplus \operatorname{Im} Q''.$$

Finally A' and B'' are invertible, and the linear pencils $\lambda B' + A'$ (respectively $\lambda B'' + A''$) is invertible for λ inside and on (respectively outside and on) the unit circle. Therefore, the spectra of the operators $A'^{-1}B'$, and $B''^{-1}A''$, satisfy

$$(2.16)\qquad \sigma(A'^{-1}B') \subseteq \Lambda_+; \qquad \sigma(B''^{-1}A'') \subseteq \Lambda_+$$

where $\Lambda_+ = \{z : |z| < 1\}$.

By (2.12) we have the equality

$$(\lambda B + A)^{-1} = U(\lambda)^* T^{-1} U(\lambda) = (\lambda^{j-i} T_{ij})_{ij=-\infty}^{\infty},$$

where we denote $T^{-1} = (T_{ij})_{ij=-\infty}^{\infty}$. By Cauchy's Theorem, this equality and (2.13) show that Q' and Q'' are block diagonal projections. Thus, we can write

$$(2.17)\qquad Q' = (\delta_{ij} Q_i')_{ij=-\infty}^{\infty}; \qquad Q'' = (\delta_{ij} Q_i'')_{ij=-\infty}^{\infty},$$

where $(Q_i')_{i=-\infty}^{\infty}$ and $(Q_i'')_{i=-\infty}^{\infty}$ are bounded sequences of projections in $\mathbb{C}^r$. We will prove that $(Q_n')_{n=-\infty}^{\infty}$ is a dichotomy for (2.1), satisfying condition (2.2).

The block decompositions (2.14) and (2.15) lead to the following decompositions

$$(2.18)\qquad B_n = \begin{pmatrix} B_n' & 0 \\ 0 & B_n'' \end{pmatrix} : \operatorname{Ker} Q_n' \oplus \operatorname{Im} Q_n' \to \operatorname{Ker} Q_{n-1}'' \oplus \operatorname{Im} Q_{n-1}'',$$

and

$$(2.19)\qquad A_n = \begin{pmatrix} A_n' & 0 \\ 0 & A_n'' \end{pmatrix} : \operatorname{Ker} Q_n' \oplus \operatorname{Im} Q_n' \to \operatorname{Ker} Q_n'' \oplus \operatorname{Im} Q_n''.$$

The invertibility of A' and B'' implies that $A'_n = A_n|_{\text{Ker}\,Q'_n} : \text{Ker}\,Q'_n \to \text{Ker}\,Q''_n$, and $B''_n = B_n|_{Im\,Q'_n} : \text{Im}\,Q'_n \to \text{Im}\,Q''_{n-1}$ $(n = 0, \pm 1, \ldots)$ are invertible, and that

$$(2.20) \qquad \sup_n \|(A_n|_{Ker\,Q'_n})^{-1}\| < \infty; \qquad \sup_n \|(B_n|_{Im\,Q'_n})^{-1}\| < \infty.$$

The invertibility of A'_n and B''_{n+1}, in turn leads to

$$(2.21) \qquad \dim \text{Ker}\,Q'_n = \dim \text{Ker}\,Q''_n; \qquad \dim \text{Im}\,Q'_{n+1} = \dim \text{Im}\,Q''_n,$$

as well as to

$$(2.22) \qquad \text{Im}(A_n|_{Ker\,Q'_n}) = \text{Ker}\,Q''_n; \qquad \text{Im}(B_{n+1}|_{Im\,Q'_{n+1}}) = \text{Im}\,Q''_n,$$

for $n = 0, \pm 1, \ldots$. It follows from these equalities that

$$(B_{n+1}|_{Im\,Q'_{n+1}}, A_n|_{Ker\,Q'_n})^{-1} = \begin{pmatrix} (B_{n+1}|_{Im\,Q'_{n+1}})^{-1} Q''_n \\ (A_n|_{Ker\,Q'_n})^{-1}(I_r - Q''_n) \end{pmatrix} \qquad (n = 0, \pm 1, \ldots).$$

Taking into account the bounds in (2.20) and the boundedness of the sequence $(Q''_n)_{n=-\infty}^{\infty}$, this equality shows that

$$(2.23) \qquad \sup_n \|(B_{n+1}|_{Im\,Q'_{n+1}}, A_n|_{Ker\,Q'_n})^{-1}\| < \infty.$$

By (2.21) $\text{Rank}\,Q'_n = \text{Rank}\,Q'_{n+1}$. Therefore, the sequence of projections $(Q'_n)_{n=-\infty}^{\infty}$ is of constant rank. In addition, (2.22) leads to the following equalities

$$(2.24) \qquad \text{Im}(B_{n+1}Q'_{n+1}) + \text{Im}(A_n(I_r - Q'_n)) = \mathbb{C}^r \qquad (n = 0, \pm 1, \ldots).$$

On the other hand, the block decomposition (2.19) and the equalities (2.22) imply that

$$(2.25) \qquad \text{Im}(A_n Q'_n) \subseteq \text{Im}\,Q''_n = \text{Im}(B_{n+1}Q'_{n+1}) \qquad (n = 0, \pm 1, \ldots),$$

while (2.18) and (2.22) lead to

$$(2.26) \qquad \text{Im}(B_{n+1}(I_r - Q'_{n+1})) \subseteq \text{Ker}\,Q''_n = \text{Im}(A_n(I_r - Q'_n)) \qquad (n = 0, \pm 1, \ldots).$$

Finally, note that (2.16) shows that there are positive numbers a and M_1, with $a < 1$, such that the following inequalities hold

$$(2.27) \qquad \|(A'^{-1}B')^j\| \le M_1 a^j; \quad \|(B''^{-1}A'')^j\| \le M_1 a^j \qquad (j = 1, 2, \ldots).$$

However, the operators $A'^{-1}B'$ and $B''^{-1}A''$ have the following form $B''^{-1}A'' = (\delta_{i,j+1} B''^{-1}_{j+1} A''_j)_{ij=-\infty}^{\infty}$ and $A'^{-1}B' = (\delta_{i+1,j} A'^{-1}_{j-1} B'_j)_{ij=-\infty}^{\infty}$. Consequently, (2.27) leads to

$$\|B''^{-1}_{n+j} A''_{n+j-1} \cdots B''^{-1}_{n+2} A''_{n+1} B''^{-1}_{n+1} A''_n\| \le M_1 a^j,$$

and

$$\|A'^{-1}_{n-j}B'_{n-j+1}\cdots A'^{-1}_{n-2}B'_{n-1}A'^{-1}_{n-1}B'_n\| \le M_1 a^j,$$

for $n = 0, \pm 1, \ldots$ and $j = 1, 2, \ldots$. The block decompositions (2.18) and (2.19) then show that

$$\|(B_{n+j}|\operatorname{Im} Q'_{n+j})^{-1}A_{n+j-1}\cdots(B_{n+1}|\operatorname{Im} Q'_{n+1})^{-1}A_nQ'_n\| \le M_1 \sup_n \|Q'_n\|a^j, \tag{2.28}$$

and

$$\begin{aligned} &\|(A_{n-j}|\operatorname{Ker} Q'_{n-j})^{-1}B_{n-j+1}\cdots(A_{n-1}|\operatorname{Ker} Q'_{n-1})^{-1}B_n(I-Q'_n)\| \\ &\quad \le M_1 \sup_n \|I-Q'_n\|a^j \end{aligned} \tag{2.29}$$

for $n = 0, \pm 1, \ldots$ and $j = 1, 2, \ldots$.

This proves that the sequence $P_n = Q'_n$ $(n = 0, \pm 1, \ldots)$ is a dichotomy for the system (2.1), which satisfies conditions (2.2)–(2.4). In fact, we have shown that $(Q'_n)_{n=-\infty}^{\infty}$ is a bounded sequence of projections in $\mathbb{C}^r$, of constant rank. Moreover, by (2.22), Q''_n is the projection onto $\operatorname{Im}(B_{n+1}Q'_{n+1})$ along $\operatorname{Im}(A_n(I_r - Q'_n))$ $(n = 0, \pm 1, \ldots)$. Finally, conditions (1.3), (1.4), (1.5)–(1.6), (2.2), and (2.3)–(2.4), follow from (2.25)–(2.26), (2.24), (2.28)–(2.29), (2.23), and (2.13)–(2.17), respectively. □

3. LINEARIZATION OF BAND OPERATORS

In this section we prove Theorem 1.1. A different method for investigating band operators, using two diagonal operators, is described at the end of this section. Two bounded operators $T_1: H_1 \to H_1$, and $T_2: H_2 \to H_2$, where H_1 and H_2 are Hilbert spaces, are equivalent if there exist two bounded and invertible operators $E, F: H_2 \to H_1$, such that $T_1F = ET_2$. We denote by I and $\tilde{I}$ the identity operators in $\ell_r^2(\mathbf{Z})$ and ℓ_r^2 respectively. We will also denote by $(\ell_r^2(\mathbf{Z}))^m$ and $(\ell_r^2)^m$ the direct sums $\overbrace{\ell_2^2(\mathbf{Z}) \oplus \cdots \oplus \ell_r^2(\mathbf{Z})}^{m}$ and $\overbrace{\ell_r^2 \oplus \cdots \oplus \ell_r^2}^{m}$. The proof of Theorem 1.1 will be based on the following lemma.

LEMMA 3.1. *Let $L = (a_{ij})_{ij=-\infty}^{\infty}$ be a bounded, $(m, 0)$ banded, block operator in $\ell_r^2(\mathbf{Z})$. Let $G = (a_{ij})_{ij=0}^{\infty}$, and let $(A_n)_{n=-\infty}^{\infty}$ and $(B_n)_{n=-\infty}^{\infty}$ be the sequences of $m \times m$ block matrices defined by*

$$A_n = \begin{pmatrix} I_r & 0 & \cdots & 0 & 0 \\ 0 & I_r & \cdots & 0 & 0 \\ \cdot & \cdot & \cdot & \cdot & \cdot \\ 0 & 0 & \cdots & I_r & 0 \\ 0 & 0 & \cdots & 0 & a_{n+m,n} \end{pmatrix}; \qquad B_n = \begin{pmatrix} 0 & \cdots & 0 & a_{n,n} \\ -I_r & \cdots & 0 & a_{n+1,n} \\ \cdot & \cdot & \cdot & \cdot \\ 0 & \cdots & 0 & a_{n+m-2,n} \\ 0 & \cdots & -I_r & a_{n+m-1,n} \end{pmatrix}. \tag{3.1}$$

Then, the operators

$$L_1 = \begin{pmatrix} L & 0 & \cdots & 0 & 0 \\ 0 & I & \cdots & 0 & 0 \\ \cdot & \cdot & \cdot & \cdot & \cdot \\ 0 & 0 & \cdots & I & 0 \\ 0 & 0 & \cdots & 0 & I \end{pmatrix} \quad \text{and} \quad G_1 = \begin{pmatrix} G & 0 & \cdots & 0 & 0 \\ 0 & \tilde{I} & \cdots & 0 & 0 \\ \cdot & \cdot & \cdot & \cdot & \cdot \\ 0 & 0 & \cdots & \tilde{I} & 0 \\ 0 & 0 & \cdots & 0 & \tilde{I} \end{pmatrix},$$

which act in $(\ell_r^2(\mathbb{Z}))^m$ *and* $(\ell_r^2)^m$, *are respectively equivalent to the operators* $L_2 = (\delta_{ij}B_j + \delta_{i-1,j}A_j)_{ij=-\infty}^{\infty}$ *and* $G_2 = (\delta_{ij}B_j + \delta_{i-1,j}A_j)_{ij=0}^{\infty}$, *which act in* $\ell_{rm}^2(\mathbb{Z})$ *and* ℓ_{rm}^2.

We postpone the proof of this lemma, and begin by deriving Theorem 1.1 from it.

PROOF OF THEOREM 1.1. Define the operators $L' = (a_{i-m_2,j})_{ij=-\infty}^{\infty}$ and $G' = (a_{i-m_2,j})_{ij=0}^{\infty}$. Then L' is an $(m_1+m_2, 0)$ banded block operator. Apply Lemma 3.1 to L', with $m = m_1 + m_2$. Then, the operators

$$L_1' = \begin{pmatrix} L' & 0 & \cdots & 0 & 0 \\ 0 & I & \cdots & 0 & 0 \\ \cdot & \cdot & \cdot & \cdot & \cdot \\ 0 & 0 & \cdots & I & 0 \\ 0 & 0 & \cdots & 0 & I \end{pmatrix} \quad \text{and} \quad G_1' = \begin{pmatrix} G' & 0 & \cdots & 0 & 0 \\ 0 & \tilde{I} & \cdots & 0 & 0 \\ \cdot & \cdot & \cdot & \cdot & \cdot \\ 0 & 0 & \cdots & \tilde{I} & 0 \\ 0 & 0 & \cdots & 0 & \tilde{I} \end{pmatrix},$$

are respectively equivalent to $L_2' = (\delta_{ij}B_j + \delta_{i-1,j}A_j)_{ij=-\infty}^{\infty}$ and $G_2' = (\delta_{ij}B_j + \delta_{i-1,j}A_j)_{ij=0}^{\infty}$. Here $(A_n)_{n=-\infty}^{\infty}$, and $(B_n)_{n=-\infty}^{\infty}$ are defined by (1.8). Define $L_3' = (\delta_{ij}A_j + \delta_{i+1,j}B_j)_{ij=-\infty}^{\infty}$. We now apply Theorem 2.1 to L_3'. It follows that L_3' is invertible if and only if the system (1.7) admits a dichotomy satisfying (1.9). However, the operators L_3', L_2' and L_1' are equivalent; L_1' is invertible if and only if L' is invertible; and L' is equivalent to L. Hence, L_3' is invertible if and only if L is invertible. This proves the first part of the theorem.

Assume that L is invertible. Then L_3' is invertible. By Theorem 2.2, the operator $G_3' = (\delta_{ij}A_j + \delta_{i+1,j}B_j)_{ij=0}^{\infty}$ is Fredholm and

$$\text{index}(G_3') = p \tag{3.2}$$

where p is the rank of the dichotomy of the system (1.7). Define the block shift operator $S = (\delta_{i-1,j}I_{r(m_1+m_2)})_{ij=0}^{\infty}$. Then $G_2' = SG_3' + (\delta_{i,0}\delta_{j,0}B_0)_{ij=0}^{\infty}$. Therefore, G_2' is Fredholm, and

$$\text{index } G_2' = \text{index } S + \text{index } G_3' = -r(m_1+m_2) + \text{index } G_3'. \tag{3.3}$$

However G_2' is equivalent to G_1'. Therefore G_1', and hence G' too, is a Fredholm operator, and the following equalities hold

$$\text{index } G' = \text{index } G_1' = \text{index } G_2'. \tag{3.4}$$

Finally, since G' is Fredholm, then G is Fredholm, and the respective indices are related via

$$\text{index}\, G = \text{index}\, G' + m_2 r. \tag{3.5}$$

It follows from (3.2)–(3.5) that $\text{index}\, G = p - m_1 r$. □

We now show that Theorem 1.1 continues to hold if the sequences $(A_n)_{n=-\infty}^{\infty}$ and $(B_n)_{n=-\infty}^{\infty}$ defined by (1.8), are replaced respectively by $(\overline{A}_n)_{n=-\infty}^{\infty}$ and $(\overline{B}_n)_{n=-\infty}^{\infty}$ defined by (1.11). Let $L = (a_{ij})_{ij=-\infty}^{\infty}$ be a bounded (m_1, m_2) banded block matrix. Then $L^* = (a_{ji}^*)_{ij=-\infty}^{\infty}$ is an (m_2, m_1) banded operator, and L is invertible if and only if L^* is invertible. By Theorem 1.1, L^* is invertible if and only if the system

$$\overline{\overline{B}}_{n+1} x_{n+1} = \overline{\overline{A}}_n x_n \qquad (n = 0, \pm 1, \ldots), \tag{3.6}$$

admits a dichotomy $(\overline{\overline{P}}_n)_{n=-\infty}^{\infty}$, satisfying

$$\sup_n \|(\overline{\overline{B}}_{n+1}|_{Im\, \overline{\overline{P}}_{n+1}}, \overline{\overline{A}}_n|_{Ker\, \overline{\overline{P}}_n})^{-1})\| < \infty. \tag{3.7}$$

Here, $\overline{\overline{A}}_n$ and $\overline{\overline{B}}_n$ $(n = 0, \pm 1, \ldots)$ are given by

$$\overline{\overline{A}}_n = \begin{pmatrix} I_r & 0 & \cdots & 0 & 0 \\ 0 & I_r & \cdots & 0 & 0 \\ \cdot & \cdot & \cdot & \cdot & \cdot \\ 0 & 0 & \cdots & I_r & 0 \\ 0 & 0 & \cdots & 0 & a_{n,n+m_2}^* \end{pmatrix}; \qquad \overline{\overline{B}}_n = \begin{pmatrix} 0 & \cdots & 0 & a_{n,n-m_1}^* \\ -I_r & \cdots & 0 & a_{n,n-m_1+1}^* \\ \cdot & \cdot & \cdot & \cdot \\ 0 & \cdots & 0 & a_{n,n+m_2-2}^* \\ 0 & \cdots & -I_r & a_{n,n+m_2-1}^* \end{pmatrix}$$

However, the existence of a dichotomy for (3.6), satisfying (3.7), is, by Theorem 2.1, equivalent to the invertibility of $T = (\delta_{ij} \overline{\overline{A}}_j + \delta_{i+1,j} \overline{\overline{B}}_j)_{ij=-\infty}^{\infty}$. On the other hand, T is invertible if and only if the operator $T_1 = (\delta_{ij} \overline{\overline{B}}_j^* + \delta_{i+1,j} \overline{\overline{A}}_{j-1}^*)_{ij=-\infty}^{\infty}$ is invertible. We now apply Theorem 2.1 to the operator T_1. Taking into account the equalities

$$\overline{\overline{A}}_n^* = \overline{B}_{n+1} \quad \text{and} \quad \overline{\overline{B}}_n^* = \overline{A}_n \qquad (n = 0, \pm 1, \ldots),$$

it follows that T_1 is invertible if and only if the system

$$\overline{B}_{n+1} x_{n+1} = \overline{A}_n x_n \qquad (n = 0, \pm 1, \ldots) \tag{3.8}$$

admits a dichotomy $(\overline{P}_n)_{n=-\infty}^{\infty}$, satisfying

$$\sup_n \|(\overline{B}_{n+1}|_{Im\, \overline{P}_{n+1}}, \overline{A}_n|_{Ker\, \overline{P}_n})^{-1}\| < \infty. \tag{3.9}$$

This proves the first part of Theorem 1.1. Now assume that L is invertible. Then L^*, T, and T_1 are invertible. Denote $G = (a_{ij})_{ij=0}^{\infty}$, $G^* = (a_{ji}^*)_{ij=0}^{\infty}$, $\overline{\overline{G}} = (\delta_{ij} \overline{\overline{A}}_j + \delta_{i+1,j} \overline{\overline{B}}_j)_{ij=0}^{\infty}$,

$\overline{\overline{G}}_1 = (\delta_{ij}\overline{\overline{B}}_j^* + \delta_{i+1,j}\overline{\overline{A}}_{j-1}^*)_{ij=0}^{\infty}$, and $S = (\delta_{i+1,j}I_{r(m_1+m_2)})_{ij=0}^{\infty}$. Then G, G^*, $\overline{\overline{G}}$, $\overline{\overline{G}}_1$ and S and Fredholm operators. It is clear that

$$\text{index}\, G = -\,\text{index}\, G^*. \tag{3.10}$$

Denote by $\overline{\overline{p}}$ and $\overline{p}$ the respective ranks of the dichotomies $(\overline{\overline{P}}_n)_{n=-\infty}^{\infty}$ and $(\overline{P}_n)_{n=-\infty}^{\infty}$ of (3.6) and (3.8). By Theorem 1.1 applied to L^* we have

$$\text{index}\, G^* = \overline{\overline{p}} - m_2 r, \tag{3.11}$$

while Theorem 2.2 applied to T leads to

$$\text{index}\, \overline{\overline{G}} = \overline{\overline{p}}.$$

Now note that $\overline{\overline{G}}_1 = \overline{\overline{G}}^* S + (\delta_{i,0}\delta_{j,0}\overline{\overline{B}}_0^*)_{ij=0}^{\infty}$. Therefore, taking into account the previous equality, we obtain

$$\text{index}\, \overline{\overline{G}}_1 = -\,\text{index}\, \overline{\overline{G}} + \text{index}\, S = -\overline{\overline{p}} + r(m_1 + m_2). \tag{3.12}$$

However, Theorem 2.2 applied to T_1 implies that

$$\text{index}\, \overline{\overline{G}}_1 = \overline{p}. \tag{3.13}$$

It follows from (3.12) and (3.13) that $\overline{\overline{p}} = r(m_1 + m_2) - \overline{p}$. On the other hand, (3.10) and (3.11) lead to $\text{index}\, G = m_2 r - \overline{\overline{p}}$. Hence $\text{index}\, G = \overline{p} - m_1 r$. This proves Theorem 1.1, in this case too.

We now turn to the proof of Lemma 3.1.

PROOF OF LEMMA 3.1. We will only prove that G_1 is equivalent to G_2, because the proof that L_1 is equivalent to L_2 is entirely similar.

Let $S = (\delta_{i,j+1}I_r)_{ij=0}^{\infty}$ be the shift operator in ℓ_r^2. We define $m+1$ block diagonal operators in ℓ_r^2, by $D_k = (\delta_{ij}a_{j+k,j})_{ij=0}^{\infty}$ $(k = 0, 1, \ldots, m)$. Then,

$$G = S^m D_m + \cdots + SD_1 + D_0.$$

We also consider the following operators which act in $(\ell_r^2)^m$,

$$C = \begin{pmatrix} S & 0 & \cdots & 0 & D_0 \\ -\tilde{I} & S & \cdots & 0 & D_1 \\ \cdot & \cdot & \cdot & \cdot & \cdot \\ 0 & 0 & \cdots & S & D_{m-2} \\ 0 & 0 & \cdots & -\tilde{I} & SD_m + D_{m-1} \end{pmatrix},$$

and

$$E = \begin{pmatrix} \tilde{I} & -S & \cdots & 0 & 0 \\ 0 & \tilde{I} & \cdots & 0 & 0 \\ \cdot & \cdot & \cdot & \cdot & \cdot \\ 0 & 0 & \cdots & \tilde{I} & -S \\ 0 & 0 & \cdots & 0 & \tilde{I} \end{pmatrix}; \qquad F = \begin{pmatrix} F_1 & -\tilde{I} & \cdots & 0 & 0 \\ F_2 & 0 & \cdots & 0 & 0 \\ \cdot & \cdot & \cdot & \cdot & \cdot \\ F_{m-1} & 0 & \cdots & 0 & -\tilde{I} \\ \tilde{I} & 0 & \cdots & 0 & 0 \end{pmatrix}, \tag{3.14}$$

where $F_i = S^{m-i}D_m + S^{m-i-1}D_{m-1} + \cdots + SD_{i+1} + D_i$ $(i = 0, 1, \ldots, m)$. The operators E and F are invertible. In fact, it is clear that

$$E^{-1} = \begin{pmatrix} \tilde{I} & S & \cdots & S^{m-2} & S^{m-1} \\ 0 & \tilde{I} & \cdots & S^{m-3} & S^{m-2} \\ \cdot & \cdot & \cdot & \cdot & \cdot \\ 0 & 0 & \cdots & \tilde{I} & S \\ 0 & 0 & \cdots & 0 & \tilde{I} \end{pmatrix}; \qquad F^{-1} = \begin{pmatrix} 0 & 0 & \cdots & 0 & \tilde{I} \\ -\tilde{I} & 0 & \cdots & 0 & F_1 \\ \cdot & \cdot & \cdot & \cdot & \cdot \\ 0 & 0 & \cdots & 0 & F_{m-2} \\ 0 & 0 & \cdots & -\tilde{I} & F_{m-1} \end{pmatrix}.$$

In addition, we have

$$EG_1 = \begin{pmatrix} G & -S & \cdots & 0 & 0 \\ 0 & \tilde{I} & \cdots & 0 & 0 \\ \cdot & \cdot & \cdot & \cdot & \cdot \\ 0 & 0 & \cdots & \tilde{I} & -S \\ 0 & 0 & \cdots & 0 & \tilde{I} \end{pmatrix},$$

$$CF = \begin{pmatrix} SF_1 + D_0 & -S & \cdots & 0 & 0 \\ -F_1 + SF_2 + D_1 & \tilde{I} & \cdots & 0 & 0 \\ \cdot & \cdot & \cdot & \cdot & \cdot \\ -F_{m-2} + SF_{m-1} + D_{m-2} & 0 & \cdots & \tilde{I} & -S \\ -F_{m-1} + SD_m + D_{m-1} & 0 & \cdots & 0 & \tilde{I} \end{pmatrix}.$$

However, $F_{m-1} = SD_m + D_{m-1}$, $F_i = SF_{i+1} + D_i$ $(i = 0, \ldots, m-1)$, and $SF_1 + D_0 = G$. Hence

$$EG_1 = CF. \tag{3.15}$$

Thus G_1 is equivalent to C. Before we proceed, we make a definition.

DEFINITION 3.2. *Let $U: (\ell_r^2)^m \to \ell_{rm}^2$ be the following unitary transformation. For $x = \left((x_{1,i})_{i=0}^{\infty}, \ldots, (x_{m,i})_{i=0}^{\infty}\right)^T \in (\ell_r^2)^m$, where $x_{j,i} \in \mathbb{C}^r$ $(j = 1, \ldots, m;\ i = 0, 1, \ldots)$*, set $Ux = (z_i)_{i=0}^{\infty}$, where $z_i = (x_{1,i}, \ldots, x_{m,i})^T$ $(i = 0, 1, \ldots)$.

We now show that the following equality holds

$$UC = G_2U. \tag{3.16}$$

Let $x = \left((x_{1,i})_{i=0}^{\infty}, \ldots, (x_{m,i})_{i=0}^{\infty}\right)^T$, where $x_{ji} \in \mathbb{C}^r$ $(j = 1, \ldots, m;\ i = 0, 1, \ldots)$ be an arbitrary vector in $(\ell_r^2)^m$. Denote $x_j = (x_{ji})_{i=0}^{\infty} \in \ell_r^2$ $(j = 1, \ldots, m)$. The definition of C leads to

$$Cx = \begin{pmatrix} Sx_1 + D_0x_m \\ -x_1 + Sx_2 + D_1x_m \\ \vdots \\ -x_{m-2} + Sx_{m-1} + D_{m-2}x_m \\ -x_{m-1} + SD_mx_m + D_{m-1}x_m \end{pmatrix}.$$

Denote

$$Cx = \left((y_{1,i})_{i=0}^{\infty}, \ldots, (y_{m,i})_{i=0}^{\infty}\right)^T. \tag{3.17}$$

Expanding the expression for Cx given above, we obtain that the vectors $y_{ji} \in \mathbb{C}^r$ are given by the following formulas

$$\begin{cases} y_{1,0} = a_{0,0}x_{m,0} \\ y_{2,0} = -x_{1,0} + a_{1,0}x_{m,0} \\ \quad\vdots \\ y_{m-1,0} = -x_{m-2,0} + a_{m-2,0}x_{m,0} \\ y_{m,0} = -x_{m-1,0} + a_{m-1,0}x_{m,0}, \end{cases}$$

and

$$\begin{cases} y_{1,i} = x_{1,i-1} + a_{i,i}x_{m,i} \\ y_{2,i} = -x_{1,i} + x_{2,i-1} + a_{i+1,i}x_{m,i} \\ \quad\vdots \\ y_{m-1,i} = -x_{m-2,i} + x_{m-1,i-1} + a_{i+m-2,i}x_{m,i} \\ y_{m,i} = -x_{m-1,i} + a_{i+m-1,i-1}x_{m,i-1} + a_{i+m-1,i}x_{m,i} \end{cases} \qquad (i = 1, 2, \ldots).$$

It follows from the definition (3.1) of the sequences $(A_n)_{n=-\infty}^{\infty}$ and $(B_n)_{n=-\infty}^{\infty}$, that

$$(y_{1,0}, \ldots, y_{m,0})^T = B_0(x_{1,0}, \ldots, x_{m,0})^T,$$

and

$$(y_{1,i}, \ldots, y_{m,i})^T = A_{i-1}(x_{1,i-1}, \ldots, x_{m,i-1})^T + B_i(x_{1,i}, \ldots, x_{m,i})^T \qquad (i = 1, 2, \ldots).$$

Denote $y_i = (y_{1,i}, \ldots, y_{m,i})^T$, and $z_i = (x_{1,i}, \ldots, x_{m,i})^T$, for $(i = 0, 1, \ldots)$. Then the above formulas lead to

$$y_0 = B_0 z_0,$$

and

$$y_i = A_{i-1}z_{i-1} + B_i z_i \qquad (i = 1, 2, \ldots).$$

Since $G_2 = (\delta_{ij}B_j + \delta_{i-1,j}A_j)_{ij=0}^{\infty}$, then

$$(y_i)_{i=0}^{\infty} = G_2(z_i)_{i=0}^{\infty}. \tag{3.18}$$

However, it follows from (3.17) that $UCx = (y_i)_{i=0}^{\infty}$, while it is clear that $Ux = (z_i)_{i=0}^{\infty}$. Hence (3.18) leads to

$$UCx = G_2Ux.$$

Since x is arbitrary, this implies (3.16). Hence C is equivalent to G_2. It was shown above that G_1 is equivalent to C. Thus, G_1 is equivalent to G_2. □

For future reference, we state here the following corollary of the preceding proof.

COROLLARY (OF PROOF) 3.3. *Let* $G = (a_{ij})_{ij=0}^{\infty}$ *be a bounded,* $(m,0)$ *banded block operator in* ℓ_r^2. *Let* $S = (\delta_{i,j+1}I_r)_{ij=0}^{\infty}$, *and* G_1 *be the operator*

$$G_1 = \begin{pmatrix} G & 0 & \cdots & 0 & 0 \\ 0 & \tilde{I} & \cdots & 0 & 0 \\ \cdot & \cdot & \cdot & \cdot & \cdot \\ 0 & 0 & \cdots & \tilde{I} & 0 \\ 0 & 0 & \cdots & 0 & \tilde{I} \end{pmatrix}. \tag{3.19}$$

Then,

$$G_1 = E^{-1}U^{-1}G_2UF. \tag{3.20}$$

Here $G_2 = (\delta_{ij}B_j + \delta_{i-1,j}A_j)_{ij=0}^{\infty}$, *where* $(A_n)_{n=0}^{\infty}$ *and* $(B_n)_{n=0}^{\infty}$ *are defined by* (3.1), E *and* F *are the invertible operators given by* (3.14), *and* U *is defined in Definition* 3.2.

PROOF. It is enough to note that the preceding proof depends only on a_{ij} for $i,j \geq 0$. Equality (3.20) follows from (3.15) and (3.16). □

Another reduction of band operators to two diagonal operators is the following. Let $L = (a_{ij})_{ij=-\infty}^{\infty}$ be an (m_1, m_2) banded, bounded block operator in $\ell_r^2(\mathbb{Z})$. Define block matrices A_n, and B_{n+1} $(n = 0, \pm 1, \dots)$ via

$$A_n = (a_{ij})_{i=nm, j=nm-m_1}^{(n+1)m-1,(n+1)m-m_1-1}; \qquad B_{n+1} = (a_{ij})_{i=nm, j=(n+1)m-m_1}^{(n+1)m-1;(n+2)m-m_1-1},$$

where $m = m_1 + m_2$. Then L is equivalent to the two diagonal operator $L_1 = (\delta_{ij}A_j + \delta_{i+1,j}B_j)_{ij=-\infty}^{\infty}$, which acts in $\ell_{mr}^2(\mathbb{Z})$. The results of Section 2 can be applied to L_1, thus giving theorems which correspond to those obtained in this section.

4. UNIQUENESS OF THE DICHOTOMY

The uniqueness of the dichotomy, is an immediate consequence of the next result.

PROPOSITION 4.1. *Assume that the system*

$$B_{n+1}P_{n+1} = A_nP_n \tag{4.1}$$

admits a dichotomy $(P_n)_{n=-\infty}^{\infty}$. *Then, for every integer* k,

$$\begin{aligned} \operatorname{Im} P_k = \{x_k \in \mathbb{C}^r \colon \exists x_j \in \mathbb{C}^r \ (j = k+1, k+2, \dots), \text{ such that} \\ B_{n+1}x_{n+1} = A_nx_n \ (n = k, k+1, \dots), \text{ and } \lim_{n \to +\infty} x_n = 0\}, \end{aligned} \tag{4.2}$$

and

$$\text{(4.3)}\qquad \begin{aligned} \operatorname{Ker} P_k = \{x_k \in \mathbb{C}^r \colon \exists x_j \in \mathbb{C}^r \ (j = k-1, k-2, \ldots), \text{ such that} \\ B_{n+1}x_{n+1} = A_n x_n \ (n = k-1, k-2, \ldots), \text{ and } \lim_{n\to-\infty} x_n = 0\}. \end{aligned}$$

PROOF. Since the proofs of (4.3) and (4.2) are entirely symmetric, we prove only (4.2). Assume that $x_k \in \operatorname{Im} P_k$. Define

$$x_{k+j} = (B_{k+j}|_{Im\, P_{k+j}})^{-1} A_{k+j-1} \cdots (B_{k+1}|_{Im\, P_{k+1}})^{-1} A_k x_k \qquad (j = 1, 2, \ldots).$$

Since $x_k \in \operatorname{Im} P_k$, then x_{k+j} $(j = 1, 2, \ldots)$ is well defined. It is clear that $B_{n+1}x_{n+1} = A_n x_n$ $(n = k, k+1, \ldots)$. Finally, inequality (1.5), taking into account that $x_k = P_k x_k$, leads to $\lim_{n\to+\infty} x_n = 0$.

Conversely, assume that the sequence $(x_n)_{n=k}^{\infty}$ satisfies $B_{n+1}x_{n+1} = A_n x_n$ $(n = k, k+1, \ldots)$ and $\lim_{n\to+\infty} x_n = 0$. Denote

$$y_n = P_n x_n \quad \text{and} \quad z_n = (I_r - P_n)x_n \qquad (n = k, k+1, \ldots).$$

Then we have

$$B_{n+1}y_{n+1} + B_{n+1}z_{n+1} = A_n y_n + A_n z_n \qquad (n = k, k+1, \ldots).$$

Hence

$$\text{(4.4)}\qquad B_{n+1}y_{n+1} - A_n y_n = A_n z_n - B_{n+1}z_{n+1} \qquad (n = k, k+1, \ldots).$$

Since $y_n \in \operatorname{Im} P_n$ and $z_n \in \operatorname{Ker} P_n$ $(n = k, k+1, \ldots)$, the relations (1.3) show that the left hand side of equality (4.4) belongs to $\operatorname{Im} B_{n+1}P_{n+1}$, while the right hand side belongs to $\operatorname{Im} A_n(I_r - P_n)$. However, (1.4) and (1.2) imply that $\operatorname{Im}(B_{n+1}P_{n+1}) \cap \operatorname{Im}(A_n(I_r - P_n)) = \{0\}$ for every integer n. Therefore, (4.4) leads to

$$\text{(4.5)}\qquad B_{n+1}y_{n+1} = A_n y_n \quad \text{and} \quad B_{n+1}z_{n+1} = A_n z_n \qquad (n = k, k+1, \ldots).$$

On the other hand, the mappings $B_{n+1}|_{Im\, P_{n+1}}$ and $A_n|_{Ker\, P_n}$ are invertible, $y_{n+1} \in \operatorname{Im} P_{n+1}$, $z_n \in \operatorname{Ker} P_n$, and $A_n y_n$ (respectively $B_{n+1}z_{n+1}$) is in the range of $B_{n+1}|_{Im\, P_{n+1}}$ (respectively $A_n|_{Ker\, P_n}$) by (1.3). Therefore, (4.5) implies that

$$\text{(4.6)}\qquad y_{n+1} = (B_{n+1}|_{Im\, P_{n+1}})^{-1} A_n y_n \qquad (n = k, k+1, \ldots),$$

and,

$$\text{(4.7)}\qquad z_n = (A_n|_{Ker\, P_n})^{-1} B_{n+1} z_{n+1} \qquad (n = k, k+1, \ldots).$$

It follows from (4.6) that

$$y_{k+j} = (B_{k+j}|_{Im\, P_{k+j}})^{-1} A_{k+j-1} \cdots (B_{k+1}|_{Im\, P_{k+1}})^{-1} A_k y_k \qquad (j = 1, 2, \ldots).$$

Since $y_k \in \operatorname{Im} P_k$, the inequalities in (1.5) lead to $\lim_{n\to+\infty} y_n = 0$. However, $z_n = x_n - y_n$ $(n = k, k+1, \ldots)$, and $\lim_{n\to+\infty} x_n = 0$ by assumption. Hence, $\lim_{n\to+\infty} z_n = 0$.

On the other hand, (4.7) leads to

$$z_k = (A_k|_{Ker\, P_k})^{-1} B_{k+1} \cdots (A_{k+j}|_{Ker\, P_{k+j-1}})^{-1} B_{k+j} z_{k+j} \qquad (j = 1, 2, \ldots).$$

Since $z_{k+j} \in \operatorname{Ker} P_{k+j}$, then (1.6) implies that $\|z_k\| \leq M a^j \|z_{k+j}\|$ $(j = 1, 2, \ldots)$. But $\lim_{n\to+\infty} z_n = 0$, and $0 < a < 1$. Therefore, letting $j \to +\infty$ in the last inequality we obtain $z_k = 0$. Since $z_k = (I_r - P_k)x_k$, then $x_k \in \operatorname{Im} P_k$. □

The following result is an immediate consequence of the previous lemma.

COROLLARY 4.2. *The system*

$$B_{n+1} x_{n+1} = A_n x_n \qquad (n = 0, \pm 1, \ldots),$$

admits at most one dichotomy. □

5. REGULAR BAND MATRICES

If $L = (a_{ij})_{ij=-\infty}^{\infty}$ is a regular (m_1, m_2) banded block matrix, then the matrices A_n, B_n, $\overline{A}_n$, and $\overline{B}_n$ defined by (1.8) and (1.11) are invertible. Thus, in this case, we are led to consider systems $B_{n+1}x_{n+1} = A_n x_n$ $(n = 0, \pm 1, \ldots)$ in which A_n and B_n $(n = 0, \pm 1, \ldots)$ are invertible matrices. The following result gives a description of dichotomy of systems in the case where B_n is invertible for every integer n.

PROPOSITION 5.1. *Let*

$$B_{n+1} x_{n+1} = A_n x_n \qquad (n = 0, \pm 1, \ldots), \tag{5.1}$$

be a system, with B_n invertible for every integer n. A bounded sequence of projections $(P_n)_{n=-\infty}^{\infty}$ in $\mathbb{C}^r$, of constant rank, is a dichotomy for the system (5.1) if and only if the following commutation relations hold

$$P_{n+1} C_n = C_n P_n \qquad (n = 0, \pm 1, \ldots) \tag{5.2}$$

where $C_n = B_{n+1}^{-1} A_n$ $(n = 0, \pm 1, \ldots)$, and for some positive numbers $a < 1$ and M,

$$\|C_{n+j-1} \cdots C_n P_n x\| \leq M a^j \|P_n x\|, \tag{5.3}$$

$$\|C_{n+j-1} \cdots C_n (I_r - P_n) x\| \geq \frac{1}{M a^j} \|(I_r - P_n) x\|, \tag{5.4}$$

for each $x \in \mathbb{C}^r$; $n = 0, \pm 1, \ldots$; $j = 1, 2, \ldots$.

PROOF. Assume that $(P_n)_{n=-\infty}^{\infty}$ is a dichotomy for the system 5.1. Let n be an integer. The inclusions (1.3) and the invertibility of B_{n+1} lead to

$$\operatorname{Im}(B_{n+1}^{-1} A_n P_n) \subset \operatorname{Im}(P_{n+1}); \qquad \operatorname{Im}(I_r - P_{n+1}) \subset \operatorname{Im}\big(B_{n+1}^{-1} A_n (I_r - P_n)\big). \tag{5.5}$$

Since $\operatorname{Rank} P_n = \operatorname{Rank} P_{n+1}$, the right hand side inclusion implies that $\operatorname{Im}(I_r - P_{n+1}) = \operatorname{Im}(B_{n+1}^{-1}A_n(I_r - P_n))$. Hence, $C_n(I_r - P_n) = (I_r - P_{n+1})C_n(I_r - P_n)$, which is equivalent to $P_{n+1}C_n = P_{n+1}C_nP_n$. However, the left hand side of (5.5) leads to $C_nP_n = P_{n+1}C_nP_n$. Hence, the equality (5.2) holds. It is clear that (5.3) follows from (1.5). We now prove (5.4). Let x, n and j be as in (5.4). Denote $x_0 = (I_r - P_n)x$, and $x_k = C_{n+k-1}\cdots C_n(I_r - P_n)x$ $(k = 1,\dots,j)$. By (5.2), we have $x_k \in \operatorname{Ker} P_{n+k}$ $(k = 0,\dots,j)$. Moreover, the equality $x_{k+1} = C_{n+k}x_k$ implies $B_{n+k+1}x_{k+1} = A_{n+k}x_k$, and therefore $x_k = (A_{n+k}|_{Ker\, P_{n+k}})^{-1}B_{n+k+1}x_{k+1}$ $(k = 0,\dots,j-1)$. Hence, $x_0 = (A_n|_{Ker\, P_n})^{-1}B_{n+1}\cdots(A_{n+j-1}|_{Ker\, P_{n+j-1}})^{-1}B_{n+j}x_j$. Taking into account that $x_j \in \operatorname{Ker} P_{n+j}$, it follows from (1.6) that $\|x_0\| \le Ma^j\|x_j\|$. Thus, $\|(I_r - P_n)x\| \le Ma^j\|C_{n+j-1}\cdots C_n(I_r - P_n)x\|$. This proves (5.4).

Conversely, assume that $(P_n)_{n=-\infty}^{\infty}$ is a bounded sequence of projections in $\mathbb{C}^r$, of constant rank, and such that (5.2)–(5.4) hold. Let n be an integer. By (5.2),

$$\operatorname{Im}(C_n(I_r - P_n)) \subset \operatorname{Im}(I_r - P_{n+1}). \tag{5.6}$$

On the other hand, it follows from (5.4), with $j = 1$, that $\operatorname{Ker} C_n \cap \operatorname{Im}(I_r - P_n) = \{0\}$. Therefore, $\dim \operatorname{Im}(C_n(I_r - P_n)) = r - p$, where $p = \operatorname{Rank} P_n$. Since $r - p$ is also equal to $\operatorname{Rank}(I_r - P_{n+1})$, we obtain from (5.6) that

$$\operatorname{Im}(C_n(I_n - P_n)) = \operatorname{Im}(I_r - P_{n+1}).$$

Hence

$$\operatorname{Im}(A_n(I_r - P_n)) = \operatorname{Im}(B_{n+1}(I_r - P_{n+1})). \tag{5.7}$$

The equality (1.4) follows from the last equality and the invertibility of B_{n+1}. Equality (5.2) also implies that $\operatorname{Im}(C_nP_n) \subset \operatorname{Im} P_{n+1}$. Thus,

$$\operatorname{Im}(A_nP_n) \subset \operatorname{Im}(B_{n+1}P_{n+1}). \tag{5.8}$$

The equalities in (1.3) are consequences of (5.7) and (5.8).

By (5.3) we obtain

$$\|B_{n+j}^{-1}A_{n+j-1}\cdots B_{n+1}^{-1}A_nP_n\| \le Ma^j \sup_k \|P_k\|.$$

Therefore, (1.5) holds with the number M being replaced by $M \sup_k \|P_k\|$. We now prove that (1.6) holds. Let $y \in \mathbb{C}^r$ be arbitrary, and $j = 1, 2, \dots$. Denote

$$z = (A_{n-j}|_{Ker\, P_{n-j}})^{-1}B_{n-j+1}\cdots(A_{n-1}|_{Ker\, P_{n-1}})^{-1}B_n(I_r - P_n)y. \tag{5.9}$$

Since $z \in \operatorname{Im}(A_{n-j}|_{Ker\, P_{n-j}})^{-1}$, then $z \in \operatorname{Ker} P_{n-j}$. It is also clear that

$$(I_r - P_n)y = C_{n-1}\cdots C_{n-j}z = C_{n-1}\cdots C_{n-j}(I_r - P_{n-j})z.$$

By (5.4) we have $\|(I_r - P_n)y\| \geq \frac{1}{Ma^j}\|(I_r - P_{n-j})z\| = \frac{1}{Ma^j}\|z\|$. Hence $\|z\| \leq Ma^j\|(I_r - P_n)y\| \leq Ma^j\|y\| \sup_k \|I_r - P_k\|$. Recalling (5.9) we have, taking into account the arbitrariness in y,

$$\|(A_{n-j}|_{Ker\, P_{n-j}})^{-1}B_{n-j+1}\cdots(A_{n-1}|_{Ker\, P_{n-1}})^{-1}B_n(I_r - P_n)\| \leq Ma^j \sup_k \|I_r - P_k\|.$$

This proves (1.6) with the bound M replaced by $M \sup_k \|I_r - P_k\|$. □

Since conditions (5.2)–(5.4) involve only C_n $(n = 0, \pm1, \ldots)$, the following corollary is an immediate consequence of the preceding proposition.

COROLLARY 5.2. *The sequence* $(P_n)_{n=-\infty}^{\infty}$ *is a dichotomy for the system* $B_{n+1}x_{n+1} = A_n x_n$ $(n = 0, \pm1, \ldots)$, *where every* B_n $(n = 0, \pm1, \ldots)$ *is invertible, if and only if it is a dichotomy for the system* $x_{n+1} = B_{n+1}^{-1}A_n x_n$ $(n = 0, \pm1, \ldots)$. □

Another consequence of Proposition 5.1 is that, in the case when $B_n = I_r$ $(n = 0, \pm1, \ldots)$, the definition of dichotomy given here coincides with the definition given in [1; Section 2]. It was shown in [1; Section 2] that if we also assume that A_n is invertible for $n = 0, \pm1, \ldots$, then the definition of dichotomy coincides with the classical definition given earlier in [6].

In the case of regular band matrices, Theorem 1.1 leads to the following theorem, which is related to the results in [2].

THEOREM 5.3. *Let* $L = (a_{ij})_{ij=-\infty}^{\infty}$ *be a bounded,* (m_1, m_2) *banded, regular block matrix. Then* L *defines an invertible operator in* $\ell_r^2(\mathbf{Z})$, *if and only if the system*

$$x_{n+1} = C_n x_n \qquad (n = 0, \pm1, \ldots), \tag{5.10}$$

where

$$C_n = \begin{pmatrix} 0 & -I_r & \cdots & 0 \\ 0 & 0 & \cdots & 0 \\ \cdot & \cdot & \cdot & \cdot \\ 0 & 0 & \cdots & -I_r \\ a_{n,n+m_2}^{-1}a_{n,n-m_1} & a_{n,n+m_2}^{-1}a_{n,n-m_1+1} & \cdots & a_{n,n+m_2}^{-1}a_{n,n+m_2-1} \end{pmatrix}$$

$(n = 0, \pm1, \ldots)$, *admits a dichotomy* $(P_n)_{n=-\infty}^{\infty}$ *such that*

$$\sup_n\{\|P_{n+1}(0, \ldots, I_r)^T a_{n,n+m_2}^{-1}\|, \|(I_{r(m_1+m_2)} - P_n)(I_r, 0, \ldots, 0)^T a_{n,n-m_1}^{-1}\|\} < \infty. \tag{5.11}$$

If L *is invertible, then* $G = (a_{ij})_{ij=0}^{\infty}$ *defines a Fredholm operator in* ℓ_r^2, *and*

$$\operatorname{index} G = p - m_1 r, \tag{5.12}$$

where p *is the rank of the dichotomy. Finally, in the case when one of the conditions*

$$\sup_n \|a_{n,n-m_1}^{-1}\| < \infty, \tag{5.13}$$

or

$$\sup_n \|a_{n,n+m_2}^{-1}\| < \infty, \tag{5.14}$$

holds, and the system (5.10) *admits a dichotomy, then inequality* (5.11) *is automatically satisfied.*

PROOF. Let the matrices $\overline{A}_n$ and $\overline{B}_n$ $(n = 0, \pm 1, \ldots)$ be defined by (1.11). Note that $C_n = \overline{B}_{n+1}^{-1}\overline{A}_n$ $(n = 0, \pm 1, \ldots)$. It follows from Corollary 5.2 that $(P_n)_{n=-\infty}^{\infty}$ is a dichotomy for the system

$$\overline{B}_{n+1}x_{n+1} = \overline{A}_n x_n \qquad (n = 0, \pm 1, \ldots), \tag{5.15}$$

if and only if it is a dichotomy for (5.10).

Assume now that $(P_n)_{n=-\infty}^{\infty}$ is a dichotomy for (5.15), and hence for (5.10). The inclusions (1.3), applied to $\overline{A}_n$ and $\overline{B}_{n+1}$ lead to $\operatorname{Im} P_n \subset \operatorname{Im}(\overline{A}_n^{-1}\overline{B}_{n+1}P_{n+1})$, and $\operatorname{Im}(I_{r_1} - P_{n+1}) \subset \operatorname{Im}(\overline{B}_{n+1}^{-1}\overline{A}_n(I_{r_1} - P_n))$ $(n = 0, \pm 1, \ldots)$, where $r_1 = r(m_1 + m_2)$. Since $\operatorname{Rank} P_n = \operatorname{Rank} P_{n+1}$, it follows that $\operatorname{Im} P_n = \operatorname{Im}(\overline{A}_n^{-1}\overline{B}_{n+1}P_{n+1})$ and $\operatorname{Im}(I_{r_1} - P_{n+1}) = \operatorname{Im}(\overline{B}_{n+1}^{-1}\overline{A}_n(I_{r_1} - P_n))$ $(n = 0, \pm 1, \ldots)$. Hence, $(I_{r_1} - P_n)\overline{A}_n^{-1}\overline{B}_{n+1}P_{n+1} = 0$, and $P_{n+1}\overline{B}_{n+1}^{-1}\overline{A}_n(I_{r_1} - P_n) = 0$ $(n = 0, \pm 1, \ldots)$. The last two equalities imply that

$$\begin{pmatrix} P_{n+1}\overline{B}_{n+1}^{-1} \\ (I_{r_1} - P_n)\overline{A}_n^{-1} \end{pmatrix} (\overline{B}_{n+1}|_{Im\, P_{n+1}}, \overline{A}_n|_{Ker\, P_n}) = \begin{pmatrix} I_p & 0 \\ 0 & I_{r_1 - p} \end{pmatrix} \qquad (n = 0, \pm 1, \ldots).$$

Since $\dim(\operatorname{Im} P_{n+1} \oplus \operatorname{Ker} P_n) = r_1$, it follows that

$$(\overline{B}_{n+1}|_{Im\, P_{n+1}}, \overline{A}_n|_{Ker\, P_n})^{-1} = \begin{pmatrix} P_{n+1}\overline{B}_{n+1}^{-1} \\ (I_{r_1} - P_n)\overline{A}_n^{-1} \end{pmatrix} \qquad (n = 0, \pm 1, \ldots). \tag{5.16}$$

It is clear that $\overline{A}_n^{-1}$ and $\overline{B}_{n+1}^{-1}$ $(n = 0, \pm 1, \ldots)$ have the following form,

$$\overline{A}_n^{-1} = \begin{pmatrix} a_{n,n-m_1}^{-1} & 0 & \cdots & 0 & 0 \\ 0 & I_r & \cdots & 0 & 0 \\ \cdot & \cdot & \cdot & \cdot & \cdot \\ 0 & 0 & \cdots & I_r & 0 \\ 0 & 0 & \cdots & 0 & I_r \end{pmatrix} \begin{pmatrix} a_{n,n-m_1+1} & \cdots & a_{n,n+m_2-1} & I_r \\ -I_r & \cdots & 0 & 0 \\ \cdot & \cdot & \cdot & \cdot \\ 0 & \cdots & 0 & 0 \\ 0 & \cdots & -I_r & 0 \end{pmatrix},$$

and

$$\overline{B}_{n+1}^{-1} = \begin{pmatrix} I_r & 0 & \cdots & 0 & 0 \\ 0 & I_r & \cdots & 0 & 0 \\ \cdot & \cdot & \cdot & \cdot & \cdot \\ 0 & 0 & \cdots & I_r & 0 \\ 0 & 0 & \cdots & 0 & a_{n,n+m_2}^{-1} \end{pmatrix}.$$

Since $\sup_{ij} \|a_{ij}\| < \infty$, and $\sup_n \|P_n\| < \infty$, these formulas show that (5.11) is equivalent to the condition

$$\sup_n \{\|(I_{r_1} - P_n)\overline{A}_n^{-1}\|, \ \|P_{n+1}\overline{B}_{n+1}^{-1}\|\} < \infty. \tag{5.17}$$

It is clear by equality (5.16), that (5.17) is equivalent to

$$\sup_n \|(\overline{B}_{n+1}|_{Im\, P_{n+1}}, \overline{A}_n|_{Ker\, P_n})^{-1}\| < \infty. \tag{5.18}$$

Hence, (5.11) is equivalent to (5.18).

Summarizing, we have shown that $(P_n)_{n=-\infty}^{\infty}$ is a dichotomy for (5.15) satisfying (5.18), if and only if it is a dichotomy for (5.10) satisfying (5.11). The first two parts of the theorem follow from this, and from Theorem 1.1, applied to $(\overline{A}_n)_{n=-\infty}^{\infty}$ and $(\overline{B}_n)_{n=-\infty}^{\infty}$.

We now prove the last statement of the theorem. Assume that the system (5.10) admits the dichotomy $(P_n)_{n=-\infty}^{\infty}$, and that one of the conditions (5.13) or (5.14) holds. By Proposition 5.1, with $B_n = I_{r_1}$ and $A_n = C_n$, the matrices P_n and C_n $(n = 0, \pm 1, \ldots)$ satisfy conditions (5.2)–(5.4). Let y be an arbitrary vector in $\mathbb{C}^r$ and n be an integer. By (5.2) we have

$$C_n(I_{r_1} - P_n)C_n^{-1} = I_{r_1} - P_{n+1}. \tag{5.19}$$

In addition, (5.4) with $j = 1$ and $x = C_n^{-1}y$, leads to

$$\|C_n(I_{r_1} - P_n)C_n^{-1}y\| \geq \frac{1}{Ma}\|(I_{r_1} - P_n)C_n^{-1}y\|.$$

This inequality and (5.19) lead to

$$\|(I_{r_1} - P_n)C_n^{-1}y\| \leq Ma\|C_n(I_{r_1} - P_n)C_n^{-1}y\| = Ma\|(I_{r_1} - P_{n+1})y\|.$$

Hence, $\|(I_{r_1} - P_n)C_n^{-1}y\| \leq Ma \sup_n \|I_{r_1} - P_n\|\|y\|$. Since y is arbitrary, this implies

$$\sup_n \|(I_{r_1} - P_n)C_n^{-1}\| < \infty. \tag{5.20}$$

In addition, (5.2) and (5.3) lead to

$$\sup_n \|P_{n+1}C_n\| \leq \sup_n \|C_n P_n\| \leq Ma \sup_n \|P_n\| < \infty. \tag{5.21}$$

If (5.13) holds, then (5.21) and the boundedness of $(P_n)_{n=-\infty}^{\infty}$ imply that

$$\sup_n \{\|P_{n+1}C_n(I_r, 0, \ldots, 0)^T a_{n,n-m_1}^{-1}\|, \|(I_{r_1} - P_n)(I_r, 0, \ldots, 0)^T a_{n,n-m_1}^{-1}\|\} < \infty. \tag{5.22}$$

On the other hand, if (5.14) holds, then (5.20) leads to

$$(5.23)\quad \sup_n\{\|P_{n+1}(0,\ldots,I_r)^T a_{n,n+m_2}^{-1}\|, \|(I_{r_1}-P_n)C_n^{-1}(0,\ldots,0,I_r)^T a_{n,n+m_2}^{-1}\|\} < \infty.$$

Finally, the definition of C_n implies that

$$C_n(I_r,0,\ldots,0)^T a_{n,n-m_1}^{-1} = (0,\ldots,0,I_r)^T a_{n,n+m_2}^{-1}.$$

Therefore, each one of the inequalities (5.22) or (5.23) implies (5.11). □

6. FREDHOLM OPERATORS WITH BAND CHARACTERISTIC

In this section we study the following class of operators. Let $G = (a_{ij})_{ij=0}^{\infty}$, and $L = (b_{ij})_{ij=-\infty}^{\infty}$, be (m_1, m_2) banded block operators. We say that L is an extension of G, if $a_{ij} = b_{ij}$ $(i,j \geq 0)$. A bounded, (m_1,m_2) banded operator $G = (a_{ij})_{ij=0}^{\infty}$, is called a Fredholm operator with band characteristic (m_1,m_2), if there exists a bounded and invertible, (m_1,m_2) banded operator $L = (a_{ij})_{ij=-\infty}^{\infty}$, which is an extension of G.

The following proposition gives another characterization of Fredholm operators with band characteristic. We denote by M_n $(n = 0,1,\ldots)$, the subspace of ℓ_r^2 consisting of all the vectors $(x_i)_{i=0}^{\infty} \in \ell_r^2$ $(x_i \in \mathbb{C}^r;\ i = 0,1,\ldots)$ such that $0 = x_n = x_{n+1} = \cdots$. Clearly, $M_0^{\perp} = \ell_r^2$, and $(M_n)^{\perp} = \{(x_i)_{i=0}^{\infty} \in \ell_r^2 : x_0 = x_1 = \cdots = x_{n-1} = 0\}$ $(n = 1,2,\ldots)$. Here and in the sequel, for a subspace S of a Hilbert space H, we denote by $S^{\perp}$ the orthogonal complement of S in H.

PROPOSITION 6.1. *Let $G = (a_{ij})_{ij=0}^{\infty}$ be a bounded, (m_1,m_2) banded, block operator in ℓ_r^2. Then G is a Fredholm operator with (m_1,m_2) band characteristic, if and only if the following conditions hold:*

$$(6.1)\qquad G \text{ is Fredholm,}$$

and

$$(6.2)\qquad \operatorname{Ker} G \cap (M_{m_2})^{\perp} = \{0\}, \quad \text{and} \quad (\operatorname{Im} G)^{\perp} \cap (M_{m_1})^{\perp} = \{0\}.$$

PROOF. Assume first that G is Fredholm with (m_1,m_2) band characteristic. Let $L = (a_{ij})_{ij=-\infty}^{\infty}$ be a bounded and invertible, (m_1,m_2) banded extension of G, Since G is a direct summand of a finite dimensional perturbation of L, then G is Fredholm. Let $x = (x_i)_{i=0}^{\infty} \in \operatorname{Ker} G \cap (M_{m_2})^{\perp}$. Since $x \in (M_{m_2})^{\perp}$, then $x_0 = x_1 = \cdots = x_{m_2-1} = 0$. Hence, the vector $y = (\cdots,0,0,x_0,x_1,\ldots) \in \ell_r^2(\mathbb{Z})$ (here x_0 has index 0), is contained in $\operatorname{Ker} L$. Thus, $y = 0$, and therefore $x = 0$. This proves the left hand side equality in (6.2). Consider now $G^* = (a_{ji})_{ij=0}^{\infty}$ and $L^* = (a_{ji}^*)_{ij=-\infty}^{\infty}$. These operators are (m_2,m_1) banded, and L^* is invertible. Hence G^* is a Fredholm operator with (m_2,m_1)

band characteristic. By the first part of the proof, $\operatorname{Ker} G^* \cap (M_{m_1})^\perp = \{0\}$. Since $\operatorname{Ker} G^* = (\operatorname{Im} G)^\perp$, this proves the right hand side equality of (6.2).

Conversely, assume that G is a bounded, (m_1, m_2) banded, block operator, and that (6.1) and (6.2) hold. Denote $d_1 = \dim(\operatorname{Im} G)^\perp$, and $d_2 = \dim \operatorname{Ker} G$. By the right hand side equality in (6.2), $(\operatorname{Im} G + M_{m_1})^\perp = (\operatorname{Im} G)^\perp \cap (M_{m_1})^\perp = \{0\}$. However, $\operatorname{Im} G + M_{m_1}$ is closed, because $\operatorname{Im} G$ is closed by (6.1), and M_{m_1} is finite dimensional. Hence, $\operatorname{Im} G + M_{m_1} = \ell_r^2$.

Let $y_1, \ldots, y_{d_1}$ be d_1 linearly independent vectors in $(\operatorname{Im} G)^\perp$. There exist vectors $z_i \in \operatorname{Im} G$, and $w_i \in M_{m_1}$ $(i = 1, \ldots, d_1)$, such that $y_i = z_i + w_i$. Assume that $\sum_{i=1}^{d_1} \alpha_i w_i \in \operatorname{Im} G$, where $\alpha_1, \ldots, \alpha_{d_1}$ are complex numbers. Since $z_i \in \operatorname{Im} G$ $(i = 1, \ldots, d_1)$, then $\sum_{i=1}^{d_1} \alpha_i y_i = \sum_{i=1}^{d_1} \alpha_i z_i + \sum_{i=1}^{d_1} \alpha_i w_i \in \operatorname{Im} G$. But $y_i \in (\operatorname{Im} G)^\perp$ $(i = 1, \ldots, d_1)$, and therefore $\sum_{i=1}^{d_1} \alpha_1 y_i = 0$. Since $y_1, \ldots, y_{d_1}$ are linearly independent, then $\alpha_1 = \alpha_2 = \cdots = \alpha_{d_1} = 0$. This argument shows, in particular, that $w_1, \ldots, w_{d_1}$ are linearly independent; and that $\operatorname{Im} G \cap \operatorname{sp}(w_1, \ldots, w_{d_1}) = \{0\}$. Denote $N_1 = \operatorname{sp}(w_1, \ldots, w_{d_1})$. The above construction leads to

$$\text{(6.3)} \qquad \dim N_1 = d_1; \quad N_1 \subset M_{m_1}; \quad \operatorname{Im} G \cap N_1 = \{0\}.$$

Taking into account that $\operatorname{Ker} G$ is closed, it follows from the left hand side equality in (6.2) that $\left((\operatorname{Ker} G)^\perp + M_{m_2}\right)^\perp = \operatorname{Ker} G \cap (M_{m_2})^\perp = \{0\}$. However, $(\operatorname{Ker} G)^\perp$ is closed and M_{m_2} is finite dimensional, hence $(\operatorname{Ker} G)^\perp + M_{m_2}$ is closed. Therefore,

$$\text{(6.4)} \qquad (\operatorname{Ker} G)^\perp + M_{m_2} = \ell_r^2.$$

Let $x_1, \ldots, x_{d_2}$ be d_2 linearly independent vectors in $\operatorname{Ker} G$. By (6.4) there exist vectors $s_i \in M_{m_2}$, and $t_i \in (\operatorname{Ker} G)^\perp$, such that $x_i = t_i + s_i$ $(i = 1, \ldots, d_2)$. Assume that for some complex numbers $\alpha_1, \ldots, \alpha_{d_2}$, the following equality holds $\langle \sum_{i=1}^{d_2} \alpha_i x_i, \sum_{i=1}^{d_2} \alpha_i s_i \rangle = 0$. Since $x_i \in \operatorname{Ker} G$ and $t_j \in (\operatorname{Ker} G)^\perp$, then $\langle x_i, t_j \rangle = 0$ $(i, j = 1, \ldots, d_2)$. Therefore

$$\left\langle \sum_{i=1}^{d_2} \alpha_i x_i, \sum_{i=1}^{d_2} \alpha_i x_i \right\rangle = \left\langle \sum_{i=1}^{d_2} \alpha_i x_i, \sum_{i=1}^{d_2} \alpha_i t_i \right\rangle + \left\langle \sum_{i=1}^{d_2} \alpha_i x_i, \sum_{i=1}^{d_2} \alpha_i s_i \right\rangle = 0.$$

Hence, $\sum_{i=1}^{d_2} \alpha_i x_i = 0$. But $x_1, \ldots, x_{d_2}$ are linearly independent, and therefore $\alpha_1 = \cdots = \alpha_{d_2} = 0$. First, it follows from this argument that $s_1, \ldots, s_{d_2}$ are linearly independent. In addition, if $x \in \operatorname{Ker} G \cap \left(\operatorname{sp}(s_1, \ldots, s_{d_2})\right)^\perp$, then $x = \sum_{i=1}^{d_2} \alpha_i x_i$ for some $\alpha_i \in \mathbb{C}$ $(i = 1, \ldots, d_2)$. Hence $\langle \sum_{i=1}^{d_2} \alpha_i x_i, \sum_{i=1}^{d_2} \alpha_i s_i \rangle = \langle x, \sum_{i=1}^{d_2} \alpha_i s_i \rangle = 0$, which forces $x = 0$. Hence $\operatorname{Ker} G \cap \operatorname{sp}(s_2, \ldots, s_{d_2})^\perp = \{0\}$. Denote $N_2 = \operatorname{sp}(s_1, \ldots, s_{d_i})$, then we have:

$$\text{(6.5)} \qquad \dim N_2 = d_2; \quad N_2 \subset M_{m_2}; \quad \operatorname{Ker} G \cap (N_2)^\perp = \{0\}.$$

Using a normalization process, we can find two linearly independent sets of vectors $\{v_i\}_{i=1}^{d_1}$, and $\{u_i\}_{i=1}^{d_2}$, such that

$$\text{(6.6)} \qquad \operatorname{span}(v_i)_{i=1}^{d_1} = N_1, \quad \text{and} \quad \operatorname{span}(u_i)_{i=1}^{d_2} = N_2,$$

and the following conditions are satisfied

$$v_i \subset M_{m_1-[(i-1)/r]} \qquad (i = 1, \ldots, d_1), \tag{6.7}$$

and

$$u_i \subset M_{m_2-[(i-1)/r]} \qquad (i = 1, \ldots, d_2). \tag{6.8}$$

Here $[(i-1)/r]$ denotes the greatest integer less or equal to $(i-1)/r$, and r is the size of the blocks.

We now leave the block structure of G, and consider the following operator in ℓ^2

$$G_1 = \begin{pmatrix} & & u_{d_2}^* \\ & 0 & \vdots \\ & & u_1^* \\ v_{d_1} & \cdots v_1 & G \end{pmatrix},$$

where 0 denotes the null matrix of size $d_2 \times d_1$. It is clear that G_1 is a Fredholm operator, and that,

$$\operatorname{index} G_1 = \operatorname{index} G + d_1 - d_2 = 0. \tag{6.9}$$

We now prove that

$$\operatorname{Ker} G_1 = \{0\}. \tag{6.10}$$

Let $x = (x_i)_{i=1}^{\infty} \in \operatorname{Ker} G_1$, where $x_i \in \mathbb{C}$ $(i = 1, 2, \ldots)$. Denote $y = (x_{i+d_1})_{i=1}^{\infty}$. Then $x \in \operatorname{Ker} G_1$ implies that

$$\langle y, u_i \rangle = 0 \qquad (i = 1, \ldots, d_2), \tag{6.11}$$

and

$$\sum_{i=1}^{d_1} x_i v_{d_1-i+1} + Gy = 0. \tag{6.12}$$

Here $\langle \cdot, \cdot \rangle$ denotes the inner product in ℓ^2. By (6.12), $\sum_{i=1}^{d_1} x_i v_{d_1-i+1} \in \operatorname{Im} G$. Then, (6.3) and (6.6) lead to $\sum_{i=1}^{d_1} x_i v_{d_1-i+1} = 0$. Since the set $(v_i)_{i=1}^{d_1}$ is linearly independent, then

$$x_1 = \cdots = x_{d_1} = 0. \tag{6.13}$$

Equality (6.12) implies now that $y \in \operatorname{Ker} G$. By (6.11) we have $y \in \left(\operatorname{span}(u_i)_{i=1}^{d_2}\right)^{\perp}$. Hence, (6.5) and (6.6) imply $y = 0$. Thus, $(x_{i+d_1})_{i=1}^{\infty} = 0$. This equality and (6.13) leads to $x = 0$. This proves (6.10).

It follows from (6.9) and (6.10) that G_1 is invertible in ℓ_r^2. Denote $I = (\delta_{ij})_{ij=-\infty}^{-1}$. Then the following block operator

$$L_1 = \begin{pmatrix} I & 0 \\ 0 & G_1 \end{pmatrix}$$

is invertible in $\ell^2(\mathbb{Z})$. Let $S = (\delta_{i,j+1})_{ij=-\infty}^{\infty}$. Then $L = S^{*d_2} L_1 S^{d_1}$ is an invertible extension of G. By (6.3) and (6.5), $d_1 \leq m_1 r$, and $d_2 \leq m_2 r$. these inequalities, and (6.7)–(6.8), imply that the $r \times r$ block subdivision of L is (m_1, m_2) banded. Hence G is a Fredholm operator with (m_1, m_2) band characteristic. □

It follows from the above proposition, that the class of Fredholm operators with band characteristic contains all regular band operators which are Fredholm. On the other hand, it is well known (see [4]), that every Fredholm Toeplitz operator also belongs to that class.

The following result is concerned with the behaviour of Fredholm operators with $(m, 0)$ band characteristic under linearization.

LEMMA 6.2. *Let $G = (a_{ij})_{ij=0}^{\infty}$ be a bounded $(m, 0)$ banded operator in ℓ_r^2. Let A_n and B_n $(n = 0, 1, \ldots)$ be given by (3.1). Then G is a Fredholm operator in ℓ_r^2 with $(m, 0)$ band characteristic, if and only if $G_2 = (\delta_{ij} B_j + \delta_{i-1,j} A_j)_{ij=0}^{\infty}$ is a Fredholm operator with $(1, 0)$ band characteristic in ℓ_{rm}^2.*

PROOF. By Proposition 6.1, G is a Fredholm operator with $(m, 0)$ band characteristic if and only if

(6.14) $$G \text{ is Fredholm,}$$

(6.15) $$\operatorname{Ker} G = \{0\},$$

and

(6.16) $$(\operatorname{Im} G)^{\perp} \cap (M_m)^{\perp} = \{0\}.$$

On the other hand, the same proposition shows that G_2 is a Fredholm operator with $(1, 0)$ band characteristic in ℓ_{rm}^2, if and only if the following conditions hold.

(6.17) $$G_2 \text{ is Fredholm,}$$

(6.18) $$\operatorname{Ker} G_2 = \{0\},$$

and

(6.19) $$(\operatorname{Im} G_2)^{\perp} \cap \left(M_1(\ell_{rm}^2)\right)^{\perp} = \{0\}.$$

Here $M_1(\ell^2_{rm})$ denotes the subspace of ℓ^2_{rm} consisting of all the vectors $(x_i)_{i=0}^\infty \in \ell^2_{rm}$ ($x_i \in \mathbb{C}^{rm}$; $i = 0, 1, \ldots$), such that $0 = x_1 = x_2 = \cdots$. We will prove the lemma by showing that (6.14), (6.15), and (6.16) are respectively equivalent to (6.17), (6.18), and (6.19).

We will use Corollary 3.3 and its notation. By (3.20) we have

$$G_1 = E^{-1}U^{-1}G_2UF, \tag{6.20}$$

where G_1, E, F, and U are given by (3.19), (3.14) and Definition (3.2). The operators E and F are invertible, and $U: (\ell^2_r)^m \to \ell^2_{rm}$ is unitary.

It is clear from (6.20) that G_1 is Fredholm if and only if G_2 is Fredholm. Since G is Fredholm if and only if G_1 is Fredholm, this shows that (6.14) and (6.17) are equivalent. In addition, (6.20) again shows that $\operatorname{Ker} G_2 = \{0\}$ if and only if $\operatorname{Ker} G_1 = \{0\}$. However $\operatorname{Ker} G_1 = \{0\}$ is equivalent to $\operatorname{Ker} G = \{0\}$. Hence (6.15) and (6.18) are equivalent.

We now show that (6.16) and (6.19) are equivalent. Define an isometry $J: \ell^2_r \to (\ell^2_r)^m$ by $Jx = (x, \overbrace{0, \ldots, 0}^{m-1})^T$ ($x \in \ell^2_r$). It is clear that

$$(\operatorname{Im} G_1)^\perp = J\big((\operatorname{Im} G)^\perp\big). \tag{6.21}$$

On the other hand, (6.20) and the invertibility of E, F, and U, imply that $\operatorname{Im} G_2 = UE(\operatorname{Im} G_1)$. Therefore $y \in (\operatorname{Im} G_2)^\perp$ if and only if $E^*U^*y \in (\operatorname{Im} G_1)^\perp$. Hence,

$$(\operatorname{Im} G_2)^\perp = UE^{*-1}(\operatorname{Im} G_1)^\perp.$$

By (6.21), this leads to

$$(\operatorname{Im} G_2)^\perp = UE^{*-1}J(\operatorname{Im} G)^\perp. \tag{6.22}$$

We now prove the following statement.

(6.23) A vector $x \in \ell^2_r$ satisfies $UE^{*-1}Jx \in \big(M_1(\ell^2_{rm})\big)^\perp$ if and only if $x \in (M_m)^\perp$.

Let $x \in \ell^2_r$. Note that,

$$E^{*-1} = \begin{pmatrix} \tilde{I} & 0 & \cdots & 0 \\ S^* & \tilde{I} & \cdots & 0 \\ \cdot & \cdot & \cdot & \cdot \\ S^{*(m-1)} & S^{*(m-2)} & \cdots & \tilde{I} \end{pmatrix}.$$

Hence,

$$E^{*-1}Jx = (x, S^*x, \ldots, S^{*(m-1)}x)^T.$$

Denote $x = (x_i)_{i=0}^{\infty}$ ($x_i \in \mathbb{C}^r$; $i = 0, 1, \ldots$). Then the preceding equality leads to

$$E^{*-1}Jx = \left((x_i)_{i=0}^{\infty}, (x_{i+1})_{i=0}^{\infty}, \ldots, (x_{i+m-1})_{i=0}^{\infty}\right)^T.$$

Therefore, by the definition of $U: (\ell_r^2)^m \to \ell_{rm}^2$, we have

$$UE^{*-1}Jx = \left((x_i, x_{i+1}, \ldots, x_{i+m-1})^T\right)_{i=0}^{\infty}.$$

Hence $UE^{*-1}Jx \in \left(M_1(\ell_{rm}^2)\right)^{\perp}$ if and only if $(x_0, x_1, \ldots, x_{m-1})^T = 0$. But this is equivalent to $x \in (M_m)^{\perp}$. Thus, (6.23) holds.

Assume now that (6.16) holds, and let $y \in (\operatorname{Im} G_2)^{\perp} \cap \left(M_1(\ell_{rm}^2)\right)^{\perp}$. By (6.22) there exists a vector $x \in (\operatorname{Im} G)^{\perp}$ such that $y = UE^{*-1}Jx$. Since $y \in \left(M_1(\ell_{rm}^2)\right)^{\perp}$, (6.23) shows that $x \in (M_m)^{\perp}$. Thus, $x \in (\operatorname{Im} G)^{\perp} \cap (M_m)^{\perp}$, and by (6.16), $x = 0$. Therefore $y = 0$. This proves that (6.19) follows from (6.16). Conversely assume (6.19), and let $x \in (\operatorname{Im} G)^{\perp} \cap (M_m)^{\perp}$. By (6.22) and (6.23) $UE^{*-1}Jx \in (\operatorname{Im} G_2)^{\perp} \cap \left(M_1(\ell_{rm}^2)\right)^{\perp}$. Hence (6.19) leads to $UE^{*-1}Jx = 0$. But $\operatorname{Ker} UE^{*-1}J = \{0\}$. Hence $x = 0$. Thus, (6.19) implies (6.16). □

In order to treat Fredholm operators with band characteristic we need to define dichotomy for one sided systems. A system

$$B_{n+1}x_{n+1} = A_n x_n \qquad (n = 0, 1, \ldots), \tag{6.24}$$

where A_n and B_{n+1} ($n = 0, 1, \ldots$) are $r \times r$ complex matrices, and $(x_n)_{n=0}^{\infty}$ is a sequence of vectors in $\mathbb{C}^r$, is called a one sided system. The difference between these systems and the one considered above is that, here, n is restricted to the nonnegative integers.

The definition of dichotomy for one sided systems is analogous to the case considered up to now. We consider bounded sequences $(P_n)_{n=0}^{\infty}$ of projections in $\mathbb{C}^r$, which satisfy the following conditions:

$$\operatorname{Rank} P_n \quad (n = 0, 1, \ldots) \quad \text{is constant,} \tag{6.25}$$

$$\operatorname{Im}(A_n P_n) \subset \operatorname{Im}(B_{n+1}P_{n+1}) \quad \text{and} \quad \operatorname{Im}\big(B_{n+1}(I_r - P_{n+1})\big) \subset \operatorname{Im}\big(A_n(I_r - P_n)\big), \tag{6.26}$$

for $n = 0, 1, \ldots$, and

$$\operatorname{Im}(B_{n+1}P_{n+1}) + \operatorname{Im}\big(A_n(I_r - P_n)\big) = \mathbb{C}^r \qquad (n = 0, 1, \ldots). \tag{6.27}$$

Since $\operatorname{Rank} P_n$ ($n = 0, 1, \ldots$) is constant, (6.27) implies that the restricted mappings $A_n|_{Ker\, P_n}: \operatorname{Ker} P_n \to \operatorname{Im}\big(A_n(I_r - P_n)\big)$, and $B_{n+1}|_{Im\, P_{n+1}}: \operatorname{Im} P_{n+1} \to \operatorname{Im}(B_{n+1}P_{n+1})$ are invertible for $n = 0, 1, \ldots$. We now define dichotomy for one sided systems. A bounded sequence of projections $(P_n)_{n=0}^{\infty}$ with properties (6.25)–(6.27) above, is called a dichotomy for the system (6.24), if there exist positive numbers $a < 1$ and M, such that the following inequalities hold

$$(6.28) \qquad \|(B_{n+j}|_{Im\, P_{n+j}})^{-1}A_{n+j-1}\cdots(B_{n+1}|_{Im\, P_{n+1}})^{-1}A_nP_n\| \le Ma^j$$

for $n = 0, 1, \ldots;\ j = 1, 2, \ldots$, and

$$(6.29) \qquad \|(A_{n-j}|_{Ker\, P_{n-j}})^{-1}B_{n-j+1}\cdots(A_{n-1}|_{Ker\, P_{n-1}})^{-1}B_n(I_r - P_n) \le Ma^j$$

for $n = 0, 1, \ldots;\ j = 1, \ldots, n$. Note that in (6.29) j takes only the values $1, \ldots, n$. We call the integer $p = \operatorname{Rank} P_n$ $(n = 0, 1, \ldots)$ the rank of the dichotomy. The next result connects the dichotomy for one sided systems, to the corresponding notion for two sided systems.

PROPOSITION 6.3. *Let*

$$(6.30) \qquad B_{n+1}x_{n+1} = A_nx_n \qquad (n = 0, 1, \ldots),$$

be a one sided system, where $(A_n)_{n=0}^{\infty}$ and $(B_n)_{n=1}^{\infty}$ are sequences of $r \times r$ matrices. A sequence of projections $(P_n)_{n=0}^{\infty}$ is a dichotomy for the system (6.30), if and only if there exists $r \times r$ complex matrices $A_{-1}, A_{-2}, \ldots$, $B_0, B_{-1}, \ldots$, and projections $P_{-1}, P_{-2}, \ldots$ in $\mathbb{C}^r$, such that $(P_n)_{n=-\infty}^{\infty}$ is a dichotomy for the system

$$(6.31) \qquad B_{n+1}x_{n+1} = A_nx_n \qquad (n = 0, \pm 1, \ldots).$$

Moreover, if the dichotomy $(P_n)_{n=0}^{\infty}$ of (6.30) satisfies

$$(6.32) \qquad \sup_{n=0,1,\ldots} \|(B_{n+1}|_{Im\, P_{n+1}}, A_n|_{Ker\, P_n})^{-1}\| < \infty,$$

then the sequences $(A_n)_{n=-\infty}^{-1}$, $(B_n)_{n=-\infty}^{0}$, and $(P_n)_{n=-\infty}^{-1}$ can be chosen so that

$$(6.33) \qquad \sup_{n=0,\pm 1,\ldots} \|(B_{n+1}|_{Im\, P_{n+1}}, A_n|_{Ker\, P_n})^{-1}\| < \infty.$$

PROOF. Assume that $(P_n)_{n=0}^{\infty}$ is a dichotomy for the one sided system (6.30). Define $P_n = P_0$, $A_n = 2(I_r - P_0)$ $(n = -1, -2, \ldots)$, and $B_n = I_r$ $(n = 0, -1, \ldots)$. It is clear that $\operatorname{Rank} P_n = p$ $(n = 0, \pm 1, \ldots)$, where p is the rank of the dichotomy $(P_n)_{n=0}^{\infty}$. Moreover $(P_n)_{n=-\infty}^{\infty}$ is bounded because $(P_n)_{n=0}^{\infty}$ is bounded. In addition, (1.3) and (1.4) hold for $n = 0, 1, \ldots$ by (6.26) and (6.27), and for $n = -1, -2, \ldots$, by the above definition. Condition (1.5) holds for $n = 0, 1, \ldots;\ j = 1, 2, \ldots$ by (6.28). If $n = -1, -2, \ldots$ then $A_nP_n = 0$. Therefore (1.5) holds also in the case when $n = -1, -2, \ldots;$ $j = 1, 2, \ldots$. This proves (1.5). We now show that

$$(6.34) \qquad \|(A_{n-j}|_{Ker\, P_{n-j}})^{-1}B_{n-j+1}\cdots(A_{n-1}|_{Ker\, P_{n-1}})^{-1}B_n(I_r - P_n)\| \le M_1a_1^j$$

for $n = 0, \pm 1, \ldots;\ j = 1, 2, \ldots$, where $a_1 = \max\left(\frac{1}{2}, a\right)$ and $M_1 = \max(\|I_r - P_0\|, M)$. If $j \le n$, then (6.34) follows from (6.29). Assume $n < j$. If $n \le 0$, then the term on the left hand side of (6.34) is equal to $\frac{1}{2^j}\|I_r - P_0\|$ which is less or equal than $M_1 a_1^j$. Finally, if $n > 0$, then

$$\begin{aligned}
&\|(A_{n-j}|_{Ker\,P_{n-j}})^{-1}B_{n-j+1}\cdots(A_{n-1}|_{Ker\,P_{n-1}})^{-1}B_n(I_r - P_n)\| \\
&= \frac{1}{2^{j-n}}\|(A_0|_{Ker\,P_0})^{-1}B_1\cdots(A_{n-1}|_{Ker\,P_{n-1}})^{-1}B_n(I_r - P_n)\| \\
&\leq \frac{1}{2^{j-n}}a^n M \leq M_1 a_1^j,
\end{aligned}$$

where we made use of (6.29) with $j = n$. Thus (6.34) holds for $n = 0, \pm 1, \ldots; j = 1, 2, \ldots$. This proves that $(P_n)_{n=-\infty}^{\infty}$ is a dichotomy for (6.31). Since $(B_{n+1}|_{Im\,P_{n+1}}, A_n|_{Ker\,P_n})$ is constant for $n = -1, -2, \ldots$, then (6.32) implies (6.33). This proves the proposition in one direction. The proof in the other direction is self-evident. □

REMARK 6.4. Note that, in the case when

$$\sup_{n=0,1,\ldots}\{\|A_n\|, \|B_n\|\} < \infty, \tag{6.35}$$

then, the above choice for $A_{-1}, A_{-2}, \ldots, B_0, B_{-1}, \ldots$ leads to

$$\sup_{n=0,\pm 1,\ldots}\{\|A_n\|, \|B_n\|\} < \infty. \tag{6.36}$$

COROLLARY 6.5. *If the one sided system*

$$B_{n+1}x_n = A_n x_n \qquad (n = 0, 1, \ldots), \tag{6.37}$$

admits the dichotomy $(P_n)_{n=0}^{\infty}$, *then*

$$\begin{aligned}
\operatorname{Im} P_k = \{x_k \in \mathbb{C}^r : \exists x_j \in \mathbb{C}^r\ (j = k+1, k+2, \ldots),\ \text{such that} \\
B_{n+1}x_{n+1} = A_n x_n\ (n = k, k+1, \ldots),\ \text{and}\ \lim_{n\to+\infty} x_n = 0\}.
\end{aligned} \tag{6.38}$$

All the dichotomies of the one sided system (6.37) *have the same rank.*

PROOF. The first statement follows from the previous proposition, and Proposition 4.1. The second statement is a consequence of the first. □

A definition of dichotomy for one sided systems of the form $x_{n+1} = A_n x_n$ $(n = 0, 1, \ldots)$ was given in [1; Section 6], and, earlier in [6], in the case when each A_n is invertible. It is shown in Section 5 above, that the definition of dichotomy of two sided systems given in the introduction, coincide with the corresponding definition in [1; Section 2]. Hence, Proposition 6.3 above and Proposition 6.1 of [1] show that, the definitions of dichotomy for one sided systems too, are equivalent. In particular, it follows from the example in [1; Section 6], that dichotomics for one sided systems are not necessarily unique.

For upper triangular band matrices, the connection between Fredholm operators with band characteristic, and dichotomy, is given by the following result.

THEOREM 6.6. *Let $G = (a_{ij})_{ij=0}^{\infty}$ be a bounded, $(0,m)$ banded, block operator in ℓ_r^2. Then G is a Fredholm operator with $(0,m)$ band characteristic if and only if the one sided system*

$$\overline{B}_{n+1}x_{n+1} = \overline{A}_n x_n \qquad (n = 0,1,\ldots), \tag{6.39}$$

where $\overline{A}_0, \overline{A}_1, \ldots$ and $\overline{B}_1, \overline{B}_2, \ldots$ are block matrices of order m given by

$$\begin{aligned} \overline{A}_n &= \begin{pmatrix} 0 & -I_r & \cdots & 0 \\ 0 & 0 & \cdots & 0 \\ \cdot & \cdot & \cdot & \cdot \\ 0 & 0 & \cdots & -I_r \\ a_{n,n} & a_{n,n+1} & \cdots & a_{n,n+m-1} \end{pmatrix}; \\ \overline{B}_{n+1} &= \begin{pmatrix} I_r & 0 & \cdots & 0 & 0 \\ 0 & I_r & \cdots & 0 & 0 \\ \cdot & \cdot & \cdot & \cdot & \cdot \\ 0 & 0 & \cdots & I_r & 0 \\ 0 & 0 & \cdots & 0 & a_{n,n+m} \end{pmatrix} \qquad (n = 0,1,\ldots), \end{aligned} \tag{6.40}$$

admits a dichotomy $(P_n)_{n=0}^{\infty}$ such that

$$\sup_{n=0,1,\ldots} \|(\overline{B}_{n+1}|_{Im\,P_{n+1}}, \overline{A}_n|_{Ker\,P_n})^{-1}\| < \infty. \tag{6.41}$$

Moreover, if G is Fredholm with $(0,m)$ band characteristic, then

$$\operatorname{index} G = p, \tag{6.42}$$

where p is the rank of the dichotomy.

PROOF. It follows directly from the definition of Fredholm operators with band characteristic, that G is Fredholm with $(0,m)$ band characteristic, if and only if G^* is Fredholm with $(m,0)$ band characteristic. We apply Lemma 6.2 to G^*. Then, G^* is Fredholm with $(m,0)$ band characteristic, if and only if $G_2 = (\delta_{ij}B_j + \delta_{i-1,j}A_j)_{ij=0}^{\infty}$ is a Fredholm operator with $(1,0)$ band characteristic in ℓ_{rm}^2. Here, $(A_n)_{n=0}^{\infty}$ and $(B_n)_{n=0}^{\infty}$, are given by

$$\begin{aligned} A_n &= \begin{pmatrix} I_r & 0 & \cdots & 0 & 0 \\ 0 & I_r & \cdots & 0 & 0 \\ \cdot & \cdot & \cdot & \cdot & \cdot \\ 0 & 0 & \cdots & I_r & 0 \\ 0 & 0 & \cdots & 0 & a_{n,n+m}^* \end{pmatrix}; \\ B_n &= \begin{pmatrix} 0 & \cdots & 0 & a_{n,n}^* \\ -I_r & \cdots & 0 & a_{n,n+1}^* \\ \cdot & \cdot & \cdot & \cdot \\ 0 & \cdots & 0 & a_{n,n+m-2}^* \\ 0 & \cdots & -I_r & a_{n,n+m-1}^* \end{pmatrix} \qquad (n = 0,1,\ldots). \end{aligned} \tag{6.43}$$

By definition, G_2 is a Fredholm operator with $(1,0)$ band characteristic if and only if, there exist two bounded sequences of $m \times m$ block matrices, $(A_n)_{n=-\infty}^{-1}$, and $(B_n)_{n=-\infty}^{-1}$, such that the operator $L_2 = (\delta_{ij}B_j + \delta_{i-1,j}A_j)_{ij=-\infty}^{\infty}$ is invertible. But L_2 is invertible if and only if $L_2^* = (\delta_{ij}B_j^* + \delta_{i+1,j}A_i^*)_{ij=-\infty}^{\infty}$ is invertible. We now apply Theorem 2.1 to L_2^*. Then, L_2^* is invertible if and only if the system

$$A_n^* x_{n+1} = B_n^* x_n \qquad (n = 0, \pm 1, \ldots), \tag{6.44}$$

admits a dichotomy $(P_n)_{n=-\infty}^{\infty}$ such that

$$\sup_n \|(A_n^*|_{Im\, P_{n+1}}, B_n^*|_{Ker\, P_n})^{-1}\| < \infty. \tag{6.45}$$

It follows from Proposition 6.3, and Remark 6.4, that the existence of bounded sequences of matrices $(A_n)_{n=-\infty}^{-1}$ and $(B_n)_{n=-\infty}^{-1}$ such that the last condition holds, is equivalent to the statement that the one sided system

$$A_n^* x_{n+1} = B_n^* x_n \qquad (n = 0, 1, \ldots), \tag{6.46}$$

admits a dichotomy $(P_n)_{n=0}^{\infty}$, satisfying

$$\sup_{n=0,1,\ldots} \|(A_n^*|_{Im\, P_{n+1}}, B_n^*|_{Ker\, P_n})^{-1}\| < \infty. \tag{6.47}$$

However, note that $A_n^* = \overline{B}_{n+1}$, and $B_n^* = \overline{A}_n$ $(n = 0, 1, \ldots)$. This proves the first part of the theorem.

If G is Fredholm with $(0, m)$ band characteristic, let $L = (a_{ij})_{ij=-\infty}^{\infty}$ be a bounded and invertible extension which is $(0, m)$ banded. Application of Theorem 1.1 with $(\overline{A}_n)_{n=-\infty}^{\infty}$ and $(\overline{B}_n)_{n=-\infty}^{\infty}$ given by (1.11), and with $(m_1, m_2) = (0, m)$, and of Corollary 6.5, yields (6.42). □

In order to give criteria for non-triangular Fredholm operator with band characteristic we fix the following setting. Let $G = (a_{ij})_{ij=0}^{\infty}$ be a bounded (m_1, m_2) banded block operator in ℓ_r^2. Define block matrices A_n, B_n, $\overline{A}_n$, $\overline{B}_n$ of order $m_1 + m_2$ by

$$A_n = \begin{pmatrix} 0 & -I_r & \cdots & 0 \\ 0 & 0 & \cdots & 0 \\ \cdot & \cdot & \cdot & \cdot \\ 0 & 0 & \cdots & -I_r \\ a_{n,n+m_2}^* & a_{n+1,n+m_2}^* & \cdots & a_{n+m_1+m_2-1,n+m_2}^* \end{pmatrix};$$

$$B_{n+1} = \begin{pmatrix} I_r & 0 & \cdots & 0 & 0 \\ 0 & I_r & \cdots & 0 & 0 \\ \cdot & \cdot & \cdot & \cdot & \cdot \\ 0 & 0 & \cdots & I_r & 0 \\ 0 & 0 & \cdots & 0 & a_{n+m_1+m_2,n+m_2}^* \end{pmatrix}$$

and

$$\overline{A}_n = \begin{pmatrix} 0 & -I_r & \cdots & 0 \\ 0 & 0 & \cdots & 0 \\ \cdot & \cdot & \cdot & \cdot \\ 0 & 0 & \cdots & -I_r \\ a_{n+m_1,n} & a_{n+m_1,n+1} & \cdots & a_{n+m_1,n+m_1+m_2-1} \end{pmatrix};$$

$$\overline{B}_{n+1} = \begin{pmatrix} I_r & 0 & \cdots & 0 & 0 \\ 0 & I_r & \cdots & 0 & 0 \\ \cdot & \cdot & \cdot & \cdot & \cdot \\ 0 & 0 & \cdots & I_r & 0 \\ 0 & 0 & \cdots & 0 & a_{n+m_1,n+m_1+m_2} \end{pmatrix},$$

for $n = 0, 1, \ldots$. Then we call the system $B_{n+1}x_{n+1} = A_n x_n$ $(n = 0, 1, \ldots)$, the first companion system of G, and the system $\overline{B}_{n+1}x_{n+1} = \overline{A}_n x_n$ $(n = 0, 1, \ldots)$, the second companion system of G. Let us also agree to call a dichotomy $(\overline{\overline{P}}_n)_{n=0}^{\infty}$ of an arbitrary system $\overline{\overline{B}}_{n+1}x_{n+1} = \overline{\overline{A}}_n x_n$ $(n = 0, 1, \ldots)$, a normal dichotomy, if

$$\sup_{n=0,1,\ldots} \|(\overline{\overline{B}}_{n+1}|_{Im\,\overline{\overline{P}}_{n+1}}, \overline{\overline{A}}_n|_{Ker\,\overline{\overline{P}}_n})^{-1}\| < \infty.$$

The following result holds.

THEOREM 6.7. *Let $G = (a_{ij})_{ij=0}^{\infty}$ be a bounded, (m_1, m_2) banded, block operator in ℓ_r^2. Then the following statements are equivalent:*

I) *The operator G is a Fredholm operator with (m_1, m_2) band characteristic.*

II) *The first companion system of G admits a normal dichotomy, and $(\operatorname{Im} G)^{\perp} \cap (M_{m_1})^{\perp} \subset (M_{m_1+m_2})^{\perp}$.*

III) *The second companion system of G admits a normal dichotomy, and $\operatorname{Ker} G \cap (M_{m_2})^{\perp} \subset (M_{m_1+m_2})^{\perp}$.*

IV) *Both the first, and the second companion systems admit normal dichotomies.*

In the course of the proof of Theorem 6.7, we will use the following lemma.

LEMMA 6.8. *With the same notation as in Theorem 6.7, if the second companion system of G admits a normal dichotomy, then G is Fredholm, and*

$$\operatorname{Ker} G \cap (M_{m_1+m_2})^{\perp} = \{0\}, \quad \text{and} \quad (\operatorname{Im} G)^{\perp} \cap (M_{m_1})^{\perp} = \{0\}. \tag{6.48}$$

PROOF. Assume that the second companion system of G admits a normal dichotomy. Define $G_1 = (a_{i+m_1,j})_{ij=0}^{\infty}$. Then G_1 is a $(0, m_1 + m_2)$ banded operator. We can apply Theorem 6.6 to G_1, because the corresponding system (6.39) is the second companion system to G. It follows that G_1 is Fredholm operator with $(0, m_1 + m_2)$ band characteristic. We now apply Proposition 6.1 to G_1, then G_1 is Fredholm, and the following equalities hold

$$\operatorname{Ker} G_1 \cap (M_{m_1+m_2})^{\perp} = \{0\}, \quad \text{and} \quad (\operatorname{Im} G_1)^{\perp} = \{0\}. \tag{6.49}$$

Since G_1 is Fredholm, then G is Fredholm. Furthermore, assume that $x = (x_i)_{i=0}^{\infty} \in (\operatorname{Im} G)^{\perp} \cap (M_{m_1})^{\perp}$. Then $0 = x_0 = x_1 = \cdots = x_{m_1-1}$. These equalities and $x \in (\operatorname{Im} G)^{\perp}$, imply that $(x_{i+m_1})_{i=0}^{\infty} \in (\operatorname{Im} G_1)^{\perp}$. By (6.49), $(x_{i+m_1})_{i=0}^{\infty} = 0$, and hence $x = 0$. This proves the right hand side equality in (6.48).

Finally note that if $x \in \operatorname{Ker} G \cap (M_{m_1+m_2})^{\perp}$, then $x \in \operatorname{Ker} G_1$. Thus, (6.49) implies that the left hand side equality in (6.48) holds. □

PROOF OF THEOREM 6.7. We first prove that I) and III) are equivalent.

If III) holds, then, by Lemma 6.8, G is Fredholm and (6.48) holds. The left equality in (6.48) combined with the inclusion in III) gives, $\operatorname{Ker} G \cap (M_{m_2})^{\perp} \in \operatorname{Ker} G \cap (M_{m_1+m_2})^{\perp} = \{0\}$. Hence, the left hand side equality in (6.2) holds. However, the right hand side equality of (6.48) leads to the corresponding equality in (6.2). Hence, (6.1) and (6.2) hold. Thus, Proposition 6.1 shows that I) holds.

Conversely, assume that I) holds. We first use Proposition 6.1. The inclusion in III) follows immediately from the left hand side equality in (6.2). However, by definition, there exists a bounded and invertible block (m_1, m_2) banded extension $L = (a_{ij})_{ij=-\infty}^{\infty}$ of G. Then, $G_1 = (a_{i+m_1,j})_{ij=0}^{\infty}$ is a Fredholm operator with $(0, m_1+m_2)$ band characteristic. Applying Theorem 6.6 to G_1 we obtain that the second companion system of G admits a normal dichotomy. Therefore III) holds. Hence I) and III) are equivalent.

Consider now the (m_2, m_1) band operator G^*. Note that G is a Fredholm operator with (m_1, m_2) band characteristic if and only if G^* is a Fredholm operator with (m_2, m_1) band characteristic. This can be seen from the definitions directly, and follows also from Proposition 6.1. Note also that the first (respectively second) companion system to G is equal to the second (respectively first) companion system to G^*. In addition, $\operatorname{Ker} G^* = (\operatorname{Im} G)^{\perp}$. It follows from this, that the equivalence of I) and III) for the operator G^*, implies that I) and II) are equivalent.

In particular, statement I) implies II) and III), and therefore, IV) also.

We conclude the proof by showing that IV) implies I). Assume that IV) holds. It follows from this and the above relations between G and G^*, that the second companion systems of G, and G^*, admit normal dichotomies. We apply Lemma 6.8 to G and G^* separately. It follows that G is Fredholm, and the right hand side equality in (6.48) leads to

$$(\operatorname{Im} G^*)^{\perp} \cap (M_{m_2})^{\perp} = \{0\}, \quad \text{and} \quad (\operatorname{Im} G)^{\perp} \cap (M_{m_1})^{\perp} = \{0\}.$$

Since $\operatorname{Ker} G = (\operatorname{Im} G^*)^{\perp}$, then the last two equalities imply that (6.2) holds. But, we have shown that G is Fredholm. Therefore, we can apply Proposition 6.1, and hence I) holds. □

REFERENCES

[1] A. Ben-Artzi and I. Gohberg, Inertia theorems for nonstationary discrete systems and dichotomy, *Linear Algebra and Its Applications* **120**, 95–138 (1989).

[2] A. Ben-Artzi and I. Gohberg, Fredholm properties of band matrices and dichotomy, in Operator Theory: Advances and Applications, Vol. 32, Topics in Operator Theory, Constantin Apostol Memorial Issue, Birkhäuser Verlag, 1988.

[3] Ch.V. Coffman and J.J. Schäffer, Dichotomies for linear difference equations, *Math. Annalen* **172**, 139–166 (1967).

[4] I.C. Gohberg and I.A. Feldman, Convolution equations and projection methods for their solutions, Transl. Math. Monographs, Vol. 41, Amer. Math. Soc., Providence, R.I., 1974.

[5] I. Gohberg and M.A. Kaashoek, Block Toeplitz operators with rational symbols, in Operator Theory: Advances and Applications, Vol. 35, Contribution to Operator Theory and Its Applications, pp. 385–440, Birkhäuser Verlag, 1988.

[6] I. Gohberg, M.A. Kaashoek, and F. van Schagen, Non-compact integral operators with semi-separable kernels and their discrete analogous: inversion and Fredholm properties, *Integral Equations and Operator Theory* **7**, 642–703 (1984).

[7] J.J. Schäffer, A note on systems of linear difference equations, *Math. Annalen* **177**, 23–30 (1968).

[8] J.J. Schäffer, Linear difference equations: closedness of covariant sequences, *Math. Annalen* **187**, 69–76 (1970).

A. Ben-Artzi
Department of Mathematics
University of California, San Diego
La Jolla, California 92093, U.S.A.

I. Gohberg
School of Mathematical Sciences
The Raymond and Beverly Sackler Faculty of Exact Sciences
Tel-Aviv University
Tel-Aviv, Ramat-Aviv 69989, Israel

Operator Theory:
Advances and Applications, Vol. 50

INVERTIBILITY OF SYSTEMS OF TOEPLITZ OPERATORS

Marisela Domínguez

We give a matricial extension of the Widom - Devinatz theorem concerning the invertibility of Toeplitz operators, in terms of the matricially weakly positive measures.

1. INTRODUCTION

The aim of this paper is to explore the properties of the so-called matricially weakly positive measures and relate them with the invertibility of systems of Toeplitz operators.

This notion of matricial weak positivity is discussed in greater detail in [6] and it is closely related with the notion of weak positivity introduced by Cotlar and Sadosky [4] (see also [2]). It [7] this notion was used to give matricial extensions of the theorems of Helson - Szegö and Helson - Sarason [9,10,11], by studing two weighted inequalities for the Hilbert Transform, with matricial weights and for functions whose Fourier Transform vanishes in a fixed interval, and by characterizing the spectral density of a strongly mixing stationary gaussian multivariate process.

In the scalar case, Widom [24,25] and Devinatz [5] gave necessary and sufficient conditions for the invertibility of a Toeplitz operator in terms of its symbol (see also [17]). Arocena and Cotlar [1] established a relation between this results and weak positivity.

Here we study the inversion of systems of Toeplitz operators and we obtain a matricial extension of the Widom - Devinatz theorem using matricially weakly positive measures.

It should be noted that many of the concepts and preliminary facts presented in this paper, concerning systems of Toeplitz operators, appeared in some papers of Pousson [19,20].

Pousson and Rabindranathan [21] have derived results concerning the invertibility of Toeplitz operators relative to non-simple shifts. With the aid of the Sz. Nagy - Foias [14] lifting theorem Page [16] established some criteria for the invertibility of Toeplitz operators, which include and in fact extend the results of Pousson and Rabindranathan.

2. THE SPACE OF MATRICES

Let q be a positive integer.

Let C^q ={1xq complex vectors}, C^q is a Hilbert space with the Euclidean scalar product

$$\langle v,w\rangle_q = v\, w^* = \sum_{k=1}^{q} v_k\overline{w_k}$$

where $v,w\in C^q$, $v=(v_1,\dots,v_q)$, $w=(w_1,\dots,w_q)$ and w^* is the adjoint of w.

Let C_{qxq} ={complex qxq matrices}. C_{qxq} is a Hilbert space with scalar product

$$\langle\langle A,B\rangle\rangle_q = \mathrm{tr}\ (AB^*) = \sum_{i;k=1}^{q} a_{ik}\overline{b_{ik}}$$

where $A,B\in C_{qxq}$, $A=(a_{ik})$, $B=(b_{ik})$ and B^* is the adjoint of B.

The basic proposition on the topological structure of the space of the matrices is as follows [26, p. 118]: The space of qxq matrices is a Banach algebra under the usual algebraic operations and either of the norms

$$|A|_E = \sqrt{\mathrm{tr}\ (AA^*)}$$

(Euclidean norm, for the proof that this is actually a norm see [23, p.125]),

$$\text{or}\quad |A|_B = \sup_{|v|\neq 0} |Av|/|v|$$

(Banach norm) where $|v|$ denotes the Euclidean length of a q-dimensional vector with components v_i : $|v|^2 = \sum_{i=1}^{q} |v_i|^2$.

The two norms define equivalent topologies in view of the inequalities

$$|A|_B \le |A|_E \le \sqrt{q}\ |A|_B .$$

Also the following inequality is true [27, p. 96]

$$|AB|_E \le |A|_B\, |B|_E \qquad \text{and} \qquad |AB|_E \le |A|_E\, |B|_B .$$

Let I be the identity matrix in C_{qxq}. For $i,k,m,n\in\{1,\dots,q\}$ let I_{ik} be the matrix given by $(I_{ik})_{mn} = \delta_{mi}\,\delta_{nk}$.

LEMMA 2.1. Given $A,B,D\in C_{qxq}$ with $A = (a_{ik})$, $B = (b_{ik})$ and $D = (D_{ik})$, we consider the following matrices:

$$A_{ik} = a_{ik}\, I_{ik}\ , \qquad B_{Lj} = b_{Lj}\, I_{Lj}\ ,$$

$$A_i = \sum_{k=1}^{q} a_{ik}\, I_{ik} \quad \text{and} \quad B_L = \sum_{j=1}^{q} b_{Lj}\, I_{Lj}$$

(i) $$\langle\langle A_{ik}\, D\ ,\ B_{Lj}\rangle\rangle_q = \begin{cases} a_{ik} d_{kj} \overline{b_{Lj}} & \text{if } i=L \\ 0 & \text{if } i\neq L \end{cases}$$ and

(ii) $$\langle\langle A_i D, B_L\rangle\rangle_q = \begin{cases} \langle [a_{i1}\dots a_{iq}]\ D\ ;\ [b_{i1}\dots b_{iq}]\rangle_q & \text{if } i=L \\ 0 & \text{if } i\neq L \end{cases}$$

PROOF.

(i) $$B_{Lj}{}^* = \overline{b_{Lj}}\, I_{jL}\ ,$$

$$D\, B_{Lj}{}^* = \sum_{m=1}^{q} d_{mj}\, \overline{b_{Lj}}\, I_{mL}\ ,$$

$$A_{ik}\, D\, B_{Lj}{}^* = a_{ik}\, d_{kj}\, \overline{b_{Lj}}\, I_{iL}\ .$$

Therefore

$$\langle\langle A_{ik}.D\ ;\ B_{Lj}\rangle\rangle_q = \operatorname{tr}\,(A_{ik}\, D\, B_{Lj}{}^*) = a_{ik}\, d_{kj}\, \overline{b_{Lj}}\, \delta_{iL}\ .$$

(ii) This assertion follows from (i) since

$$A_i = \sum_{k=1}^{q} A_{ik} \text{ and } B_L = \sum_{j=1}^{q} B_{Lj}\ .$$

3. MATRIX - VALUED FUNCTIONS

Let $T = [0,2\pi]$ and $1 \le p < \infty$. Let L^p and L^∞ be the usual Lebesgue spaces on T, H^p the space of functions in L^p whose Poisson extensions into the unit circle are analytic and $H_o{}^{p-}$ the space of functions in L^p with mean value zero and whose Poisson extensions out of the unit circle are analytic.

For $p > 0$, let

$$L^p{}_{qxq} = \{qxq \text{ matricial valued functions } (h_{ik}) \text{ with } h_{ik}\in L^p\}\ ,$$

$$H^p{}_{qxq} = \{qxq \text{ matricial valued functions } (h_{ik}) \text{ with } h_{ik}\in H^p\}\ ,$$

$$H_o{}^p{}_{qxq}{}^- = \{qxq \text{ matricial valued functions } (h_{ik}) \text{ with } h_{ik}\in H_o{}^{p-}\}\ .$$

If $f\in L^\infty{}_{qxq}$, $||f||_\infty = \text{ess sup}_{x\in T}\, |f(x)|_B$.

The inner product in $L^2{}_{qxq}$ is given by

$$\langle\langle f , g \rangle\rangle = \mathrm{tr} \int_0^{2\pi} f\, g^* \, dx .$$

and $$||f||_2 = \sqrt{\langle\langle f ; f \rangle\rangle} .$$

If $f \in L^p_{q\times q}$, then $||f||_p = \int_0^{2\pi} \mathrm{tr}\, (f\, f^*)^{p/2} \, dx$.

THEOREM 3.1. [20, p.530] *Products of the form* $\Psi_1 \cdot \Psi_2$ *where* Ψ_1 *is in the unit ball of* $H^2_{q\times q}$ *and* Ψ_2 *is in the unit ball of* $H_o{}^p{}_{q\times q}$ *are dense in the open unit ball of* $H_o{}^1{}_{q\times q}$.

We will use the following notations:

$e_n : T \to C$, $e_n(x) = e^{inx}$,

$R = \{\text{real numbers}\}$

$Z = \{\text{integer numbers}\}$

$Z_1 = \{n \in Z /\ n \geq 0\}$

$Z_2 = \{n \in Z /\ n < 0\}$

$$P(C_{q\times q}) = \left\{ \phi : T \to C_{q\times q} \ / \ \phi = \sum_n e_n \hat{\phi}(n) \text{ and } \hat{\phi} : Z \to C_{q\times q} \text{ has finite support} \right\}$$

$P_1(C_{q\times q}) = \{\phi \in P(C_{q\times q});\ \hat{\phi}(n) = 0 \text{ for all } n \in Z_2\}$,

$P_2(C_{q\times q}) = \{\phi \in P(C_{q\times q});\ \hat{\phi}(n) = 0 \text{ for all } n \in Z_1\}$.

$P(C_{q\times q})$, $P_1(C_{q\times q})$ and $P_2(C_{q\times q})$ are dense in $L^2_{q\times q}$, $H^2_{q\times q}$ and $H_o{}^2{}_{q\times q}{}^-$ respectively.

DEFINITION. $\eta \in H^p_{q\times q}$ is outer if

$$\frac{1}{2\pi} \int_0^{2\pi} \log |\det \eta| \, dx = \log |\det \hat{\eta}(0)| > -\infty .$$

THEOREM 3.2. [8, p.193] *If* $w \in L^1_{q\times q}$ *is positive semi-definite, hermitian, and*

$$\int_0^{2\pi} \log \det w \, dx > -\infty$$

then w *has a factorization* $w = \eta\eta^*$ *where* $\eta \in H^2_{q\times q}$ *is outer (and* η^{-1} *exists).*

Even more, we have the following property:

PROPOSITION 3.3. *Let* $w \in L^1_{q\times q}$ *be hermitian, positive semi-definite and suppose* $\log \det w$ *is summable. Then we have the following properties:*

(a) There exists two factorizations of w; *that is,*

$$w = \sigma^* \sigma = \delta \delta^*$$

where σ, $\delta \in H^2_{qxq}$, σ *and* δ *are both outer functions.*

(b) Let $f = \delta^{-1} \sigma^* = \delta^* \sigma^{-1}$ *then f is unitary and*

$$w = \delta f \sigma.$$

THEOREM 3.4. [20, p. 531] *Suppose* $f \in L^\infty_{qxq}$. *The linear functional* $L(.)$ *on* L^1_{qxq} *defined for all* $\varphi \in L^1_{qxq}$ *by*

$$L(\varphi) = \mathrm{tr} \int_0^{2\pi} \varphi f \, dx$$

is bounded and $||L|| = ||f||_\infty$. *All bounded linear functionals on* L^1_{qxq} *have this form.*

4. SYSTEMS OF TOEPLITZ OPERATORS

DEFINITION. A Toeplitz operator T_f: $H^2 \to H^2$ is determined by its symbol $f \in L^\infty$, according to the formula $T_f(h) = P_+(fh)$ for all $h \in H^2$, where P_+: $L^2 \to H^2$ is the projection.

DEFINITION. Let $f \in L^\infty_{qxq}$, a system of Toeplitz operators is T_f: $H^2_{qxq} \to H^2_{qxq}$ defined by $T_f(h) = P_+(fh)$ for all $h \in H^2_{qxq}$, where P_+: $L^2_{qxq} \to H^2_{qxq}$ is the projection.

Let $P_- = I - P_+$.

PROPOSITION 4.1.

(1) $||T_f|| \leq ||f||_\infty$.

(2) $(T_f)^* = T_{f^*}$. *Hence* T_f *is invertible if and only if* T_{f^*} *is invertible and* $(T_{f^*})^{-1} = ((T_f)^{-1})^*$.

(3) $T_f(h) = f h$ *if* $f \in H^\infty_{qxq}$.

(4) If $g \in H^\infty_{qxq}$ $T_{fg} = T_f T_g$ *and* $T_{g^* f} = T_{g^*} T_f$.

(5) Let $f \in H^\infty_{qxq}$ *then:* T_f *is invertible if and only if* $f^{-1} \in H^\infty_{qxq}$ *in which case* $(T_f)^{-1} = T_{f^{-1}}$.

(6) If f is unitary and T_f *is invertible then* $f = \eta^* \gamma$ *where* η, η^{-1}, γ, $\gamma^{-1} \in H^2_{qxq}$ *and* $\eta \eta^* = (\gamma \gamma^*)^{-1}$.

(7) If T_f *is invertible then* $f^{-1} \in L^\infty_{qxq}$.

PROOF.

(1) The result follows from the inequality that relates the

Euclidean norm of a product of matrices with the Banach norm and the Euclidean norm of the factors.

$$||T_f|| = \sup_{||h||_2=1} ||T_f(h)||_2 = \sup_{||h||_2=1} ||P_+(fh)||_2$$
$$\leq \sup_{||h||_2=1} ||fh||_2 \leq ||f||_\infty .$$

for a proof of the last inequality see [27, p.98].

(2) This assertion is obvious.

(3) This assertion is also evident.

The proofs of the other results are available in the literature: (4), (5) and (6) are proved in [19, p.607], (7) is proved in [19, p.606].

We are interested in the question under what circumstances a system of Toeplitz operators is invertible. The following lemmas shows that in certain cases it is enough to consider T_f for the case where f is unitary

LEMMA 4.2. *Let* $f \in L^\infty_{qxq}$ *and* $f^{-1} \in L^\infty_{qxq}$, let $|f| = \sqrt{f^*f}$.

(a) Then $|f|^{-1}$ *has the factorization* $|f|^{-1} = \eta\ \eta^*$ *where* $\eta \in H^2_{qxq}$ *is outer.*

(b) $\eta^*\ |f|\ \eta = I$.

(c) $\eta^*\ f\ \eta$ *is unitary, if* f *is normal.*

(d) T_f *is invertible if and only if* $T_{\eta^* f \eta}$ *is invertible.*

PROOF.

(a) If we prove that $g \in L^1_{qxq}$, g is positive semi-definite, hermitian and

$$\int_0^{2\pi} \log \det |f|^{-1}\, dx > -\infty$$

theh the result will follow from Theorem 3.2.

$f \in L^\infty_{qxq}$ implies $f^* \in L^\infty_{qxq}$, thus $f^*f,\ ff^*,\ \sqrt{ff^*} \in L^\infty_{qxq}$. $|f|$ is semi-positive definite implies that $|f|^{-1}$ is positive semi-definite. f^*f is hermitian implies that $|f|^{-1}$ is hermitian.

We have that

$$(f\ |f|^{-1}\ f^*)^2 = f\ |f|^{-1}\ |f|^2\ |f|^{-1}\ f^* = ff^*.$$

then

$$f\ |f|^{-1}\ f^* = \sqrt{ff^*} .$$

and

$$|f|^{-1} = f^{-1} \sqrt{ff^*}\ (f^*)^{-1} .$$

As $f^{-1}\in L^{\infty}_{qxq}$ we have that $|f|^{-1}\in L^{\infty}_{qxq}$, therefore $|f|^{-1}\in L^{1}_{qxq}$.

We have that $\det|f|\in L^{\infty}$. So there exists a such that $\det(|f|^{-1}) = (\det|f|)^{-1} > a > 0$. It follows that

$$\int_0^{2\pi} \log\det|f|^{-1}\,dx > -\infty$$

(b) This assertion is obvious.

(c) If f is normal

$$f\,|f|^{-1} f^{*} = |f|.$$

Therefore

$$(\eta^{*} f \eta)(\eta^{*} f \eta)^{*} = \eta^{*} f \eta \eta^{*} f^{*} \eta = \eta^{*} f\,|f|^{-1} f^{*} \eta = \\ = \eta^{*}\sqrt{ff^{*}}\,\eta = \eta^{*}|f|\,\eta = I.$$

(d) $f^{-1}\in L^{\infty}_{qxq}$ clearly implies $\eta\in L^{\infty}_{qxq}$. It follows readily from $\eta\in H^{2}_{qxq}$ that $\eta\in H^{\infty}_{qxq}$. Thus from the Proposition 4.1. (4)

$$T_{\eta^{*}f\eta} = T_{\eta^{*}}T_{f}T_{\eta}.$$

It can also be proved that $\eta^{-1}\in H^{\infty}_{qxq}$. Proposition 4.1. (5) and (2) imply that T_{η} is invertible and thus $T_{\eta^{*}}$ is invertible. Therefore T_f is invertible if and only if $T_{\eta^{*}f\eta}$ is invertible. And the proof is complete.

LEMMA 4.3. *Let* $f\in L^{\infty}_{qxq}$. T_f *is invertible if and only if* $f^{-1}\in L^{\infty}_{qxq}$ *and* $T_{\eta^{*}f\eta}$ *is invertible, where* $\eta\in H^{2}_{qxq}$ *is outer and* $|f|^{-1} = \eta\,\eta^{*}$. *If f is normal then* $\eta^{*} f \eta$ *is unitary.*

PROOF. The result follows from Proposition 4.1. (7) and from the Lemma 4.2.

This Lemma shows that if f is normal, it is enough to consider T_f for the case where f is unitary.

5. MATRIX - VALUED MEASURES AND MATRICIAL POSITIVITY

Let

$$M_{qxq} = \{C_{qxq} \text{ - valued hermitian finite Borel measures } \mu \text{ in } T\}.$$

DEFINITION. Let $\mu\in M_{qxq}$ μ is positive if and only if for every borel set $\Delta\subset T$, for all q-dimensional row vector $v = [v_1,\dots,v_q]$

$$\langle v.\mu(\Delta), v\rangle_q = \sum_{k=1}^{q}\sum_{j=1}^{q} v_k\,\mu_{kj}(\Delta)\,\overline{v_j} \geq 0.$$

Let $\mu\in M_{qxq}$ be fixed, μ positive and $\mathrm{tr}\mu = \mathrm{trace}(\mu)$, we define [22, p.294].

$$L^2_{qxq}(\mu) = \left\{ \phi : T \to C_{qxq} \text{ such that } \int_0^{2\pi} \phi(t) d\mu(t) \phi^*(t) \text{ exists} \right\}$$

where:

$$\left(\int_0^{2\pi} \phi \, d\mu \, \psi^* \right)_{11} = \int_0^{2\pi} \left(\phi \frac{d\mu}{dtr\mu} \psi^* \right)_{11} dtr\mu,$$

and

$$\frac{d\mu}{dtr\mu} = \left(\frac{d\mu_{kj}}{dtr\mu} \right)_{k,j=1,\dots,q}$$

By a theorem of Rosenberg and Rozanov [12, p.362; 22, p.295], $L^2_{qxq}(\mu)$ is complete under the inner product

$$<<\phi , \psi>>_\mu = tr \int_0^{2\pi} \phi \, d\mu \, \psi^* .$$

Let $\mu \in M_{qxq}$ then $P(C_{qxq}) \subset L^2_{qxq}(\mu)$.

Denote $||\phi||_\mu^2 = <<\phi , \phi>>_\mu$. And denote by $H^2_{qxq}(\mu)$ and $H_o{}^2_{qxq}{}^-(\mu)$ the subspaces of $L^2_{qxq}(\mu)$ generated by $P_1(C_{qxq})$ and $P_2(C_{qxq})$ respectively.

To avoid unnecessary notation when $d\mu = w(x)dx$ we shall use $L^2_{qxq}(w)$, $<<\phi , \psi>>_w$, $||\phi||_w$, $H^2_{qxq}(w)$ and $H_o{}^2_{qxq}{}^-(w)$ instead of $L^2_{qxq}(\mu)$, $<<\phi , \psi>>_\mu$, $||\phi||_\mu$, $H^2_{qxq}(\mu)$ and $H_o{}^2_{qxq}{}^-(\mu)$.

Once $<<\phi , \psi>>_\mu$ is defined for any positive $\mu \in M_{qxq}$, it can be defined for a general $\mu \in M_{qxq}$ in the same way that this is done for scalar valued measures.

Now we will introduce the notions of matricial positivity. The connections between these new notions and the well known notions of positivity are discussed in greater detail in [6].

DEFINITION. Let $\mu \in M_{qxq}$ μ is matricially positive if for every borel set $\Delta \subset T$, for all $A \in C_{qxq}$

$$<<A.\mu(\Delta),A>>_q \geq 0 .$$

PROPOSITION 5.1. Let $\mu \in M_{qxq}$, μ is positive if and only if μ is matricially-positive.

PROOF. The result follows from Lemma 2.1. (ii).

$(\rightarrow)$ Let $A \in C_{qxq}$, $A = (a_{ik})$ and $A_i = \sum_{k=1}^{q} a_{ik} I_{ik}$

$$<<A.\mu(\Delta),A>>_q = \sum_{i;L=1}^{q} <<A_i.\mu(\Delta);A_L>>_q$$
$$= \sum_{i=1}^{q} <[a_{i1}...a_{iq}]\ \mu(\Delta);\ [a_{i1}...a_{iq}]>_q \geq 0 .$$

$(\leftarrow)$ Let $v \in C^q$, $v = [v_1, \dots, v_q]$. Define $A_q = (a_{ik})$ where

$$a_{ik} = \begin{cases} v_k & \text{if } i=q \\ 0 & \text{if } i\neq q \end{cases}$$

then

$$<v.\mu(\Delta),v>_q = <<A_q.\mu(\Delta),A_q>>_q \geq 0 .$$

$(\mu_{\alpha\beta})_{\alpha,\beta=1,2}$ will allways be an hermitian matrix $\begin{pmatrix} \mu_{11} & \mu_{12} \\ \mu_{21} & \mu_{22} \end{pmatrix}$ with $\mu_{\alpha\beta} \in M_{qxq}$, for $\alpha,\beta=1,2$.

DEFINITION. $(\mu_{\alpha\beta})_{\alpha,\beta=1,2}$ is positive if for all $\Delta \subset T$, for all $v_1, v_2 \in C^q$

$$\sum_{\alpha=1}^{2} \sum_{\beta=1}^{2} <v_\alpha.\mu_{\alpha\beta}(\Delta),v_\beta>_q \geq 0.$$

DEFINITION. $(\mu_{\alpha\beta})_{\alpha,\beta=1,2}$ is matricially positive if for all $\Delta \subset T$, for all $A_1, A_2 \in C_{qxq}$

$$\sum_{\alpha=1}^{2} \sum_{\beta=1}^{2} <<A_\alpha.\mu_{\alpha\beta}(\Delta);A_\beta>>_q \geq 0.$$

PROPOSITION 5.2. $(\mu_{\alpha\beta})_{\alpha,\beta=1,2}$ is positive if and only if $(\mu_{\alpha\beta})_{\alpha,\beta=1,2}$ is matricially positive.

PROOF. The result follows from Lemma 2.1. (ii).

$(\rightarrow)$ Let $A^1, A^2 \in C_{qxq}$, $A^1 = (a^1_{ik})$, $A^2 = (a^2_{ik})$, consider

$$A^1_i = \sum_{k=1}^{q} a^1_{ik} I_{ik} \qquad \text{and} \qquad A^2_L = \sum_{j=1}^{q} a^2_{Lj} I_{Lj}$$

then for $\alpha,\beta = 1,2$

$$<<A^\alpha\mu_{\alpha\beta}(\Delta),A^\beta>>_q = \sum_{i;L=1}^{q} <<A^\alpha{}_i\mu_{\alpha\beta}(\Delta),A^\beta{}_L>>_q =$$
$$= \sum_{i=1}^{q} <[a^\alpha{}_{i1}...a^\alpha{}_{iq}]\ \mu_{\alpha\beta}(\Delta),\ [a^\beta{}_{i1}...a^\beta{}_{iq}]>_q$$

Thus

$$\sum_{\alpha=1}^{2}\sum_{\beta=1}^{2} \langle\langle A^{\alpha}\mu_{\alpha\beta}(\Delta), A^{\beta}\rangle\rangle_q$$

$$= \sum_{i=1}^{q}\sum_{\alpha=1}^{2}\sum_{\beta=1}^{2} \langle [a^{\alpha}_{i1}\ldots a^{\alpha}_{iq}]\mu_{\alpha\beta}(\Delta) \ , \ [a^{\beta}_{i1}\ldots a^{\beta}_{iq}]\rangle_q \geq 0.$$

$(\leftarrow)$ For $\alpha=1,2$ let $v^{\alpha}\in C^q$, $v^{\alpha} = [v^{\alpha}_1, \ldots, v^{\alpha}_q]$. Define $A^{\alpha}_q = (a^{\alpha}_{ik})$ where

$$a^{\alpha}_{ik} = \begin{cases} v^{\alpha}_k & \text{if } i=q \\ 0 & \text{if } i\neq q \end{cases}$$

Thus

$$\sum_{\alpha=1}^{2}\sum_{\beta=1}^{2} \langle v^{\alpha}.\mu_{\alpha\beta}(\Delta) \ , \ v^{\beta}\rangle_q = \sum_{\alpha=1}^{2}\sum_{\beta=1}^{2} \langle\langle A^{\alpha}_q.\mu_{\alpha\beta}(\Delta) \ , \ A^{\beta}_q\rangle\rangle_q \geq 0.$$

This finishes the proof.

For given $(\mu_{\alpha\beta})_{\alpha,\beta=1,2}$, for every $\phi_1,\phi_2\in P(C_{qxq})$, let

$$\mathcal{M}_{qxq}(\phi_1,\phi_2) = \sum_{\alpha=1}^{2}\sum_{\beta=1}^{2} \langle\langle\phi_{\alpha} \ , \ \phi_{\beta}\rangle\rangle_{\mu_{\alpha\beta}} .$$

PROPOSITION 5.3. [6] *The following conditions are equivalent:*

(a) $(\mu_{\alpha\beta})_{\alpha,\beta=1,2}$ *is matricially positive.*

(b) $\mathcal{M}_{qxq}(\phi_1,\phi_2) \geq 0$ *for all* $\phi_1,\phi_2\in P(C_{qxq})$.

(c) μ_{11} *and* μ_{22} *are matricially positive and for every* $\phi_1,\phi_2\in P(C_{qxq})$

$$|\langle\langle\phi_1 \ , \ \phi_2\rangle\rangle_{\mu_{12}}|^2 \leq \langle\langle\phi_1 \ , \ \phi_1\rangle\rangle_{\mu_{11}} \langle\langle\phi_2 \ , \ \phi_2\rangle\rangle_{\mu_{22}} .$$

PROOF.

(a) $\to$ (b) Let Π any subdivision of T and $\phi_1,\phi_2\in P(C_{qxq})$

$$\mathcal{M}_{qxq}(\phi_1,\phi_2) = \sum_{\alpha=1}^{2}\sum_{\beta=1}^{2} \langle\langle\phi_{\alpha}, \ \phi_{\beta}\rangle\rangle_{\mu_{\alpha\beta}}$$

$$= \sum_{\alpha=1}^{2}\sum_{\beta=1}^{2} \operatorname{tr}\int_0^{2\pi} \phi_{\alpha} \, d\mu_{\alpha\beta} \, \phi_{\beta}^*$$

$$= \sum_{\alpha=1}^{2}\sum_{\beta=1}^{2} \lim_{\Pi} \sum_{\Delta_j\in\Pi} \operatorname{tr}(\phi_{\alpha}(x_j) \ \mu_{\alpha\beta}(\Delta_j) \ \phi_{\beta}^*(x_j))$$

$$= \lim_{\Pi} \sum_{\Delta_j\in\Pi}\sum_{\alpha=1}^{2}\sum_{\beta=1}^{2} \langle\langle\phi_{\alpha}(x_j) \ \mu_{\alpha\beta}(\Delta_j) \ , \ \phi_{\beta}(x_j)\rangle\rangle_q \geq 0.$$

(b) $\to$ (a) Let $\Delta \subset T$, $A_1,A_2 \in C_{qxq}$ then there exist two decreasing sequences of continuous functions from T on C_{qxq} $(\phi_{1,n})_n$ and $(\phi_{2,n})_n$

such that
$\lim_{n\to\infty} \phi_{1,n} = A_1 \chi_\Delta$ and $\lim_{n\to\infty} \phi_{2,n} = A_2 \chi_\Delta$ a.e.

For every n, $\phi_{1,n}$ and $\phi_{2,n}$ can be aproximated by functions of $P(C_{qxq})$, so using the hypothesis we obtain $\mathcal{M}_{qxq}(\phi_{1,n}, \phi_{2,n}) \geq 0$. From the Lebesgue convergence theorem it follows that

$$\sum_{\alpha=1}^{2} \sum_{\beta=1}^{2} <<A_\alpha . \mu_{\alpha\beta}(\Delta); A_\beta>>_q = \mathcal{M}_{qxq}(A_1 \chi_\Delta, A_2 \chi_\Delta) \geq 0.$$

(b) $\leftrightarrow$ (c) Let $\phi_1, \phi_2 \in P(C_{qxq})$ and $\lambda_1, \lambda_2 \in C$. Changing ϕ_1 by $\lambda_1\phi_1$ and ϕ_2 by $\lambda_2\phi_2$, the inequality in (b) becomes

$$\sum_{\alpha=1}^{2} \sum_{\beta=1}^{2} \lambda_\alpha \overline{\lambda_\beta} <<\phi_\alpha , \phi_\beta>>_{\mu_{\alpha\beta}} \geq 0$$

which is the positive definiteness of

$$\begin{pmatrix} <<\phi_1 ; \phi_1>>_{\mu_{11}} & <<\phi_1 ; \phi_2>>_{\mu_{12}} \\ <<\phi_2 ; \phi_1>>_{\mu_{21}} & <<\phi_2 ; \phi_2>>_{\mu_{22}} \end{pmatrix}$$

and is equivalent to (c).

This finishes the proof.

DEFINITION. $(\mu_{\alpha\beta})_{\alpha,\beta=1,2}$ is matricially weakly positive if $\mathcal{M}_{qxq}(\phi_1, \phi_2) \geq 0$ for all $(\phi_1, \phi_2) \in P_1(C_{qxq}) x P_2(C_{qxq})$.

PROPOSITION 5.4. [6] *The following conditions are equivalent*:

(a) $(\mu_{\alpha\beta})_{\alpha,\beta=1,2}$ *is matricially weakly positive.*

(b) μ_{11} *and* μ_{22} *are matricially positive and for every* $(\phi_1, \phi_2) \in P_1(C_{qxq}) x P_2(C_{qxq})$

$$|<<\phi_1 , \phi_2>>_{\mu_{12}}|^2 \leq <<\phi_1 , \phi_1>>_{\mu_{11}} <<\phi_2 , \phi_2>>_{\mu_{22}} .$$

PROOF. Let $(\phi_1, \phi_2) \in P_1(C_{qxq}) x P_2(C_{qxq})$ and $\lambda_1, \lambda_2 \in C$. Changing ϕ_1 by $\lambda_1\phi_1$ and ϕ_2 by $\lambda_2\phi_2$, the inequality

$$\mathcal{M}_{qxq}(\phi_1, \phi_2) \geq 0$$

for all $(\phi_1, \phi_2) \in P_1(C_{qxq}) x P_2(C_{qxq})$, becomes

$$\sum_{\alpha=1}^{2} \sum_{\beta=1}^{2} \lambda_\alpha \overline{\lambda_\beta} <<\phi_\alpha , \phi_\beta>>_{\mu_{\alpha\beta}} \geq 0$$

for all $(\phi_1, \phi_2) \in P_1(C_{qxq}) x P_2(C_{qxq})$, which is the positive definiteness of

$$\begin{pmatrix} <<\phi_1 ; \phi_1>>_{\mu_{11}} & <<\phi_1 ; \phi_2>>_{\mu_{12}} \\ <<\phi_2 ; \phi_1>>_{\mu_{21}} & <<\phi_2 ; \phi_2>>_{\mu_{22}} \end{pmatrix}$$

for all $(\phi_1,\phi_2)\in P_1(C_{qxq})xP_2(C_{qxq})$, and is equivalent to

$$|<<\phi_1 , \phi_2>>_{\mu_{12}}|^2 \leq <<\phi_1 , \phi_1>>_{\mu_{11}} <<\phi_2 , \phi_2>>_{\mu_{22}} .$$

This finishes the proof.

6. A MATRICIAL EXTENSION OF THE WIDOM - DEVINATZ THEOREM

In the rest of this section f will be asumed to be unitary. The Widom - Devinatz theorem on invertibility of the Toeplitz operators amounts to the equivalence of some assertions which are connected with an arbitrary unitary ($|f|=1$ a.e.) symbol f.

The following concepts are necessary for formulating this theorem.

DEFINITION. Given $f\in L^{\infty}_{qxq}$ and $\mu\in M_{qxq}$, $\mu \geq 0$ let $\rho(f,\mu)$ be the cosine of the angle between the subspaces of $L^2_{qxq}(\mu)$: $H^2_{qxq}(\mu).f$ and $H_o^2{}_{qxq}^-(\mu)$

$$\rho(f,\mu) = \sup_{\substack{\phi_1\in H^2_{qxq}(\mu);\ \phi_2\in H_o^2{}_{qxq}^-(\mu) \\ <<\phi_1;\phi_1>>_\mu = <<\phi_2;\phi_2>>_\mu = 1}} |<<\phi_1 f ; \phi_2>>_\mu| .$$

It is useful to note that the value $\rho(f,\mu)$ does not change if this sup is taken over functions from $P_1(C_{qxq})$ and $P_2(C_{qxq})$. ($P(C_{qxq})$ is dense in $L^2_{qxq}(w)$ provided $\det w \neq 0$ [20, p.532]).

The subspaces $f.H^2_{qxq}(\mu)$ and $H_o^2{}_{qxq}^-(\mu)$ are at positive angle provided $\rho(f,\mu) < 1$.

If $d\mu = w(x)dx$, then we will write $\rho(f,w)$ instead of $\rho(f,\mu)$.

When μ is the spectral measure of a multivariate stationary process $\rho(I,\mu)$ is the maximal correlation coefficient of the process [7,9,10,11,12,18].

PROPOSITION 6.1. *Let* $f\in L^{\infty}_{qxq}$, $\mu\in M_{qxq}$, $\mu \geq 0$ *and* $r\in(0,1)$.

$$Q = \begin{pmatrix} r\,\mu & f\,\mu \\ \mu^*\, f^* & r\,\mu \end{pmatrix}$$

is matricially weakly positive if and only if $\rho(f,\mu) \leq r$.

PROOF. $\rho(f,\mu) \leq r$ if and only if for all $(\phi_1,\phi_2) \in P_1(C_{qxq})xP_2(C_{qxq})$

$$\left|<<\phi_1 f ; \phi_2>>_\mu\right|^2 \leq (r <<\phi_1;\phi_1>>_\mu\ r <<\phi_2;\phi_2>>_\mu)$$

if and only if Q is matricially weakly positive. This finishes the proof.

The Hilbert transform H (also called the harmonic conjugate), is defined linearly in $P(C_{qxq})$ by

$$H(\phi) = -i\phi \quad \text{if } \phi \in P_1(C_{qxq}) \quad \text{and} \quad H(\phi) = +i\phi \text{ if } \quad \phi \in P_2(C_{qxq}).$$

THEOREM 6.2. (Widom - Devinatz) [5,15,17,24,25] *Let* $f \in L^\infty$ *, a unitary function. Then the following conditions are equivalent:*

(a) T_f *is invertible.*

(b) $\mathrm{dist}(f\ , H^\infty) < 1$ *and* $\mathrm{dist}(\overline{f}\ , H^\infty) < 1$.

(c) $f = c\overline{g}/g$, *where* $c \in C$, $|c| = 1$, *g is an outer function and* $w = |g|^2 = gfg$ *satisfy* $\rho(1,w) < 1$.

(d) $f = c\overline{g}/g$, *where* $c \in C$, $|c| = 1$, *g is an outer function and* $w = |g|^2 = gfg$ *is such that for the Hilbert transform* H

$$\int_0^{2\pi} |H\phi|^2 \, w \, dx \quad \le \quad a \int_0^{2\pi} |\phi|^2 \, w \, dx$$

for all $\phi \in P(C_{1x1})$, $a > 0$.

(e) $f = \exp(i(c + Hu + v))$, *where* $c \in R$, u *and* v *are real valued functions in* L^∞ *and* $||v||_\infty < \pi/2$.

In this theorem the function $|g|^2$ satisfyes the so-called Helson - Szegö condition, i.e., the possibility of writing $|g|^2$ in the form $\exp(u + Hv)$ where u and v real valued functions in L^∞ and $||v||_\infty < \pi/2$.

In general the problem of characterizing the measures μ such that the subspaces of $L^2_{qxq}(\mu)$ generated by $P_1(C_{qxq})$ and $P_2(C_{qxq})$ are at positive angle is equivalent to a weighted inequality problem for the Hilbert Transform and is studied in detail in [10] for q=1, and in [7,18] for other q.

Arocena and Cotlar [1, p.93-94] established a connection between weakly positive matrices and invertibility of systems operators for the scalar case. The next theorem gives a matricial extension of their result.

THEOREM 6.3. *Let* $f \in L^\infty_{qxq}$, $f \neq 0$, f *unitary. Then the following conditions are equivalent:*

(a) T_f *is left - invertible.*

(b) There exists $r \in (0,1)$ *such that*

$$\begin{pmatrix} r.I.dx & f.dx \\ f^*.dx & r.I.dx \end{pmatrix}$$

is matricially weakly positive.

(c) $\rho(f,dx) \leq r$ *for some* $r \in (0,1)$.

(d) $\operatorname{dist}(f\ ,\ H^{\infty}_{qxq}) \leq r$ *for some* $r \in (0,1)$.

PROOF.

(a) → (b) We first show that for $r > 0$, the following conditions are equivalent:

(1) $|\langle\langle\phi_1 f,\phi_2\rangle\rangle|^2 \leq r||\phi_1||_2^2||\phi_2||_2^2$ for all $(\phi_1,\phi_2)\in P_1(C_{qxq})xP_2(C_{qxq})$

(2) $|\langle\langle f\psi_1,\psi_2\rangle\rangle|^2 \leq r||\psi_1||_2^2||\psi_2||_2^2$ for all $(\psi_1,\psi_2)\in P_1(C_{qxq})xP_2(C_{qxq})$

In fact, considere $\phi_1,\phi_2 \in P(C_{qxq})$.

Let $\psi_1 = e_{-1}\phi_2^*$ and $\psi_2 = e_{-1}\phi_1^*$ then $\phi_2\in P_2(C_{qxq})$ if and only if $\psi_1\in P_1(C_{qxq})$, $||\phi_2||_2 = ||\psi_1||_2$, $\phi_1\in P_1(C_{qxq})$ if and only if $\psi_2\in P_2(C_{qxq})$ and $||\phi_1||_2 = ||\psi_2||_2$.

The equivalence between (1) and (2) follows.

Now suppose that T_f is left - invertible. As $P_1(C_{qxq}) \subset H^2_{qxq}$ there exists $\varepsilon\in(0,1)$ such that for all $\psi_1\in P_1(C_{qxq})$

$$\varepsilon\ ||\psi_1||_2^2 \leq ||T_f\ \psi_1||_2^2$$

$$||\psi_1||_2^2 = ||f\ \psi_1||_2^2 = ||P_+(f\ \psi_1)||_2^2 + ||P_-(f\ \psi_1)||_2^2$$

$$\geq \varepsilon\ ||\psi_1||_2^2 + ||P_-(f\ \psi_1)||_2^2 .$$

Thus

$$||P_-(f\ \psi_1)||_2^2 \leq (1 - \varepsilon)\ ||\psi_1||_2^2 .$$

Let $\psi_2\in P_2(C_{qxq})$ then

$$|\langle\langle f\ \psi_1\ ,\ \psi_2\rangle\rangle|^2 = |\langle\langle P_-(f\ \psi_1)\ ,\ \psi_2\rangle\rangle|^2$$

$$\leq (1 - \varepsilon)\ ||\psi_1||_2^2\ ||\psi_2||_2^2 .$$

Thus for every $(\phi_1,\phi_2)\in P_1(C_{qxq})xP_2(C_{qxq})$

$$|\langle\langle\phi_1\ f,\phi_2\rangle\rangle|^2 \leq (1 - \varepsilon)\ ||\phi_1||_2^2||\phi_2||_2^2 .$$

Therefore

$$\begin{pmatrix} \sqrt{1 - \varepsilon}.I.dx & f.dx \\ f^*.dx & \sqrt{1 - \varepsilon}.I.dx \end{pmatrix}$$

is matricially weakly positive.

(b) → (a) It is proved in a similar way going backwards.

(b) ↔ (c) See Proposition 6.1.

(c) ↔ (d) We have that

$$\rho(f,dx) = \sup_{\substack{\psi_1 \in H^2_{qxq}\ ;\ \psi_2 \in H_o{}^2_{qxq} \\ ||\psi_1|| = ||\psi_2|| = 1}} \left| tr \int_0^{2\pi} \psi_1 \, f \, \psi_2 \, dx \right|$$

From Theorem 3.1. it follows that

$$\rho(f,dx) = \sup_{\substack{\varphi \in H_o{}^1_{qxq} \\ ||\varphi||=1}} \left| tr \int_0^{2\pi} \varphi \, f \, dx \right|$$

Theorem 3.4. implies that

$$tr \int_0^{2\pi} \varphi \, f \, dx$$

defines a bounded linear functional on L^1_{qxq} which, when restricted to $H_o{}^1_{qxq}$ has norm $\rho(f,dx)$.

Suppose $g \in L^\infty_{qxq}$ is an annihilator of $H_o{}^1_{qxq}$; that is, for all $\varphi \in H_o{}^1_{qxq}$

$$tr \int_0^{2\pi} \varphi \, g \, dx = 0$$

Necessarily, $g \in H^\infty_{qxq}$. Moreover, all elements of H^∞_{qxq} are annihilators of $H_o{}^1_{qxq}$.

We now apply a corollary of the Hahn - Banach theorem, (see [3]) and obtain

$$\rho(f,dx) = \inf_{g \in H^\infty_{qxq}} ||f - g||_\infty = \text{dist } (f \ , \ H^\infty_{qxq})$$

And this finishes the proof of the theorem.

THEOREM 6.4. *Let* $w \in L^1_{qxq}$ *be hermitian, positive semi-definite and suppose* log det w *is summable. Then we have the following properties:*

(1) there exists two factorizations of w*; that is,*

$$w = \sigma^* \, \sigma = \delta \, \delta^*$$

where σ, $\delta \in H^2_{qxq}$, σ *and* δ *are both outer functions.*

(2) Let $f = \delta^{-1} \sigma^* = \delta^* \sigma^{-1}$ *then* f *is unitary and*

$$w = \delta \, f \, \sigma.$$

(3) $w^{-1} \in L^\infty_{qxq}$, *is hermitian positive semidefinite,*

(4) $w^{-1} = (\delta^{-1})^* \, \delta^{-1} = \sigma^{-1} \, (\sigma^{-1})^*$

(5) $f^* = \sigma \, (\delta^{-1})^* = (\sigma^{-1})^* \, \delta$, f^* *is unitary and*

$$w^{-1} = \sigma^{-1} f^* \delta^{-1} .$$

And the following conditions are equivalent to the conditions (a), (b), (c) and (d) of the last theorem:

(e) $\rho(I,w) < r$ *for some* $r \in (0,1)$.

(f) There exists $r \in (0,1)$ *such that*

$$\begin{pmatrix} r.w.dx & w.dx \\ w^*.dx & r.w.dx \end{pmatrix}$$

is matricially weakly positive.

(g) For the Hilbert transform H

$$\mathrm{tr} \int_0^{2\pi} (H\phi)\ w\ dx\ (H\phi)^* \quad \leq \quad a \ \ \mathrm{tr} \int_0^{2\pi} \phi\ w\ dx\ \phi^*$$

for all $\phi \in P(C_{qxq})$, *where* $a > 0$.

PROOF.

(c) $\leftrightarrow$ (e) First observe that the set of all products $\phi\ \delta$ where $\phi \in H^2_{qxq}(w)$ and $||\phi||_w = 1$, is dense in the unit ball of H^2_{qxq}

In fact, let $\phi \in H^2_{qxq}(w)$ then there exists a sequence $\{\phi_n\}_{n\geq 1} \subset P(C_{qxq})$ such that $\lim_{n\to\infty} ||\ \phi_n - \phi\ ||_w = 0$.

For every $n \geq 1$, $||\ \phi_n - \phi\ ||_w = ||\ \phi_n\ \delta - \phi\ \delta\ ||_2$, therefore

$\lim_{n\to\infty} ||\ \phi_n\ \delta - \phi\ \delta\ ||_w = 0$, we also have that $\phi_n\ \delta \in H^2_{qxq}$ for every $n \geq 1$, thus $\phi\ \delta \in H^2_{qxq}$.

Moreover $||\phi||_w = ||\phi\ \delta||_2$.

Thus ϕ in the unit ball of $H^2_{qxq}(w)$ implies $\phi\ \delta$ in the unit ball of H^2_{qxq} .

A similar argument shows that the set of all products $\sigma\ \phi$ where $\phi \in H_o{}^2{}_{qxq}(w)$ and $||\phi||_w = 1$, is dense in the unit ball of $H_o{}^2{}_{qxq}$.

Therefore we have that

$$\rho(I,w) = \sup_{\substack{\phi_1 \in H^2_{qxq}(w);\ \phi_2 \in H_o{}^2{}_{qxq}{}^-(w) \\ ||\phi_1||_w = ||\phi_2||_w = 1}} \left| \mathrm{tr} \int_0^{2\pi} \phi_1\ w\ \phi_2{}^*\ dx \right|$$

$$= \sup_{\substack{\phi_1 \in H^2_{qxq}(w);\ \phi_2 \in H_o{}^2{}_{qxq}{}^-(w) \\ ||\phi_1||_w = ||\phi_2||_w = 1}} \left| \mathrm{tr} \int_0^{2\pi} \phi_1\ \delta\ f\ \sigma\ \phi_2{}^*\ dx \right|$$

$$= \sup_{\substack{\Psi_1 \in H^2_{qxq},\ \Psi_2 \in H_o^{2}{}_{qxq}^{-} \\ ||\Psi_1||_2 = ||\Psi_2||_2 = 1}} \left| \mathrm{tr} \int_0^{2\pi} \Psi_1\, f\, \Psi_2^*\, dx \right|$$

$$= \rho(f, dx)$$

(e) $\leftrightarrow$ (f) See Proposition 6.1.

(f) $\leftrightarrow$ (g) See [7].

So the proof is complete.

COROLLARY 6.5. *Let* $f \in L^\infty_{qxq}$, $f \neq 0$, f *unitary. Then the following conditions are equivalent:*

(a) T_f *is invertible.*

(b) There exists $r \in (0,1)$ *such that*

$$\begin{pmatrix} r.I.dx & f.dx \\ f^*.dx & r.I.dx \end{pmatrix} \text{ and } \begin{pmatrix} r.I.dx & f^*.dx \\ f.dx & r.I.dx \end{pmatrix}$$

are matricially weakly positive.

(c) $\rho(f, dx) \leq r$ *and* $\rho(f^*, dx) \leq r$ *for some* $r \in (0,1)$.

(d) $\mathrm{dist}(f\ ,\ H^\infty_{qxq}) < r$ *and* $\mathrm{dist}(f^*\ ,\ H^\infty_{qxq}) < r$ *for some* $r \in (0,1)$.

COROLLARY 6.6. *Let* $w \in L^1_{qxq}$ *be hermitian, positive semi-definite and suppose* log det w *is summable. Then there exists two factorizations of* w; *that is,* $w = \sigma^* \sigma = \delta\, \delta^*$ *where* σ, $\delta \in H^2_{qxq}$, σ *and* δ *are both outer functions. Let*

$$f = \delta^{-1} \sigma^* = \delta^* \sigma^{-1}$$

then f is unitary and

$$w = \delta\, f\, \sigma.$$

And the following properties are equivalent to the conditions (a), (b), (c) and (d) of the last corollary:

(e) $\rho(I, w) < r$ *and* $\rho(I, w^{-1}) < r$ *for some* $r \in (0,1)$.

(f) There exists $r \in (0,1)$ *such that*

$$\begin{pmatrix} r.w.dx & w.dx \\ w^*.dx & r.w.dx \end{pmatrix} \text{ and } \begin{pmatrix} r.w.dx & w^{-1}.dx \\ (w^{-1})^*.dx & r.w.dx \end{pmatrix}$$

are matricially weakly positive.

(g) For the Hilbert transform H

$$\mathrm{tr} \int_0^{2\pi} (H\phi)\, w\, dx\, (H\phi)^* \quad \leq \quad a \quad \mathrm{tr} \int_0^{2\pi} \phi\, w\, dx\, \phi^*$$

and

$$\mathrm{tr} \int_0^{2\pi} (H\phi)\, w^{-1}\, dx\, (H\phi)^* \quad \leq \quad a \quad \mathrm{tr} \int_0^{2\pi} \phi\, w^{-1}\, dx\, \phi^*$$

for all $\phi \in P(C_{qxq})$, *where* $a > 0$.

REMARK. Let f unitary such that T_f is invertible, in view of Proposition 4.1. (6) we have that $f = \eta^* \gamma = (\gamma^* \eta)^{-1}$ where $\eta, \eta^{-1}, \gamma, \gamma^{-1} \in H^2_{qxq}$ and $\eta \eta^* = (\gamma \gamma^*)^{-1}$. Let $w = \eta f \gamma^{-1}$ then $w = \eta \eta^* = (\gamma \gamma^*)^{-1}$. We have that w is hermitian, positive semi-definite and log det w is summable, thus we can apply the last corollary. Therefore Corollary 6.6. gives a representation theorem for those unitary f for which T_f is invertible.

REFERENCES

[1] AROCENA, R., and M. COTLAR; Generalized Toeplitz kernels, Hankel forms and Sarason's commutation theorem; *Acta Científica Venezolana*, Vol. **33**, (1982), 89 - 98.

[2] AROCENA, R., M. COTLAR and C. SADOSKY; Weighted inequalities in L^2 and lifting properties; *Adv. Math, Sup.Studies*, V.**7A**, (1981), 95 - 128.

[3] BONSALL, F. F.; Dual extremum problems in the theory of functions; *J. London Math. Soc.* **31** (1956), 105 - 110.

[4] COTLAR, M. and C. SADOSKY; The Helson-Szegö theorem and a related class of modified Toeplitz kernels, *Proc. Symp. Pure Math. AMS.* **35**: I (1979), 383 - 407.

[5] DEVINATZ, A. Toeplitz operators on H^2 spaces. *Trans. Amer. Math. Soc.* **112**, 304 - 317, 1964.

[6] DOMINGUEZ, M.; Different kinds of positivity;(submitted).

[7] DOMINGUEZ, M.; (in press). A matricial extension of the Helson-Sarason theorem and a characterization of some multivariate linearly completely regular processes; *Journal of Multivariate Analysis*, **31**; (1989).

[8] HELSON, H. and D. LOWDENSLAGER. Prediction theory and Fourier series in several variables; *Acta Math.* **99**, (1958), 165 - 202.

[9] HELSON, H., D. SARASON; Past and Future; *Math. Scand.*, **21**, (1967), 5-16.

[10] HELSON, H., and G. SZEGÖ; A problem in prediction theory; *Am. Math. Pura Appl.* **51**, (1960), 107 - 138.

[11] IBRAGIMOV, I.A. and Y.A.. ROZANOV; Gaussian random processes; Applications of Mathematics, **9**, Springer Verlag, New York, (1978).

[12] MASANI, P.; Recent trends in multivariate prediction theory. *Multivariate Analysis* (P. R. Krishnaiah, Ed.), Ac. Press, New York; (1966); 351 - 382.

[13] MASANI, P.; Shift invariant spaces and prediction theory, *Acta Math.* **107** (1962), 275 - 290.

[14] Sz. NAGY, B. and C. FOIAS; Dilations des Commutants d'operateurs, *Comptes Rendus* , **266**, (1968), 32 - 42.

[15] NIKOL'SKII, N. K.; Treatise on the shift operator, *Springer Verlag*, 1986.

[16] PAGE, L.; Applications of the Sz. Nagy and Foias lifting theorem; *Indiana University Math. Journal* , **20**, 2, (1970), 135 - 145.

[17] PELLER, V.V. and S.V. HRUSCEV; Hankel operators, best approximations and stationary gaussian processes. LOMI preprint E-4-81 Leningrad, (1981).

[18] POURAHMADI, M.; A matricial extension of the Helson - Szegö Theorem and its application in multivariate prediction; *Journal of multivariate analysis* **16**; 265 - 275, (1985).

[19] POUSSON, H.R. Systems of Toeplitz operators on H^2. *Proceedings American Mathematical Society*, **19**; (1968); 603 - 608.

[20] POUSSON , H.R. Systems of Toeplitz operators on H^2, II. *Trans. Amer. Math. Soc.* **133**, (1968); 527 - 536.

[21] RABINDRANATHAN, M.; Generalized Toeplitz operators, J. Math. Mech., **19**, (1969), 195 - 206.

[22] ROSENBERG, M.; The square integrability of matrix valued functions with respect to a non negative hermitian measure; *Duke Math. J.* **31**, (1964), 291 - 298.

[23] WEDDERBURN, J. H. M.; Lectures on matrices. Amer. Math. Soc. Colloq. Publ., vol. **17**, Amer. Math. Soc., Providence, R. I., 1934.

[24] WIDOM, H.; Inversion of Toeplitz matrices II, *Illinois J. Math.* **4** (1960), 88 - 89.

[25] WIDOM, H.; Inversion of Toeplitz matrices III. *Notices Amer. Math. Soc.* **7**, (1960), 63. (Abstract 564 - 246).

[26] WIENER, N. and P. MASANI; The prediction theory of multivariate stochastic processes, Part I. *Acta Math.* **98**, (1957), 111 - 150.

[27] WIENER, N. and P. MASANI; The prediction theory of multivariate stochastic processes, Part II. *Acta Math.* **99**, (1958), 93 - 137.

Facultad de Ciencias, Universidad Central de Venezuela.

Mailing address:

Apartado Postal 47.159,

Caracas 1041 - A,

Venezuela.

Operator Theory:
Advances and Applications, Vol. 50

A HERMITE THEOREM FOR MATRIX POLYNOMIALS

Harry Dym*

An analogue of the Hermite theorem for the number of zeros in a half plane for a scalar polynomial is obtained for a class of $m \times m$ matrix polynomials by (finite dimensional) reproducing kernel Krein space methods. The paper, which is largely expository, is partially modelled on an earlier paper with N.J. Young which developed similar analogues of the Schur-Cohn theorem for matrix polynomials. More complete results in a somewhat different formulation have been obtained by Lerer and Tismenetsky by other methods. New proofs of some recent results on the distribution of the roots of certain matrix polynomials which are associated with invertible Hermitian block Hankel and block Toeplitz matrices are presented as an application of the main theorem.

CONTENTS

---

* The author wishes to express his thanks to Renee and Jay Weiss for endowing the chair which supported this research.

1. INTRODUCTION

In a recent paper [DY] with Nicholas Young, finite dimensional reproducing kernel Krein spaces were used to establish a Schur Cohn theorem for matrix polynomials. One of the purposes of the present note is to explain how to adapt the methods of that paper to obtain the analogue for matrix polynomials of the corresponding theorem of Hermite, which gives information on the number of zeros in the open upper half plane $\mathbb{C}_+$ rather than in the open unit disc $\mathbb{D}$. The adaptation is not totally routine because the Hardy space $H^2(\mathbb{C}_+)$ of the halfplane is not closed under multiplication by polynomials and therefore a number of steps in the proof of the main theorem of [DY] must be modified. We shall focus on these and touch lightly on those steps which, roughly speaking, amount to little more than a routine passage from the disc to the halfplane which can be effected almost automatically by invoking the changes of notation indicated in the following table, wherein $\mathbb{T}$ denotes the unit circle, $\mathbb{R}$ the real numbers, H^2_m the space of $m \times 1$ vector valued functions with components in H^2, and the superscript $*$ denotes conjugate transpose for a matrix and complex conjugate for a number. In addition we shall discuss some applications of the main theorem in Sections 5 and 6.

	$\mathbb{T}$	$\mathbb{R}$
$F^{\#}(\lambda)$	$F(1/\lambda^*)^*$	$F(\lambda^*)^*$
$\rho_\omega(\lambda)$	$1-\lambda\omega^*$	$-2\pi i(\lambda-\omega^*)$
$b_\omega(\lambda)$	$(\lambda-\omega)/(1-\lambda\omega^*)$	$(\lambda-\omega)/(\lambda-\omega^*)$
$\langle f,g\rangle$	$\frac{1}{2\pi}\int_0^{2\pi} g(e^{i\theta})^* f(e^{i\theta})d\theta$	$\int_{-\infty}^{\infty} g(\lambda)^* f(\lambda)d\lambda$
H^2_m	$H^2_m(\mathbb{D})$	$H^2_m(\mathbb{C}_+)$
allpass	analytic and unitary at every point $\lambda \in \mathbb{T}$	analytic and unitary at every point $\lambda \in \mathbb{R}$

Table 1

Recall that: A rational $m \times m$ matrix valued function F is said to be *nonsingular* if its determinant is not identically equal to zero. A point $\alpha \in \mathbb{C}$ is said to be a *pole* of a rational $m \times m$ matrix valued function F if one or more entries in F has a pole at α (or equivalently if F admits a matrix Laurent expansion $F(\lambda) = \Sigma_{j=-\nu}^{\infty} F_j(\lambda - \alpha)^j$ with matrix coefficients F_j where $\nu \geq 1$ and $F_{-\nu}$ is nonzero). A point $\alpha \in \mathbb{C}$ is said to be a *zero* of a nonsingular rational $m \times m$ matrix valued function $F(\lambda)$ if there exists an $m \times 1$ vector valued polynomial f such that (1) Ff is analytic at α, (2) $(Ff)(\alpha) = 0$, and (3) $f(\alpha) \neq 0$. If F is already analytic at α, then it suffices to choose $f = f(\alpha)$ constant. Moreover, it is readily checked that a point $\alpha \in \mathbb{C}$ is a zero of a nonsingular rational $m \times m$ matrix function F if and only if it is a pole for F^{-1}.

The need for care in the definition of a zero of a rational matrix valued function arises because even an allpass function can have both a zero and a pole at the same point $\alpha \in \mathbb{C}$ (neither of which appear in the determinant) as is illustrated by the example

$$F(\lambda) = \text{diag}\{b_\alpha(\lambda), b_\alpha(\lambda)^{-1}\} .$$

If F is a matrix polynomial, then $\delta_j(F)$ will denote the degree of the j'th column of F, i.e., the degree of the highest power of λ occurring in the j'th column with nonzero vector coefficient.

Throughout this paper

$$N(\lambda) = N_0 + N_1\lambda + \ldots + N_k\lambda^k$$

will denote an $m \times m$ matrix polynomial with column degrees

$$\delta_j(N) = d_j , \qquad j = 1, \ldots, m , \tag{1.1}$$

so that

$$k = \max\{d_j, \ j = 1, \ldots, m\} , \tag{1.2}$$

and we shall let

$$\delta(\lambda) = \text{diag}\{(\lambda + i)^{d_1}, \ldots, (\lambda + i)^{d_m}\} \tag{1.3}$$

and

$$\begin{aligned} \Delta(\lambda) &= \delta^{\#}(\lambda)\delta(\lambda)^{-1} \\ &= \text{diag}\left\{\left(\frac{\lambda - i}{\lambda + i}\right)^{d_1}, \ldots, \left(\frac{\lambda - i}{\lambda + i}\right)^{d_m}\right\} . \end{aligned} \tag{1.4}$$

N is said to be *column reduced* if $N(\lambda)\delta(\lambda)^{-1}$ tends to an invertible matrix as $\lambda \to \infty$, or equivalently if det $N(\lambda)$ is a polynomial of degree

$$d = \sum_{j=1}^{m} d_j . \tag{1.5}$$

We shall say that an $m \times m$ matrix polynomial D is a *reflection* of N if

(1) D is nonsingular,

(2) ND^{-1} is allpass (i.e., analytic and unitary on $\mathbb{R}$), and

(3) N and D are right coprime.

We shall see below that a necessary condition for N to admit a reflection is that its zeros are off $\mathbb{R}$ (see Theorem 2.2), whereas (by Theorem 2.1) a sufficient condition is that its zeros are nonconjugate with respect to $\mathbb{R}$ (i.e., if α is a zero of N then α^* is not a zero of N).

The notation $\mathbb{C}_-$ for the open lower half plane, $\delta_\pm(F)$ [resp. $\delta_0(F)$] for the number of zeros of a matrix polynomial F (counting multiplicities) in $\mathbb{C}_\pm$ [resp. $\mathbb{R}$] and $\mu_\pm(B)$ [resp. $\mu_0(B)$] for the number of eigenvalues of a matrix B (counting multiplicities) in $\mathbb{C}_\pm$ [resp. $\mathbb{R}$] will also be used.

The main result of this paper is now formulated as Theorem 1.1. It is perhaps well to bear in mind that if the matrix polynomial $N(\lambda)$ which appears therein has an invertible top coefficient N_k, then $N(\lambda)$ is automatically column reduced and the matrix E is just the $km \times km$ identity matrix.

THEOREM 1.1. *Let $N(\lambda) = N_0 + \lambda N_1 + \ldots + \lambda^k N_k$ be a nonsingular column reduced $m \times m$ matrix polynomial with column degrees $d_1, \ldots, d_m$ (so that $k = \max d_j$) which admits a reflection*

$$D(\lambda) = D_0 + D_1\lambda + \ldots + D_k\lambda^k$$

and let

$$E = \operatorname{diag}\{E_0, \ldots, E_{k-1}\} ,$$

where E_s, $s = 0, \ldots, k-1$, is the $m_s \times m$ matrix which is obtained from the identity matrix I_m by deleting the t'th row if $d_t \leq s$, $t = 1, \ldots, m$. Then the $d \times d$ matrix

$$P = \frac{iE}{2\pi}\left\{\begin{bmatrix} N_1^* & N_2^* & \dots & N_k^* \\ N_2^* & & & \\ \vdots & & & O \\ N_k^* & & & \end{bmatrix}\begin{bmatrix} N_0 & N_1 & \dots & N_{k-1} \\ 0 & N_0 & & \\ & & & \vdots \\ O & & & N_0 \end{bmatrix} - \begin{bmatrix} D_1^* & D_2^* & \dots & D_k^* \\ D_2^* & & & \\ \vdots & & & O \\ D_k^* & & & \end{bmatrix}\begin{bmatrix} D_0 & D_1 & \dots & D_{k-1} \\ 0 & D_0 & & \\ & & & \vdots \\ O & & & D_0 \end{bmatrix}\right\} E^* \tag{1.6}$$

is invertible and Hermitian, $\delta_0(N) = \delta_0(D) = \mu_0(P) = 0$ *and* $\delta_\pm(N) = \delta_\mp(D) = \mu_\pm(P)$.

The proof of this theorem is presented in Section 4. More complete results in a somewhat different formulation have been obtained by other methods by Lerer and Tismenetsky [LT].

We remark that the block matrix $B = -2\pi i E^* P E$ is the Bezoutian (as defined by Anderson and Jury [AJ]) of the matrix polynomials $N^\#, N, D^\#, D$, as follows readily from (4.6) and (4.9). The matrix $P = G^{-1}$ is the inverse of the Gram matrix of the finite dimensional reproducing kernel Krein space $\mathcal{K}(\Theta)$ which is introduced in Section 3. Thus, if the top coefficient N_k of $N(\lambda)$ is invertible, then $d_1 = \dots = d_m = k$, $E = I_{mk}$ and hence $(i/2\pi)B = P = G^{-1}$ is the inverse of the Gram matrix of a reproducing kernel Krein space. If $(i/2\pi)B$ is not invertible, then presumably it is a generalized inverse of the Gram matrix of a structured indefinite inner product space of the type introduced and referred to as i^3 spaces in [AD2]. Although a number of details still have to be checked out, the recent work of Wimmer [W] which identifies Bezoutians as generalized inverses gives additional weight to this conjecture.

Unfortunately, Theorem 1.1 is not easy to implement because of the need to calculate the reflection D. Difficulties of the same sort apply to the results of [LT] also. The case of scalar polynomials is a happy exception because if $m = 1$, then $D = N^\#$ and hence the $k \times k$ matrix

$$P = \frac{i}{2\pi}\left\{\begin{bmatrix} N_1^* & N_2^* & \dots & N_k^* \\ N_2^* & & & \\ \vdots & & O & \\ N_k^* & & & \end{bmatrix}\begin{bmatrix} N_0 & N_1 & \dots & N_{k-1} \\ 0 & N_0 & & \\ & & & \vdots \\ O & & & N_0 \end{bmatrix}\right.$$

$$\left. - \begin{bmatrix} N_1 & N_2 & \dots & N_k \\ N_2 & & & \\ \vdots & & O & \\ N_k & & & \end{bmatrix}\begin{bmatrix} N_0^* & N_1^* & \dots & N_{k-1}^* \\ 0 & N_0^* & & \\ & & & \vdots \\ O & & & N_0^* \end{bmatrix}\right\} E^*$$

is expressable totally in terms of the coefficients of N and Theorem 1.1 then reduces to the classical theorem of Hermite. For a statement and proof of the refined version (a less precise form appears to be due to Fujiwara [Fuj]), see Theorem 2.4 of Heinig and Rost [HR] and/or Section 3 of Krein and Naimark [KN]. Another approach which seems to be rather natural in terms of the present development is sketched in Section 5.

2. REFLECTIONS OF MATRIX POLYNOMIALS

In this section we shall present sufficient conditions for the existence of a reflection.

We shall say that the zeros $\alpha_1, \dots, \alpha_n$ of N are nonconjugate with respect to $\mathbb{R}$ if $\alpha_i - \alpha_j^* \neq 0$ for $i, j = 1, \dots, n$.

THEOREM 2.1. *Let N be a nonsingular $m \times m$ matrix polynomial whose zeros are nonconjugate with respect to $\mathbb{R}$. Then N admits a reflection D.*

PROOF. The proof can be broken into two steps the first of which consists of constructing a nonsingular matrix polynomial D such that ND^{-1} is allpass on $\mathbb{R}$. This can be done by successive extraction just as in Lemma 2.1 of [DY] except that now the formula given there for $b_\omega(\lambda)$ should be replaced by its counterpart for the upper half plane as indicated in the table in Section 1. Correspondingly, factors such as $(1 - \alpha_j^* \lambda)$ which appear in the proof of Lemma 2.1 of [DY] get replaced by $\lambda - \alpha_j^*$. The supplementary assumption that N has no conjugate zeros (which has not been used to this point) then guarantees that N and D are right coprime, just as in [DY]. ∎

THEOREM 2.2. *Let N be a nonsingular $m \times m$ matrix polynomial which* *admits a reflection D and let $\Theta = ND^{-1}$. Then:*

(1) *$N, D, N^{\#}$ and $D^{\#}$ are invertible on $\mathbb{R}$.*

(2) *$\Theta(\lambda)$ tends to a unitary constant matrix $\Theta(\infty)$ as $\lambda \to \infty$.*

(3) *$\delta_j(D) = \delta_j(N)$ for $j = 1, \ldots, m$.*

(4) *D is column reduced if and only if N is.*

PROOF. Suppose first that $N(\alpha)\xi = 0$ for some $\alpha \in \mathbb{R}$ and $\xi \in \mathbb{C}^m$. Then

$$D(\alpha)^* D(\alpha)\xi = N(\alpha)^* N(\alpha)\xi = 0 .$$

But this implies that

$$D(\alpha)\xi = 0 ,$$

which is not compatible with the assumption that N and D are right coprime unless $\xi = 0$. Thus N has no real zeros, nor does D by the very same argument. The same holds true for $N^{\#}$ and $D^{\#}$ since their zeros are just the conjugates of the zeros of N and D, respectively.

Now, since N and D are nonsingular, there exist a pair of finite Blaschke-Potapov products Ψ_1 and Ψ_2 such that $\Psi_1^{-1}N$ and $\Psi_2^{-1}D$ are matrix polynomials which have no zeros in $\bar{\mathbb{C}}_+$. Thus

$$\Psi_3 = \Psi_1^{-1}N(\Psi_2^{-1}D)^{-1} = \Psi_1^{-1}\Theta\Psi_2$$

is a rational allpass function with no zeros or poles in $\bar{\mathbb{C}}_+$. The formula

$$\Psi_3^{\#}(\lambda)\Psi_3(\lambda) = I_m \tag{2.1}$$

implies further that Ψ_3 has no zeros or poles in $\bar{\mathbb{C}}_-$. Consequently Ψ_3 is a matrix polynomial which, in view of (2.1), must in fact be a unitary constant matrix. Therefore, since Ψ_1 and Ψ_2 are just finite products of elementary factors of the form $I_m + \{b_\alpha(\lambda) - 1\}\xi\xi^*$ with $\alpha \in \mathbb{C}_+$ and ξ a unit vector in $\mathbb{C}^m$, it is readily seen that $\Theta(\lambda)$ tends to a unitary constant as $\lambda \to \infty$.

Next, the formula

$$\begin{aligned}\lim_{\lambda\to\infty} N(\lambda)\delta(\lambda)^{-1} &= \lim_{\lambda\to\infty} \Theta(\lambda)D(\lambda)\delta(\lambda)^{-1} \\ &= \Theta(\infty)\lim_{\lambda\to\infty} D(\lambda)\delta(\lambda)^{-1}\end{aligned}$$

implies that the last limit on the right has nonzero columns since $\Theta(\infty)$ is unitary. Therefore $\delta_j(D) = \delta_j(N)$ for $j = 1, \ldots, m$, by the very same formula, and D is clearly column reduced if and only if N is. ∎

3. THE REPRODUCING KERNEL KREIN SPACE $\mathcal{K}(\Theta)$

In this section we introduce a finite dimensional reproducing kernel Krein space $\mathcal{K}(\Theta)$ for every $m \times m$ rational allpass function Θ on $\mathbb{R}$. The statement means that there is an $m \times m$ matrix valued function $K_\omega(\lambda)$ defined on $\Omega \times \Omega$ such that, for every choice of $f \in \mathcal{K}(\Theta)$, $v \in \mathbb{C}^m$ and $\omega \in \Omega$,

(1) $K_\omega v \in \mathcal{K}(\Theta)$ and

(2) $[f, K_\omega v]_{\mathcal{K}} = v^* f(\omega)$

wherein Ω denotes the domain of analyticity of Θ in $\mathbb{C}_+$ and $[\ ,\ ]_{\mathcal{K}}$ denotes the indefinite inner product for $\mathcal{K} = \mathcal{K}(\Theta)$.

The fact that the reproducing kernel for such a space is unique and can be written in a number of different ways (which must therefore match) will be used to advantage in the sequel.

The main result is adapted from Theorem 6.6 of [AD1], which is valid in both the disc and the upper half plane. We shall, however, formulate the result directly in terms of Θ as in [DY] rather than in terms of the inner factors Ψ_1 and Ψ_2 which appear in the left inner coprime factorization $\Theta = \Psi_1^{-1}\Psi_2$ as was done in [AD1].

THEOREM 3.1. *Let Θ be a rational $m \times m$ allpass function and let Ω denote its domain of analyticity in $\mathbb{C}_+$. Then*

(1) *The space*

$$\mathcal{K}(\Theta) = (H_m^2 + \Theta H_m^2) \ominus (H_m^2 \cap \Theta H_m^2)$$

is a finite dimensional subspace of $L_m^2(\mathbb{R})$ consisting of $m \times 1$ vector valued rational functions which are analytic in Ω.

(2) *The intersection $\mathcal{K}_1 \cap \mathcal{K}_2$ of the subspaces*

$$\mathcal{K}_1 = (H_m^2 + \Theta H_m^2) \ominus H_m^2$$

and

$$\mathcal{K}_2 = (H_m^2 + \Theta H_m^2) \ominus \Theta H_m^2$$

is equal to zero.

(3) *The space* $\mathcal{K} = \mathcal{K}(\Theta)$ *is the direct sum of* $\mathcal{K}_1$ *and* $\mathcal{K}_2$*:*

$$\mathcal{K} = \mathcal{K}_1 \dot{+} \mathcal{K}_2 \ .$$

(4) *The space* $\mathcal{K} = \mathcal{K}(\Theta)$ *endowed with the indefinite inner product*

$$[f,g]_{\mathcal{K}} = \langle f_2, g_2 \rangle - \langle f_1, g_1 \rangle \ ,$$

where the f_i *and* g_i *are the components of* f *and* g *respectively relative to the direct sum decomposition in (3) is a finite dimensional reproducing kernel Krein space with reproducing kernel*

$$K_\omega(\lambda) = \frac{I_m - \Theta(\lambda)\Theta(\omega)^*}{\rho_\omega(\lambda)} \tag{3.1}$$

(5) *The MacMillan degree of* Θ *is equal to the dimension of* $\mathcal{K}(\Theta)$.

PROOF. The proof is much the same as the proof of Theorem 3.1 of [DY]. ∎

4. GRAM MATRICES AND THE DISTRIBUTION OF ZEROS

In this section we shall obtain a second description of the space $\mathcal{K}(\Theta)$ for $\Theta = ND^{-1}$, when D is a reflection of the given nonsingular column reduced matrix polynomial N. This will yield a second formula for the reproducing kernel in terms of a basis for $\mathcal{K}(\Theta)$ and the Gram matrix of this basis. A formula for the Gram matrix in terms of the coefficients of the polynomials N and D will then be obtained by choosing a good basis (the analogue of the expedient basis of [DY]) and then matching the two expressions for the reproducing kernel.

THEOREM 4.1. *Let* N *be a nonsingular column reduced* $m \times m$ *matrix polynomial with column degrees* $d_1, \ldots, d_m$ *and reflection* D. *Let* $\Theta = ND^{-1}$ *and let* $\delta(\lambda)$ *and* $\Delta(\lambda)$ *be defined as in (1.3) and (1.4), respectively. Then:*

(1) $N, D, N^\#$ *and* $D^\#$ *are all invertible on* $\mathbb{R}$.

(2)
$$H_m^2 + \Theta H_m^2 = (N^\#)^{-1}\delta H_m^2. \tag{4.1}$$

(3)
$$H_m^2 \cap \Theta H_m^2 = N\delta^{-1} H_m^2. \tag{4.2}$$

(4)
$$\mathcal{K}(\Theta) = (N^\#)^{-1}\delta(H_m^2 \ominus \Delta H_m^2). \tag{4.3}$$

PROOF. (1) has already been established in Theorem 2.2.

Next, let

$$\delta^{-1}N^{\#} = B_1Q_1 \quad \text{and} \quad \delta^{-1}D^{\#} = B_2Q_2 \ ,$$

where B_1 [resp. B_2] is a Blaschke-Potapov product which contains all the zeros of $\delta^{-1}N^{\#}$ [resp. $\delta^{-1}D^{\#}$] in $\mathbb{C}_+$. Then, in view of (1), Q_1 and Q_2 have no zeros in $\bar{\mathbb{C}}_+$. Moreover, since $B_1(\infty)$, $B_2(\infty)$ and $\Theta(\infty)$ are unitary, N is column reduced and

$$\lim_{\lambda\to\infty} \delta(\lambda)^{-1}N^{\#}(\lambda) = \lim_{\lambda\to\infty} \delta(\lambda)^{-1}D^{\#}(\lambda)\Theta^{\#}(\lambda) \ ,$$

it follows that $Q_1(\lambda)$ and $Q_2(\lambda)$ tend to finite invertible limits as $\lambda \to \infty$ also. Thus Q_1 and Q_2 are invertible in the Wiener algebra $(\mathcal{W}_{m\times m})_+$ of $m \times m$ matrix valued functions of the form

$$C + \int_0^\infty e^{i\lambda t}F(t)dt \ ,$$

where C is a constant $m \times m$ matrix and $F \in L^1_{m\times m}$ on the half line $t \geq 0$. In particular, this guarantees that

$$Q_1H^2_m = H^2_m \quad \text{and} \quad Q_2H^2_m = H_m$$

(i.e., Q_1 and Q_2 are outer). Furthermore, since N and D are right coprime, it is readily checked first that $N^{\#}$ and $D^{\#}$ are left coprime and then that B_1 and B_2 are left inner coprime. Thus, by Theorem 14.10 of Fuhrmann [Fu] (adapted to $\mathbb{C}_+$), there exist a pair of $m \times m$ matrix valued functions A_1 and A_2 in $H^\infty_{m\times m}$ such that

$$B_1(\lambda)A_1(\lambda) + B_2(\lambda)A_2(\lambda) = I_m$$

for every point $\lambda \in \mathbb{C}_+$. Consequently,

$$\begin{aligned} \delta^{-1}(N^{\#}H^2_m + D^{\#}H^2_m) &= B_1H^2_m + B_2H^2_m \\ &\supset B_1A_1H^2_m + B_2A_2H^2_m \\ &\supset (B_1A_1 + B_2A_2)H^2_m = H^2_m \ . \end{aligned}$$

But this in turn implies that

$$H^2_m + \Theta H^2_m \supset (N^{\#})^{-1}\delta H^2_m \ . \tag{4.4}$$

The selfevident inclusion

$$\begin{aligned} N^{\#}H^2_m + D^{\#}H^2_m &= \delta\{\delta^{-1}N^{\#}H^2_m + \delta^{-1}D^{\#}H^2_m\} \\ &\subset \delta H^2_m \end{aligned}$$

leads readily to the opposite inclusion of (4.4), which serves to complete the proof of (4.1).

To obtain (3), let us first express

$$N\delta^{-1} = Q_3B_3 \quad \text{and} \quad D\delta^{-1} = Q_4B_4 \ ,$$

where B_3 and B_4 are Blaschke-Potapov products which contain the upper half plane roots of $N\delta^{-1}$ and $D\delta^{-1}$, respectively. Then, since N and D are column reduced, it follows much as in the proof of (2) that Q_3 and Q_4 are invertible in $\bar{\mathbb{C}}_+ \cup \{\infty\}$ and hence that

$$Q_3H_m^2 = H_m^2 \quad \text{and} \quad Q_4H_m^2 = H_m^2 \ .$$

Next, since N and D are right coprime, it is readily checked that B_3 and B_4 are right inner coprime. Thus, by another application of Theorem 14.10 of [Fu], there exist a pair of $m \times m$ matrix valued functions A_3 and A_4 in $H_{m\times m}^\infty$ such that

$$A_3(\lambda)B_3(\lambda) + A_4(\lambda)B_4(\lambda) = I_m$$

for every point $\lambda \in \mathbb{C}_+$.

Now, if $f \in H_m^2 \cap ND^{-1}H_m^2$, then $f = ND^{-1}g$ for some $g \in H_m^2$. Therefore, since

$$\begin{aligned} ND^{-1} &= N\delta^{-1}(D\delta^{-1})^{-1} \\ &= Q_3B_3(Q_4B_4)^{-1} \ , \end{aligned}$$

it follows that

$$\begin{aligned} f &= Q_3B_3B_4^{-1}Q_4^{-1}g \\ &= Q_3B_3(A_3B_3 + A_4B_4)B_4^{-1}Q_4^{-1}g \\ &= Q_3B_3\{A_3Q_3^{-1}f + A_4Q_4^{-1}g\} \\ &\in Q_3B_3H_m^2 \ . \end{aligned}$$

This proves that

$$H_m^2 \cap ND^{-1}H_m^2 \subset N\delta^{-1}H_m^2 \ . \tag{4.5}$$

On the other hand,

$$\begin{aligned} ND^{-1}H_m^2 &= N\delta^{-1}(D\delta^{-1})H_m^2 \\ &= N\delta^{-1}B_4^{-1}Q_4^{-1}H_m^2 \\ &= N\delta^{-1}B_4^{-1}H_m^2 \\ &\supset N\delta^{-1}H_m^2 \ , \end{aligned}$$

which, when combined with (4.5), serves to complete the proof of (4.2).

Finally, upon combining (4.1) and (4.2) with (1) of Theorem 3.1, it follows that $f \in \mathcal{K}(\Theta)$ if and only if $f = (N^{\#})^{-1}\delta g$ for some $g \in H_m^2$ and

$$0 = \langle f, N\delta^{-1}h \rangle$$

for every choice of $h \in H_m^2$. But this in turn implies that

$$\begin{aligned} 0 &= \langle N^{\#} f, \delta^{-1}h \rangle \\ &= \langle \delta g, \delta^{-1}h \rangle \\ &= \langle g, \Delta h \rangle \end{aligned}$$

for every choice of $h \in H_m^2$ and hence that

$$g \in H_m^2 \ominus \Delta H_m^2 \; .$$

The rest is plain. ∎

THEOREM 4.2. *Let N be a nonsingular column reduced $m \times m$ matrix polynomial with column degrees $d_1, \ldots, d_m$ and reflection D. Let φ_i, $i = 1, \ldots, d$, be a basis for the $d = \Sigma_{j=1}^m d_j$ dimensional Hilbert space $H_m^2 \ominus \Delta H_m^2$, let*

$$f_j = (N^{\#})^{-1}\delta\varphi_j \; , \qquad j = 1, \ldots, d \; ,$$

and let G denote the Gram matrix of the basis $f_1, \ldots, f_d$ of $\mathcal{K}(\Theta)$, i.e., the $d \times d$ matrix with ij entry

$$G_{ij} = [f_j, f_i]_{\mathcal{K}} \; , \qquad i,j = 1, \ldots, d \; .$$

Then G is invertible and

$$\frac{N^{\#}(\lambda)N^{\#}(\omega)^* - D^{\#}(\lambda)D^{\#}(\omega)^*}{\rho_\omega(\lambda)} = \sum_{i,j=1}^{d} \delta(\lambda)\varphi_i(\lambda)(G^{-1})_{ij}\varphi_j(\omega)^*\delta(\omega)^* \tag{4.6}$$

for every choice of λ and ω in $\mathbb{C}$.

PROOF. By Theorem 4.1, $f_1, \ldots, f_d$ is a basis for the reproducing kernel Krein space $\mathcal{K}(\Theta)$. Therefore, since such a space contains no nonzero isotropic vectors, the Gram matrix G is invertible and it is readily checked by direct calculation that

$$K_\omega(\lambda) = \sum_{i,j=1}^{d} f_i(\lambda)(G^{-1})_{ij} f_j(\omega)^* \tag{4.7}$$

is a reproducing kernel for $\mathcal{K}(\Theta)$. Thus, as a reproducing kernel Krein space admits only one reproducing kernel, the right hand sides of the two formulas (3.1) and (4.7) for $K_\omega(\lambda)$ must agree:

$$\frac{I_m - \Theta(\lambda)\Theta(\omega)^*}{\rho_\omega(\lambda)} = \sum_{i,j=1}^{d} f_i(\lambda)(G^{-1})_{ij} f_j(\omega)^* \tag{4.8}$$

for every choice of λ and ω in Ω, the domain of analyticity of Θ in $\mathbb{C}_+$. Formula (4.6) drops out upon multiplying both sides of (4.8) on the left by $N^{\#}(\lambda)$ and on the right by $N^{\#}(\omega)$. ∎

The next step is to choose a convenient basis for $H_m^2 \ominus \Delta H_m^2$. We shall use the basis

$$\left\{ \frac{\lambda^s}{(\lambda+i)^{d_t}} e_t \ , \quad s = 0, \ldots, d_t - 1 \ , \quad t = 1, \ldots, m \right\}$$

with the understanding that no entries are counted for the standard basis vector e_t of $\mathbb{C}^m$ if $d_t = 0$. This basis will be labelled $\varphi_1, \ldots, \varphi_d$ according to the order in which the elements $\delta(\lambda)\varphi_1(\lambda), \ldots, \delta(\lambda)\varphi_d(\lambda)$ appear in the sequence

$$e_1, \ldots, e_m; \ \lambda e_1, \ldots, \lambda e_m; \ \ldots \ ; \lambda^{k-1} e_1, \ldots, \lambda^{k-1} e_m$$

(just as in [DY]). This is meaningful because

$$\delta(\lambda) \frac{\lambda^s}{(\lambda+i)^{d_t}} e_t = \lambda^s e_t \ .$$

THEOREM 4.3. *Let N be a nonsingular column reduced $m \times m$ matrix polynomial with column degrees $d_1, \ldots, d_m$ and reflection D. Then the Gram matrix G of the basis $f_j = (N^{\#})^{-1}\delta\varphi_j$, $j = 1, \ldots, d$, of $\mathcal{K}(\Theta)$, when $\delta\varphi_1, \ldots, \delta\varphi_d$ are chosen as described just above, is equal to the inverse of the matrix P which is described in Theorem 1.1.*

PROOF. Because of the chosen ordering, the functions $\delta\varphi_j$, $j = 1, \ldots, d$, fall naturally into $k = \max_j(d_j)$ blocks, numbered 0 to $k-1$ according to the power of λ which appears. Let E_s, $s = 0, \ldots, k-1$, be the matrix which is obtained from the identity matrix I_m by deleting the t'th row if $d_t \leq s$; E_s is an $m_s \times m$ matrix, where m_s is equal to the number of entries in the sequence $\delta\varphi_1, \ldots, \delta\varphi_d$ which are of the form $\lambda^s e$ for some $e \in \mathbb{C}^m$. In fact E_s^* is just the stacked array of the column vectors appearing in this list as multipliers of λ^s. Corresponding to this division of the basis $f_1, \ldots, f_d$, G^{-1} can be viewed as a $k \times k$ block matrix with blocks g^{pq}, $p, q = 0, \ldots, k-1$, of size $m_p \times m_q$. We shall also write $G^{-1} = [\gamma_{ij}]$, $i, j = 1, \ldots, d$. Then the right hand side of (4.6) can be reexpressed in the form

$$\sum_{i,j=1}^{d} \delta(\lambda)\varphi_i(\lambda)\gamma_{ij}\varphi_j(\omega)^*\delta(\omega)^* = \sum_{p,q=0}^{k-1} \lambda^p E_p^* g^{pq} E_q \omega^{*q} , \tag{4.9}$$

whereas the left hand side of (4.6) is equal to

$$\frac{N^{\#}(\lambda) - N^{\#}(\omega^*)}{-2\pi i(\lambda - \omega^*)} N(\omega^*) + \frac{D^{\#}(\omega^*) - D^{\#}(\lambda)}{-2\pi i(\lambda - \omega^*)} D(\omega^*)$$

since $N^{\#}N = D^{\#}D$.

Now for any matrix polynomial

$$A(\lambda) = \sum_{j=0}^{k} A_j \lambda^j ,$$

it is readily checked that

$$\frac{A(\lambda) - A(\mu)}{\lambda - \mu} = \sum_{i=0}^{k-1} \lambda^i \sum_{j=i}^{k-1} A_{j+1} \mu^{j-i}$$

and hence that the coefficient of $\lambda^s \omega^{*t}$ in

$$\frac{N^{\#}(\lambda) - N^{\#}(\omega^*)}{\lambda - \omega^*} N(\omega^*)$$

is equal to the coefficient of ω^{*t} in

$$\left(\sum_{j=s}^{k-1} N_{j+1}^* \omega^{*j-s} \right) \sum_{i=0}^{k} N_i \omega^{*i} .$$

But this is equal to

$$[N_{s+1}^* \quad N_{s+2}^* \quad \cdots \quad] \begin{bmatrix} N_t \\ N_{t-1} \\ \vdots \end{bmatrix}$$

for as long as the indicated entries make sense. More precisely, the coefficient of $\lambda^s \omega^{*t}$, $s, t = 0, \ldots, k-1$, in

$$\frac{N^{\#}(\lambda) - N^{\#}(\omega^*)}{\lambda - \omega^*} N(\omega^*)$$

is the st block entry of the following matrix product:

$$\begin{bmatrix} N_1^* & N_2^* & \dots & N_k^* \\ N_2^* & N_3^* & & 0 \\ \vdots & \vdots & & \vdots \\ N_k^* & 0 & \dots & 0 \end{bmatrix} \begin{bmatrix} N_0 & N_1 & \dots & N_{k-1} \\ 0 & N_0 & \dots & N_{k-2} \\ \vdots & \vdots & & \vdots \\ 0 & 0 & \dots & N_0 \end{bmatrix} .$$

The rest is straightforward since a similar formula applies to the polynomial involving $D^{\#}$ and the matrices E_s^* are left invertible. ∎

We are now able to complete the proof of the main theorem.

PROOF OF THEOREM 1.1. Let $\hat{N} = \delta^{-1}N^{\#}$ and $\hat{D} = \delta^{-1}D^{\#}$. Then (4.6) can be reexpressed as

$$\frac{\hat{N}(\lambda)\hat{N}(\omega)^* - \hat{D}(\lambda)\hat{D}(\omega)^*}{\rho_\omega(\lambda)} = Z(\lambda)G^{-1}Z(\omega)^* , \tag{4.10}$$

where

$$Z = [\varphi_1 \cdots \varphi_d]$$

is the $m \times d$ matrix valued function with columns $\varphi_1, \dots, \varphi_d$. Since $\hat{N} \in (\mathcal{W}_{m\times m})_+$ and $\hat{N}$ is invertible at every point λ in the extended line $\mathbb{R} \cup \{\infty\}$ (by the statement and proof of Theorem 4.1), it follows just as in the proof of Lemma 6.2 of [D2] that

$$\dim \ker\{\underline{\underline{p}}\hat{N}^* \Big|_{H_m^2}\} = \delta_+(N^{\#}) ,$$

in which $\underline{\underline{p}}$ denotes the orthogonal projection of L_m^2 onto H_m^2. Moreover, since the present (4.10) is equivalent to (6.3) of [D2] with $X = [\hat{N} \quad \hat{D}]$, the proof of Theorem 6.2 of [D2] can easily be adapted to show that

$$\left(\sum_{s=1}^k b_s u_s\right)^* G^{-1} \sum_{t=1}^k b_t u_t = -\left\| \underline{\underline{p}}\hat{D}^* \sum_{t=1}^k b_t g_t \right\|^2 , \tag{4.11}$$

in which $g_1, \dots, g_k$ is a basis for the kernel of the operator $\underline{\underline{p}}\hat{N}^* \Big|_{H_m^2}$ and $u_1, \dots, u_k$ is a set of vectors in $\mathbb{C}^d$ which is generated by $g_1, \dots, g_k$ just as in Theorem 6.1 of [D2]. Furthermore, since $\hat{N}$ and $\hat{D}$ are left inner coprime, there exist a pair of functions A and B in $H_{m\times m}^\infty$ such that

$$\hat{N}(\lambda)A(\lambda) + \hat{D}(\lambda)B(\lambda) = I_m$$

in $\mathbb{C}_+$, by Theorem 14.10 of [Fu]. Thus

$$\begin{aligned} g &= \underline{\underline{p}}A^*\hat{N}^*g + \underline{\underline{p}}B^*\hat{D}^*g \\ &= \underline{\underline{p}}A^*\underline{\underline{p}}\hat{N}^*g + \underline{\underline{p}}B^*\underline{\underline{p}}\hat{D}^*g \end{aligned}$$

for every choice of $g \in H^2_m$. But this clearly implies that $\underline{\underline{p}}\hat{D}^*g$ and $\underline{\underline{p}}\hat{N}^*g$ cannot vanish simultaneously if g is nonzero. Therefore the right hand side of (4.11) is strictly negative and hence the span of the vectors $u_1, \ldots, u_k$ is a strictly negative k dimensional subspace of $\mathbb{C}^d$ with respect to the indefinite inner product induced by G^{-1}. Thus

$$k = \delta_+(N^\#) = \delta_-(N) \leq \mu_-(G^{-1}) = \mu_-(G) \tag{4.12}$$

and, similarly,

$$\delta_+(D^\#) = \delta_-(D) \leq \mu_+(G^{-1}) = \mu_+(G) \; . \tag{4.13}$$

Next, upon reexpressing (4.10) in terms of $\hat{N}_-$, $\hat{D}_-$ and Z_-, where $F_-(\lambda) = \delta(\lambda)^{-1}\delta(-\lambda)F(-\lambda)$, it follows in much the same way that

$$\delta_+(D) = \delta_+(\hat{D}_-) \leq \mu_-(G) \tag{4.14}$$

and

$$\delta_+(N) = \delta_+(\hat{N}_-) \leq \mu_+(G) \; . \tag{4.15}$$

Therefore, just as in the proof of Theorem 6.2 of [D2], it follows that

$$d = \delta_-(N) + \delta_+(N) \leq \mu_-(G) + \mu_+(G) = d$$

and hence that equality prevails in (4.12) and (4.15). Since D also had d zeros by Theorem 2.2, the same argument works with D in place of N and yields equality in (4.13) and (4.14) also.

The identification of G with P^{-1}, which was established in the last theorem, does the rest. ∎

5. THE REFINED HERMITE THEOREM FOR SCALAR POLYNOMIALS

In this section we shall indicate how to obtain the refined Hermite theorem for scalar polynomials from Theorem 1.1, by an iterative procedure which reduces the general case to the case covered by Theorem 1.1. In principle this reduction method should work for matrix polynomials also, but this has yet to be explored. In case of emergency, the strongest available results may be found in [LT].

To begin, let $c(\lambda)$ denote the greatest common divisor of the now scalar polynomials $N(\lambda)$ and $N^\#(\lambda)$ and suppose that c is normalized so that

$$c(\lambda) = (\lambda - \beta_1) \cdots (\lambda - \beta_s) = c^\#(\lambda) \; .$$

Then the proper substitute for formula (4.6) can now be expressed as

$$c(\lambda)\left\{\frac{b^{\#}(\lambda)b^{\#}(\omega)^* - b(\lambda)b(\omega)^*}{\rho_\omega(\lambda)}\right\}c(\omega)^* = \varphi(\lambda)\Gamma_0\varphi(\omega)^* \ , \tag{5.1}$$

where

$$b(\lambda) = N(\lambda)c(\lambda)^{-1} = a(\lambda-\alpha_1)\cdots(\lambda-\alpha_t) \ , \quad a \neq 0$$

$$\varphi(\lambda) = [1 \ \ \lambda \ \ \cdots \ \ \lambda^{n-1}]$$

$$\Gamma_k = \begin{bmatrix} \gamma_{kk} & \cdots & \gamma_{k,n-1} \\ \vdots & & \vdots \\ \gamma_{n-1,k} & \cdots & \gamma_{n-1,n-1} \end{bmatrix} \ , \qquad k = 0,\ldots,n-1 \ ,$$

$s+t=n$ and b and $b^{\#}$ are coprime.

Formula (5.1) clearly implies that

$$\varphi(\beta_1)\Gamma_0 = 0$$

and hence that

$$c_j(\lambda)\left\{\frac{b^{\#}(\lambda)b^{\#}(\omega)^* - b(\lambda)b(\omega)^*}{\rho_\omega(\lambda)}\right\}c_j(\omega)^* = \varphi_j(\lambda)\Gamma_0\varphi_j(\omega)^* \tag{5.2}$$

for $j = 1$, where

$$c_1(\lambda) = \frac{c(\lambda)}{\lambda-\beta_1} \quad \text{and} \quad \varphi_1(\lambda) = \frac{\varphi(\lambda)-\varphi(\beta_1)}{\lambda-\beta_1} \ .$$

But this procedure can be iterated. If $s > 1$, then (5.2) implies that

$$\varphi_1(\beta_2)\Gamma_0 = 0$$

and hence that (5.2) holds for $j = 2$ with

$$c_j(\lambda) = \frac{c_{j-1}(\lambda)}{\lambda-\beta_j} \quad \text{and}$$

$$\varphi_j(\lambda) = \frac{\varphi_{j-1}(\lambda)-\varphi_{j-1}(\beta_j)}{\lambda-\beta_j} \ .$$

The iteration continues until $j = s$ and hence, since the row vectors $\varphi(\beta_1), \varphi_1(\beta_2), \ldots, \varphi_{s-1}(\beta_s)$ are linearly independent members of the left null space of Γ_0, it follows that

$$\mu_0(\Gamma_0) \geq s \ . \tag{5.3}$$

Moreover, since $c_s(\lambda) = 1$, and the first j entries of the row vector $\varphi_j(\lambda)$ are equal to zero for $j = 1, \ldots, s$, it follows readily upon writing out the step by step calculations in detail, that

$$\frac{b^{\#}(\lambda)b^{\#}(\omega)^* - b(\lambda)b(\omega)^*}{\rho_\omega(\lambda)} = \varphi_s(\lambda)\Gamma_0\varphi_s(\omega)^*$$
$$= [1 \ \ \lambda \ \cdots \ \lambda^{t-1}]U_t(\beta_s)\cdots U_t(\beta_1)\Gamma_s U_t(\beta_1)^* \cdots U_t(\beta_s)^*[1 \ \ \omega \ \cdots \ \omega^{t-1}]^* \ ,$$

where $U_t(\alpha)$ is the $t \times t$ upper triangular Toeplitz matrix with top row equal to $[1 \ \ \alpha \ \cdots \ \alpha^{t-1}]$. But now as b and $b^{\#}$ are coprime and the Toeplitz matrices are invertible, it follows from Theorem 1.1 and the Sylvester inertia theorem that Γ_s is invertible and

$$\delta_{\pm}(b) = \mu_{\pm}(\Gamma_s) \ .$$

Consequently,

$$\delta_{\pm}(N) = \delta_{\pm}(b) + \delta_{\pm}(c)$$
$$= \mu_{\pm}(\Gamma_s) + \delta_{\pm}(c)$$

and

$$\delta_0(N) = \delta_0(c) = s - \delta_+(c) - \delta_-(c) \ .$$

Therefore, since

$$\mu_{\pm}(\Gamma_0) \geq \mu_{\pm}(\Gamma_s) \tag{5.4}$$

(by Schur complements and the Sylvester inertia theorem), it follows that

$$n = \mu_0(\Gamma_0) + \mu_-(\Gamma_0) + \mu_+(\Gamma_0)$$
$$\geq s + \mu_-(\Gamma_s) + \mu_+(\Gamma_s)$$
$$= s + t = n$$

and hence that equality prevails in (5.3) and (5.4). Thus, since

$$\delta_+(c) = \delta_-(c) \tag{5.5}$$

(which is equal to the number of complex conjugate pairs at zeros in N counting multiplicities), we see that

$$\delta_{\pm}(N) = \mu_{\pm}(\Gamma_0) + \delta_{\pm}(c) \tag{5.6}$$

and

$$\delta_0(N) = \mu_0(\Gamma_0) - 2\delta_+(c) \ . \tag{5.7}$$

But this is precisely the statement of the refined version of Hermite's theorem. As an extra benefit the preceding argument also serves to identify the matrix

$$-2\pi i\{U_t(\beta_s)\cdots U_t(\beta_1)\Gamma_s\, U_t(\beta_1)^*\cdots U_t(\beta_s)^*\}$$

as the Bezoutian for the polynomials $b^{\#}$ and b.

6. ON APPLICATIONS

In general Theorem 1.1 is, as we have already remarked, not easy to apply because of the difficulty of finding a reflection D for the given polynomial N. Nevertheless, there are some happy exceptions apart from the scalar case: Theorem 6.2 of [D2] may be obtained from Theorem 1.1, and Theorem 11.1 of [D1] may be obtained from its Schur Cohn counterpart in [DY]. The former deals with a pair of matrix polynomials associated with an invertible block Hankel matrix whereas the latter, which deals with a pair of matrix polynomials associated with a block Toeplitz matrix, is equivalent to a theorem which was first established (more or less simultaneously by different methods) by Alpay and Gohberg [AG] and Gohberg and Lerer [GL]. In both cases it turns out the matrix polynomials of interest are simply related to a matrix polynomial N and its reflection D.

We begin with a paraphrase of Theorem 6.2 of [D2].

THEOREM 6.1. *Let H_n be an invertible Hermitian block Hankel matrix with blocks of size $m \times m$ and inverse*

$$\Gamma_n = [\gamma_{ij}^{(n)}]\ , \qquad i,j = 0,\ldots,n\ .$$

Suppose further that $\gamma_{nn}^{(n)}$ is definite (which guarantees the invertibility of H_{n-1}) and let

$$C_n(\lambda) = \sum_{j=0}^{n} \gamma_{jn}^{(n)}\lambda^j\ ,$$

$$P_n(\lambda) = \lambda C_n(\lambda) - C_{n-1}(\lambda) + iC_n(\lambda)\ ,$$

and

$$Q_n(\lambda) = \lambda C_n(\lambda) - C_{n-1}(\lambda) - iC_n(\lambda)\ .$$

Then

$$\begin{array}{rcll} \delta_+(P_n) & = & \mu_-(H_n) & [\text{resp. } \mu_+(H_n)] \\ \delta_-(P_n) & = & \mu_+(H_n) & [\text{resp. } \mu_-(H_n)] \\ \delta_+(Q_n) & = & \mu_+(H_n) & [\text{resp. } \mu_-(H_n)] \\ \delta_-(Q_n) & = & \mu_-(H_n) & [\text{resp. } \mu_+(H_n)] \end{array}$$

if $\gamma_{nn}^{(n)}$ is positive [resp. negative] definite.

PROOF. Suppose first that $\gamma_{nn}^{(n)}$ is positive definite. Then the polynomials $E = E_n^K$ and $F = F_n^K$ which are defined in Lemma 9.1 of [D2] are invertible on $\mathbb{R}$ by Step 1 of Theorem 9.2, and left coprime by (9.24); furthermore $E^{-1}F$ is allpass by (9.8) (all the references are to [D2]). Moreover, E and F are both matrix polynomials of the form $A_0 + A_1\lambda + \cdots + A_{n+1}\lambda^{n+1}$ with A_{n+1} invertible. Thus, $E, F, E^\#$ and $F^\#$ are all column reduced (with $d_j = n+1$ for $j = 1, \ldots, m$). Consequently the polynomials $N = E^\#$ and $D = F^\#$ satisfy the hypotheses of Theorem 1.1, and (as follows by comparing the present (4.6) and (4.9) with (9.21) of [D2]), $2\pi P = \Gamma_n$ and the matrix E which appears in (1.6) is equal to the identity. This completes the proof in case $\gamma_{nn}^{(n)} > 0$ since in this instance

$$P_n = E_n\{2\gamma_{nn}^{(n)}\}^{\frac{1}{2}} , \qquad Q_n = F_n\{2\gamma_{nn}^{(n)}\}^{\frac{1}{2}}$$

and E_n and F_n are just E and F with $K = 0$.

The proof for $\gamma_{nn}^{(n)} < 0$ goes through in much the same way except that now one works with the normalization factor $\{-\gamma_{nn}^{(n)}\}^{\frac{1}{2}}$. ∎

From now on ρ_ω and $F^\#$ are defined according to the first column of the Table in Section 1 and $\delta_+(F)$ [resp. $\delta_-(F)$] will denote the number of zeros of the matrix polynomial F inside $\mathbb{D}$ [resp. outside $\bar{\mathbb{D}}$].

Our next objective is to explain how to extract the theorem of Alpay, Gohberg and Lerer from Theorem 1.1 of [DY]. To this end it is convenient to paraphrase the latter in the special case that the top coefficient of the matrix polynomial of interest is invertible as the next theorem.

THEOREM 6.2. *Let*

$$N(\lambda) = N_0 + \cdots + N_k\lambda^k \quad \text{and} \quad D(\lambda) = D_0 + \cdots + D_k\lambda^k$$

be $m \times m$ matrix polynomials such that N_k and D_0 are both invertible. Suppose further that N and D are right coprime and that ND^{-1} is allpass (i.e., analytic and unitary on

$\mathbb{T}$) *and let*

$$P = \left\{ \begin{bmatrix} N_k^* & 0 & & O \\ N_{k-1}^* & N_k^* & & \\ \vdots & \vdots & & \\ N_1^* & N_2^* & \cdots & N_k^* \end{bmatrix} \begin{bmatrix} N_k & N_{k-1} & \cdots & N_1 \\ 0 & N_k & \cdots & N_2 \\ & & & \vdots \\ O & & & N_k \end{bmatrix} - \begin{bmatrix} D_k^* & 0 & & O \\ D_{k-1}^* & D_k^* & & \\ \vdots & & & \\ D_1^* & D_2^* & \cdots & D_k^* \end{bmatrix} \begin{bmatrix} D_k & D_{k-1} & \cdots & D_1 \\ 0 & D_k & \cdots & D_2 \\ & & & \vdots \\ O & & & D_k \end{bmatrix} \right\} .$$

Then P is invertible and $\delta_\pm(N) = \mu_\pm(P)$.

Now, on to the application:

THEOREM 6.3. *Let H_n be an invertible Hermitian block Toeplitz matrix with blocks of size $m \times m$ and inverse*

$$\Gamma_n = [\gamma_{ij}^{(n)}] , \qquad i,j = 0,\ldots,n .$$

Suppose further that $\gamma_{nn}^{(n)}$ is definite (which guarantees the invertibility of H_{n-1}) and let

$$A_n(\lambda) = \sum_{i=0}^{n} \lambda^i \gamma_{i0}^{(n)} \quad and \quad C_n(\lambda) = \sum_{i=0}^{n} \lambda^i \gamma_{in}^{(n)} .$$

Then

$$\delta_\pm(\lambda C_n) = \mu_\pm(H_n) \qquad [\text{resp. } \delta_\pm(\lambda C_n) = \mu_\mp(H_n)]$$

if $\gamma_{nn}^{(n)}$ is positive [resp. negative] definite.

PROOF. Suppose first that $\gamma_{nn}^{(n)}$ is positive definite. Then H_{n-1} is invertible (see e.g. Lemma 3.1 of [D2]) and $\gamma_{00}^{(n)}$ is positive definite (see e.g. Lemma 4.2 of [D1]). Now let

$$N(\lambda) = \{\gamma_{nn}^{(n)}\}^{-\frac{1}{2}} \{\lambda^* C_n(\lambda^*)\}^* \quad \text{and} \quad D(\lambda) = \{\gamma_{00}^{(n)}\}^{-\frac{1}{2}} A_n(\lambda^*)^* .$$

Then clearly both $D(0)$ and the top coefficient of $N(\lambda)$ are invertible. Next it follows readily from formula (4.19) of [D1] (which is just the generating function form of the

Gohberg-Heinig formula) that $A_n(\lambda)$ and $\lambda C_n(\lambda)$ are left coprime. Therefore N and D are clearly right coprime. The same formula also implies that

$$N(\lambda)^* N(\lambda) = D(\lambda)^* D(\lambda) \quad \text{for} \quad \lambda \in \mathbb{T}$$

and therefore, since $N(\lambda)$ and $D(\lambda)$ are invertible for every point $\lambda \in \mathbb{T}$ (see e.g. Corollary 2 to Theorem 6.1 of [D1]), ND^{-1} is allpass. Thus Theorem 6.2 is applicable and implies that

$$\delta_\pm(N) = \mu_\pm(P) = \mu_\pm(P^{-1}) \; .$$

But this completes the proof (when $\gamma_{nn}^{(n)} > 0$) since

$$\delta_\pm(N) = \delta_\pm(\lambda C_n(\lambda))$$

and in this setting P is just the Gohberg-Heinig formula for $\Gamma_n' = [\gamma_{n-i,n-j}^{(n)}]$.

The proof when $\gamma_{nn}^{(n)} < 0$ is carried out in much the same way except that now the normalizing factors $\{\gamma_{jj}^{(n)}\}^{\frac{1}{2}}$ are replaced by $\{-\gamma_{jj}^{(n)}\}^{\frac{1}{2}}$. ∎

Theorem 6.3 is equivalent to the theorem of Alpay, Gohberg and Lerer. The same methods can be used to show that the zeros of the matrix polynomial

$$\lambda^{n+1} A_n^\#(\lambda) = \lambda \sum_{i=0}^{n} \lambda^{n-i} \gamma_{0i}^{(n)}$$

are distributed in exactly the same way with respect to $\mathbb{T}$ as those of $\lambda C_n(\lambda)$:

THEOREM 6.4. *If H_n is an invertible Hermitian block Toeplitz matrix, then*

$$\delta_\pm(\lambda^{n+1} A_n^\#(\lambda)) = \mu_\pm(H_n) \quad [\text{resp.} \;\; \mu_\mp(H_n)]$$

if $\gamma_{nn}^{(n)}$ is positive [resp. negative] definite.

PROOF. If $\gamma_{nn}^{(n)} > 0$, take

$$N(\lambda) = \{\gamma_{00}^{(n)}\}^{-\frac{1}{2}} \lambda^{n+1} A_n^\#(\lambda) \;\; \text{and} \;\; D(\lambda) = \{\gamma_{nn}^{(n)}\}^{-\frac{1}{2}} \lambda^n C_n^\#(\lambda) \; .$$

The rest is easily adapted from the proof of Theorem 6.3. ∎

Theorem 6.4 is equivalent to Theorem 6.3 because the polynomial $\lambda^{n+1} A_n^\#(\lambda)$ based on $\Gamma_n = [\gamma_{ij}^{(n)}]$ is equal to the polynomial $\{\lambda^* C_n(\lambda^*)\}^*$ based on $\Gamma_n' = [\gamma_{n-i,n-j}^{(n)}]$.

REFERENCES

[AD1] D. Alpay and H. Dym, *On applications of reproducing kernel spaces to the Schur algorithm and rational J unitary factorization*, in: I. Schur Methods in Operator Theory and Signal Processing (I. Gohberg, ed.), Operator Theory: Advances and Applications **OT18**, Birkhäuser, Basel, 1986, pp. 89-159.

[AD2] D. Alpay and H. Dym, *Structured invariant spaces of vector valued rational functions, Hermitian matrices and a generalization of the Iohvidov laws*, Linear Algebra Appl., in press.

[AG] D. Alpay and I. Gohberg, *On orthogonal matrix polynomials*, in: Orthogonal Matrix-valued Polynomials and Applications (I. Gohberg, ed.), Operator Theory: Advances and Applications, **OT34**, Birkhäuser, Basel, 1988, pp. 79-135.

[AJ] B.D.O. Anderson and E.I. Jur, *Generalized Bezoutian and Sylvester matrices in multivariable linear control*, IEEE Trans. Autom. Control, **AC21** (1976), 551-556.

[D1] H. Dym, *Hermitian block Toeplitz matrices, orthogonal polynomials, reproducing kernel Pontryagin spaces, interpolation and extension*, in: Orthogonal Matrix-valued Polynomials and Applications (I. Gohberg, ed.), Operator Theory: Advances and Applications, **OT34**, Birkhäuser, Basel, 1988, pp. 79-135.

[D2] H. Dym, *On Hermitian block Hankel matrices, matrix polynomials, the Hamburger moment problem, interpolation and maximum entropy*, Integral Equations Operator Theory, **12** (1989), 757-812.

[DY] H. Dym and N.J. Young, *A Schur-Cohn theorem for matrix polynomials*, Proc. Edinburgh Math. Soc., in press.

[Fu] P.A. Fuhrmann, *Linear Systems and Operators in Hilbert Space*, McGraw Hill, New York, 1981.

[Fuj] M. Fujiwara, *Uber die algebraischen Gleichungen, deren Wurzeln in einem Kreise oder in einer Halbebene liegen*, Math. Zeit. **24** (1926), 161-169.

[GL] I. Gohberg and L. Lerer, *Matrix generalizations of M.G. Krein theorems on orthogonal polynomials*, in: Orthogonal Matrix-valued Polynomials and Applications (I. Gohberg, ed), Operator Theory: Advances and Applications, **OT34** Birkhäuser, Basel, 1988, pp. 137-202.

[HR] G. Heinig and K. Rost, *Algebraic Methods for Toeplitz-like Matrices and Operators*, Operator Theory: Advances and Applications **OT13**, Birkhäuser, Basel, 1984.

[KN] M.G. Krein and M.A. Naimark, *The method of symmetric and Hermitian forms in the theory of the separation of the roots of algebraic equations*, Linear and Multilinear Algebra, **10** (1981), 265-308.

[LT] L. Lerer and M. Tismenetsky, *The Bezoutian and the eigenvalue-separation problem for matrix polynomials*, Integral Equations Operator Theory, **5** (1982), 386-445.

[W] H. Wimmer, *Bezoutians of polynomial matrices*, Linear Algebra Appl., **122/123/124** (1989), 475-487.

Department of Theoretical Mathematics
The Weizmann Institute of Science
Rehovot 76100, Israel

Operator Theory:
Advances and Applications, Vol. 50

ON A SINGULARLY PERTURBED EIGENVALUE PROBLEM IN THE THEORY OF ELASTIC RODS

L.S. Frank

A singularly perturbed eigenvalue problem appearing in the theory of elastic rods is considered. The least eigenvalue λ_0^ε of the corresponding operator turns out to be exponentially decreasing as the small parameter ε vanishes, λ_0^ε being strictly positive for each $\varepsilon > 0$. Usual techniques based either on the parametrix constructions or on rescaling and stretching of variables fail to produce asymptotic formulae for λ_0^ε and the associated eigenfunction $\psi_0^\varepsilon(x)$ in the case considered. Classical geometrical optics approach is used here in order to derive asymptotic formulae for λ_0^ε and $\psi_0^\varepsilon(x)$ as $\varepsilon \to +0$.

INTRODUCTION.

Singularly perturbed eigenvalue problems for ordinary differential operators affected by the presence of a small positive parameter is one of the classical topics in the singular perturbation theory which goes back to the work by Lord Rayleigh [12], where the following problem was considered:

$$\left\{\begin{array}{l} \varepsilon^2 u^{(iv)} - u'' = \lambda u, \; x \in (0,1) \\ u(0) = u'(0) = u(1) = u'(1) = 0, \end{array}\right. \tag{0.1}$$

and the following asymptotic formula for the eigenvalue λ_n^ε of this problem was established:

$$\lambda_n^\varepsilon = \pi^2 n^2 + 4\pi^2 n^2 \varepsilon + O(\varepsilon^2), \varepsilon \to 0, n = 1, 2, \ldots. \tag{0.2}$$

The operator

$$L_\varepsilon := \varepsilon^2 (d/dx)^4 - (d/dx)^2, L_\varepsilon : D_{L_\varepsilon} \to L_2(0,1)$$

with the domain

$$D_{L_\varepsilon} := \{u \in H_4(0,1), u(0) = u'(0) = u(1) = u'(1) = 0\},$$

is self-adjoint $\forall \varepsilon > 0$ and so it is for the reduced operator

$$L_0 = -(d/dx)^2, L_0 : D_{L_0} \to L_2(0,1)$$

with the domain

$$D_{L_0} = \{u \in H_2(0,1), u(0) = u(1) = 0\}.$$

Similar abstract self-adjoint problems were investigated in [9].

Not necessarily self-adjoint ordinary differential operators of the form:

$$L_\varepsilon = \varepsilon^{2(n-m)} Q + P$$

with ord$Q = 2n >$ ord$P = 2m$ were considered in [11], where an assumption of strong ellipticity of L_ε (see [14]) is made and boundary conditions are considered which are a specific case of more general coercive boundary conditions for operators with a small parameter (see [1], [3]).

Very essential in [11] is the assumption that the boundary operators associated with the reduced problem for $L_0 = P$ have their orders$<$ ord$P = 2m$.

The method used in [11] is closely related to the one in [16] and goes back to the classical geometrical optics asymptotic method applied in the specific one dimensional situation in [16].

The method introduced in [15] has the advantage to be applicable also in the case of elliptic partial differential operators. However, its realization in specific situations requires a considerable amount of technical work for deriving asymptotic expansions and for proving their convergence as the parameter vanishes.

The reduction method for coercive singular perturbations sketched in [2] and developed in [6], [7] (see also [5], [8], [17]) allows to derive in a simple way asymptotic formulae for the eigenvalues and eigenfunctions of coercive singular perturbations (see [4]) in the case when the perturbation shifts the spectrum of the reduced problem to distances which are of order of some positive power of the small parameter. Since only the principal symbols of the coercive singular perturbations are used for producing a singularly perturbed operator which reduces a given coercive singular perturbation to a regular one, the reduction method based upon such a construction can not be applied in the situations when the shift of the spectrum as a result of the perturbation is exponentially small when the parameter vanishes. Neither is the method in [15] applicable in this situation for the same reason as the reduction method mentioned above.

Yet problems of this type appear in a natural way in the theory of elastic rods. Such a problem is considered here and is analyzed directly by using the classical geometrical optics approach. The singularly perturbed eigenvalue problem in the interval $U = (0,1)$ considered here is neither self-adjoint nor it satisfies the conditions in [11], since for the corresponding reduced operator one of the boundary conditions on $\partial U = \{0,1\}$ has the same order as the one of the reduced operator in the interval U. Asymptotic formulae are derived and justified for the least eigenvalue and the associated eigenfunction of the coercive singular perturbation in the theory of elastic rods in the case when the rod is subjected to a large (rescaled dimensionless) longitudinal pulling out force and has one of its end points clamped and the other one free.

1. STATEMENT OF THE PROBLEM

The following singularly perturbed boundary value problem describes an elastic rod at the equilibrium state in the presence of a large (rescaled dimensionless) pulling out force when one of its end points is clamped and the other one is free (see, for instance, [10]):

$$\varepsilon^2 D_x^2(q^2(x)D_x^2u(x)) + D_x^2u(x) = f(x),\ x \in U = (0,1), \tag{1.1}$$

$$B_j(x', D_x)u(x') = \varphi_j(x'),\ j = 1,2,\ x' \in \partial U = \{0,1\}, \tag{1.2}$$

where $D_x = -id/dx$, $q(x) > 0, \forall x \in U, q \in C^\infty(\overline{U}), \varepsilon \in (0,\varepsilon_0], \varepsilon_0 \ll 1$ (ε is proportional to $T^{-1/2}$ with T the dimensionless parameter characterizing the pulling out longitudinal force) and the boundary operators $B_j(x, D_x)$ are given as follows:

$$B_1(x, D_x) = (1-x) - xD_x^2, B_2(x, D_x) = B_1(x, D_x)\partial_n, \tag{1.3}$$

with $\partial_n = (-1)^{x'}d/dx$ the inward normal derivative.

Thus, the boundary conditions have the form:

$$u(0) = \varphi_1(0), u'(0) = \varphi_2(0), u''(1) = \varphi_1(1), u'''(1) = \varphi_2(1).$$

where the upper dashes stand for the derivatives with respect to $x :' = d/dx$.

We associate with (1.1)-(1.3) the following singularly perturbed column-operator:

$$\mathcal{A}^\varepsilon := (\pi_U r^\varepsilon D_x^2, \pi_{\partial U}B_1, \pi_{\partial U}B_2)^T, \tag{1.4}$$

where π_U and $\pi_{\partial U}$ are the restriction operators (traces of continuous functions) to $U = (0,1)$ and $\partial U = \{0,1\}$, respectively, the differential operator $r^\varepsilon(x, D_x)$ is defined as follows:

$$r^\varepsilon(x, D_x) = \varepsilon^2 D_x^2 q^2(x) + 1, \tag{1.5}$$

and the upper T stands for the column-vector which is the transpose of the corresponding row-vector.

One associates with $\mathcal{A}^\varepsilon$ its reduced operator $\mathcal{A}^0$ defined as follows:

$$\mathcal{A}^0 := (\pi_U D_x^2, \pi_{\partial U} B_1)^T, \tag{1.6}$$

the corresponding reduced problem being stated in an obvious way:

$$\mathcal{A}^0 u^0 = (f, \varphi_1)^T. \tag{1.7}$$

Considering the boundary value problem:

$$\begin{aligned} &(\varepsilon^2 D_x^4 + D_x^2 - \lambda) u^\varepsilon(x) = 0, \ x \in U \\ &B_j(x', D_x) u(x') = \varphi_j(x'), \ j = 1, 2, \ x' \in \partial U \end{aligned}$$

with $B_j(x, D_x)$ defined by (1.3) and with given $\lambda \in (0, \pi^2)$, it is readily seen that

$$\lim_{\varepsilon \to +0} u^\varepsilon(x) = u^0(x)$$

where $u^0(x)$ is the solution of the problem:

$$\begin{aligned} &(D_x^2 - \lambda) u^0(x) = 0, \ x \in U \\ &B_1(x', D_x) u^0(x') = \varphi_1(x'), \ x' \in \partial U. \end{aligned}$$

Of course, the same is true if one considers inhomogeneous perturbed and reduced equations in U with a smooth second member $f(x)$ instead of zero.

This justifies the some-what formal definition (1.6) of the reduced operator $\mathcal{A}^0$ associated with the perturbed one defined by (1.4), (1.5), (1.3).

It is readily seen that the kernel of $\mathcal{A}^\varepsilon$ is trivial (while for the reduced operator it consists of all functions Cx with C any constant).

Indeed, introducing $v^\varepsilon = D_x^2 u^\varepsilon$ with $u^\varepsilon \in \ker \mathcal{A}^\varepsilon$, (i.e. $\mathcal{A}^\varepsilon u^\varepsilon = (0,0,0)^T$), one gets for v^ε the following problem:

$$\begin{cases} r^\varepsilon v^\varepsilon = 0, \ x \in U, \\ v^\varepsilon(1) = 0, D_x v^\varepsilon(1) = 0, \end{cases} \tag{1.8}$$

so that $v^\varepsilon(x) \equiv 0, \ \forall x \in \overline{U}$.

Furthermore, one finds for u^ε:

$$\begin{cases} D_x^2 u^\varepsilon = 0, \ x \in U \\ u^\varepsilon(0) = 0, D_x u^\varepsilon(0) = 0, \end{cases} \tag{1.9}$$

so that $u^\varepsilon(x) \equiv 0,\ \forall x \in \overline{U}$.

As a consequence of such a situation, it is impossible to factorize $\mathcal{A}^\varepsilon$ by $\mathcal{A}^0$, i.e. to find an operator R^ε (which would be a 3×2 matrix-operator), such that $\mathcal{A}^\varepsilon = R^\varepsilon \mathcal{A}^0$, given that $\ker \mathcal{A}^\varepsilon = \{0\}$ and $\ker \mathcal{A}^0 = \{Span\ x\}$. Also, as indicates (1.8), even if the inverse operator $(\mathcal{A}^\varepsilon)^{-1}$ exists, its norm grows exponentially, i.e. as $\exp(\gamma/\varepsilon)$ for $\varepsilon \to +0$ with some constant $\gamma > 0$. In other words, the eigenvalue $\lambda_0^0 = 0$ of the reduced operator (with x the associated eigenfunction) is shifted for its perturbation $\mathcal{A}^\varepsilon$ given by (4), to some value $\lambda_0^\varepsilon = O(\exp(-\gamma/\varepsilon))$ with some constant $\gamma > 0$, as $\varepsilon \to +0$.

We are going to find asymptotic formulae for λ_0^ε and the associated eigenfunction of $\mathcal{A}^\varepsilon$.

A direct method based upon the classical geometrical optics approach will be applied for deriving these asymptotic formulae.

Thus, consider the eigenvalue problem:

$$\pi_U(r^\varepsilon D_x^2 - \lambda_0^\varepsilon)\psi_0^\varepsilon = 0,\ \pi_{\partial U} B_j(x, D_x)\psi_0^\varepsilon = 0,\ j = 1, 2, \tag{1.10}$$

where $B_j(x, D_x)$ are given by (1.3).

REMARK 1.1. Singularly perturbed operator $\mathcal{A}^\varepsilon$ defined by (1.3)-(1.5) is a coercive singular perturbation (see [3]) and so it is also for the following perturbation $\mathcal{A}_\mu^\varepsilon$ of $\mathcal{A}^\varepsilon$:

$$\mathcal{A}_\mu^\varepsilon := (\pi_U(r^\varepsilon D_x^2 + \mu^2), \pi_{\partial U} B_1, \pi_{\partial U} B_2)^T,\ \mu > 0.$$

Since the reduced operator

$$\mathcal{A}_\mu^0 := (\pi_U(D_x^2 + \mu^2), \pi_{\partial U} B_1)^T,\ \mu > 0$$

is invertible, so it is also for $\mathcal{A}_\mu^\varepsilon$, $\forall \varepsilon \in (0, \varepsilon_0]$, provided that $\varepsilon_0 > 0$ is sufficiently small, and, moreover, $(\mathcal{A}_\mu^\varepsilon)^{-1}$ is uniformly bounded with respect to $\varepsilon \in (0, \varepsilon_0]$ as a linear operator in corresponding Sobolev type spaces (see [5], [6], [7], [17]).

An unusual feature of $\mathcal{A}_\mu^\varepsilon, \mu \geq 0$, is the boundary operators $B_k(x', D_x), k = 1, 2$, whose orders are different at $x' = 0$ and $x' = 1 : 0 = \mathrm{ord} B_1(0, D_x) < \mathrm{ord} B_1(1, D_x) = 2$ and $1 = \mathrm{ord} B_2(0, D_x) < \mathrm{ord} B_2(1, D_x) = 3$.

2. ASYMPTOTIC SOLUTIONS

Here two different types of asymptotic solutions to the differential equation in (1.10) with $\lambda \in \mathbf{C}$ instead of λ_0^ε will be constructed.

We denote by $q(x)$, $q : \mathbf{R} \to \mathbf{R}$ a smooth extension of $q \in C^\infty(\overline{U})$ to $\mathbf{R}$ such that $q(x) \geq q_0 > 0$, $\forall x \in \mathbf{R}$ and $q(x) = q_\infty + q_1(x)$ with $q_1 \in C_0^\infty(\mathbf{R})$ (see, for instance, [13] where the possibility of such an extension is shown).

Introduce the notation:

$$L(\varepsilon, \lambda, x, \partial_x) := \varepsilon^2 \partial_x^2 q^2(x) \partial_x^2 - \partial_x^2 - \lambda \tag{2.1}$$

where $\partial_x = d/dx$, $\varepsilon \in (0, \varepsilon_0]$, $\lambda \in \mathbf{C}$, $x \in \mathbf{R}$ and $q(x)$, $q \in C^\infty(\mathbf{R})$ is extended as indicated above.

We start with the following

PROPOSITION 2.1. *The equation*

$$L(\varepsilon, \lambda, x, \partial_x) u_\lambda^\varepsilon(x) = 0, \; x \in \mathbf{R}, \tag{2.2}$$

has the following formal asymptotic solutions:

$$u_j(\varepsilon, \lambda, x) \sim \sum_{k \geq 0} \varepsilon^{2k} u_{jk}(\lambda, x), \quad j = 1, 2 \tag{2.3}$$

and

$$u_j(\varepsilon, \lambda, x) \sim \exp(-Q_j(x)/\varepsilon) \sum_{k \geq 0} \varepsilon^k u_{jk}(\lambda, x), \quad j = 3, 4 \tag{2.4}$$

where

$$u_{10}(\lambda, x) = \lambda^{-1/2} \sin(\lambda^{1/2} x), u_{20}(\lambda, x) = \cos(\lambda^{1/2} x), \tag{2.5}$$

$$Q_3(x) = \int_0^x (q(y))^{-1} dy, Q_4(x) = \int_x^1 (q(y))^{-1} dy, u_{j0}(\lambda, x) = (q(x))^{1/2}, j = 3, 4, \tag{2.6}$$

and where $u_{jk}(\lambda, x)$, $k > 0$, $1 \leq j \leq 4$ *are defined recursively as follows:*
(i) *for* $j = 1, 2$:

$$(\partial_x^2 + \lambda) u_{jk}(\lambda, x) = \partial_x^2 q^2 \partial_x^2 u_{j,k-1}(\lambda, x), u_{jk}(\lambda, 0) = \partial_x u_{jk}(\lambda, 0) = 0, k = 1, 2, ..., \tag{2.7}$$

(ii) *for* $j = 3, 4$:

$$\begin{cases} u_{jk}(\lambda, x) \equiv 0, k < 0, u_{j0}(\lambda, x) = (q(x))^{1/2}, \\ L_q(u_{jk})(\lambda, x) = ((L_q q(x) L_q + q(x)(\partial_x^2 - \lambda)) u_{j,k-1})(\lambda, x) - \\ - ((L_q q^2(x) \partial_x^2 + q^2(x) \partial_x^2 q(x) L_q) u_{j,k-2})(\lambda, x) + \\ + (q^2(x) \partial_x^2 q^2(x) \partial_x^2 u_{j,k-3})(\lambda, x), k = 0, 1, ..., \\ u_{3,k}(\lambda, 0) = 0, u_{4,k}(\lambda, 1) = 0, k = 1, 2, ..., \end{cases} \tag{2.8}$$

with

$$L_q(u) := \partial_x u + q(x)\partial_x(u/q(x)). \tag{2.9}$$

PROOF. An elementary computation shows that for the formal asymptotic solutions $u_j(\varepsilon,\lambda,x), j=1,2$, defined by (2.3), (2.7) holds:

$$L(\varepsilon,\lambda,x,\partial_x)u_j(\varepsilon,\lambda,x) = O(\varepsilon^\infty), \varepsilon \to +0, j=1,2.$$

We show briefly that formally one has:

$$L(\varepsilon,\lambda,x,\partial_x)u_3(\varepsilon,\lambda,x) = \rho(\varepsilon,\lambda,x)\exp(-Q_3(x)/\varepsilon),$$

where uniformly with respect to $x \in \mathbf{R}$ and $|\lambda| \leq r < \infty$ holds:

$$\rho(\varepsilon,\lambda,x) = O(\varepsilon^\infty), \varepsilon \to +0. \tag{2.10}$$

Indeed, rewriting $L(\varepsilon,\lambda,x,\partial_x)$ in the form:

$$L(\varepsilon,\lambda,x,\partial_x) = \partial_x^2 r(x,\varepsilon\partial_x) - \lambda, r(x,\varepsilon\partial_x) = q^2(x)(\varepsilon\partial_x)^2 - 1, \tag{2.11}$$

an elementary straightforward computation shows that

$$r(x,\varepsilon\partial_x)u_3(\varepsilon,\lambda,x) \sim \varepsilon^2\exp(-Q_3(x)/\varepsilon)\sum_{k\geq 0}\varepsilon^k(-q(x)L_q(u_{3,k+1}) + q^2(x)\partial_x^2 u_{3,k}) \tag{2.12}$$

Substitution of (2.12) into (2.2) yields:

$$\begin{aligned} &L(\varepsilon,\lambda,x,\partial_x)u_3(\varepsilon,\lambda,x) \sim \\ &\sim \exp(-Q_3(x)/\varepsilon)\sum_{k\geq 0}\varepsilon^k((q(x))^{-2}v_{k+2}(\lambda,x) - (q(x))^{-1}L_q(v_{k+1}) + \partial_x^2 v_k(\lambda,x), \end{aligned} \tag{2.13}$$

where we have denoted

$$v_{k+2}(\lambda,x) := q(x)L_q(u_{3,k+1}))(\lambda,x) + q^2(x)\partial_x^2 u_{3,k}(\lambda,x), k=0,1,... \tag{2.14}$$

Using (2.14), (2.8), it is readily seen, that

$$(q(x))^{-2}v_{k+2}(\lambda,x) - (q(x))^{-1}(L_q(v_{k+1}))(\lambda,x) + \partial_x^2 v_k(\lambda,x) \equiv 0, k \geq 0, \tag{2.15}$$

where, of course, $v_k(\lambda, x) \equiv 0, k = 0, 1$, as a consequence of (2.14), (2.8) and (2.6), the latter having as a consequence the identity: $(L_q(u_{j,0}))(x) \equiv 0$. The same argument applies to $u_4(\varepsilon, \lambda, x)$. ◇

We are going to use the classical construction due to T. Carleman, in order to produce functions $v_j(\varepsilon, \lambda, x)$ which are C^∞ in variables $(\varepsilon, x) \in (0, \varepsilon_0] \times \overline{U}$, analytic in $\lambda \in \mathbf{C}$ and have the following properties: $v_j(\varepsilon, \lambda, x), 1 \leq j \leq 4$, admit an asymptotic expansion by the formal series representing the corresponding $u_j(\varepsilon, \lambda, x), 1 \leq j \leq 4$.

LEMMA 2.2. *Let $\varphi_k \in C^\infty(\overline{U}), k \geq 0$ and $\mu \in [\mu_0, \infty)$.*
Then there exists a function $\varphi(\mu, x), \varphi \in C^\infty([\mu_0, \infty) \times U)$ such that the following asymptotic relations hold:

$$\partial_\mu^\alpha \partial_x^\beta \varphi(\mu, x) \sim \sum_{k \geq 0} (\partial_\mu^\alpha \mu^{-k}) \partial_x^\beta \varphi_k(x), \mu \to +\infty, \forall \alpha, \beta, \tag{2.16}$$

uniformly with respect to $x \in \overline{U}$, i.e. for each integer $N > 0$ one has:

$$|\partial_\mu^\alpha \partial_x^\beta \varphi(\mu, x) - \sum_{0 \leq k < N} (\partial_\mu^\alpha \mu^{-k}) \partial_x^\beta \varphi_k(x)| \leq C_{\alpha,\beta,N} \mu^{-(N+\alpha)}, \tag{2.17}$$

where the constant $C_{\alpha,\beta,N}$ may depend only on its subscripts.

PROOF. Let $\chi \in C^\infty(\mathbf{R})$ be such that $\chi(t) \equiv 1$ for $|t| \geq 1$ and $\chi(t) \equiv 0$ for $|t| \leq 1/2$. Define

$$\varphi(\mu, x) = \sum_{k \geq 0} \chi(\delta_k \mu) \mu^{-k} \varphi_k(x) \tag{2.18}$$

where the sequence $\delta_k \downarrow 0$ for $k \to \infty$ will be chosen later on.

Notice that for each $\mu \geq \mu_0$ given the series on the right hand side of (2.18) contains only a finite number non-vanishing terms, i.e. it is convergent for each given $\mu \geq \mu_0$. The numbers $\delta_k \downarrow 0, k \to \infty$, are chosen to satisfy the inequalities.

$$|\partial_\mu^\alpha \partial_x^\beta (\chi(\delta_k \mu) \mu^{-k} \varphi_k(x)| \leq \mu^{1-\alpha-k},\ 0 \leq \alpha + \beta \leq k\ \forall \mu \geq \mu_0,\ \forall x \in \overline{U}. \tag{2.19}$$

Thus, for $\forall N > 0$ and for $\alpha + \beta \leq N$, one finds:

$$|\sum_{k \geq N+2} \partial_\mu^\alpha \partial_x^\beta (\chi(\delta_k \mu) \mu^{-k} \varphi_k(x)| \leq \sum_{k \geq N+1} \mu^{-1-\alpha-k} = O(\mu^{-1-\alpha-N}), \text{ as } \mu \to \infty. \tag{2.20}$$

Thus, (2.20) yields:

$$\partial_\mu^\alpha \partial_x^\beta \varphi(\mu, x) = \partial_\mu^\alpha \partial_x^\beta \sum_{0 \leq k \leq N+1} \chi(\delta_k \mu) \mu^{-k} \varphi_k(x) +$$
$$+ O(\mu^{-1-\alpha-N}) = \partial_\mu^\alpha \partial_x^\beta \sum_{0 \leq k \leq N+1} \mu^{-k} \varphi_k(x) + O(\mu^{-1-\alpha-N}), \text{ as } \mu \to \infty.$$ ◇

Now define the functions

$$v_j(\varepsilon,\lambda,x)=\sum_{k\geq 0}\chi(\delta_k\varepsilon^{-2})\varepsilon^{2k}u_{jk}(\lambda,x),\ j=1,2 \tag{2.21}$$

and

$$v_j(\varepsilon,\lambda,x)=\exp(-Q_j(x)/\varepsilon)\sum_{k\geq 0}\chi(\delta_k\varepsilon^{-1})\varepsilon^{k}u_{jk}(\lambda,x),\ j=3,4 \tag{2.22}$$

where u_{jk}, $j=1,2$ and u_{jk}, $j=3,4$ are defined by (2.7) and (2.8) respectively. ◇

THEOREM 2.3. *Equation* (2.2) *has a fundamental system of solutions* $w_j(\varepsilon,\lambda,x), 1\leq j\leq 4$ *such that:*
(i) w_j *are* C^∞ *in variables* $(\varepsilon,x)\in(0,\varepsilon_0]\times\overline{U}$ *and analytic in variable* $\lambda\in\mathbf{C}$;
(ii) *for any* $\alpha\geq 0,\beta\geq 0$ *and* $N>0$ *given the following asymptotic relations hold:*

$$(\varepsilon^3\partial_\varepsilon)^\alpha\partial_x^\beta(w_j(\varepsilon,\lambda,x)-v_j(\varepsilon,\lambda,x))=O(\varepsilon^{2\alpha+2N}), j=1,2 \tag{2.23}$$

and

$$(\varepsilon^2\partial_\varepsilon)^\alpha\partial_x^\beta(w_j(\varepsilon,\lambda,x)-v_j(\varepsilon,\lambda,x))=O(\varepsilon^{\alpha+N}\exp(-Q_j(x)(\varepsilon)),\ j=1,2, \tag{2.24}$$

or, equivalently,

$$(\varepsilon^2\partial_\varepsilon)^\alpha\partial_x^\beta(\exp(Q_j(x)/\varepsilon)(w_j(\varepsilon,\lambda,x)-v_j(\varepsilon,\lambda,x)))=O(\varepsilon^{\alpha+N}), \tag{2.25}$$

the asymptotic formulae (2.23), (2.25) *being valid uniformly with respect to* $x\in\overline{U}$ *and* $\lambda\in\mathbf{C}$, $|\lambda|\leq r<\infty$.

PROOF. We shall briefly sketch the proof emphasizing the main ideas and constructions and omitting technical details which can be easily verified.

Let us start with the solutions $w_j(\varepsilon,\lambda,x)$, $j=1,2$ of the first type. Using a smooth extension of q onto $\mathbf{R}$ (with the properties indicated above), we may consider the equation

$$L(\varepsilon,\lambda,x,\partial_x)w_j(\varepsilon,\lambda,x)=0,\ x\in\mathbf{R},\ j=1,2. \tag{2.26}$$

As a consequence of Proposition 2.1 and Lemma 2.2, one has for the asymptotic solutions $v_j(\varepsilon,\lambda,x)$, $j=1,2$ defined by (2.3), (2.7), (2.21):

$$L(\varepsilon,\lambda,x,\partial_x)v_j(\varepsilon,\lambda,x)=\delta_j(\varepsilon,\lambda,x),\quad j=1,2 \tag{2.27}$$

where $\delta_j(\varepsilon,\lambda,x)$ is $O(\varepsilon^\infty)$, as $\varepsilon \to 0$, with all its derivatives, i.e. in the topology of $C^\infty((0,\varepsilon_0]\times\mathbf{R})$; furthermore as a consequence of the construction, $\lambda \to \delta(\varepsilon,\lambda,x)$ is an entire function of $\lambda \in \mathbf{C}$.

We shall seek the solution w_j of (2.26) such that

$$w_j(\varepsilon,\lambda,0) = v_j(\varepsilon,\lambda,0), \partial_x w_j(\varepsilon,\lambda,0) = \partial_x v_j(\varepsilon,\lambda,0),\ j=1,2. \tag{2.28}$$

Introducing $Y_j(\varepsilon,\lambda,x) = v_j(\varepsilon,\lambda,x) - w_j(\varepsilon,\lambda,x)$, $j=1,2$, one finds:

$$\begin{cases} (r^\varepsilon(x,D_x)D_x^2 - \lambda)Y_j(\varepsilon,\lambda,x) = \delta_j(\varepsilon,\lambda,x),\ x\in\mathbf{R} \\ Y_j(\varepsilon,\lambda,0) = 0,\ \partial_x Y_j(\varepsilon,\lambda,0) = 0, \quad j=1,2, \end{cases} \tag{2.29}$$

where $r^\varepsilon(x,D_x) = \varepsilon^2 D_x^2 q^2(x) + 1$ is an elliptic singular perturbation of order (0,0,2).

As a consequence of Theorem 2.2.1 in [6], $r^\varepsilon(x,D_x)$ has an inverse $s^\varepsilon(x,D_x)$ which is an elliptic singular perturbation of order (0,0,-2). Hence, one may rewrite (2.29) equivalently in the following fashion:

$$Y_j(\varepsilon,\lambda,x) - \lambda\int_0^x (x-y)s^\varepsilon(y,D_y)Y_j(\varepsilon,\lambda,y)dy = \rho_j(\varepsilon,\lambda,x), \tag{2.30}$$

with

$$\rho_j(\varepsilon,\lambda,x) = \int_0^x (x-y)s^\varepsilon(y,D_y)\delta_j(\varepsilon,\lambda,y)dy,\ j=1,2$$

still being $O(\varepsilon^\infty)$ in the topology of $C^\infty_{\varepsilon,x}$ and entire functions of $\lambda\in\mathbf{C}$.

Volterra integral equations (2.30) have well-defined solutions $Y_j(\varepsilon,\lambda,x)$, $j=1,2$, which possess the same properties, as $\rho_j(\varepsilon,\lambda,x)$, i.e.

$$Y_j(\varepsilon,\lambda,x) = O(\varepsilon^\infty),\ j=1,2, \tag{2.31}$$

in the topology of $C^\infty((0,\varepsilon_0]\times\mathbf{R})$ and are entire functions of $\lambda\in\mathbf{C}$.

Next, we consider the solutions $w_j(\varepsilon,\lambda,x)$, $j=3,4$ of the second type. Again, using a smooth extension of $q(x)$ onto $\mathbf{R}$ (as indicated above), one may consider the equation:

$$L(\varepsilon,\lambda,x,\partial_x)w_j(\varepsilon,\lambda,x) = 0,\ x\in\mathbf{R},\ j=3,4. \tag{2.32}$$

Again, as a consequence of Proposition 2.1 and Lemma 2.2 one has for the asymptotic solutions $v_j(\varepsilon,\lambda,x)$ defined by (2.4), (2.5), (2.8), (2.22):

$$\exp(Q_j(x)/\varepsilon)L(\varepsilon,\lambda,x,\partial_x)v_j(\varepsilon,\lambda,x) = \delta_j(\varepsilon,\lambda,x),\ j=3,4, \tag{2.33}$$

where again $\delta_j(\varepsilon,\lambda,x) = O(\varepsilon^\infty)$ in the topology of $C^\infty_{\varepsilon,x}$ and are entire functions of $\lambda \in \mathbf{C}$.

We shall seek solutions $w_j(\varepsilon,\lambda,x)$ of (2.32) which satisfy the conditions:

$$w_j(\varepsilon,\lambda,0) = v_j(\varepsilon,\lambda,0), \quad j = 3,4,$$

so that for

$$Y_j(\varepsilon,\lambda,x) := \exp(Q_j(x)/\varepsilon)(v_j(\varepsilon,\lambda,x) - w_j(\varepsilon,\lambda,x)), \tag{2.34}$$

one gets the following problem:

$$\begin{cases} \exp(Q_j(x)/\varepsilon)L(\varepsilon,\lambda,x,\partial_x)(\exp(-Q_j(x)/\varepsilon)Y_j(\varepsilon,\lambda,x)) = \delta_j(\varepsilon,\lambda,x), x \in \mathbf{R}, \\ Y_j(\varepsilon,\lambda,0) = 0, \quad j = 3,4. \end{cases} \tag{2.35}$$

A straightforward computation shows that

$$\varepsilon \exp(Q_j(x)/\varepsilon)L(\varepsilon,\lambda,x,\partial_x)\exp(-Q_j(x)/\varepsilon) = M(x,\varepsilon D_x)\partial_x + R(\varepsilon,\lambda,x,\partial_x) \tag{2.36}$$

where

$$M(x,\eta) := -iq^2(x)\eta^3 + 4q(x)\eta^2 + 5i\eta - 6/q(x), \tag{2.37}$$

$M(x,\varepsilon D_x)$ is an elliptic perturbation of order (0,0,3) and $R(\varepsilon,\lambda,x,\partial_x)$ is a differential singular perturbation whose order is at most (0,0,2).

Hence, again as a consequence of Theorem 2.2.1 in [6] $M(x,\varepsilon D_x)$ has an inverse $N(x,\varepsilon,D_x)$ which is an elliptic singular perturbation of order (0,0,-3).

Therefore one may rewrite (2.35) in the following equivalent fashion:

$$Y_j(\varepsilon,\lambda,x) + \int_0^x N(\varepsilon,\lambda,D_y) \circ R(\varepsilon,\lambda,y,\partial_y)Y_j(\varepsilon,\lambda,y)dy = \rho_j(\varepsilon,\lambda,x), \quad j = 3,4, \tag{2.38}$$

where $\rho_j(\varepsilon,\lambda,x)$ have the same properties as $\delta_j(\varepsilon,\lambda,x)$.

Thus, the well-defined solutions Y_j of Volterra integral equations with

$$\text{ord } N(\varepsilon,\lambda,D_y) \circ R(\varepsilon,\lambda,y,\partial_y) \leq (0,0,-1)$$

have the same properties as $\delta_j(\varepsilon,\lambda,x)$, i.e.

$$Y_j(\varepsilon,\lambda,x) = O(\varepsilon^\infty), \quad j = 3,4,$$

in the topology of $C^\infty((0,\varepsilon_0] \times \mathbf{R})$ and are entire functions of $\lambda \in \mathbf{C}$. ◇

REMARK 2.4. The results in Theorem 2.3 are similar to the ones in [11], [16]. However their proof given here and based on the reduction theory developed in [6], [7] is quite different and much simpler.

3. ASYMPTOTIC FORMULAE

Here asymptotic formulae for the eigenvalues and the eigenfunctions of the operator $\mathcal{A}^\varepsilon$ defined by (1.3)-(1.5) are derived by means of Proposition 2.1 and Theorem 2.3.

Let $w_j(\varepsilon,\lambda,x)$, $1 \le j \le 4$, be the fundamental system of solutions of equation (2.2) which has been constructed in the proof of Theorem 2.3. Introduce the matrix

$$\mathcal{D}(\varepsilon,\lambda) := \begin{Vmatrix} w_1(\varepsilon,\lambda,0), & w_2(\varepsilon,\lambda,0), & \varepsilon w_3(\varepsilon,\lambda,0), & \varepsilon^3 w_4(\varepsilon,\lambda,0) \\ w_1'(\varepsilon,\lambda,0), & w_2'(\varepsilon,\lambda,0), & \varepsilon w_3'(\varepsilon,\lambda,0), & \varepsilon^3 w_4'(\varepsilon,\lambda,0) \\ w_1''(\varepsilon,\lambda,1), & w_2''(\varepsilon,\lambda,1), & \varepsilon w_3''(\varepsilon,\lambda,1), & \varepsilon^3 w_4''(\varepsilon,\lambda,1) \\ w_1'''(\varepsilon,\lambda,1), & w_2'''(\varepsilon,\lambda,1), & \varepsilon w_3'''(\varepsilon,\lambda,1), & \varepsilon^3 w_4'''(\varepsilon,\lambda,1) \end{Vmatrix}, \tag{3.1}$$

where the upper dash stands for the derivative with respect to x $:'= d/dx$.

Obviously, the eigenvalues of $\mathcal{A}^\varepsilon$ defined by (1.3)-(1.5) are the zeros of the equation:

$$F(\varepsilon,\lambda) := \det \mathcal{D}(\varepsilon,\lambda) = 0, \quad \varepsilon \in (0,\varepsilon_0]. \tag{3.2}$$

$F(\varepsilon,\lambda)$ being an entire function of $\lambda \in \mathbf{C}$ for each $\varepsilon \in (0,\varepsilon_0]$, equation (3.2) has isolated zeros λ_n^ε, $n = 0,1,\ldots$ such that $|\lambda_n^\varepsilon| \to \infty$ for $n \to \infty$.

As a consequence of Proposition 2.1 and Theorem 2.3 one has:

$$F(0,\lambda) : = \lim_{\varepsilon\to+0} F(\varepsilon,\lambda) =$$

$$= \det \begin{Vmatrix} O, & 1, & O, & O \\ 1, & O, & -(q(0))^{-1/2}, & O \\ -\lambda^{1/2}\sin\lambda^{1/2}, & -\lambda\cos\lambda^{1/2}, & O, & O \\ -\lambda\cos\lambda^{1/2}, & \lambda^{3/2}\sin\lambda^{1/2}, & O, & (q(1))^{-5/2} \end{Vmatrix} = \tag{3.3}$$

$$= (q(0))^{-1/2}(q(1))^{-5/2}\lambda^{1/2}\sin\lambda^{1/2}.$$

The zeros

$$\lambda_n^0 = \pi^2 n^2, \quad n = 0,1,\ldots \tag{3.4}$$

of $F(0,\lambda)$ are nothing else but the eigenvalues of the reduced operator $\mathcal{A}^0$ defined by (1.6), (1.3).

LEMMA 3.1. *The following asymptotic formula holds uniformly with respect to* λ *on each compact set in* $\mathbf{C}$:

$$\begin{aligned} F(\varepsilon,\lambda) &= F(0,\lambda) - \\ &- \varepsilon(q(0))^{-1/2}(q(1))^{-5/2}(\lambda\cos\lambda^{1/2} \sum_{x'\in\partial U} \pi_{\partial U} q(x') + \\ &+ \lambda^{1/2}\sin\lambda^{1/2} \sum_{x'\in\partial U} \pi_{\partial U}\partial_n q(x')) + O(\varepsilon^2), \end{aligned} \tag{3.5}$$

where $\partial_n = (-1)^{x'} d/dx$ *is the inward normal derivative at* $x' \in \partial U$.

PROOF. As a consequence of Proposition 2.1 and Theorem 2.3 one has the following asymptotic relations uniformly with respect to $x \in \overline{U}$ and λ belonging to each compact set in $\mathbf{C}$:

$$
(3.6) \quad \begin{cases} w_1(\varepsilon,\lambda,x) = \lambda^{-1/2}\sin(\lambda^{1/2}x) + O(\varepsilon^2), \\ w_2(\varepsilon,\lambda,x) = \cos(\lambda^{1/2}x) + O(\varepsilon^2), \\ w_3(\varepsilon,\lambda,x) = \exp(-\varepsilon^{-1}\int_0^x (q(y))^{-1}dy)((q(x))^{1/2} + \varepsilon u_{31}(\lambda,x) + O(\varepsilon^2)), \\ w_4(\varepsilon,\lambda,x) = \exp(-\varepsilon^{-1}\int_x^1 (q(y)^{-1}dy)((q(x))^{1/2} + \varepsilon u_{41}(\lambda,x) + O(\varepsilon^2)), \end{cases}
$$

where $u_{j1}(\lambda,x)$, $j = 3,4$, are defined by (2.8).

Inserting (3.6) into (3.1) and using (3.3), one gets (3.5), while keeping all the terms with ε^k, $k = 0,1$ and neglecting the ones which are $O(\varepsilon^2)$, as $\varepsilon \to +0$. ◇

THEOREM 3.2. *The eigenvalues* λ_n^ε, $n = 0,1,...$ *of* $\mathcal{A}^\varepsilon$ *defined by (1.3)-(1.5), are real and for each* $n = 0,1,...$ *the following asymptotic formulae are valid:*

$$
(3.7) \quad \lambda_n^\varepsilon = \pi^2 n^2 (1 - 2\varepsilon \sum_{x' \in \partial U} \pi_{\partial U} q(x')) + O(\varepsilon^2).
$$

PROOF. As a consequence of (3.3), one has for $F_\lambda = (\partial/\partial\lambda)F$:

$$
(3.8) \quad \begin{cases} F_\lambda(0,\pi^2 n^2) = (1/2)(-1)^n (q(0))^{-1/2}(q(1))^{-5/2} \neq 0, n = 1,2,... \\ F_\lambda(0,0) = (q(0))^{-1/2}(q(1))^{-5/2} \neq 0. \end{cases}
$$

Furthermore, as a consequence of (3.5), $F_\varepsilon(\varepsilon,\lambda) = (\partial/\partial\varepsilon)F(\varepsilon,\lambda)$ is continuous on $[0,\varepsilon_0] \times \mathbf{C}$ for each $\varepsilon_0 > 0$ and $F_{\varepsilon\varepsilon}(\varepsilon,\lambda) = (\partial/\partial\varepsilon)^2 F(\varepsilon,\lambda)$ is bounded on $(0,\varepsilon_0] \times K$ for each $\varepsilon_0 > 0$ and each compact $K \in \mathbf{C}$.

Since $\lambda_n^0 = \pi^2 n^2$, $n = 0,1,...$ are real and $\lambda_n^0 > 0$, $n = 1,2,...$, the implicit functions theorem implies that the zeros $\lambda_n^\varepsilon = \lambda_n(\varepsilon)$ of $F(\varepsilon,\lambda)$ defined by (3.2), are real and continuously differentiable with respect to $\varepsilon \in [0,\varepsilon_0]$ with any $\varepsilon_0 > 0$; moreover, one has for $\varepsilon_0 > 0$ sufficiently small: $\lambda_n^\varepsilon > 0$, $n = 1,2,...$, $\forall \varepsilon \in [0,\varepsilon_0]$.

Furthermore, for μ_n,

$$
(3.9) \quad \mu_n := (d/d\varepsilon)\lambda_n(\varepsilon)|_{\varepsilon=0}, \quad n = 0,1,...
$$

one finds using (3.5):

$$
\mu_n = -F_\varepsilon(0,\pi^2 n^2)(F_\lambda(0,\pi^2 n^2))^{-1} = -2\pi^2 n^2 \sum_{x' \in \partial U} \pi_{\partial U} q(x'), \quad n = 0,1,...
$$

and that proves (3.7). ◇

REMARK 3.3. In fact, using Proposition 2.1 and Theorem 2.3, one gets the conclusion that $F(\varepsilon,\lambda)$ defined by (3.1), (3.2), is infinitely differentiable with respect to $\varepsilon \geq 0$, being an entire function of $\lambda \in \mathbf{C}$ with all its derivatives with respect to $\varepsilon \geq 0$. Thus, the eigenvalues $\lambda_n^\varepsilon = \lambda_n(\varepsilon)$, $n = 0, 1, \ldots$ of $\mathcal{A}^\varepsilon$ defined by (1.3)-(1.5) are C^∞ in $\varepsilon \geq 0$, and one has for each integer $N > 0$ and each $n = 1, 2, \ldots$

$$\lambda_n(\varepsilon) = \sum_{0 \leq k < N} \lambda_n^k \varepsilon^k + O(\varepsilon^N), \quad \varepsilon \to +0,$$

where, of course,

$$\lambda_n^0 = \pi^2 n^2, \lambda_n^1 = -2\pi^2 n^2 \sum_{x' \in \partial U} \pi_{\partial U} q(x'),$$

and all λ_n^k, $k \geq 2$, can be defined recursively using Proposition 2.1.

Furthermore, obviously $\lambda_0^\varepsilon = \lambda_0(\varepsilon) = O(\varepsilon^N)$, $\forall N > 0$, and an asymptotic formula for λ_0^ε (which is exponentially small as $\varepsilon \to +0$) will be exhibited later on. ◇

Now an asymptotic formula for $\lambda_0^\varepsilon = \lambda_0(\varepsilon)$ will be derived.

Using (2.7), one finds for $F(\varepsilon,\lambda)$ defined by (3.1), (3.2):

$$F(\varepsilon, 0) = \varepsilon^4 (w_3'''(\varepsilon,0,1) w_4''(\varepsilon,0,1) - w_3''(\varepsilon,0,1) w_4'''(\varepsilon,0,1)) \tag{3.10}$$

and Proposition 2.1 and Theorem 2.3 yield with any integer $N > 0$:

$$F(\varepsilon,0) = -2(q(1))^{-4} \delta_\varepsilon \Big(\sum_{0 \leq k < N} \gamma_k \varepsilon^{k-1} + O(\varepsilon^N)\Big), \quad \varepsilon \to 0, \tag{3.11}$$

where $\gamma_0 = 1$ and γ_k, $k > 0$, are computed recursively using (2.8), and where we have denoted:

$$\delta_\varepsilon := \exp\Big(-\varepsilon^{-1} \int_0^1 (q(y))^{-1} dy\Big). \tag{3.12}$$

The same argument yields for $F_\lambda(\varepsilon, 0)$:

$$\begin{aligned} F_\lambda(\varepsilon,0) = \det \begin{Vmatrix} 0, & 1, & \varepsilon w_3(\varepsilon,0,0), & 0 \\ 0, & 0, & \varepsilon w_3'(\varepsilon,0,0), & 0 \\ -1, & 0, & 0 & \varepsilon^3 w_4''(\varepsilon,0,1) \\ -1, & 0, & 0 & \varepsilon^3 w_4'''(\varepsilon,0,1) \end{Vmatrix} + \\ + \det \begin{Vmatrix} 0, & 0, & \varepsilon w_3(\varepsilon,0,0), & 0 \\ 1, & 0, & \varepsilon w_3'(\varepsilon,0,0), & 0 \\ 0, & -1, & 0, & \varepsilon^3 w_4''(\varepsilon,0,1) \\ 0, & 0, & 0, & \varepsilon^3 w_4'''(\varepsilon,0,1) \end{Vmatrix} + O(\varepsilon^{-2}\delta_\varepsilon), \end{aligned} \tag{3.13}$$

since, as a consequence of (3.6), one has for $\varepsilon \to +0$:

$$\partial_\lambda^k \partial_x^p w_3(\varepsilon,0,1) = O(\varepsilon^{-p}\delta_\varepsilon), \partial_\lambda^k \partial_x^p w_4(\varepsilon,0,0) = O(\varepsilon^{-p}\delta_\varepsilon), \quad \forall\, p \geq 0, \quad \forall\, k \geq 0. \tag{3.14}$$

Computing the determinants on the right hand side of (3.13) and using (3.14), one finds:

$$\begin{aligned} F_\lambda(\varepsilon,0) &= -\varepsilon^4 (w_3'(\varepsilon,0,0)(w_4'''(\varepsilon,0,1) - w_4''(\varepsilon,0,1)) + \\ &+ w_4'''(\varepsilon,0,1) w_3(\varepsilon,0,1)) + O(\varepsilon^{-2}\delta_\varepsilon) = \\ &= (q(0))^{-1/2}(q(1))^{-5/2} \sum_{0 \leq k < N} c_k \varepsilon^k + O(\varepsilon^N) \end{aligned} \tag{3.15}$$

with any integer $N > 0$, where $c_0 = 1$ and c_k, $k > 0$, can be computed recursively using (2.8).

Thus, one has for each $\varepsilon \geq 0$ fixed:

$$F(\varepsilon,\lambda) = F(\varepsilon,0) + \lambda F_\lambda(\varepsilon,0) + O(\lambda^2), \text{ as } \lambda \to 0 \tag{3.16}$$

and for $|\lambda| \leq \varepsilon^{-2}\delta_\varepsilon$ one finds:

$$F(\varepsilon,\lambda) = F(\varepsilon,0) + \lambda F_\lambda(\varepsilon,0) + O(\varepsilon^{-4}\delta_\varepsilon^2). \tag{3.17}$$

Hence, one gets for the zero λ_0^ε of $F(\varepsilon,\lambda)$ in the interval $|\lambda| \leq \varepsilon^{-2}\delta_\varepsilon$ the following asymptotic formula with any integer $N > 0$:

$$\begin{aligned} \lambda_0^\varepsilon &= -F(\varepsilon,0)(F_\lambda(\varepsilon,0))^{-1} + O(\varepsilon^{-4}\delta_\varepsilon^2) = \\ &= 2(q(0))^{1/2}(q(1))^{-3/2}\delta_\varepsilon\Big(\sum_{0 \leq k < N} b_k \varepsilon^{k-1} + O(\varepsilon^N)\Big), \quad \varepsilon \to +0, \end{aligned} \tag{3.18}$$

where $b_0 = 1$, b_k, $k > 0$, can be computed recursively and where, of course, (3.11), (3.15), (3.17) have been used in order to derive (3.18).

Besides, for $\varepsilon \in (0,\varepsilon_0]$ with ε_0 sufficiently small, one finds using (3.11) and (3.15), (3.17):

$$F(\varepsilon,0) < 0, \quad F(\varepsilon,\varepsilon) > 0, \quad \forall \varepsilon \in (0,\varepsilon_0],$$

so that, in fact, $\lambda_0^\varepsilon \in (0,\varepsilon)$.

Thus, summarizing, we have proved the following

THEOREM 3.4. *The least eigenvalue λ_0^ε of $\mathcal{A}^\varepsilon$ defined by (1.3)-(1.5) is strictly positive for $\forall \varepsilon \in (0,\varepsilon_0]$ with $\varepsilon_0 > 0$ sufficiently small and, moreover, the following asymptotic formula holds for λ_0^ε with any integer $N > 0$:*

$$\lambda_0^\varepsilon = 2(q(0))^{1/2}(q(1))^{-3/2}\delta_\varepsilon\Big(\sum_{0 \leq k < N} b_k \varepsilon^{k-1} + O(\varepsilon^N)\Big), \text{ as } \varepsilon \to +0, \tag{3.19}$$

where $b_0 = 1$, b_k, $k > 0$, can be computed recursively using (2.8) and where δ_ε is given by (3.12).

Next, we are going to prove the following

THEOREM 3.5. *For the eigenfunction $\psi_0^\varepsilon(x)$ of $\mathcal{A}^\varepsilon$ (given by (1.3)-(1.5)) associated with the least eigenvalue λ_0^ε the following asymptotic formula holds:*

$$\begin{aligned}\psi_0^\varepsilon(x) &= xC_1(\varepsilon) + \varepsilon C_2(\varepsilon) + \varepsilon C_3(\varepsilon) w_3(\varepsilon, 0, x) + \\ &+ \lambda_0^\varepsilon((-x^3/6)C_1(\varepsilon) - (\varepsilon x^2/2)C_2(\varepsilon) + \\ &+ \varepsilon C_3(\varepsilon)\partial_\lambda w_3(\varepsilon, 0, x) - \varepsilon^2 w_4(\varepsilon, 0, x)) + \\ &+ O((\lambda_0^\varepsilon)^2),\end{aligned} \tag{3.20}$$

where the coefficients $C_k(\varepsilon)$, $1 \le k \le 3$, are C^∞ functions in $\varepsilon \in [0, \varepsilon_0]$ admitting the asymptotic expansions:

$$C_k(\varepsilon) = \sum_{0 \le p < N} c_{kp}\varepsilon^p + O(\varepsilon^N), \tag{3.21}$$

with any integer $N > 0$, and for the functions $w_3(\varepsilon, 0, x)$, $\partial_\lambda w_3(\varepsilon, 0, x)$, $w_4(\varepsilon, 0, x)$ the asymptotic expansions (2.4) with $u_{jk}(\lambda, x)$ defined by (2.8) are valid.

PROOF. We seek an eigenfunction $\psi_0^\varepsilon(x)$ of $\mathcal{A}^\varepsilon$ associated with the least eigenvalue λ_0^ε in the form:

$$\begin{aligned}\psi_0^\varepsilon(x) &= C_1(\varepsilon)w_1(\varepsilon, \lambda_0^\varepsilon, x) + \varepsilon C_2(\varepsilon) w_2(\varepsilon, \lambda_0^\varepsilon, x) + \varepsilon C_3(\varepsilon) w_3(\varepsilon, \lambda_0^\varepsilon, x) - \\ &- \varepsilon^2 \lambda_0^\varepsilon w_4(\varepsilon, \lambda_0^\varepsilon, x),\end{aligned} \tag{3.22}$$

where $w_k(\varepsilon, \lambda, x)$, $1 \le k \le 4$, is the fundamental system of solutions for equation (2.2) constructed in the proof of Theorem 2.3.

The boundary conditions for $\psi_0^\varepsilon(x)$ yield an overdetermined system for the coefficients $C_k(\varepsilon)$, $1 \le k \le 3$, which has non-trivial solutions, since $F(\varepsilon, \lambda_0^\varepsilon) = 0$, $\forall \varepsilon \in (0, \varepsilon_0]$ with $F(\varepsilon, \lambda)$ defined by (3.2).

Thus, neglecting the equation for $C_k(\varepsilon)$, $1 \le k \le 3$, which results from the boundary condition $\partial_x^3 \psi_0^\varepsilon(\varepsilon, \lambda_0^\varepsilon, 1) = 0$ and rewriting the boundary conditions $\psi_0^\varepsilon(\varepsilon, \lambda_0^\varepsilon, 0) = 0$, $\partial_x^2 \psi_0^\varepsilon(\varepsilon, \lambda_0^\varepsilon, 1) = 0$ in the equivalent form: $\varepsilon^{-1}\psi_0^\varepsilon(\varepsilon, \lambda_0^\varepsilon, 0) = 0$, $(\lambda_0^\varepsilon)^{-1}\partial_x \psi_0^\varepsilon(\varepsilon, \lambda_0^\varepsilon, 1) = 0$, one gets for $C(\varepsilon) = (C_1(\varepsilon), C_2(\varepsilon), C_3(\varepsilon))$ the linear system:

$$A(\varepsilon)C(\varepsilon) = g(\varepsilon), \tag{3.23}$$

where

$$A(\varepsilon) := \begin{Vmatrix} 0, & w_2(\varepsilon, \lambda_0^\varepsilon, 0), & w_3(\varepsilon, \lambda_0^\varepsilon, 0) \\ w_1'(\varepsilon, \lambda_0^\varepsilon, 0), & 0, & \varepsilon w_3'(\varepsilon, \lambda_0^\varepsilon, 0) \\ (\lambda_0^\varepsilon)^{-1} w_1''(\varepsilon, \lambda_0^\varepsilon, 1), & \varepsilon(\lambda_0^\varepsilon)^{-1} w_2''(\varepsilon, \lambda_0^\varepsilon, 1), & \varepsilon(\lambda_0^\varepsilon)^{-1} w_3''(\varepsilon, \lambda_0^\varepsilon, 1) \end{Vmatrix} \tag{3.24}$$

and

$$g(\varepsilon) := (\varepsilon\lambda_0^\varepsilon w_4(\varepsilon,\lambda_0^\varepsilon,0), \varepsilon^2\lambda_0^\varepsilon w_4'(\varepsilon,\lambda_0^\varepsilon,0), \varepsilon^2 w_4''(\varepsilon,\lambda_0^\varepsilon,1)). \tag{3.25}$$

Indeed, as a consequence of Proposition 2.1, Lemma 2.2 and Theorem 2.3, one has:

$$w_1(\varepsilon,\lambda,0) = w_2'(\varepsilon,\lambda,0) = 0.$$

Using again Proposition 2.1, Theorem 2.3, and (3.19) one easily finds:

$$A(\varepsilon) = A(0) + \varepsilon B(\varepsilon), \tag{3.26}$$

where

$$A(0) := \begin{Vmatrix} 0, & 1, & (q(0))^{1/2} \\ 1, & 0, & -(q(0))^{-1/2} \\ -1, & 0, & 2(q(0))^{-1/2} \end{Vmatrix} \tag{3.27}$$

and where $B(\varepsilon)$ is a 3×3 matrix which is infinitely differentiable with respect to $\varepsilon \in [0,\varepsilon_0]$ and whose asymptotic expansion for $\varepsilon \to +0$ can be found explicitly,

$$B(\varepsilon) \sim \sum_{k\geq 0} \varepsilon^k B_k,$$

using Proposition 2.1 and Theorem 2.3.

Since $A(0)$ is invertible, the matrix $A(\varepsilon)$ defined by (3.24) is invertible, as well, for $\forall\varepsilon \in [0,\varepsilon_0]$ with ε_0 sufficiently small, and, moreover, $(A(\varepsilon))^{-1}$ is an infinitely differentiable matrix-function of $\varepsilon \in [0,\varepsilon_0]$.

Furthermore, taking the first two terms in the Taylor's expansion of $w_k(\varepsilon,\lambda,x)$, $1 \leq k \leq 3$, with respect to λ around the point $\lambda = 0$, and noticing that $w_1(\varepsilon,0,x) \equiv x$, $\partial_\lambda w_1(\varepsilon,0,x) = -x^3/6$, $w_2(\varepsilon,0,x) \equiv 1$, $\partial_\lambda w_2(\varepsilon,0,x) = -x^2/2$ (as a consequence of (2.5), (2.7), Lemma 2.2. and Theorem 2.3), one gets asymptotic formula (3.20) with $C_k(\varepsilon)$, $1 \leq k \leq 3$, admitting asymptotic expansion (3.21). ◇

REMARK 3.6. An easy computation yields for the coefficients c_{kp} on the right hand side of (3.21):

$$c_{10} = 2(q(1))^{-3/2}, c_{20} = -2q(0)(q(1))^{-3/2}, c_{30} = 2(q(0))^{1/2}(q(1))^{-3/2}. \tag{3.28}$$

Besides, as a consequence of (2.8), one also has:

$$\begin{cases} w_3(\varepsilon,0,x) = ((q(x))^{1/2} + O(\varepsilon))\exp(-\varepsilon^{-1}\int_0^x (q(y))^{-1}dy), \\ \partial_\lambda w_3(\varepsilon,0,x) = O(\varepsilon)\exp(-\varepsilon^{-1}\int_0^x (q(y))^{-1}dy), \\ w_4(\varepsilon,0,x) = ((q(x))^{1/2} + O(\varepsilon))\exp(-\varepsilon^{-1}\int_x^1 (q(y))^{-1}dy), \end{cases} \tag{3.29}$$

so that keeping only the main terms, one gets the following asymptotic expansion for $\psi_0^\varepsilon(x)$:

$$\psi_0^\varepsilon(x) = 2(q(1))^{-3/2}(1+O(\varepsilon))(x - \varepsilon q(0) + \varepsilon(q(0)q(x))^{1/2}\exp(-\varepsilon^{-1}\int_0^x (q(y))^{-1}dy)).$$

Moreover, the last formula may also be rewritten in the following equivalent form:

$$\psi_0^\varepsilon(x) = 2(q(1))^{-3/2}(1+O(\varepsilon))(x - \varepsilon q(0)(1-\exp(-x/(\varepsilon q(0))))), \tag{3.30}$$

since freezing the coefficient $q(x)$ at $x = 0$ brings over an error, which is of order $O(\varepsilon^2\exp(-x/(\varepsilon q(0)))$.

Formula (3.30) can, of course, be also derived using the reduction method mentioned above in the introduction. ◇

Next, an asymptotic formula will be derived for an eigenfunction $\psi_n^\varepsilon(x)$ associated with an eigenvalue λ_n^ε, for which (2.7) is valid with a given integer $n > 0$.

Seeking $\psi_n^\varepsilon(x)$ in the form:

$$\begin{aligned}\psi_n^\varepsilon(x) = \pi n w_1(\varepsilon,\lambda_n^\varepsilon,x) + \varepsilon C_2(\varepsilon)w_2(\varepsilon,\lambda_n^\varepsilon,x) + \varepsilon C_3(\varepsilon)w_3(\varepsilon,\lambda_n^\varepsilon,x) +\\ + \varepsilon^3 C_4(\varepsilon)w_4(\varepsilon,\lambda_n^\varepsilon,x),\end{aligned} \tag{3.31}$$

the same argument as above in the case of $\psi_0^\varepsilon(x)$ yields a linear system for $C(\varepsilon) = (C_2(\varepsilon), C_3(\varepsilon), C_4(\varepsilon))$ with a 3×3 matrix $A(\varepsilon)$ which is infinitely differentiable with respect to $\varepsilon \in [0,\varepsilon_0]$ and such that

$$A(0) = \left\| \begin{matrix} 1, & (q(0))^{1/2}, & 0 \\ 0, & -(q(0))^{-1/2}, & 0 \\ 0, & 0, & (q(1))^{-5/2} \end{matrix} \right\|, \tag{3.32}$$

the second member $g(\varepsilon)$ of this linear system being

$$g(\varepsilon) = (0, -\pi n w_1'(\varepsilon,\lambda_n^\varepsilon,0), -\pi n w_1'''(\varepsilon,\lambda_n^\varepsilon,1)), \tag{3.33}$$

a C^∞ function of $\varepsilon \in [0,\varepsilon_0]$.

Therefore, the coefficients $C_k(\varepsilon), 2 \le k \le 4$ on the right hand side of (3.31) are well-defined C^∞-functions of $\varepsilon \in [0,\varepsilon_0]$, provided that ε_0 is sufficiently small. Besides, $C_k(\varepsilon)$ again admit asymptotic expansion:

$$C_k(\varepsilon) \sim \sum_{p\ge 0} c_{kp}\varepsilon^p. \tag{3.34}$$

It is readily seen that

$$C_2(0) = -q(0)\pi n, C_3(0) = (q(0))^{1/2}\pi n, C_4(0) = (q(1))^{5/2}(-1)^n(\pi n)^3. \tag{3.35}$$

Thus, freezing again $q(x)$ at $x = 0$ for w_3 and at $x = 1$ for w_4, and using the fact that $\lambda_n^\varepsilon - \lambda_n^0 = O(\varepsilon)$, (3.31), (3.35) yield the following asymptotic formula for $\psi_n^\varepsilon(\lambda)$:

$$\begin{aligned}\psi_n^\varepsilon(x) = \{&\sin \pi n x + \varepsilon q(0)(\exp(-x/(\varepsilon q(0))) - \cos \pi n x) + \\ &+ (-1)^n(\varepsilon q(1)\pi n)^3 \exp(-(1-x)/(\varepsilon q(1))\}(1 + O(\varepsilon))\end{aligned} \tag{3.36}$$

Also using (3.31) and Taylor's expansions in λ around $\lambda = \pi^2 n^2$ of $w_k(\varepsilon, \lambda_n^\varepsilon, x)$, $1 \leq k \leq 4$, one gets for $\psi_n^\varepsilon(x)$ a full asymptotic expansion in the form:

$$\begin{aligned}\psi_n^\varepsilon \sim &\sum_{p\geq 0} \varepsilon^p u_{1p}(x) + \\ &\exp(-\varepsilon^{-1}\int_0^x (q(y))^{-1} dy) \sum_{p\geq 1} \varepsilon^p u_{2p}(x) + \\ &+ \exp(-\varepsilon^{-1}\int_x^1 (q(y))^{-1} dy) \sum_{p\geq 3} \varepsilon^p u_{3p}(x),\end{aligned} \tag{3.37}$$

where, of course,

$$\begin{aligned}&u_{10}(x) = \sin \pi n x, u_{11}(x) = -\varepsilon q(0) \cos \pi n x \\ &u_{21}(x) = (q(0)q(x))^{1/2}, u_{33}(x) = (q(1))^{5/2}(q(x))^{1/2},\end{aligned}$$

and all other coefficients $u_{pk}(x)$ can be computed recursively.

REMARK 3.7. Of course, using the asymptotic formulae for $w_j(\varepsilon, \lambda, x)$, $1 \leq j \leq 4$, in Theorem 3.1, one can establish asymptotic formulae for λ_0^ε and $\psi_0^\varepsilon(x)$ in (1.10) up to an error term $O(\varepsilon^\infty \delta_\varepsilon)$ with δ_ε given by (3.3), as $\varepsilon \to +0$.

For other eigenvalues λ_k^ε, $k = 1, 2, ...$, of (2.1) one may apply the reduction method developed in [6], [7] as well, the procedure being very similar to the one in [4], since one has for the eigenvalues λ_k^ε, $k = 1, 2, ...$, of (1.10):

$$\lambda_k^\varepsilon = \pi^2 k^2 + O(\varepsilon), \ \varepsilon \to 0,$$

as it was the case for $\mathcal{A}^\varepsilon$ in [4] (i.e. $\lambda_k^\varepsilon - \lambda_k^0$ for $k \geq 1$ is no longer exponentially small, as $\varepsilon \to +0$, so that the parametrix used previously in the construction of a reducing operator for coercive singular perturbations $\mathcal{A}^\varepsilon$ may be used again with necessary minor modifications).

REMARK 3.8. The following argument can be used in order to get heuristically the first term in the asymptotic expansion of the least eigenvalue λ_0^ε.

As a consequence of Theorem 2.3, one has the following asymptotic formulae for the fundamental system of solutions $w_k(\varepsilon, \lambda, x)$, $1 \leq k \leq 4$ of equation (2.2):

$$\begin{aligned}
w_1(\varepsilon,\lambda,x) &= \lambda^{-1/2}\sin(\lambda^{1/2}x) + O(\varepsilon^2),\\
w_2(\varepsilon,\lambda,x) &= \cos(\lambda^{1/2}x) + O(\varepsilon^2)\\
w_3(\varepsilon,\lambda,x) &= (1+O(\varepsilon))(q(x))^{1/2}\exp(-\varepsilon^{-1}\int_0^x (q(y))^{-1}dy)\\
w_4(\varepsilon,\lambda,x) &= (1+O(\varepsilon))(q(x))^{1/2}\exp(-\varepsilon^{-1}\int_x^1 (q(y))^{-1}dy).
\end{aligned}$$

Using only the first terms in the asymptotic expansions for $w_k(\varepsilon, \lambda, x)$, $1 \leq k \leq 4$ and attempting to satisfy for a solution $u(\varepsilon, \lambda, x)$ of (2.2) the boundary conditions: $u(\varepsilon, \lambda, 0) = \partial_x u(\varepsilon, \lambda, 0) = 0$, one gets for $u(\varepsilon, \lambda, x)$ the asymptotic representation

$$\begin{aligned}
u(\varepsilon,\lambda,x) \sim{}& -\lambda^{-1/2}\sin(\lambda^{1/2}x) + \varepsilon q(0)\cos(\lambda^{1/2}x) -\\
&- \varepsilon(q(0)q(x))^{1/2}(\exp(-\varepsilon^{-1}\int_0^x (q(y))^{-1}dy) +\\
&+ C_\varepsilon \exp(-\varepsilon^{-1}\int_x^1 (q(y))^{-1}dy)).
\end{aligned}$$

Trying to cancel the leading term in the asymptotic expansion for $\partial_x^3 u(\varepsilon, \lambda, 1)$, one gets the conclusion that the only reasonable choice for the constant C_ε on the right hand side of the last formula is:

$$C_\varepsilon = \delta_\varepsilon = \exp(-\varepsilon^{-1}\int_0^1 (q(y))^{-1}dy).$$

Furthermore, one also realizes that the only way to satisfy asymptotically the boundary conditions $\partial_x^k u(\varepsilon, \lambda, 1) = 0, k = 2, 3$, is to have $\lambda = \lambda(\varepsilon) \to 0$ as $\varepsilon \to +0$.

Afterwards, replacing $\lambda^{-1/2}\sin(\lambda^{1/2}x)$ and $\cos(\lambda^{1/2}x)$ by $x - \lambda x^3/6$ and $1 - \lambda x^2/2$, respectively and attempting to satisfy the boundary condition: $\partial_x^2 u(\varepsilon, \lambda, 1) = 0$, one gets in this heuristic way for λ_0^ε the first term in the asymptotic formula given by (3.19) and for the associated eigenfunction $\psi_0^\varepsilon(x)$ the first terms in the asymptotic formula given by (3.20).

The last term on the right hand side of (3.20), which is of order $O(\delta_\varepsilon^2)$ at $x = 0$, seems to be redundant; however it is not so, since at $x = 1$ this term is comparable with all other terms in the asymptotic formulae for $\partial_x^k \psi_0^\varepsilon(x)|_{x=1}$, $k = 2, 3, \ldots$. But away from the point $x = 1$ and, especially in the neighbourhood of $x = 0$ only the three first terms on the right hand side of (3.20) are relevant for the asymptotic behaviour of $\psi_0^\varepsilon(x)$ as $\varepsilon \to +0$. Of course, freezing $q(x)$ in the exponential term at the point $x = 0$, one gets an asymptotic formula for $\psi_0^\varepsilon(x)$ valid in a neighbourhood of $x = 0$,

$$\psi_0^\varepsilon(x) = -x + \varepsilon q(0)(1 - \exp(-x/(\varepsilon q(0))) + \varepsilon^2 w_\varepsilon$$

with $\sup_{\varepsilon} \max_{x} |\partial_x^k w_\varepsilon(x)| < \infty$, $k = 0, 1, 2$, which may be also derived by using a reducing operator S^ε, constructed in the same way as in [6], [7], i.e. by using the parametrix constructions where only the principal symbol of the coercive singular perturbation is involved. ◇

REMARK 3.9. It is readily seen that the reduced problem $\mathcal{A}^0 u^0 = (f, \varphi_1)^T$ is solvable iff the following condition is satisfied:

$$f(1) + \varphi_1(1) = 0,$$

the solution $u^0(x)$ being well-defined up to the additive term Cx, which is the solution of the homogeneous problem $\mathcal{A}^0 x = (0, 0)^T$.

In other words, the index $\kappa(0)$ of the reduced problem is zero: $\kappa(0) = 0$.

Notice that so it is also for the perturbed problem $\mathcal{A}^\varepsilon$, i.e. $\kappa(\varepsilon) = 0$, $\forall \varepsilon \in (0, \varepsilon_0]$, since for $\forall \varepsilon \in (0, \varepsilon_0]$ with ε_0 sufficiently small the perturbed problem has a well-defined solution for any data $(f, \varphi_1, \varphi_2)^T$ (sufficiently smooth), i.e. $\dim \ker \mathcal{A}^\varepsilon = \dim \operatorname{coker} \mathcal{A}^\varepsilon = 0$.

The stability of the index $\kappa(\varepsilon)$ of elliptic boundary value problems with respect to coercive singular perturbations is also a consequence of the general reduction procedure indicated in [6], [7], [17]. ◇

4. ADJOINT OPERATOR.

We identify the singular perturbation $\mathcal{A}^\varepsilon$ given by (1.4), (1.3) with the differential operator $L^\varepsilon = L(\varepsilon, 0, x, \partial_x) = \varepsilon^2 \partial_x^2 q^2(x) \partial_x^2 - \partial_x^2$ considered as an unbounded operator in $L_2(U)$ with the domain D_{L^ε},

$$D_{L^\varepsilon} := \{u \in H_4(U), \pi_{\partial U} B_k u(x') = 0,\ k = 1, 2\}, \tag{4.1}$$

where $H_4(U)$ is the Sobolev space of order 4.

Denote by $(L^\varepsilon)^*$ the adjoint of L^ε. The partial integration implies that $(L^\varepsilon)^*$ is the same differential operator $L(\varepsilon, 0, x, \partial_x)$ with the domain $D_{(L^\varepsilon)^*}$ defined as follows:

$$D_{(L^\varepsilon)^*} := \{u \in H_4(U), \pi_{\partial U} {}^t B_k^\varepsilon(x', \partial_x) u(x') = 0,\ k = 1, 2\}, \tag{4.2}$$

where

$$\begin{cases} {}^t B_1^\varepsilon(x, \partial_x) = (x - 1) + x(\varepsilon^2 q^2(x) \partial_x^2 - 1), \\ {}^t B_2^\varepsilon(x, \partial_x) = (1 - x) q^4(x) \partial_x q^{-2}(x) + x \partial_x (1 - \varepsilon^2 q^2(x) \partial_x^2). \end{cases} \tag{4.3}$$

Hence, one may consider the adjoint operator $(L^\varepsilon)^* : D_{(L^\varepsilon)^*} \to L_2(U)$ of $L^\varepsilon : D_{L^\varepsilon} \to L_2(U)$ as a restriction (to homogeneous boundary conditions) of the operator $(\mathcal{A}^\varepsilon)^*$ defined as follows:

$$(\mathcal{A}^\varepsilon)^* := (\pi_U r^\varepsilon D_x^2, \pi_{\partial U}{}^t B_1^\varepsilon, \pi_{\partial U}{}^t B_2^\varepsilon)^T. \tag{4.4}$$

Since the least eigenvalue λ_0^ε of $\mathcal{A}^\varepsilon$ defined by (1.4) is real (and strictly positive for $\varepsilon \in (0, \varepsilon_0]$), it coincides with the least eigenvalue of $(\mathcal{A}^\varepsilon)^*$ defined by (4.4), (4.3).

We are going to exhibit an asymptotic formula for the eigenfunction $\varphi_0^\varepsilon(x)$ of $(\mathcal{A}^\varepsilon)^*$ associated with the least eigenvalue $\lambda_0^\varepsilon > 0$.

Using Theorem 2.3 and the same argument as above for the eigenfunction ψ_0^ε of $\mathcal{A}^\varepsilon$ associated with λ_0^ε, one gets the following asymptotic formula for $\varphi_0^\varepsilon(x)$:

$$\begin{aligned} \varphi_0^\varepsilon(x) \sim\ & \delta_\varepsilon(2(q(0)q(1))^{1/2}x - 2(q(0))^{1/2}+ \\ & + (q(x))^{1/2}\exp(-\varepsilon^{-1}\int_0^x (q(y))^{-1}dy))+ \\ & + (q(x))^{1/2}\exp(-\varepsilon^{-1}\int_x^1 (q(y))^{-1}dy). \end{aligned} \tag{4.5}$$

Neglecting terms which are $O(\delta_\varepsilon)$ uniformly with respect to $x \in \overline{U}$, one gets the following simplified (and less accurate) asymptotic formula for the eigenfunction $\varphi_0^\varepsilon(x)$ of the adjoint problem:

$$\varphi_0^\varepsilon(x) \sim (q(x))^{1/2}\exp(-\varepsilon^{-1}\int_x^1 (q(y))^{-1}dy). \tag{4.6}$$

Furthermore, freezing $q(x)$ at the point $x = 1$, one gets for $\varphi_0^\varepsilon(x)$ an asymptotic formula which may be established using the reduction procedure from [6], [7], as well:

$$\varphi_0^\varepsilon(x) \sim (q(1))^{1/2}\exp(-(1-x)/(\varepsilon q(1))). \tag{4.7}$$

For the normalized eigenfunction, i.e. φ_0^ε such that $|\langle\varphi_0^\varepsilon, \psi_0^\varepsilon\rangle| = 1$ with ψ_0^ε the corresponding eigenfunction of $\mathcal{A}^\varepsilon$ given by (3.2), (3.3), one finds, using (4.7):

$$\varphi_0^\varepsilon(x) \sim (\varepsilon q(1))^{-1}\exp(-(1-x)/(\varepsilon q(1))). \tag{4.8}$$

Note that the normalized eigenfunction converges to the Dirac's mass $\delta(1-x)$ at the point $x = 1$.

The adjoint operator $(\mathcal{A}^\varepsilon)^*$ defined by (4.4) is not a coercive singular perturbation, the operator $\pi_U r^\varepsilon D_x^2$ however being still an elliptic singular perturbation (see [3] for the definition of coercive singular perturbations).

One may wonder what should be like the reduced operator $(\mathcal{A}^0)^*$ associated with $(\mathcal{A}^\varepsilon)^*$ given by (4.4). The formally defined reduced operator $(\mathcal{A}^0)^*$ is the one, where only the first boundary operator ${}^tB_1^\varepsilon(x, \partial_x)$ in (4.3) is kept and afterwards one sets $\varepsilon = 0$ in both $\pi_U r^\varepsilon D_x^2$ and $\pi_{\partial U}{}^tB_1^\varepsilon$, the latter yielding

$$(\mathcal{A}^0)^* := (\pi_U D_x^2, \pi_{\partial U} 1)^T. \tag{4.9}$$

In order to realize that formally defined operator (4.9) is indeed the reduced operator for $(\mathcal{A}^\varepsilon)^*$ given by (4.4), one has to consider a suitable perturbation of $(\mathcal{A}^\varepsilon)^*$ by lower order operators which shift the spectrum of $(\mathcal{A}^\varepsilon)^*$ away from zero. For instance, considering the boundary value problem (with $q(x) \equiv 1$ for simplicity):

$$\begin{cases} \pi_U(\varepsilon^2 D_x^2 + D_x^2 - \lambda)u_\lambda^\varepsilon = 0 \\ \pi_{\partial U}{}^tB_k^\varepsilon u_\lambda^\varepsilon(x') = \varphi_k(x'),\ 1 \le k \le 2, \end{cases} \tag{4.10}$$

with $\lambda \in \mathbf{C}, 0 < |\lambda| < \pi^2$ and $\varphi_k(x'), k = 1, 2$ given and independent of ε, it is readily seen that there exists the pointwise limit:

$$u_\lambda^0(x) = \lim_{\varepsilon \to o} u_\lambda^\varepsilon(x),\ \forall x \in [0, 1) \tag{4.11}$$

and, moreover, $u_\lambda^0(x)$ is the solution of the boundary value problem

$$\pi_U(D_x^2 - \lambda)u_\lambda^0(x) = 0,\ \pi_{\partial U} u_\lambda^0(x') = \varphi_1(x').$$

The same is true if one considers an inhomogeneous equation in (4.10) with a smooth second member $f(x)$ which does not depend on ε.

A somewhat surprizing situation is the fact that the reduced operator $(\mathcal{A}^0)^*$ for $(\mathcal{A}^\varepsilon)^*$ defined by (4.4) has no longer $\lambda = 0$ as its eigenvalue, while for $\mathcal{A}^0$ defined by (1.6), (1.3) zero is an eigenvalue with Cx as the associated eigenfunctions.

In fact, the solution $u_\lambda^\varepsilon(x)$ to (4.10) contains a singular part which converges (as $\varepsilon \to +0$) to $\gamma_\varphi \delta(1-x)$ with some constant γ_φ depending on $\varphi_1(x')$ and $\varphi_2(x')$, where $\delta(1-x)$ is the Dirac's δ-function. An easy computation shows also that

$$\lim_{\varepsilon \to 0} \operatorname{Res} u_\lambda^\varepsilon(x)|_{\lambda=0} = \gamma_\varphi \delta(1-x) = \varphi_0^0(x)$$

so that, to some extent, it would be natural to consider $\delta(1-x)$ as an "eigenfunction" of $(\mathcal{A}^0)^*$.

Let us consider again the operator $(\mathcal{A}_\lambda^\varepsilon)^*$ associated with (4.10):

$$(\mathcal{A}_\lambda^\varepsilon)^* := (\pi_U(\varepsilon^2 D_x^4 + D_x^2 - \lambda), \pi_U{}^tB_1^\varepsilon, \pi_{\partial U}{}^tB_2^\varepsilon)^T,\ 0 < \lambda < \pi^2,$$

where, of course, ${}^tB_k^\varepsilon(x,\partial_x)$ are defined by (4.3) with $q(x)\equiv 1$, i.e.

$$ {}^tB_1^\varepsilon(x,\partial_x)=(x-1)+x(\varepsilon^2\partial_x^2-1),\, {}^tB_2^\varepsilon=(1-x)\partial_x+x\partial_x(1-\varepsilon^2\partial_x^2). $$

It is readily seen, that the function

$$ u_\lambda^\varepsilon(x)=\varepsilon^{-1}(1+\varepsilon\lambda x/2)\exp(-(1-x)/\varepsilon),\ 0<\lambda<\pi^2, $$

is the solution of the boundary value problem

$$ (\mathcal{A}_\lambda^\varepsilon)^* u_\lambda^\varepsilon(x)=(f^\varepsilon(x),\varphi_1^\varepsilon,\varphi_2^\varepsilon)^T, $$

with

$$ (4.12)\qquad \begin{cases} f^\varepsilon(x)=(-\lambda^2 x/2)\exp(-(1-x)/\varepsilon),\ \sup\limits_{0<\varepsilon\le 1}\max\limits_{x\in\overline{U}}|f^\varepsilon(x)|<\infty \\ \varphi_1^\varepsilon(x')=(1-x')\varepsilon^{-1}\exp(-\varepsilon^{-1})+\varepsilon\lambda x',\ \sup\limits_{0<\varepsilon\le 1}\max\limits_{x'\in\partial U}|\varphi_1^\varepsilon(x')|<\infty \\ \varphi_2^\varepsilon(x')=(1-x')\varepsilon^{-2}(1+\varepsilon\lambda/2)\exp(-\varepsilon^{-1})+\lambda x', \\ \sup\limits_{0<\varepsilon\le 1}\max\limits_{x'\in\partial U}|\varphi_2^\varepsilon(x')|<\infty. \end{cases} $$

Nevertheless,

$$ \max_{x\in\overline{U}} u_\lambda^\varepsilon(x)=\varepsilon^{-1}(1+\varepsilon\lambda/2)\to\infty \text{ as } \varepsilon\to+0. $$

Such a situation is impossible, for instance, for the following singular perturbation of $(\mathcal{A}_\lambda^0)^* =$
$(\pi_U(D_x^2-\lambda),\pi_{\partial U}1)^T$:

$$ \mathcal{B}_\lambda^\varepsilon := (\pi_U(\varepsilon^2 D_x^4+D_x^2-\lambda),\pi_{\partial U}1,\pi_{\partial U}\partial_n)^T,\ 0\le\lambda<\pi^2, $$

which is coercive (see [3]), so that for the solution u_λ^ε of the problem

$$ \mathcal{B}_\lambda^\varepsilon u_\lambda^\varepsilon=(f^\varepsilon,\varphi_1^\varepsilon,\varphi_2^\varepsilon)^T $$

with the data $(f^\varepsilon,\varphi_1^\varepsilon,\varphi_2^\varepsilon)^T$ satisfying (4.12), one always has the following a priori estimate (a version of the maximum principle):

$$ \sup_{0<\varepsilon\le 1}\max_{x\in\overline{U}}|u_\lambda^\varepsilon(x)|\le C(\sup_{0<\varepsilon\le 1}\max_{x\in\overline{U}}|f^\varepsilon(x)|+\sum_{1\le k\le 2}\sup_{0<\varepsilon\le 1}\max_{x'\in\partial U}|\varphi_k^\varepsilon(x')|), $$

with some constant $C>0$.

The right formulation of the reduced boundary value problem for $(\mathcal{A}^0)^*$ given by (4.9) should be as follows: for a given $f\in H_{s-2}(U)$, $s>1/2$, and $\varphi(x')$ find a solution

$u \in H_{-1}(U)$ such that $u(x)$, $x \in [0,1)$ can be extended by continuity to a function $lu(x)$, $x \in [0,1]$, $lu \in H_s$, lu being the solution of the problem:

$$(\mathcal{A}^0)^* lu = (f,\varphi)^T, \quad u \in H_{-1}(U). \tag{4.13}$$

The solution of (4.13) is not unique and can be represented in the form:

$$u(x) = \left((\mathcal{A}^0)^*\right)^{-1}(f,\varphi)^T + c\delta(1-x) \tag{4.14}$$

where $c \in \mathbb{C}$ is an arbitrary constant.

LIST OF REFERENCES:

[1] L.S. Frank, Problèmes aux limites coercifs avec un petit paramètre, C.R. Acad. Sci., Paris, t.282, Série A, 1976, p.1109-1111.

[2] L.S. Frank, Perturbazioni Singolari Ellittiche, Rendiconti del Politecnico di Milano, XLVII, 1977, pp.135-163.

[3] L.S. Frank, Coercive Singular Perturbations I: A prioiri estimates, Annali di Mat. Pura Appl. (IV), 19, 1979, pp.41-113.

[4] L.S. Frank, Perturbations Singulières Coercives IV: Problème des valeurs propres, C.R. Acad.Sci. Paris, Série I, t.301, n3, 1985, pp.69-72.

[5] L.S. Frank, Perturbations singulières coercives: Reduction à des perturbations régulières et applications, Séminaire Equations aux Dérivées Partielles 1986-1987, Centre de Mathématiques, Ecole Polytechnique, exposé no.XVIII, 26 avril 1987, pp.1-26.

[6] L.S. Frank, W.D. Wendt, Coercive Singular Perturbations II: Reduction to regular perturbations and applications, Comm. P.D.E. 7, 1982, pp.469-535.

[7] L.S. Frank, W.D. Wendt, Coercive Singular Perturbations III: Wiener-Hopf Operators, Journal d'Analyse Mathématique, Vol.43, 1983/84, pp.88-135.

[8] L.S. Frank, Sharp error estimates in the reduction procedure for the coercive singular perturbations, Asymptotic Analysis, vol. 3, no. 1, 1990, pp. .

[9] T. Kato, Perturbation Theory of semi-bounded Operators, Math. Ann., 125 (1953), pp. 435-447.

[10] L. Landau, E. Lifschitz, Théorie de l'Elasticité, MIR, Moscou 1967 (transl. from Russian).

[11] J. Moser, Singular Perturbation of Eigenvalue Problems for Linear Differential Equations of Even Order, Comm. Pure and Appl. Math., vol. VIII (1955), pp. 251-278.

[12] Rayleigh, Lord, The Theory of Sound, vol. I, London 1937.

[13] R.T. Seeley, Extension of C^{∞} functions defined on a half-space, Proc. Amer. Math. Soc., 15, (1964), pp. 625-626.

[14] M.I. Vishik, On strongly Elliptic Systems of Differential Equations, Mat. Sb. (N.S.) 29 (71) (1951), pp. 615-676.

[15] M.I. Vishik and L.A. Lyusternik, Regular Degeneration and Boundary Layer for Linear Differential Equations with Small Parameter, Uspekhi Mat. Nauk 12 no. 5 (1957), pp. 3-122, AMS Transl. (2), 20, 1962, pp. 239-364.

[16] W. Wasow, Asymptotic Expansions for Ordinary Differential Equations, Interscience, 1965.

[17] W.D. Wendt, Coercive Singularly perturbed Wiener-Hopf Operators and Applications, Ph.D.Thesis, Univ. of Nijmegen, 1983.

L.S. Frank
Department of Mathematics
Catholic University of Nijmegen
Toernooiveld
6525 ED Nijmegen
The Netherlands.

Operator Theory:
Advances and Applications, Vol. 50

Matrix Polynomials with prescribed zero structure in the finite complex plane

I. Gohberg, M.A. Kaashoek, A.C.M. Ran

Explicit formulas are deduced for all regular matrix polynomials with a prescribed zero structure in the finite complex plane and with the additional property that the inverses are analytic at infinity. The solutions are parametrized by two matrices satisfying a linear equation which is connected to feedback problems and Rosenbrock's theorem from mathematical systems theory.

0. Introduction

Let $L(\lambda)=L_0+\lambda L_1+\ldots+\lambda^l L_l$ be an $m\times m$ matrix polynomial which is regular (i.e., $\det L(\lambda)$ does not vanish identically). A pair of matrices (A, B) is said to be a *(right) zero pair* for $L(\lambda)$ if A is a square matrix whose order n is equal to the degree of the scalar polynomial $\det L(\lambda)$, B is an $n\times m$ matrix and the following conditions are satisfied:

$$\bigvee_{j=0}^{n-1} \operatorname{Im} A^j B = \operatorname{Im} (B\ AB\ \cdots\ A^{n-1}B) = \mathbb{C}^n, \tag{0.1}$$

$$BL_0+ABL_1+\ldots+A^l BL_l=0. \tag{0.2}$$

If $L(\lambda)$ is monic (i.e., the leading coefficient L_l is the identity matrix I), then the pair (H, F), where

$$H=\begin{pmatrix} 0 & I & & & \\ & & \cdot & & \\ & & & \cdot & \\ & & & & \cdot \\ & & & & I \\ -L_0 & -L_1 & \cdot & \cdot & \cdot & -L_{l-1} \end{pmatrix}, \quad F=\begin{pmatrix} 0 \\ \cdot \\ \cdot \\ \cdot \\ 0 \\ I \end{pmatrix},$$

is a right zero pair for $L(\lambda)$ and any other zero pair (A, B) for $L(\lambda)$ is obtained from this pair by similarity in the following way:

$$A=S^{-1}HS, \quad B=S^{-1}F. \tag{0.3}$$

If in (0.3) the similarity transformation S is chosen such that A has the Jordan normal form, then the columns of B are eigenvectors and generalized eigenvectors of $L(\lambda)$ associated with the zeros of $\det L(\lambda)$ (see [GLR] for further details).

The present paper concerns the inverse problem. Let A and B be matrices of sizes $n \times n$ and $n \times m$, respectively, and assume that (0.1) holds. Determine, if possible, a regular $m \times m$ matrix polynomial $L(\lambda)$ such that (A, B) is a zero pair for $L(\lambda)$. For monic polynomials the inverse problem has a strikingly simple solution. First of all the pair (A, B) has to satisfy two conditions, namely:

(i) $n = ml$ for some positive integer l,

(ii) $\det\,(B\,AB\,\ldots A^{l-1}B) \neq 0$.

If these conditions are satisfied, then the inverse problem has a unique solution (in the class of monic matrix polynomials), which is given by

$$L(\lambda) = \lambda^l I - (W_1 + \lambda W_2 + \ldots + \lambda^{l-1} W_l) A^l B, \tag{0.4}$$

where W_i are the $m \times ml$ matrices determined by

$$(B\,AB\,\ldots A^{l-1}B)^{-1} = \begin{pmatrix} W_1 \\ W_2 \\ \cdot \\ \cdot \\ \cdot \\ W_l \end{pmatrix}.$$

Moreover, in that case

$$L(\lambda)^{-1} = C(\lambda - A)^{-1}B \tag{0.5}$$

with $C = W_l$ (see [GLR], Theorem 2.4).

For arbitrary regular matrix polynomials the inverse problem has been solved in [GR], (see also [GLR], Chapter 7). From the analysis in [GR, GLR] it follows that in the non-monic case there are, in general, many solutions. In the present paper we solve the inverse problem under the additional constraint that the inverse of the matrix polynomial $L(\lambda)$ that we are looking for is required to be analytic at infinity. For the monic case, as is clear from (0.5), this extra condition is satisfied automatically.

Let us state now in more detail the problem solved in this paper. Given an $n \times n$ matrix A and an $n \times m$ matrix B such that (0.1) holds, a complex number α which is not an eigenvalue of A and an invertible matrix D, we want to construct a regular $m \times m$ matrix polynomial $L(\lambda)$ such that

(C_1) (A, B) is a zero pair for $L(\lambda)$,

(C_2) $L(\alpha)=D$,

(C_3) $L(\lambda)^{-1}$is analytic at infinity.

We shall show that condition (C_1) can also be expressed by requiring the existence of an $m\times n$ matrix C such that $\bigcap_{j=0}^{n-1}\operatorname{Ker} CA^j=(0)$ and

$$L(\lambda)^{-1}-C(\lambda-A)^{-1}B$$

is an entire function. Note that condition (C_3) together with the above will guarantee that, in fact, $L(\lambda)^{-1}-C(\lambda-A)^{-1}B$ is a constant matrix. Also condition (C_3) can be rephrased by requiring the McMillan degree of $L(\lambda)$ (see [BGK], Section 4.2) to be as small as possible.

The solution of this problem will lead us to a problem which is of interest in its own right. It can be stated as follows: given a pair of matrices (A, B) as above find an $m\times n$ matrix F and an $n\times n$ nilpotent matrix T such that

$$\begin{bmatrix} A & B \end{bmatrix}\begin{bmatrix} T \\ F \end{bmatrix}=I. \tag{0.6}$$

In fact, the general solution to the problem stated in the previous paragraph is given by

$$L(\lambda)=D+(\lambda-\alpha)\sum_{j=0}^{\omega-1}\lambda^j FT^j(\alpha-A)^{-1}BD,$$

where (T, F) is as in (0.6) and ω is the order of nilpotency of T. For this reason we shall call equation (0.6) the *parameter equation.* Several interesting properties of equation (0.6) will be analysed in this paper. In particular, we shall be interested in the possible invariant polynomials of any matrix T for which there exists an F such that (0.6) holds. This has a surprising connection with Rosenbrock's theorem [R] and its generalization by Zaballa [Z].

The paper is divided into seven sections. In the first one we shall outline some preliminary material concerning zero pairs of matrix polynomials together with some remarks on the McMillan degree. In the second section we solve the main problem as stated in the first paragraph of this introduction. The third section contains an application to an extension problem involving standard pairs. In the fourth section we start the investigation of the parameter equation. The results of this section are applied in the next one to the special case which is relevant to our main problem. Some examples are given in Section 6. In the last section we study the parameter equation for general pairs of matrices (A, B) not necessarily satisfying (0.1).

1. Preliminaries

This section concerns the notion of a zero pair. Let A be an $n \times n$ matrix and B an $n \times m$ matrix. The pair (A, B) is called a *full range* pair in case $\bigvee_{j=0}^{n-1} \operatorname{Im} A^j B = \mathbb{C}^n$. Likewise, if C is an $m \times n$ matrix, the pair (C, A) is called a *zero kernel* pair if $\bigcap_{j=0}^{n-1} \operatorname{Ker} CA^j = (0)$.

Now let $L(\lambda) = \lambda^l L_l + \lambda^{l-1} L_{l-1} + \ldots + \lambda L_1 + L_0$ be a regular $m \times m$ matrix polynomial. The full range pair (A, B) is called a *(right) zero pair* for $L(\lambda)$ if

$$A^l B L_l + A^{l-1} B L_{l-1} + \ldots + B L_0 = 0, \tag{1.1}$$

$$\det L(\lambda) \text{ is a polynomial of degree } n. \tag{1.2}$$

Such a zero pair for $L(\lambda)$ can be built from the right Jordan chains of L at each of the zeros of $\det L(\lambda)$, see [GLR], Section 7.1. In fact, there a left zero pair is built from the left Jordan chains; for a right zero pair a similar construction can be employed. The analogue of Theorem 7.7 in [GLR] for right zero pairs implies that if (A, B) is a right zero pair for $L(\lambda)$, then there is a matrix C such that (A, B, C) is a minimal system (i.e., (C, A) is a zero kernel pair) and

$$L(\lambda)^{-1} - C(\lambda - A)^{-1} B$$

is a polynomial in λ. The converse is also true.

Proposition 1.1 *The full range pair (A, B) is a zero pair for $L(\lambda)$ if and only if there exists a matrix C such that (C, A) is a zero kernel pair and*

$$L(\lambda)^{-1} - C(\lambda - A)^{-1} B = \sum_{i=0}^{k} \lambda^i A_i \tag{1.3}$$

for some A_i, $i = 0, \ldots, k$, and some k.

Proof. We only have to show that (1.3) implies (1.1), (1.2). From (1.3) and the minimality of (A, B, C) we see that L^{-1} has n poles in the finite complex plane, counting multiplicities. But this implies (1.2). To see (1.1), evaluate $C(\lambda - A)^{-1} B L(\lambda)$ from (1.3). We see that the latter expression must be a polynomial in λ. However, for large λ we obtain

$$C(\lambda - A)^{-1} B L(\lambda) = \sum_{j=0}^{\infty} \sum_{i=0}^{l} \lambda^{i-j-1} C A^j B L_i .$$

Consider the coefficient of λ^{ν}, $\nu=-1,\dots,-n$; this yields

$$CA^{-\nu-1}(BL_0+ABL_1+\dots+A^lBL_l)=0.$$

Hence $\operatorname{Im}(BL_0+ABL_1+\dots+A^lBL_l)\subset\bigcap_{j=0}^{n-1}\operatorname{Ker} CA^j=(0)$. This proves (1.1) □

The previous proposition allows us to use (1.3) as the definition of a zero pair for L. In the latter form the definition coincides with the definition of 'zero pair' for a rational matrix function, as given in, e.g., [GKLR, GK].

Let $L(\lambda)$ be a regular $m\times m$ matrix polynomial, and (A, B) a zero pair for L. Then the McMillan degree of L (see e.g., [BGK]) is given by

$$n+\operatorname{rank}\begin{bmatrix} A_1 & A_2 & \cdot & \cdot & \cdot & A_k \\ A_2 & & & & \cdot & \\ \cdot & & & \cdot & & \\ \cdot & & \cdot & & & \\ \cdot & \cdot & & & 0 & \\ A_k & & & & & \end{bmatrix},$$

where $A_1,\dots,A_k$ are such that (1.3) holds. Let us consider the problem posed in the introduction. Given a full range pair (A, B), a number $\alpha\notin\sigma(A)$ and an invertible matrix D, we are looking for a regular $m\times m$ matrix polynomial $L(\lambda)$ such that

$$(A, B) \text{ is a zero pair for } L(\lambda), \tag{1.4}$$

$$L(\alpha)=D, \tag{1.5}$$

$$L(\lambda)^{-1} \text{ is analytic at infinity}. \tag{1.6}$$

From (1.3) we see that the latter requirement is equivalent to $A_1=A_2=\dots=A_k=0$, i.e., to the requirement that the McMillan degree of L is n. Note that n is a lower bound for the McMillan degree of any matrix polynomial satisfying (1.4), (1.5). We shall show in the sequel that this lower bound can be achieved. From these remarks the relation with our earlier work [GKR1,2] is clear. There we studied rational matrix functions with prescribed pole and zero data, of smallest McMillan degree. However, we shall not use [GKR1,2] here, but instead, we shall give independent proofs of our main results.

2. Solution to the main problem

In this section we shall solve the problem posed in the introduction. Let (A, B) be a full range pair of $n\times n$ and $n\times m$ matrices, respectively, α a complex number not an eigenvalue of A, and D an invertible matrix. First we choose a basis $\{g_{jk}\}_{k=1 j=1}^{\omega_j \ \ s}$ in $\mathbb{C}^n$ such that

$$\{g_{j\omega_j}\}_{j=1}^{s} \text{ is a basis for } \operatorname{Im} B, \tag{2.1}$$

$$Ag_{j\,k+1}-g_{jk}\in \operatorname{Im} B \qquad (g_{j0}:=0). \tag{2.2}$$

The existence of such a basis follows from the Brunovsky canonical form of the pair (A, B), see, e.g., [K, W]. Using this basis we have the following result.

Theorem 2.1 *A regular matrix polynomial $L(\lambda)$ such that L has (A, B) as its zero pair, L^{-1} is analytic at infinity and $L(\alpha)=D$, can be constructed as follows. Take a basis $\{g_{jk}\}_{k=1 j=1}^{\omega_j \ \ s}$ satisfying (2.1), (2.2), $\omega_1 \geqq \ldots \geqq \omega_s$. Define*

$$T\colon \mathbb{C}^n \to \mathbb{C}^n,\ Tg_{jk}=g_{j\,k+1},\ (g_{j\omega_j+1}:=0) \quad k=1,\ldots,\omega_j. \tag{2.3}$$

Take any $F\colon \mathbb{C}^n \to \mathbb{C}^m$ such that

$$AT+BF=I, \tag{2.4}$$

and put

$$L(\lambda)=D+(\lambda-\alpha)\sum_{j=0}^{\omega_1-1}\lambda^j FT^j(\alpha-A)^{-1}BD. \tag{2.5}$$

Then L has the desired properties and

$$L(\lambda)^{-1}=D^{-1}+(\lambda-\alpha)D^{-1}F(I-\alpha T)^{-1}(A-\lambda)^{-1}B. \tag{2.6}$$

Proof. First note that the right hand side of (2.5) can be rewritten as

$$L(\lambda)=D+(\lambda-\alpha)F(I-\lambda T)^{-1}(\alpha-A)^{-1}BD. \tag{2.7}$$

Now write out the product of the right hand sides of (2.6) and (2.7); one gets

$$I+(\lambda-\alpha)F(I-\lambda T)^{-1}(\alpha-A)^{-1}B+(\lambda-\alpha)F(I-\alpha T)^{-1}(A-\lambda)^{-1}B+$$

$$+(\lambda-\alpha)^2F(I-\lambda T)^{-1}(\alpha-A)^{-1}BF(I-\alpha T)^{-1}(A-\lambda)^{-1}B.$$

Observe that

$$(I-\lambda T)^{-1}(\alpha-A)^{-1}+(I-\alpha T)^{-1}(A-\lambda)^{-1}+$$

$$+(\lambda-\alpha)(I-\lambda T)^{-1}(\alpha-A)^{-1}BF(I-\alpha T)^{-1}(A-\lambda)^{-1}=$$

$$=(I-\lambda T)^{-1}(\alpha-A)^{-1}(\alpha-\lambda)(I-AT-BF)(I-\alpha T)^{-1}(A-\lambda)^{-1},$$

which is zero by (2.4). So the product of the right hand sides of (2.5) and (2.6) indeed equals I.

Next note that from (2.2), (2.3) we have $(AT-I)g_{jk} \in \operatorname{Im} B$ for all j,k. So the existence of an F such that (2.4) holds is guaranteed.

From (2.5) we see that $L(\alpha)=D$. To see that $L(\lambda)^{-1}$ is analytic at infinity, rewrite (2.6) as

$$D^{-1}+(\lambda-\alpha)D^{-1}F(I-\alpha T)^{-1}(A-\lambda)^{-1}B=$$

$$=(D^{-1}-D^{-1}F(I-\alpha T)^{-1}B)+D^{-1}F(I-\alpha T)^{-1}(A-\alpha)(A-\lambda)^{-1}B,$$

which clearly is analytic at infinity. The realization (2.7) is observable, i.e., the pair (F,T) is a zero kernel pair. Indeed, suppose there is an $x \neq 0$ such that $Tx=\lambda_0 x$, $Fx=0$. Then $\lambda_0=0$, so $x \in \operatorname{Ker} \begin{pmatrix} T \\ F \end{pmatrix}$. But by (2.4) this kernel is (0), so $x=0$. Contradiction. Now we compute the inverse of $L(\lambda)$ using the realization (2.7):

$$L(\lambda)^{-1}=$$

$$=D^{-1}-(\lambda-\alpha)D^{-1}F(\lambda(T+(A-\alpha)^{-1}BF)-(I+\alpha(A-\alpha)^{-1}BF))^{-1}(A-\alpha)^{-1}B=$$

$$=D^{-1}-(\lambda-\alpha)D^{-1}F(\lambda(AT-\alpha T+BF)-(A-\alpha I+\alpha BF))^{-1}B=$$

$$=D^{-1}-(\lambda-\alpha)D^{-1}F(I-\alpha T)^{-1}(\lambda-A)^{-1}B,$$

which is realization (2.6) for $L(\lambda)^{-1}$. It follows that this realization is observable as well. From (2.6) we then see that (A,B) is indeed a zero pair for $L(\lambda)$. □

From the first part of the above proof one sees that the particular construction of T and F is irrelevant. More precisely, for any T and F satisfying (2.4), with T nilpotent we have that (2.7) and (2.6) are each others inverse. This leads us to the following theorem.

Theorem 2.2 *Let (A, B) be a full range pair, $\alpha \notin \sigma(A)$ a complex number, and D an invertible matrix. Then the set of regular matrix polynomials $L(\lambda)$ satisfying*

$$L(\alpha)=D, \tag{2.8}$$

$$L \text{ has } (A, B) \text{ as its zero pair}, \tag{2.9}$$

$$L(\lambda)^{-1} \text{ is analytic at infinity }, \tag{2.10}$$

is parametrized by the set of pairs of matrices T, F of sizes $n \times n$ and $m \times n$, respectively, satisfying

$$AT+BF=I \tag{2.11}$$

where T is nilpotent. More precisely, given a pair of matrices (T, F) with the above properties then

$$L(\lambda)=D+(\lambda-\alpha)\sum_{j=0}^{\omega-1}\lambda^j FT^j(\alpha-A)^{-1}BD, \tag{2.12}$$

satisfies (2.8)-(2.10), where ω is the order of nilpotency of T, i.e., $T^{\omega}=0$, $T^{\omega-1}\neq 0$. In that case

$$L(\lambda)^{-1}=D^{-1}+(\lambda-\alpha)D^{-1}F(I-\alpha T)^{-1}(A-\lambda)^{-1}B. \tag{2.13}$$

Conversely, any matrix polynomial satisfying (2.8)-(2.10) is given by (2.12) for a unique pair of matrices (T, F) solving (2.11), with T nilpotent.

Proof. In case $L(\lambda)$ is given by (2.12), then $L(\lambda)^{-1}$ is given by (2.13) as we saw above. Hence L satisfies (2.8)-(2.10), using the same arguments as in the proof of Theorem 2.1.

Conversely, suppose (2.8)-(2.10) hold. Then we can write

$$L(\lambda)^{-1}=X+C(\lambda-A)^{-1}B, \tag{2.14}$$

for some X and C, and, moreover, we can take this realization to be minimal. Rewrite (2.14) as

$$L(\lambda)^{-1}=X-C(A-\alpha)^{-1}B+(\lambda-\alpha)C(\lambda-A)^{-1}(A-\alpha)^{-1}B=$$

$$=D^{-1}+(\lambda-\alpha)C(\lambda-A)^{-1}(A-\alpha)^{-1}B,$$

because of (2.8). According to [GK] the function $L(\lambda)$ itself is then given by

$$L(\lambda)=D+(\lambda-\alpha)DC(\lambda G^{\times}-A^{\times})^{-1}(\alpha-A)^{-1}BD,$$

where $G^{\times}=I+(A-\alpha)^{-1}BDC$, $A^{\times}=A+\alpha(A-\alpha)^{-1}BDC$. Furthermore, this realization of L is also minimal. Write

$$\lambda G^{\times}-A^{\times}=E_1\begin{pmatrix}\lambda I-A_1 & 0\\ 0 & I-\lambda A_2\end{pmatrix}E_2,$$

with invertible E_1, E_2 and nilpotent A_2. As L is a polynomial by assumption we see that A_1 can have at most eigenvalue α, which is cancelled by the factor $\lambda-\alpha$. However, the minimality of the realization, the fact that the state space dimension is as small as possible, gives that this cannnot occur, i.e., the matrix A_1 is vacuous, and hence

$$\lambda G^{\times}-A^{\times}=E_1(I-\lambda A_2)E_2,$$

with A_2 nilpotent. So $A^{\times}$ is invertible and $G^{\times}A^{\times -1}$ is nilpotent. Now put $T=G^{\times}A^{\times -1}$, $F=-DCA^{\times -1}$. Then

$$L(\lambda)=D+(\lambda-\alpha)F(I-\lambda T)^{-1}(\alpha-A)^{-1}BD,$$

as required in (2.12). Furthermore,

$$AT+BF=(AG^{\times}-BDC)A^{\times -1},$$

and

$$AG^{\times}-BDC=A+\{A(A-\alpha)^{-1}-I\}BDC=A+\alpha(A-\alpha)^{-1}BDC=A^{\times},$$

so indeed (2.11) holds.

It remains to prove the unicity of the pair (T,F). To see this observe that the following realization for $L(\lambda)^{-1}$ is minimal:

$$L(\lambda)^{-1}=(D^{-1}-D^{-1}F(I-\alpha T)^{-1}B)+D^{-1}F(I-\alpha T)^{-1}(A-\alpha)(\lambda-A)^{-1}B.$$

Now assume that we have a second pair of matrices $(\tilde{T},\tilde{F})$ such that $\tilde{T}$ is nilpotent of order ν,

$$A\tilde{T}+B\tilde{F}=I,$$

and

$$L(\lambda)=D+(\lambda-\alpha)\sum_{j=0}^{\nu}\lambda^j\tilde{F}\tilde{T}^j(\alpha-A)^{-1}BD.$$

Then, if we replace in the above realization for $L(\lambda)^{-1}$ the matrices T and F by $\tilde{T}$ and $\tilde{F}$, respectively, we obtain another realization for $L(\lambda)^{-1}$. As both these realizations are minimal, there exists an invertible S such that

$$SAS^{-1}=A,\ SB=B,\ D^{-1}\tilde{F}(I-\alpha\tilde{T})^{-1}S^{-1}=D^{-1}F(I-\alpha T)^{-1}.$$

Since (A,B) is full range, we have $S=I$. Thus $\tilde{F}(I-\alpha\tilde{T})^{-1}=F(I-\alpha T)^{-1}$. Now use that

$$-A(I-\alpha T)+\alpha BF=(\alpha-A).$$

Hence

$$-A+\alpha BF(I-\alpha T)^{-1}=(\alpha-A)(I-\alpha T)^{-1}.$$

This also holds if we replace (T,F) by $(\tilde{T},\tilde{F})$. So

$$(\alpha-A)(I-\alpha T)^{-1}=(\alpha-A)(I-\alpha\tilde{T})^{-1}.$$

It follows that $T=\tilde{T}$, and hence also $F=\tilde{F}$, i.e., the pair (T,F) is indeed unique. □

We see from Theorem 2.2 that all solutions $L(\lambda)$ to (2.8)-(2.10) are parametrized by all pairs of matrices (F,T) with T nilpotent and satisfying

$$\begin{bmatrix} A & B \end{bmatrix} \begin{bmatrix} T \\ F \end{bmatrix} = I. \tag{2.15}$$

For this reason (2.15) will be called the *parameter equation.*

3. Extension to standard pairs

In this section we study extensions to standard pairs for a given full range pair. Let us start by defining the notion of a standard pair. A full range pair of matrices (Y, X) where X is $ml\times m$ and Y is $ml\times ml$, is called a *standard pair* for the monic $m\times m$ matrix polynomial $L(\lambda)=\lambda^l I+\lambda^{l-1}L_{l-1}+\ldots+L_0$ if

$$Y^l X+Y^{l-1}XL_{l-1}+\ldots+YXL_1+XL_0=0. \tag{3.1}$$

In other words, a full range pair of matrices is a standard pair for $L(\lambda)$ if it is a right zero pair for $L(\lambda)$. As already noted in the introduction the pair

$$Y=\begin{bmatrix} 0 & I & & & \\ & & \cdot & & \\ & & & \cdot & \\ & & & & \cdot \\ & & & & I \\ -L_0 & -L_1 & \cdot & \cdot & \cdot & -L_{l-1} \end{bmatrix}, \quad X=\begin{bmatrix} 0 \\ \cdot \\ \cdot \\ \cdot \\ 0 \\ I \end{bmatrix}$$

is a standard pair for $L(\lambda)$, and any other standard pair is similar to this one.

Given a full range pair of matrices (A, B), where A is $n\times n$ and B is $n\times m$, the pair (Y, X) is called an *extension* of (A, B) if there exists and Y-invariant subspace M and an invertible $S: M\rightarrow \mathbb{C}^n$ such that

$$A = SY|_M S^{-1}, \quad B = SP_M X,$$

where P_M is a projection onto M. The problem of existence of a standard pair (Y, X) which is an extension of a given full range pair (A, B) was studied in [GR], see also [GLR], Chapter 6. A description of special extensions is also given in [GLR].

The problem we shall study in this section is the following. Given is a full range pair (A, B) with A invertible. Describe all standard pairs (Y, X) which are extensions of (A, B) such that $\sigma(Y)\backslash\sigma(A) \subset \{0\}$, and such that $l \geqq \omega$, where l is such that Y is of size $ml \times ml$ (i.e., l is the degree of the monic polynomial corresponding to the standard pair (Y, X), compare formula (0.5)), and ω is the largest partial multiplicity of Y corresponding to the zero eigenvalue.

Theorem 3.1 *Let (A, B) be a full range pair of matrices, A is $n \times n$, B is $n \times m$, and A is invertible. Let (T, F) be such that*

$$T + ABF = A \tag{3.2}$$

and T is nilpotent of order l. Form

$$Y = \begin{bmatrix} 0 & I & & & \\ & & \cdot & & \\ & & & \cdot & \\ & & & & \cdot \\ & & & & I \\ FT^{l-1}AB & FT^{l-2}AB & \cdot & \cdot\ \cdot & FAB \end{bmatrix}, \quad X = \begin{bmatrix} 0 \\ \cdot \\ \cdot \\ \cdot \\ 0 \\ I \end{bmatrix} \tag{3.3}$$

Then (Y, X) is a standard pair which is an extension of (A, B) such that $\sigma(Y)\backslash\sigma(A) \subset \{0\}$ and $l \geqq \omega$, where ω is the largest partial multiplicity of Y corresponding to the zero eigenvalue.

Conversely, for any standard pair $(\tilde{Y}, \tilde{X})$ which is an extension of (A, B) such that $\sigma(\tilde{Y})\backslash\sigma(A) \subset \{0\}$ and $l \geqq \omega$, where l and ω are as above, there exist matrices T and F such that (3.2) holds, T is nilpotent of order l, and for which the pair $(\tilde{Y}, \tilde{X})$ is similar to the pair given by (3.3).

The polynomial corresponding to the standard pair $(\tilde{Y}, \tilde{X})$ is given by

$$L(\lambda) = \lambda^l I - \lambda^{l-1} FAB - \ldots - \lambda FT^{l-2}AB - FT^{l-1}AB.$$

Proof. The proof is based on the reduction of the problem to the problem solved in Section2. In fact we shall show below that the following holds. If $L(\lambda)$ is a monic matrix polynomial of degree l with standard pair $(\tilde{Y}, \tilde{X})$, and $(\tilde{Y}, \tilde{X})$ is an extension of (A, B), $\sigma(\tilde{Y})\backslash\sigma(A) \subset \{0\}$ and $l \geqq \omega$, where ω is the largest partial multiplicity of $\tilde{Y}$ corresponding to the zero eigenvalue, then

$$\tilde{L}(\lambda) = \lambda^l L(\lambda^{-1})$$

is a comonic matrix polynomial, i.e., $\tilde{L}(0) = I$, of degree l such that (A^{-1}, B) is a zero pair for $\tilde{L}$ and $\tilde{L}(\lambda)^{-1}$ is analytic at infinity. Conversely, if $\tilde{L}$ is such a comonic polynomial then $L(\lambda) = \lambda^l \tilde{L}(\lambda^{-1})$ is monic of degree l, its standard pair $(\tilde{Y}, \tilde{X})$ is an extension of (A, B) and $\sigma(\tilde{Y})\backslash\sigma(A) \subset \{0\}$ and $l \geqq \omega$.

Let us start by proving the converse part. Suppose $(\tilde{Y}, \tilde{X})$ is and extension of (A, B) with $\sigma(\tilde{Y})\backslash\sigma(A) \subset \{0\}$ and $l \geqq \omega$. Let $L(\lambda)$ be the monic polynomial with standard pair $(\tilde{Y}, \tilde{X})$. Then there is a matrix C such that $(C, \tilde{Y})$ is a null-kernel pair and

$$L(\lambda)^{-1} = C(\lambda - \tilde{Y})^{-1}\tilde{X}.$$

As $(\tilde{Y}, \tilde{X})$ is an extension of (A, B) there is an $\tilde{Y}$-invariant subspace M such that $A = S\tilde{Y}\,|_M S^{-1}$. Moreover, since A is invertible and $\sigma(\tilde{Y})\backslash\sigma(A) \subset \{0\}$, the space M is a spectral subspace of $\tilde{Y}$. Therefor, by applying a similarity transformation if necessary, we may assume

$$\tilde{Y} = \begin{pmatrix} A & 0 \\ 0 & Y_2 \end{pmatrix}, \quad \tilde{X} = \begin{pmatrix} B \\ X_2 \end{pmatrix}.$$

Decompose $C = \begin{pmatrix} C_1 & C_2 \end{pmatrix}$. The matrix Y_2 is nilpotent of order ω. Consider the matrix polynomial

$$\tilde{L}(\lambda) = \lambda^l L(\lambda^{-1}).$$

Then $\tilde{L}$ is a comonic matrix polynomial of degree $\leqq l$. Furthermore, $\tilde{L}(\lambda)^{-1}$ is analytic at infinity. Indeed,

$$\tilde{L}(\lambda)^{-1} = \lambda^{-l} L(\lambda^{-1})^{-1} =$$

$$\lambda^{-l} C^1(\lambda^{-1} - A)^{-1} B + \lambda^{-l} C^2(\lambda^{-1} - Y_2)^{-1} X_2 =$$

$$\lambda^{-l+1} C_1 A^{-1}(A^{-1} - \lambda)^{-1} B + \lambda^{-l+1} C_2(I - \lambda Y_2)^{-1} X_2 =$$

$$\lambda^{-l+1} C_1 A^{-1}(A^{-1} - \lambda)^{-1} B + \lambda^{-l+1} \sum_{j=0}^{\omega-1} \lambda^j C_2 Y_2^j X_2.$$

Since $l \geqq \omega$, we have that $\tilde{L}(\lambda)^{-1}$ is analytic at infinity. Moreover, (A^{-1}, B) is a zero pair for $\tilde{L}$. By Theorem 2.2 we have that there are matrices F, T with T nilpotent such that

$$A^{-1}T+BF=I,$$

i.e., (3.2) holds, and $\tilde{L}(\lambda)=I-\sum_{j=0}^{l-1}\lambda^{j+1}FT^{j}AB$. Then the pair given by (3.3) is also a standard pair for $L(\lambda)$, as one easily sees. As two standard pairs are similar, $(\tilde{Y}, \tilde{X})$ is similar to a pair of the form (3.3).

To prove the direct part of the theorem, suppose $\tilde{L}(\lambda)$ is an $m \times m$ matrix polynomial of degree l such that (A^{-1}, B) is a zero pair for $\tilde{L}$, $\tilde{L}(0)=I$ and $\tilde{L}(\lambda)^{-1}$ is analytic at infinity. By Theorem 2.2 such polynomials $\tilde{L}(\lambda)$ are in one-one correspondence with pairs of matrices (F, T) such that T is nilpotent of order l and (3.2) holds, i.e.,

$$A^{-1}T+BF=I.$$

Moreover, we have

$$\tilde{L}(\lambda)^{-1}=I+\lambda F(A^{-1}-\lambda)^{-1}B,$$

(cf (2.13), with $\alpha=0$ and $D=I$). Form $L(\lambda)=\lambda^{l}\tilde{L}(\lambda^{-1})$. Then $L(\lambda)$ is a monic matrix polynomial of degree l, and

$$L(\lambda)^{-1}=\lambda^{-l}\tilde{L}(\lambda^{-1})^{-1}=\lambda^{-l}+\lambda^{-l-1}F(A^{-1}-\lambda^{-1})^{-1}B=$$

$$\lambda^{-l}I+\lambda^{-l}F(\lambda A^{-1}-I)^{-1}B=\lambda^{-l}-\sum_{j=0}^{\infty}\lambda^{j-l}FA^{-j}B=$$

$$=\lambda^{-l}-\sum_{j=0}^{l-1}\lambda^{j-l}FA^{-j}B+FA^{-l+1}(\lambda-A)^{-1}B.$$

Now, let (Y, X) be a standard pair of L. Let M, respectively N be the spectral subspace of Y corresponding to its non-zero, respectively zero, eigenvalues. Then, with respect to $\mathbb{C}^{ml}=M \oplus N$ we can write $Y=\begin{bmatrix} Y_1 & 0 \\ 0 & Y_2 \end{bmatrix}$, $X=\begin{bmatrix} X_1 \\ X_2 \end{bmatrix}$ with Y_1 invertible and Y_2 nilpotent of order ω. As $\tilde{L}(\lambda)^{-1}$ is analytic at infinity we see $\omega \leqq l$. Moreover, there exist matrices C_1, C_2 such that with $C=\begin{bmatrix} C_1 & C_2 \end{bmatrix}$ we have (C, Y) is a null-kernel pair and

$$L(\lambda)^{-1}=C_1(\lambda-Y_1)^{-1}X_1+C_2(\lambda-Y_2)^{-1}X_2=$$

$$C_1(\lambda-Y_1)^{-1}X_1+\sum_{j=0}^{\omega-1}\lambda^{-j-1}C_2Y_2^{j}X_2.$$

Considering the Laurent expansion of $L(\lambda)^{-1}$ around zero we see that

$$C_1(\lambda - Y_1)^{-1}X_1 = FA^{-l+1}(\lambda - A)^{-1}B.$$

Both these realizations are minimal. Indeed, to see this we only have to show that (F, A) is a null-kernel pair. Assume $Ax = \lambda x$, $Fx = 0$. By (3.2) $Tx = Ax = \lambda x$, i.e., $\lambda = 0$ as T is nilpotent. However, A is invertible, so this implies that $x = 0$. The state space isomorphism theorem now implies the existence of an invertible matrix S such that

$$A = SY_1S^{-1}, \quad B = SX_1.$$

Hence (Y, X) is an extension of (A, B). Moreover, (Y, X) is similar to the pair (3.3) as $\tilde{L}(\lambda)$ is given by (2.11):

$$\tilde{L}(\lambda) = I - \sum_{j=0}^{l-1} \lambda^{j+1} FT^j AB.$$

So $L(\lambda) = \lambda^l I - \sum_{j=0}^{l-1} \lambda^{l-j-1} FT^j AB.$ □

4. Analysis of the parameter equation

In this section we analyse the equation (2.15) with (A, B) a given full range pair. We do not restrict our attention to nilpotent T, but allow T to have arbitrary eigenvalues. Our main interest will be in the possibilities for the invariant polynomials of T. The connection with the inverse problem of Section 2 will be discussed in the next section.

The invariant polynomials of a matrix T can be defined as follows: there exist unimodular matrix functions $E(\lambda)$ and $F(\lambda)$ such that

$$\lambda - T = E(\lambda) diag\ (p_1(\lambda), p_2(\lambda), \ldots, p_n(\lambda)) F(\lambda),$$

where $p_n | p_{n-1} | \ldots | p_1$ are scalar monic polynomials. These are called the *invariant polynomials* of T, and are uniquely determined by T. (See also [G].)

Let A and B be $n \times n$ and $n \times m$ matrices, respectively, such that (A, B) is full range. Take a basis $\{g_{jk}\}_{k=1}^{\omega_j}{}_{j=1}^{s}$ such that $\omega_1 \geqq \ldots \geqq \omega_s > 0$, and

$$\{g_{j\omega_j}\}_{j=1}^{s} \text{ is a basis for } \operatorname{Im} B, \tag{4.1}$$

$$Ag_{j\,k+1} - g_{jk} \in \operatorname{Im} B \qquad (g_{j0} := 0). \tag{4.2}$$

The numbers $\omega_1, .., \omega_s$ are fixed by A and B, and are called the *controllability indices* of the pair (A,B).

Theorem 4.1 *Given an $n\times n$ matrix A and an $n\times m$ matrix B such that (A, B) is a full range pair. Let $\{g_{jk}\}_{k=1}^{\omega_j}{}_{j=1}^{s}$ be a basis such that (4.1) and (4.2) hold. Then with respect to this basis any T for which there exists an F such that*

$$\begin{bmatrix} A & B \end{bmatrix}\begin{bmatrix} T \\ F \end{bmatrix}=I, \tag{4.3}$$

is of the form

$$T=\begin{bmatrix} T_{ij} \end{bmatrix}_{i,j=1}^{s} \tag{4.4}$$

where T_{ij} is an $\omega_i\times\omega_j$ matrix of the following form:

$$T_{ii}=\begin{bmatrix} * & * & . & . & . & * \\ 1 & 0 & & & & \\ & . & . & & & \\ & & . & . & & \\ & & & . & 0 & \\ & & & & 1 & 0 \end{bmatrix}, \quad T_{ij}=\begin{bmatrix} * & * & . & . & . & * \\ 0 & . & . & . & . & 0 \\ . & & & & & . \\ . & & & & & . \\ . & & & & & . \\ 0 & . & . & . & . & 0 \end{bmatrix}, \quad i\neq j.$$

In other words, any such T satisfies

$$Tg_{jk}-g_{j\,k+1}\in \text{span}\,\{g_{j1}\}_{j=1}^{s} \qquad (g_{j\,\omega_j+1}:=0) \tag{4.5}$$

for all j, k.

Conversely, given a T of the form (4.4) there exists an F such that (4.3) holds.

Proof. Because of (4.3) we have $ATg_{jk}-g_{jk}\in \text{Im}\ B=\ \text{span}\ \{g_{i\omega_i}\}_{i=1}^{s}$. Write $Tg_{jk}=\sum_{i=1}^{s}\sum_{l=1}^{\omega_i}\alpha_{il}g_{il}$. Then $ATg_{jk}=\sum_{i=1}^{s}\sum_{l=2}^{\omega_i}\alpha_{il}\ g_{il-1}+w$, with $w\in \text{Im}\ B$, because of (4.2). Since ATg_{jk} is of the form $g_{jk}+v$ with $v\in \text{Im}\ B$, we have $\alpha_{il}=0$ for $(i,l)\neq(j,k+1)$ with $l\geqq 2$, and $\alpha_{jk+1}=1$. So (4.5) holds.

For the converse, note that because of (4.2) and (4.5), the vector $(AT-I)g_{jk}\in$ Im B for all j, k. Hence there is an F such that (4.3) holds. □

As a consequence of Theorem 4.1 we have the following result, which is an equivalent version of Rosenbrock's theorem [R].

Theorem 4.2 *Let A be an $n \times n$ matrix and B be an $n \times m$ matrix such that (A, B) is a full range pair. Denote the controllability indices of (A, B) by $\omega_1 \geqq \ldots \geqq \omega_s$. Let $p_1,\ldots,p_n$ be monic polynomials with degrees $d_1,\ldots,d_n$ such that $p_n \,|\, p_{n-1} \,|\, \ldots \,|\, p_1$ and*

$$\sum_{j=1}^{s} d_j = n, \tag{4.6}$$

$$\sum_{j=1}^{k} \omega_j \leqq \sum_{j=1}^{k} d_j \qquad k=1,\ldots,s-1. \tag{4.7}$$

Then there exists an $n \times n$ matrix T with invariant polynomials $p_1,\ldots,p_n$ and an $m \times n$ matrix F such that

$$AT+BF = \begin{pmatrix} A & B \end{pmatrix} \begin{pmatrix} T \\ F \end{pmatrix} = I. \tag{4.8}$$

Conversely, if the pair (F, T) satisfies (4.8) then the invariant polynomials of T have degrees $d_1,\ldots,d_n$ satisfying (4.6), (4.7).

Proof. Because of Theorem 4.1 to prove the first part of the theorem it suffices to show that there exists an $n \times n$ matrix T of the form (4.4) with invariant polynomials $p_1,\ldots,p_n$. Indeed, once such a T is found the existence of F such that (4.8) holds follows from Theorem 4.1. Let B': $\mathbb{C}^s \to \mathbb{C}^n$ with respect to the standard basis $e_1,\ldots,e_s$ in $\mathbb{C}^s$ and the basis $\{g_{jk}\}_{k=1\, j=1}^{\omega_j \quad s}$ in $\mathbb{C}^n$ be given by $B'e_j = g_{j1}$. Let T_0: $\mathbb{C}^n \to \mathbb{C}^n$ be given by $T_0 g_{jk} = g_{j\,k+1}$ $(g_{j\omega_j+1}=0)$. Then (T_0, B') is full range. Moreover, the controllability indices of (T_0, B') are $\omega_1,..,\omega_s$. According to Rosenbrock's theorem, because of (4.6), (4.7), there is a matrix C such that $T_0+B'C=T$ has invariant polynomials $p_1,\ldots,p_n$. It remains to observe that any matrix of the form $T_0+B'C$ is of the form (4.4).

Conversely, suppose (F, T) satisfies (4.8). By Theorem 4.1 the matrix T is of the form (4.4). It follows that $T=T_0+B'C$ for some C, where T_0 and B' are as in the previous paragraph. But then Rosenbrock's theorem implies that the invariant polynomials $p_1,\ldots,p_n$ of T satisfy (4.6), (4.7). □

We now state the dual of Theorem 4.2 for zero kernel pairs. If (C, A) is a zero kernel pair the controllability indices of the pair (A^*, C^*) are called the *observability indices* of (C, A).

Theorem 4.3 *Let A be an $n \times n$ matrix and C be an $m \times n$ matrix such that (C, A) is* *a zero kernel pair. Denote the observability indices of (C, A) by* $\alpha_1 \geqq \ldots \geqq \alpha_s$. *Let* $p_1,\ldots,p_n$ *be monic polynomials with degrees* $d_1,\ldots,d_n$ *such that* $p_n \mid p_{n-1} \mid \ldots \mid p_1$ *and*

$$\sum_{j=1}^{s} d_j = n, \tag{4.9}$$

$$\sum_{j=1}^{k} \alpha_j \leqq \sum_{j=1}^{k} d_j, \qquad k=1,\ldots,s-1. \tag{4.10}$$

Then there exists an $n \times n$ *matrix T with invariant polynomials* $p_1,\ldots,p_n$ *and an* $n \times m$ *matrix G such that*

$$TA + GC = \begin{pmatrix} T & G \end{pmatrix} \begin{pmatrix} A \\ C \end{pmatrix} = I. \tag{4.11}$$

Conversely, if (T, G) satisfies (4.11) then the degrees of the invariant polynomials of T satisfy (4.9), (4.10).

The two theorems above can be rephrased in terms of subspaces. Let A be a given $n \times n$ matrix and M a subspace of $\mathbb{C}^n$ such that $\bigvee_{j=0}^{\infty} A^j M = \mathbb{C}^n$. Define subspaces

$$H_j = M + AM + \ldots + A^{j-1}M, \quad H_0 = (0),$$

and define numbers $\omega_1 \geqq \ldots \geqq \omega_s$ by $s = \dim M$,

$$\omega_j = \#\{k \mid \dim H_k/H_{k-1} \geqq j\}, \quad j=1,\ldots,s.$$

The numbers ω_j are called the *incoming indices* of A with respect to M (see [BGK1]). For such A and M we have the following theorem.

Theorem 4.4 *Let A and M be such that* $\bigvee_{j=0}^{\infty} A^j M = \mathbb{C}^n$, *and let* $\omega_1,\ldots,\omega_s$ *be the incoming indices of A with respect to M. Further, let* $p_1,\ldots,p_m$ *be given monic polynomials with degrees* $d_1,\ldots,d_m$ *satisfying* $p_m \mid p_{m-1} \mid \ldots \mid p_1$ *and*

$$\sum_{j=1}^{m} d_j = n, \tag{4.12}$$

$$\sum_{j=1}^{k} \omega_j \leqq \sum_{j=1}^{k} d_j \qquad j=1,\ldots,m. \tag{4.13}$$

Then there exists an $n \times n$ *matrix T with invariant polynomials* $p_1,\ldots,p_m$ *such that*

$$\operatorname{Im}(AT-I)\subset M. \tag{4.14}$$

Conversely, if T satisfies (4.14), then the degrees $d_1,\ldots,d_m$ of its invariant polynomials satisfy (4.12), (4.13).

Proof. Take any B such that $M=\operatorname{Im} B$. (For instance B can be defined by taking a basis in M, and using the basis vectors as columns of B.) Since $\bigvee_{j=0}^{\infty} A^j M = \mathbb{C}^n$, it follows that (A, B) is full range. Moreover, the incoming indices of A with respect to M are precisely the controllability indices of (A, B). With these observations the theorem easily follows from Theorem 4.2. □

To formulate the dual of the above theorem we first introduce outgoing indices. Let A be a given $n\times n$ matrix and M a subspace of $\mathbb{C}^n$ such that $\bigcap_{j=0}^{\infty} A^j M=(0)$. Define subspaces

$$K_j=M\cap AM\cap\cdots\cap A^{j-1}M,\quad K_0=\mathbb{C}^n$$

and define numbers $\alpha_1\geqq\ldots\geqq\alpha_t$ by $t=\operatorname{codim} M$,

$$\alpha_j=\#\{k \mid \dim K_{k-1}/K_k \geqq j\},\quad j=1,\ldots,t.$$

These numbers are called the *outgoing indices* of A with respect to M (see [BGK1]).

Theorem 4.5 *Let A and M be such that $\bigcap_{j=0}^{\infty} A^j M=(0)$, and let $\alpha_1,\ldots,\alpha_t$ be the outgoing indices of A with respect to M. Further, let $p_1,\ldots,p_m$ be given monic polynomials with degrees $d_1,\ldots,d_m$ satisfying $p_m \mid p_{m-1} \mid \cdots \mid p_1$ and*

$$\sum_{j=1}^{m} d_j=n, \tag{4.15}$$

$$\sum_{j=1}^{k}\alpha_j\leqq\sum_{j=1}^{k} d_j\quad j=1,\ldots,m. \tag{4.16}$$

Then there exists an $n\times n$ matrix T with invariant polynomials $p_1,\ldots,p_m$ such that

$$M\subset \operatorname{Ker}(TA-I). \tag{4.17}$$

Conversely, if T satisfies (4.17) then the degrees $d_1,\ldots,d_m$ of its invariant polynomials satisfy (4.15), (4.16).

Proof. Take any C such that $M=\operatorname{Ker} C$. Then $(C\ A)$ is observable, with observability indices $\alpha_1,..,\alpha_t$. Now apply Theorem 4.3 to obtain the above result. □

5. The pole pair at infinity

Let $L(\lambda)$ be an $m \times m$ regular matrix polynomial. A null-kernel pair of matrices (F, T) where F is $m \times n$ and T is $n \times n$ and nilpotent, will be called a *left pole pair at infinity* for $L(\lambda)$ if there exist $\tilde{D}$ and $\tilde{B}$ such that

$$L(\lambda)=\tilde{D}+\lambda F(I-\lambda T)^{-1}\tilde{B}, \tag{5.1}$$

and $(T, \tilde{B})$ is full range (compare [KMR], Section 4). Now assume that $L(\lambda)^{-1}$ is analytic at infinity, $L(\alpha)=D$, and that (A, B) is a zero pair of L. As we have seen in Section 2 there are (F, T) such that T is nilpotent, $AT+BF=I$, and with

$$L(\lambda)=D+(\lambda-\alpha)\sum_{j=0}^{\omega-1}\lambda^j FT^j(\alpha-A)^{-1}BD=$$

$$=D+(\lambda-\alpha)F(I-\lambda T)^{-1}(\alpha-A)^{-1}BD.$$

Here ω is the order of nilpotency of T.

We claim that (F, T) is a left pole pair at infinity for $L(\lambda)$. Indeed, (F, T) is a null-kernel pair. Put $\tilde{D}=L(0)$ and $\tilde{B}=(I-\alpha T)(\alpha-A)^{-1}BD$. Then $(T, \tilde{B})$ is full range. Indeed, suppose for some x we have $x\tilde{B}=0$ and $xT=\lambda x$. Then $\lambda=0$, and $x(\alpha-A)^{-1}B=0$. Put $y=x(\alpha-A)^{-1}$. Using $AT+BF=I$ it is easily seen that $\alpha yT=y$, which implies $y=0$, and hence $x=0$. Furthermore,

$$\tilde{D}+\lambda F(I-\lambda T)^{-1}\tilde{B}=$$

$$D-\alpha F(\alpha-A)^{-1}BD+\lambda F(I-\lambda T)^{-1}(I-\alpha T)(\alpha-A)^{-1}BD=$$

$$D-\alpha F(\alpha-A)^{-1}BD+\lambda F(I-\lambda T)^{-1}(I-\alpha T+(\lambda-\alpha)T)(\alpha-A)^{-1}BD=$$

$$=D+(\lambda-\alpha)F\{I+\lambda T(I-\lambda T)^{-1}\}(\alpha-A)^{-1}BD=$$

$$=D+(\lambda-\alpha)F(I-\lambda T)^{-1}(\alpha-A)^{-1}BD=L(\lambda).$$

Note that with (F, T) also any pair of the form (FS^{-1}, STS^{-1}) is a left pole pair at infinity of L, where S is an invertible matrix of size $n \times n$. Moreover, as (5.1) is a minimal realization, any left pole pair at infinity of L is similar to a pair (F, T) such that $AT+BF=I$. In the next theorem we describe what the possibilities are for the Jordan structure of the nilpotent matrix T in a pole pair at infinity.

Theorem 5.1 *Let (F, T) be a left pole pair at infinity of a regular $m \times m$ matrix* *polynomial* $L(\lambda)$, *and let* $\varkappa_1 \geqq \ldots \geqq \varkappa_r$ *be the sizes of the Jordan blocks of* T. *Let* (A, B) *be a zero pair of* L, *and denote by* $\omega_1 \geqq \ldots \geqq \omega_s$ *the controllability indices of* (A, B). *Then* $r \leqq s$ *and*

$$\sum_{j=1}^{k} \omega_j \leqq \sum_{j=1}^{k} \varkappa_j, \qquad k=1,\ldots,r. \tag{5.2}$$

Proof. By the remarks preceding the theorem we have $ASTS^{-1}+BFS^{-1}=I$ for some invertible matrix S. Note that the sizes of the Jordan blocks of T and STS^{-1} are the same. So we may as well replace (F, T) by (FS^{-1}, STS^{-1}). Then we may apply Theorem 4.2. It remains only to note that the elementary divisors of T , i.e., the polynomials $\lambda^{\varkappa_j}$, are precisely the non-constant invariant polynomials of T since T is nilpotent (see e.g. [G], Section VI 3). □

The next corollary concerns the problem to extend a given full range pair (A, B) to a standard pair. Solutions to this problem appear in [GR] (for the case when the set of eigenvalues of A may be extended by one point only) and in [GKR1]. In what follows we solve an extension problem of the type considered in [GR], Theorem 3.1, but with an additional constraint on the desired standard pair. The result is a corollary to Theorems 3.1 and 5.1.

Corollary 5.2 *Let* (A, B) *be a full range pair of matrices,* A *is* $n \times n$, B *is* $n \times m$, *and suppose* A *is invertible. Denote by* $\omega_1 \geqq \ldots \geqq \omega_s$ *the controllability indices of the pair* (A, B). *Let* (Y, X) *be a standard pair which is an extension of* (A, B) *such that* $\sigma(Y)\backslash\sigma(A) \subset \{0\}$ *and* $l \geqq \omega$, *where* Y *is of size* $lm \times lm$ *and* ω *is the largest partial multiplicity of* Y *corresponding to the zero eigenvalue. Denote by* $0 \leqq \nu_1 \leqq \ldots \leqq \nu_m$ *the partial multiplicities of* Y *corresponding to zero. (Note that some of these may be zero.) Let* r *be such that* $\nu_r < l$, $\nu_{r+1} = l$. *Then* $r \leqq s$ *and*

$$\sum_{j=1}^{k} \omega_j \leqq lk - \sum_{j=1}^{k} \nu_j, \qquad k=1,\ldots,r. \tag{5.3}$$

Proof. Suppose the standard pair (Y, X) is an extension of (A, B) as in Theorem 3.1. Let $L(\lambda)$ be the corresponding monic polynomial of degree l. Then $\tilde{L}(\lambda)=\lambda^l L(\lambda^{-1})$ is a comonic matrix polynomial of degree l, its inverse is analytic at infinity and it has (A^{-1}, B) as zero pair. By Theorem 2.2 there are a nilpotent matrix T and a matrix F such that

$$A^{-1}T+BF=I$$

and $\tilde{L}(\lambda)=I+F(I-\lambda T)^{-1}A^{-1}B$. Denote by $\varkappa_1 \geqq \ldots \geqq \varkappa_r$ the partial multiplicities of T. Considering the Smith-McMillan form of $\tilde{L}(\lambda^{-1})$ and $\lambda^l \tilde{L}(\lambda^{-1})=L(\lambda)$, we see that the partial multiplicities of $L(\lambda)$ at zero are $l-\varkappa_1,\ldots,l-\varkappa_r,l,\ldots,l$. So if we put $\varkappa_j=0$ for $j=r+1,\ldots,m$, we have $\nu_j=l-\varkappa_j$. Next, note that the controllability indices of (A^{-1}, B) are precisely those of (A, B). Then (5.3) follows from Theorem 5.1, using the fact that the partial multiplicities of L at zero are precisely the partial multiplicities of Y corresponding to its zero eigenvalue. □

We conjecture that the following converse of Corollary 5.2 holds. Let Y_1 be a nilpotent $(ml-n)\times(ml-n)$ matrix with $\dim \operatorname{Ker} Y_1 \leqq m$ and such that for its partial multiplicities $0 \leqq \nu_1 \leqq \ldots \leqq \nu_m$ we have $\nu_m \leqq l$ and (5.3) holds. Then we conjecture the existence of a matrix X_1 such that the pair

$$\left(\begin{pmatrix} A & 0 \\ 0 & Y_1 \end{pmatrix}, \begin{pmatrix} B \\ X_1 \end{pmatrix} \right)$$

is a standard pair and an extension of (A, B).

6. Examples

In this section we discuss two examples, which in some sense represent two extreme possibilities. The first example is the following. Take A to be the $n \times n$ Jordan block with zero eigenvalue and $B=(0\ldots.0\ 1)^T$. Clearly, if $AT+BF=I$ we have

$$T=\begin{pmatrix} * & \cdot & \cdot & \cdot & \cdot & \cdot & * \\ 1 & 0 & \cdot & \cdot & \cdot & \cdot & 0 \\ 0 & \cdot & \cdot & & & & \cdot \\ \cdot & \cdot & \cdot & \cdot & & & \cdot \\ \cdot & & \cdot & \cdot & \cdot & & \cdot \\ \cdot & & & \cdot & \cdot & \cdot & \cdot \\ 0 & \cdot & \cdot & \cdot & 0 & 1 & 0 \end{pmatrix}, \qquad F=\begin{pmatrix} 0\ldots.0 & 1 \end{pmatrix},$$

where *'s denote arbitrary entries. Since T is a companion matrix, it is clear that the invariant polynomials of T are all 1 except the first one which is an arbitrary monic polynomial of degree n. As $\omega_1=n$ in this case, the same statement follows easily from (4.6), (4.7). Obviously, T is nilpotent if and only if all the *'s are zero.

The second example concerns the case when B is invertible. Clearly, then there is no restriction on T, as we can solve F, given T, from $AT+BF=I$. That there is no restriction on the invariant polynomials of T is also clear form Theorem 4.2, as in this case $\omega_1=\omega_2=\ldots=\omega_n=1$.

7. Further analysis of the parameter equation

In this section we analyse the parameter equation further. In particular, we shall drop the assumption that (A, B) is a full range pair. For this more general case we study the same problem as in Section 4, namely: what can be said about the invariant polynomials of T if (F, T) is a solution of the parameter equation (2.15). As in Section 4 we do not restrict ourselves to the case when T is nilpotent.

Let A be an $n \times n$ matrix and B and $n \times m$ matrix. Denote the controllability indices of (A, B) by $\omega_1 \geqq \ldots \geqq \omega_s$ and let $t = n - (\omega_1 + \ldots + \omega_s)$. Then there is a basis $\{g_{jk}\}_{k=1 j=1}^{\omega_j \ \ s}$ in $M = \bigvee_{j=0}^{n-1} \operatorname{Im} A^j B$ and a basis $\{v_i\}_{i=1}^{t}$ for a complement of M in $\mathbb{C}^n$ such that with respect to the basis $\{g_{jk}\}_{k=1 j=1}^{\omega_j \ \ s} \cup \{v_i\}_{i=1}^{t}$ in $\mathbb{C}^n$ and the standard basis in $\mathbb{C}^m$ the matrices A and B have the following form

$$A = \left(A_{ij}\right)_{i,j=1}^{s+1}, \quad B = \left(B_{ij}\right)_{i=1 j=1}^{s+1 \ m}. \tag{7.1}$$

Here A_{ij} is an $\omega_i \times \omega_j$ matrix for i and j less than or equal to s, and B_{ij} is an $\omega_i \times 1$ matrix for $i \leqq s$, and $B_{jj} = (0 \ldots 0\ 1)^T$ for $j = 1, \ldots, s$ and $B_{ij} = 0$ for all other i, j. The matrices A_{ij} have the following form

$$A_{ii} = \begin{pmatrix} 0 & 1 & & & \\ & \cdot & \cdot & & \\ & & \cdot & \cdot & \\ & & & \cdot & \cdot \\ & & & 0 & 1 \\ * & \cdot & \cdot & \cdot & * \end{pmatrix}, \text{ for } i = 1, \ldots, s,$$

$$A_{ij} = \begin{pmatrix} 0 & \cdot & \cdot & \cdot & 0 \\ \cdot & & & & \cdot \\ \cdot & & & & \cdot \\ \cdot & & & & \cdot \\ 0 & & & & 0 \\ * & \cdot & \cdot & \cdot & * \end{pmatrix}, \text{ for } i = 1, \ldots, s \text{ and all } j \neq i.$$

Finally, $A_{s+1 j} = 0$ for $j = 1, .., s$, and $A_{s+1 s+1} = J$, where J is a matrix in Jordan canonical form. We shall refer to the matrix J as the *Jordan part* of the pair (A, B). The invariant polynomials of $(A - \lambda I \ \ B)$ are the invariant polynomials of J together with a number of invariant polynomials which are one. The next theorem describes all T 's such that

$AT+BF=I$ for some F.

Theorem 7.1 *There exists a T and an F such that*

$$AT+BF=\begin{bmatrix} A & B \end{bmatrix}\begin{bmatrix} T \\ F \end{bmatrix}=I \tag{7.2}$$

if and only if the Jordan part J of (A, B) is invertible, i.e., if and only if the uncontrollable eigenvalues of A are non-zero.

In that case any T for which there exists an F such that $AT+BF=I$ is of the following form with respect to the basis $\{g_{jk}\}_{k=1}^{\omega_j}{}_{j=1}^{s} \cup \{v_i\}_{i=1}^{t}$.

$$T=\begin{bmatrix} T_{ij} \end{bmatrix}_{i,j=1}^{s+1}, \tag{7.3}$$

where

$$T_{ii}=\begin{bmatrix} * & \cdot & \cdot & \cdot & \cdot & * \\ 1 & 0 & & & & \\ & \cdot & \cdot & & & \\ & & \cdot & \cdot & & \\ & & & \cdot & \cdot & \\ & & & & 1 & 0 \end{bmatrix} \quad \textit{for } i \leqq s,$$

$$T_{ij}=\begin{bmatrix} * & \cdot & \cdot & \cdot & \cdot & * \\ 0 & \cdot & \cdot & \cdot & \cdot & 0 \\ \cdot & & & & & \cdot \\ \cdot & & & & & \cdot \\ \cdot & & & & & \cdot \\ 0 & \cdot & \cdot & \cdot & \cdot & 0 \end{bmatrix} \quad \textit{for } i \neq j,\ i \leqq s.$$

Finally, $T_{s+1\,j}=0$ for $j=1,\dots,s$ and $T_{s+1\,s+1}=J^{-1}$.

Conversely, if T is of the form (7.3), then there exists an F such that (7.2) holds.

Proof. The first part of the theorem is easily proved by observing that (7.2) holds for some T and F if and only if $\operatorname{rank}\begin{bmatrix} A & B \end{bmatrix}=n$.

So suppose $\operatorname{rank}\begin{bmatrix} A & B \end{bmatrix}=n$ and that (F, T) satisfy (7.2). From (7.2) we see that $ATg_{jk}-g_{jk}\in \operatorname{Im} B$ for all j, k. Using an argument similar to the one used in the proof of Theorem 4.1 we obtain that the first $\sum_{i=1}^{s}\omega_i$ columns of T are as in (7.3). Further, $ATv_i-v_i\in \operatorname{Im} B$ for all $i=1,\dots,t$ by (7.2). In the same way this implies that the last t

columns of T are as in (7.3).

Conversely, from (7.1) and (7.2) we see that $(AT-I)x \in \operatorname{Im} B$ for all x. So there is an F for which $AT+BF=I$. □

Next, we state our main theorem of this section, which is the analogue of Theorem 4.2 for the case when (A, B) is not a full range pair. First we introduce the following notation for a scalar polynomial of degree d:

$$a^{\#}(\lambda)=a(\frac{1}{\lambda})\lambda^{d}\frac{1}{a_j},$$

where $a_j \neq 0$ is the first non-zero coefficient. Further, we shall use the abbreviation l.c.m. for least common multiple.

Theorem 7.2 *Let A be an $n \times n$ matrix and B an $n \times m$ matrix. Denote the controllability indices of (A, B) by $\omega_1 \geqq \dots \geqq \omega_s$, and the invariant polynomials of $(A-\lambda I \ B)$ by $a_n \,|\, a_{n-1} \,|\, \dots \,|\, a_1$. Assume* rank $(A \ B)=n$. *Let $p_n \,|\, p_{n-1} \,|\, \dots \,|\, p_1$ be given monic polynomials of degrees $d_n, \dots, d_1$, respectively. Then there exist matrices T and F such that*

$$\begin{pmatrix} A & B \end{pmatrix} \begin{pmatrix} T \\ F \end{pmatrix} = I$$

and T has invariant polynomials $p_1, \dots, p_n$ if and only if

$$p_{i+m} \,|\, a_i^{\#} \,|\, p_i \qquad i=1,\dots,m \quad (p_j:=1,\ j>n), \tag{7.4}$$

$$\sum_{i=1}^{s} d(\sigma_i)=n, \tag{7.5}$$

$$\sum_{i=1}^{k} \omega_i \leqq \sum_{i=1}^{k} d(\sigma_i), \qquad k=1,\dots,s-1, \tag{7.6}$$

where $d(\sigma_i)$ denotes the degree of the monic polynomial σ_i which is defined as follows. Put $\beta_i^j = l.c.m.\,(a_{n-i+j}^{\#}, p_{n-i+m})$, $i=1,\dots,n+j$, $j=0,\dots,m$, where $a_i^{\#}:=1$ for $i>n$, and define $\beta^j=\beta_1^j.\beta_2^j..\beta_{n+j}^j$. Then $\sigma_j=\dfrac{\beta^j}{\beta^{j-1}}$.

Proof. We suppose (A, B) is in the form (7.1) with respect to the basis $\{g_{jk}\}_{k=1\,j=1}^{\omega_j\ \ s} \cup \{v_i\}_{i=1}^{t}$. Let T_0 be the matrix in (7.3) with all *'s equal to zero, and let B': $\mathbb{C}^m \rightarrow \mathbb{C}^n$ be given by the second matrix in (7.1). More precisely, $B'e_j=g_{j1}$, $j=1,\dots,s$, $B'e_j=0$, $j>s$ where $e_1,\dots,e_m$ is the standard basis in $\mathbb{C}^m$. Note that any T of the form (7.3) is of the form $T_0+B'C$ for some C. Furthermore, for any T of the form (7.3) the controllability indices of (T, B') are $\omega_1,..,\omega_s$, and the invariant polynomials of $(T-\lambda I\ B')$ are $a_1^\#,...,a_n^\#$. Indeed, the invariant polynomials are those of J^{-1} supplemented by a number of 1's.

According to Zaballa's theorem, [Z] Theorem 2.6, there is a C such that $T=T_0+B'C$ has invariant polynomials $p_1,...,p_n$ if and only if (7.4)-(7.6) hold. Using Theorem 7.1 the theorem now follows in exactly the same way as Theorem 4.2 followed from Theorem 4.1. Indeed, suppose (7.4)-(7.6) hold. Take $T=T_0+B'C$ with invariant polynomials $p_1,...,p_n$. Such a T is of the form (7.3), and by Theorem 7.1 there is an F such that $AT+BF=I$. Conversely, suppose $AT+BF=I$, and T has invariant polynomials $p_1,...,p_n$. Apply Zaballa's theorem to (T, B') to show that (7.4)-(7.6) hold. □

Obviously, one can also state the dual version of the above theorem for a pair of matrices (C, A), where A is an $n\times n$ matrix and C is an $m\times n$ matrix. We omit the details.

References

[BGK] Bart, H., Gohberg, I., Kaashoek, M.A.: Minimal factorizations of Matrix and Operator Functions, OT 1, Birkhäuser, Basel, 1979.

[BGK1] Bart, H., Gohberg, I., Kaashoek, M.A.: Explicit Wiener-Hopf factorization and realization, in: Constructive Methods of Wiener-Hopf Factorization (eds. I. Gohberg, M.A. Kaashoek), OT 21, Birkhäuser, Basel, 1986, 235-316.

[G] Gantmacher, F.R.: The theory of matrices, Chelsea, New York, 1959.

[GK] Gohberg, I., Kaashoek, M.A.: Regular rational matrix functions with prescribed pole and zero structure, in: Topics in Interpolation Theory of Rational Matrix-valued Functions, OT 33, Birkhäuser, Basel, 1988, 109-122.

[GKLR] Gohberg, I., Kaashoek, M.A., Lerer, L., Rodman, L.: Minimal divisors of rational matrix functions with prescribed zero and pole structure, in: Topics in Operator Theory Systems and Networks, H. Dym and I. Gohberg eds., OT 12, Birkhäuser, Basel, 1984, 241-275.

[GKR1] Gohberg, I., Kaashoek, M.A., Ran, A.C.M.: Interpolation problems for rational matrix functions with incomplete data and Wiener-Hopf factorization, in: Topics in Interpolation Theory of Rational Matrix-valued Functions, OT 33, Birkhäuser, Basel, 1988, 73-108.

[GKR2] Gohberg, I., Kaashoek, M.A., Ran, A.C.M.: Regular rational matrix functions with prescribed null and pole data except at infinity, to appear.

[GLR] Gohberg, I., Lancaster, P., Rodman, L.: Matrix Polynomials, Academic Press, New York, 1982.

[GR] Gohberg, I., Rodman, L.: On the spectral structure of monic matrix polynomials and the extension problem, Linear Algebra Appl. 24(1979), 157-172.

[KMR] Kaashoek, M.A., van der Mee, C.V.M., Rodman, L.: Analytic operator functions with compact spectrum. I Spectral nodes, linearization and equivalence, Integral Equations and Operator Theory 4(1981), 504-547.

[K] Kailath, T.: Linear Systems, Prentice-Hall, Englewood Cliffs, NJ, 1980.

[R] Rosenbrock, H.H.: State space and Multivariate Theory, Nelson, London, 1970.

[W] Wonham, W.M., Linear Multivariable Control: A Geometric Approach, Springer Verlag, Berlin, 1979.

[Z] Zaballa, I.: Interlacing and majorization in invariant factors assignment problems, to appear.

I. Gohberg
School of Mathematical Sciences
Tel Aviv University
Tel Aviv, Ramat Aviv,
Israel

M.A. Kaashoek and A.C.M. Ran
Faculteit Wiskunde en Informatica
Vrije Universiteit
De Boelelaan 1081
1081 HV Amsterdam
The Netherlands

Operator Theory:
Advances and Applications, Vol. 50

ON STRUCTURED MATRICES, GENERALIZED BEZOUTIANS AND GENERALIZED CHRISTOFFEL-DARBOUX FORMULAS

Georg Heinig

Matrices are considered the entries of which fulfill a difference equation with constant coefficients and coefficient matrix D. It is shown that the inverse of such a matrix is a generalized Bezoutian in case that the coefficient matrix has rank 2. This leads to generalized Christoffel-Darboux formulas. Furthermore Moebius transformations of structured matrices are studied.

1. INTRODUCTION

In the present paper we investigate two classses of matrices: matrices with D-displacement structure and D-Bezoutians. Let us start with the definition of these classes supplemented by a few comments. Throughout the paper D will denote a fixed matrix, $D=[d_{ij}]$ $(i=0,\ldots,p;j=0,\ldots,q)$, with complex entries. A given matrix $A=[a_{ij}]$ $(i=0,\ldots,m-1;j=0,\ldots,n-1)$ is said to be D-structured iff

$$\sum_{k=0}^{p-1} \sum_{l=0}^{q-1} a_{i+k,j+l} d_{kl} = 0 \tag{1.1}$$

for all $i=0,\ldots,m-p$ and $j=0,\ldots,n-q$. The class of all D-structured matrices will be denoted by Str(D), the class of all mxn matrices in Str(D) by $Str_{mn}(D)$, and $Str_n(D):=Str_{nn}(D)$. Let us note that $Str_{mn}(D)$ is a linear space and $Str(tD)=Str(D)$ for $t \in \mathbb{C}$, $t \neq 0$.

For the first time the class Str(D) was introduced (in a slightly different form) in [HR2] and studied in [HR3] and [HR4] for the special case p=q=1. The present paper is aimed to generalize some results of these papers to cases when p,q>1. However, in order to complete the history of structured matrices we have to remark that matrices the entries of which fulfill difference equations of the form (1.1) already appeared in various papers about orthogonal polynomials and moment problems on algebraic curves. In fact, it was, in principle, already remarked by M.G.Krein [K] in 1933 that the moment matrix of a measure on an algebraic curve is structured. For further references and results we refer to [MA],[MM],[MP].

Let us explain in more detail how structured matrices occur in the theory of orthogonal polynomials and moment problems. Let K be an algebraic curve in the complex plane given by

$$K = \{ z: D(z,\bar{z}) = 0 \}$$

where

$$D(z,w) = \sum_{i=0}^{p} \sum_{j=0}^{q} d_{ij} z^i w^j$$

Furthermore let σ denote a complex measure concentrated on K. Then σ generates the sequilinear form

$$(x(z),y(z)) = \int x(z)\overline{y(z)}d\sigma$$

defined for polynomials x(z) and y(z). Clearly, if x(z) and y(z) are polynomials with degree <n and x and y are the corresponding coefficient vectors then $(x(z),y(z))= y^*A_n x$, where A_n is the Gram (or moment) matrix $[(z^i,z^j)]_0^{n-1}$. We have now

$$\sum_{k,l} a_{i+k,j+l}d_{kl} = z^i\ \bar{z}^j \int \sum_{k,l} z^k\ \bar{z}^l d_{kl}d\sigma= 0 \quad .$$

Thus $A\in Str(D)$. Structured matrices also occur in polynomial regression problems on algebraic curves (see [HR4]).

The concept of D-structured matrices generalizes the concepts of Toeplitz, Hankel and Toeplitz-plus-Hankel matrices. These classes are obtained for the following special choices of D, repectively:

$$D = [\begin{smallmatrix} 1 & 0 \\ 0 & -1 \end{smallmatrix}] , \quad D = [\begin{smallmatrix} 0 & -1 \\ 1 & 0 \end{smallmatrix}] , \quad D = \begin{bmatrix} 0 & 1 & 0 \\ -1 & 0 & -1 \\ 0 & 1 & 0 \end{bmatrix} \quad . \tag{1.2}$$

The curves K corresponding to these matrices are the real line, the unit circle and the union of real line and unit circle, respectively. For Bernoulli's lemniscate $|z^2-1| = 1$ and more general Cassin curves one has

$$D = \begin{bmatrix} 0 & 0 & 1 \\ 0 & 0 & 0 \\ 1 & 0 & -1 \end{bmatrix} \qquad D = \begin{bmatrix} a & 0 & c \\ 0 & 0 & 0 \\ b & 0 & d \end{bmatrix},$$

respectively. Let us point out that in all examples listed above one has rank $D = 2$ (this remark is important in the sequel). Further examples of curves corresponding with rank-two matrices D are equipotential curves defined by $|a(z)| = R$ and harmonic curves given by $\operatorname{Im} a(z)=0$. In these cases

$$D = aa^* - R^2ee^* \quad \text{and } D = ae^* - ea^*,$$

respectively, where a denotes the coefficient vector of $a(z)$ and e the first unit vector. Matrices of these types are considered in [MM],[MA],[MP].

In order to define the second matrix class under consideration we introduce a convenient notation. Let $l_n(z)$ be the vector $l_n(z) := [1\ z\ \dots\ z^{n-1}]$ ($z \in \mathbb{C}$) and $l_n(\infty) := [0 \dots 0\ 1]$. For $x \in \mathbb{C}^n$ we denote by $x(z)$ the polynomial $x(z) = l_n(z)^T x$. For an $m \times n$ matrix A the polynomial $A(z,w)$ is defined by

$$A(z,w) := l_m(z)^T A l_n(w).$$

$A(z,w)$ is usually called the generating function of A.

A matrix is said to be a D-Bezoutian iff there are polynomials $u_i(z)$ and $v_i(z)$ ($i=1,\dots,p+q$) such that

$$D(z,w)B(z,w) = \sum_{i=1}^{p+q} u_i(z)v_i(w) \tag{1.3}$$

The class of D-Bezoutians will be denoted by $\operatorname{Bez}(D)$, the $m \times n$ matrices in $\operatorname{Bez}(D)$ by $\operatorname{Bez}_{mn}(D)$, and $\operatorname{Bez}_n(D) := \operatorname{Bez}_{nn}(D)$. Let us note that $\operatorname{Bez}(tD) = \operatorname{Bez}(D)$ for $t \in \mathbb{C}$, $t \neq 0$, but $\operatorname{Bez}_{mn}(D)$ is, different to $\operatorname{Str}_{mn}(D)$, not a linear space. In the cases (1.2) the Bezoutians are called Toeplitz or unit circle, Hankel or real line and Toeplitz-plus-Hankel Bezoutians, respectively (see [HR1,2] for some results and references concerning these matrix classes). General Bezoutians for $p=q=1$ were introduced for the first time, as far as we know, in [LABK]. Let us note that other generalizations of the Bezoutian concept is also introduced in the papers [LT] and [GSh]. We point out that the Bezoutians concept considered there is essentially different from our definition.

Between the classes of Hankel matrices and Hankel Bezoutians there are various interconnections. The most important one is the fact that inverses of Hankel matrices are Hankel Bezoutians. This and the analogous Toeplitz result is a simple consequence of the Gohberg-Semencul theorem [GS] and is important to construct fast inversion algorithms. Now the natural question is to which extend this fact generalizes to general D-structured matrices. The cases $p=q=1$, Toeplitz-plus-Hankel and D with a cross structure were studied in [HR2-5], and the answer is positive. However, already simple examples show that in general the answer is "no". If, for instance, $D=\mathrm{diag}(1,-2,1)$ then the matrix $\mathrm{diag}(1,2,..,n)$ belongs to Str(D) but its inverse is not Bezoutian-like. A positive result is obtained in the case rank D = 2, which is our main theorem.

THEOREM 1.1. Let the matrix D fulfill the following conditions:

(a) rank $D = 2$

(b) If $D(z,w)=E(z,w)f(w)$ for polynomials $E(z,w)$ and $f(w)$ then $f(w)$ is a constant. Furthermore the last column of D is nonzero.

Then $A^{-1} \in \mathrm{Bez}(D^T)$ for all regular matrices $A \in \mathrm{Str}(D)$.

We conjecture that the theorem remains true without the assumption (b). Another conjecture is that $B^{-1} \in \mathrm{Str}(D)$ for regular matrices $B \in \mathrm{Bez}(D^T)$. At least this is true if $p=q=1$ or D has a cross structure (see [HR3,5]).

The proof of Theorem 1.1 will be given in Section 3. Section 2 contains some preliminary facts. In Section 4 Theorem 1.1 will be interpreted in the language of orthogonal polynomials on algebraic curves and formulated as a generalized Christoffel-Darboux formula.

Section 5 is dedicated to some kind of transformations that are related to the Frobenius-Fischer transformations in the theory of moment problems. In [HR3] we have shown that these transformations can be utilized to prove Theorem 1.1 for the case $p=q=1$. But they are also of independent interest. We study how they act on the classes Str(D) and Bez(D). As a particular result we obtain a similarity transformation transforming Toeplitz-plus-Hankel matrices of even order into block Hankel matrices with 2x2 blocks.

2. ELEMENTARY PROPERTIES

For fixed D, we introduce two linear transformations $T(D)$ and $T_\emptyset(D)$ defined for matrices $A=[a_{ij}]_\emptyset^{m-1}\,{}_\emptyset^{n-1}$ by

$$T(D)A = [c_{ij}]_{-p}^{m-1}\,{}_{-q}^{n-1}$$

and

$$T_\emptyset(D)A = [c_{ij}]_\emptyset^{m-p-1}\,{}_\emptyset^{n-q-1},$$

where

$$c_{ij} = \sum_{k,l} a_{i+k,j+l}d_{kl}$$

Then the following fact is obvious.

PROPOSITION 2.1. The matrix A belongs to Str(D) iff $T_\emptyset(D)A = \emptyset$.

The class Bez(D) can be characterized with the help of the transformation T. This follows from the following.

LEMMA 2.1. $(T(D)A)(z,w) = D(z^{-1},w^{-1})A(z,w)$.

Proof. By definition,

$$(T(D)A)(z,w) = \sum_{i,j,k,l} a_{i+k,j+l}d_{kl}z^iw^j$$

Hence

$$(T(D)A)(z,w) = \sum_{k,l,s,t} a_{st}d_{kl}z^{s-k}w^{t-l} = D(z^{-1},w^{-1})A(z,w).\#$$

Introduce the counteridentity J_n,

$$J_n(x_k)_\emptyset^{n-1} = (x_{n-1-k})_\emptyset^{n-1}$$

and define

$$\hat{D} := J_{p+1}DJ_{q+1} .$$

Then $\hat{D}(z,w) = z^pw^qD(z^{-1},w^{-1})$. Hence, by Lemma 2.1,

$$(T(\hat{D})A)(z,w) = z^{-p}w^{-q}D(z,w)A(z,w)$$

and the following is true.

PROPOSITION 2.2. The matrix B belongs to Bez(D) iff rank $T(\hat{D}) \le p+q$.

For a vector $f\in\mathbb{C}^{p+1}$, $f=(f_i)_\emptyset^p$ we define matrices

$$P_m(f) = \begin{bmatrix} f_0 & & & 0 \\ \vdots & \ddots & & f_0 \\ f_p & & \ddots & \vdots \\ 0 & & & f_p \end{bmatrix}$$

consisting of m columns. It is easily checked that $P_m(f)$ is the matrix of the multiplication operator $x(z) \longrightarrow f(z)x(z)$ defined for polynomials with degree $<m$ with respect to the standard basis $\{z^i\}$. The class Str(D) can be characterized with the help of this type of matrices. We do it for the case rank D = 2. Suppose that D is of the form

$$D = f_1 g_2^T - g_1 f_2^T \quad . \tag{2.1}$$

PROPOSITION 2.3. Let D be of the form (2.1). Then the matrix A belongs to Str(D) iff

$$P_{m-p}(f_1)^T A P_{n-q}(g_2) = P_{m-p}(g_1)^T A P_{n-q}(f_2). \tag{2.2}$$

Proof. Suppose that $f_k = (f_i^k)$, $g_k = (g_i^k)$ $(k=1,2)$. Then

$$\begin{aligned} T_0(D)A &= [\sum_{k,l} a_{i+k,j+l}(f_k^1 g_l^2 - g_k^1 f_l^2)]_{i,j} \\ &= [\sum_{s,t} f_{s-i}^1 a_{st} g_{t-j}^2 - g_{s-i}^1 a_{st} f_{t-j}^2]_{i,j} \end{aligned}$$

Hence $T_0(D)A=0$ iff (2.2) holds. It remains to take Proposition 2.1 into account. #

3. INVERSION OF STRUCTURED MATRICES

The main aim of this section is the proof of Theorem 1.1. We start with an elementary but important remark.

LEMMA 3.1. Let A be a regular nxn matrix and x_t the solution of the equation

$$Ax_t = 1_n(t) \tag{3.1}$$

Then

$$A^{-1}(z,t) = x_t(z).$$

That means, in order to find the generating function of the inverse matrix one has to solve the equation (3.1) for all $t \in \mathbb{C}$ or at least for some collection of t's.

In order to motivate our approach for solving (3.1) in

case that A is structured we consider first the case of a Hankel matrix. The basic idea of Hankel matrix theory approach in [HR1] is to consider instead of one Hankel matrix $H = [s_{i+j}]_{\emptyset}^{n-1}$ the family of Hankel matrices $H^k=[s_{i+j}]_{\emptyset}^{n-1-k}\ {}_{\emptyset}^{n-1+k}$ $(k=\emptyset,\pm1,\ldots)$. In the case of a regular H it suffices, for inversion purposes, to regard the matrix $H'=H^1$. A simple proof of Theorem 1.1 for the Hankel case results from the following fact:

If $Hx_t = 1_n(t)$, then u defined by $u(z)=(z-t)x_t(z)$ belongs to ker H'.

Since ker H' is two-dimensional we conclude that $x_t(z)= (v_1(t)u_1(z)-v_2(t)u_2(z))/(z-t)$, where $\{u_1,u_2\}$ is a basis of ker H' and $v_1(t)$, $v_2(t)$ are polynomials. Consequently H^{-1} is a Hankel Bezoutian.

In order to generalize this idea one has to generalize the map $H \longrightarrow H^k$, in particular $H \longrightarrow H'$. For the case $p=q=1$ this was done in [HR4], for the case of D with cross form in [HR5]. We are going now to consider the general rank two case with the only restriction (b) of Theorem 1.1. The latter condition means that the polynomials $g_2(z)$ and $f_2(z)$ are coprime and one of the highest order coefficients is nonzero. For given A from $Str_n(D)$ we define A' by the following. If x is given by

$$x = P_n(f_2)x_1 + P_n(g_2)x_2 \tag{3.2}$$

for some x_1, $x_2 \in \mathbb{C}^n$, then, by definition,

$$A'x = P_{m-p}(f_1)^T Ax_1 + P_{m-p}(g_1)^T Ax_2 \ . \tag{3.3}$$

Of course, we have to show that the definition is correct, which means that different representations of x give the same image A'x. Suppose that (3.2) holds and

$$x = P_n(f_2)x_1' + P_n(g_2)x_2' \tag{3.2'}$$

Our goal is to show that

$$P_{m-p}(f_1)^T Ax_1' + P_{m-p}(g_1)^T Ax_2' = P_{m-p}(f_1)^T Ax_1 + P_{m-p}(g_1)^T Ax_2 \tag{3.4}$$

Introducing $y_1= x_1-x_1'$, $y_2= x_2'-x_2$ we get from (3.2) and (3.2')

$$f_2(z)y_1(z) = g_2(z)y_2(z) \ . \tag{3.5}$$

Since $f_2(z)$ and $g_2(z)$ are coprime we conclude $y_1(z)=g_2(z)r_1(z)$

and $y_2(z) = f_2(z)r_2(z)$ for certain polynomials $r_1(z)$ and $r_2(z)$. Inserting into (3.5) we obtain $r_1=r_2=:r$. Thus

$$Ay_1 = AP_{m-p}(g_2)r \quad \text{and} \quad Ay_2 = AP_{m-p}(f_2)r \quad .$$

Multiplying by $P_{m-p}(f_1)^T$ and $P_{m-p}(g_1)^T$ from the left and taking Proposition 2.3 into account we get

$$P_{m-p}(f_1)^T Ay_1 = P_{m-p}(g_1)^T Ay_2,$$

which is equivalent to (3.4). Hence A' is correctly defined.

Furthermore, A' is defined for all $x \in \mathbb{C}^{n+q}$. In fact, since $f_2(z)$ and $g_2(z)$ are coprime, any polynomial $x(z)$ with degree $<n+q$ can be represented in the form

$$x(z) = f_2(z)x_1(z) + g_2(z)x_2(z)$$

with polynomials $x_1(z)$, $x_2(z)$ of degree $<n$. This representation is equivalent with (3.2). Finally we note that A' is obviously a map onto $\mathbb{C}^{m-p}$. Thus the following is true.

PROPOSITION 3.1. $\dim \ker A' = p+q$.

The main point of the proof of Theorem 1.1 will be the following fact.

LEMMA 3.2. Let $A \in Str_n(D)$ be regular. If x_t is a solution of (3.1) and $u \in \mathbb{C}^{n+q}$ is defined by $u(z) := D^T(z,t)x_t(z) = D(t,z)x_t(z)$ then $u \in \ker A'$.

Proof. We have $u(z)=(g_2(z)f_1(t)-f_2(z)g_1(t))x_t(z)$, which is equivalent to

$$u = (f_1(t)P_n(g_2) - g_1(t)P_n(f_2))x_t.$$

By the definition of A',

$$\begin{aligned} A'u &= f_1(t)P_{n-p}(g_1)^T Ax_t - g_1(t)P_{n-p}(f_1)^T Ax_t \\ &= f_1(t)P_{n-p}(g_1)^T 1_n(t) - g_1(t)P_{n-p}(f_1)^T 1_n(t) \\ &= f_1(t)g_1(t)1_{n-p}(t) - g_1(t)f_1(t)1_{n-p}(t) = \emptyset. \end{aligned}$$

Thus $u \in \ker A'$ and the lemma is proved. #

Proof of Theorem 1.1. Put $x_t(z) = A^{-1}(z,t)$. Then x_t is a solution of (3.1). According to Lemma 3.2 the vector u defined by $u(z) = D(t,z)x_t(z)$ belongs to $\ker A'$. In view of Proposition 3.1, there is a basis $\{u_1,\ldots,u_{p+q}\}$ of $\ker A'$. Hence

$$D(t,z)x_t(z) = \sum_{i=1}^{p+q} v_i(t)u_i(z) \tag{3.6}$$

for some numbers $v_i(t)$ depending on t. Since the left-hand side of (3.6) is a polynomial in t, the right-hand side is also a polynomial. Hence the functions $v_i(t)$ are in fact polynomials in t. Thus (3.6) states that A^{-1} is a D^T-Bezoutian. #

Let us point out that, analogously to the Hankel case, matrices $A'' = (A')'$, A''',... can be defined. This leads to the hope that the algebraic Hankel theory can completely be generalized to the more general classes considered here. A basic observation concerning this approach is the following one.

PROPOSITION 3.2. A' is D-structured.

Proof. In view of Proposition 2.3 we have to show that

$$P_{m-2p}(f_1)^T A' P_n(g_2) = P_{m-2p}(g_1)^T A' P_n(f_2) . \quad (3.7)$$

By the definition of A' we have

$$A' P_n(g_2) = P_{m-p}(g_1)^T A, \quad A' P_n(f_2) = P_{m-p}(f_1)^T A .$$

Since $P_{m-2p}(f_1)^T P_{m-p}(g_1)^T = P_{m-2p}(g_1)^T P_{m-p}(f_1)^T$, we conclude (3.7). #

4. GENERALIZED CHRISTOFFEL-DARBOUX FORMULA

In this section we restrict ourselves to the case rank D = 2 and D hermitian. Let K denote the algebraic curve defined by D, $K = \{z : D(z,\bar{z}) = 0\}$. Clearly, K is not empty iff D is indefinite. In this case D admits a representation

$$D = ff^* - gg^* \qquad (f,g \in \mathbb{C}^{p+1}).$$

Hence

$$K = \{ z \in \mathbb{C} : |R(z)| = 1 \}, \quad (4.1)$$

where $R(z) := f(z)/g(z)$.

Let λ be a real-valued measure concentrated on K and $(.,.)$ the corresponding (indefinite) inner product. A polynomial $x(z)$ with degree n is said to be orthogonal with respect to λ iff $(x(z), z^i) = 0$ for $i=0,\dots,n-1$. Obviously, $x(z)$ is orthogonal iff $A_n x = r_n e_n$, where $A_n = [a_{ij}]_0^{n-1}$, $a_{ij} = (z^i, z^j)$, $e_n = [0 \dots 0\ 1]$ and $r_n = (x(z), z^n)$.

We assume now that λ is such that all A_k $(k=0,1,2,\dots)$ are regular. This is true, in particular, if λ is positive. If this condition is fulfilled then for any k there exist exactly one monic orthogonal polynomial, which will be denoted by $x_k(z)$.

We form now an upper triangular matrix X_n consisting of the coefficient vectors of $x_k(z)$ supplemented by zeros to an n+1 dimensional vector. Then

$$A_n X_n = R_n Y_n,$$

where $R_n = \mathrm{diag}(r_\emptyset,\dots,r_n)$, $r_k=(x_k(z),z^k)$ and a lower triangular matrix Y_n. Since A_n is hermitian we have $Y_n^{-1} = X_n^*$. Hence

$$A_n^{-1} = X_n R_n^{-1} X_n^* .$$

This is equivalent with

$$A_n^{-1}(z,w) = \sum_{k=\emptyset}^{n} \frac{1}{r_k} x_k(z)\bar{x}_k(w) .$$

Comparing this formula with Theorem 1.1 we get the following formula of Christoffel-Darboux type.

THEOREM 4.1. Let K be a curve defined by (4.2) and λ a measure concentrated on K such that the Gram matrices A_k $(k=\emptyset,\dots,n)$ are regular. Then the monic orthogonal polynomials $x_k(z)$ fulfill an equality

$$\sum_{k=\emptyset}^{n} \frac{1}{r_k} x_k(z)\bar{x}_k(w) = \sum_{i=1}^{2n} u_i(z)v_i(w)/D(w,z) ,$$

where $\{u_i\}$ and $\{v_i\}$ are bases of the kernel of the operator A' defined in Section 3.

Theorem 4.1 generalizes, to some extend, the classical Christoffel-Darboux formulas, which are the formulas for K being the real line or the unit circle. However, the following questions are still open:

1) What are the relations between the u_i and the v_i? (It is clear that a given basis $\{u_i\}$ uniquely defines the basis $\{v_i\}$.)

2) How the u_i and v_i are related to the $x_k(z)$? (In the classical cases the u_i and v_i are orthogonal polynomials or reflections of them.)

3) How the u_i and v_i can be computed recursively? (Some results in this direction are contained in [MA] and [MM].)

As noted in the introduction, examples of curves meeting the condition of Theorem 4.1 are lemniscates or more general equipotential curves, harmonic curves, but also families of orthogonal circles. An ellipse fulfils the assumption only if it is a circle, and it seems that there is no Christoffel-Darboux formula for the general ellipses.

5. MOEBIUS TRANSFORMATIONS

We associate any 2x2 regular matrix $h = \begin{bmatrix} a & c \\ b & d \end{bmatrix}$ with the linear fractional transformation $h(z) = (az+b)/(cz+d)$ defined on the Riemann sphere and with nxn matrices $M_n(h)$ defined by

$$M_n(h)x = y \ , \ y(z) = x(h(z))(cz+d)^{n-1} \quad (x,y \in \mathbb{C}^n)$$

The map $h \longrightarrow M_n(h)$ is a group homomorphism from $GL(\mathbb{C}^2)$ into $GL(\mathbb{C}^n)$. We note the following relation:

$$M_n(h)^T 1_n(z) = 1_n(h(z))(cz+d)^{n-1}. \tag{5.1}$$

We are going to study the action of Moebius transformations in the classes Str(D). First of all we consider rank-one matrices of the form

$$L_{mn}(t,s) = [t^i s^j]_{\emptyset}^{m-1}{}_{\emptyset}^{n-1} = 1_m(t)1_n(s)^T.$$

Obviously $L_{mn}(t,s)$ belongs to Str(D) iff $D(t,s)=\emptyset$. From (5.1) we conclude the following relation.

PROPOSITION 5.1. Suppose that $h_i = \begin{bmatrix} a_i & c_i \\ b_i & d_i \end{bmatrix}$ $(i=1,2)$.

Then

$$M_m(h_1)^T L_{mn}(t,s) M_n(h_2) = c\ L_{mn}(h_1(t),h_2(s)) \ ,$$

where

$$c = (c_1 t+d_1)^{m-1}(c_2 s+d_2)^{n-1}.$$

PROPOSITION 5.2 If $L := L_{mn}(t,s)$ belongs to Str(D) then $L' := M_m(h_1)^T L M_n(h_2)$ belongs to Str(D'), where

$$D' := M_{p+1}(h_1^{-1}) D M_{q+1}(h_2^{-1})^T \tag{5.2}$$

Proof. Since $L \in Str(D)$ we have $D(t,s)=\emptyset$. Furthermore, for some constant k,

$$D'(h_1(t),h_2(s)) = k 1_{p+1}(t)^T M_{p+1}(h_1) D' M_{q+1}(h_2)^T 1_{q+1}(s)$$
$$= k\ 1_{p+1}(t)^T D 1_{q+1}(s).$$

Hence $D'(h_1(t),h_2(s))=\emptyset$. In view of Proposition 5.1 this implies $L' \in Str(D')$. #

The relation $M_m(h_1)^T L M_n(h_2) \in Str(D')$ extends by linearity to the span of all matrices $L_{mn}(t,s)$ with $D(t,s)=\emptyset$ which will be denoted by S(D). The question is whether S(D) coincides with Str(D). The answer is provided by the following.

THEOREM 5.1. If the polynomial D(z,w) is square-free in the polynomial ring $\mathbb{C}[z,w]$, then any matrix $A \in Str_{mn}(D)$ can be represented as linear combination of matrices $L_{mn}(t,s)$, and vice versa.

Proof. First we have to show that S(D)=Str(D) if the assumption of the theorem is satisfied. Since obviously $S(D) \subseteq Str(D)$ it suffices to prove that each linear functional vanishing on S(D) also vanishes on Str(D). Suppose that F is a linear functional on $\mathbb{C}^{m \times n}$,

$$FX = \sum_{i,j} f_{ij} x_{ij} \qquad (X=[x_{ij}]_{0\ \ 0}^{m-1\ n-1})$$

We identify F with the matrix $[f_{ij}]$. Now F vanishes on S(D) iff $F(z,w)=0$ for all $(z,w) \in N(D)$, where

$$N(D) := \{(z,w): D(z,w)=0\}.$$

By Hilbert's Nullstellensatz we have

$$F(z,w)^r = D(z,w)E(z,w)$$

for some integer r and polynomial E(z,w). Since D(z,w) is assumed to be square-free and we conclude that F(z,w) admits a representation

$$F(z,w)=D(z,w)E(z,w) \ . \tag{5.3}$$

On the other hand, a matrix A belongs to Str(D), by definition, iff the functionals corresponding to the polynomials $z^i w^j D(z,w)$ vanish on A ($i=0,\dots,m-p; j=0,\dots,n-q$). That means, any functional F satisfying (5.3) vanishes on Str(D). Thus S(D)=Str(D).

Vice versa, if D(z,w) is not square-free, then there exists an F such that $F(z,w)=0$ for $(z,w) \in N(D)$, but F(z,w) is not of the form (5.3). Such an F vanishes on S(D) but not on Str(D). That means $S(D) \neq Str(D)$. #

As a consequence of Proposition 5.2 and Theorem 5.1 we have now the following.

THEOREM 5.2. Let D be such that D(z,w) is square-free. Then the transformation

$$A \longrightarrow M_m(h_1)^T A M_n(h_2) \tag{5.4}$$

maps $Str_{mn}(D)$ onto $Str_{mn}(D')$, where D' is defined by (5.2).

Transformations of the form (5.4) occurred for the first

time in classical papers of Frobenius and Fischer (see [I]) to transform Hankel into Toeplitz forms. They were investigated in more detail in [I], [HR1,3]. In [HR3] it is shown that in the case of a 2x2 matrix D any pair of classes Str(D), Str(D′) are related by some transformation (5.4). This is, of course, not true in the general case. Nevertheless, these transformations seem to be of interest also in the case p,q>2 as the following application shows.

A matrix $[a_{ij}]$ is said to be a 2-Hankel matrix if $a_{ij} = a_{i+2,j-2}$ for all indices for which this equality makes sense. In case that the numbers of columns and rows are even 2-Hankel matrices can be regarded as block Hankel matrices with 2x2 blocks. 2-Hankel matrices and Toeplitz-plus-Hankel matrices are D-structured with D(z,w) beeing

$$D_1(z,w) = z^2 - w^2 \quad \text{and} \quad D_2(z,w) = (z-w)(1-zw)$$

respectively. Putting $h := \begin{bmatrix} a & 1 \\ -a & 1 \end{bmatrix}$ for some $a \neq 0$ we have $h(z) = a(z-1)/(z+1)$. After an elementary computation one obtains

$$M_3(h)D_1M_3(h)^T = -4a\ D_2 .$$

Applying Theorem 5.2 we get the following result.

COROLLARY 5.1. Let h be defined by (5.6). Then the transformation

$$A \longrightarrow M_m(h)^T A M_n(h)$$

maps the class of mxn Toeplitz-plus-Hankel matrices onto the class of mxn 2-Hankel matrices.

For completeness let us show how Moebius transformations act between the classes Bez(D).

THEOREM 5.3. Suppose that $h_1, h_2 \in GL(\mathbb{C}^2)$ are as in Proposition 5.1. The transformation

$$B \longrightarrow M_m(h_1)BM_n(h_2)^T$$

maps $Bez_{mn}(D)$ onto $Bez_{mn}(D')$, where

$$D' = M_{p+1}(h_1)DM_{q+1}(h_2)^T . \tag{5.5}$$

Proof. Suppose that $B \in Bez_{mn}(D)$. Then

$$D(z,w)B(z,w) = \sum_{i=1}^{p+q} u_i(z)v_i(w)$$

for certain polynomials $u_i(z)$ and $v_i(w)$. Replacing z by $h_1(z)$ and w by $h_2(w)$ we get

$$D'(z,w)B'(z,w) = \sum_{i=1}^{p+q} u_i(h_1(z))v_i(h_2(w))(c_1z+d_1)^{p+m-1} * \\ *(c_2w+d_2)^{q+n-1};$$

where D' is given by (5.5) and $B' = M_m(h_1)BM_n(h_2)$. Hence $B' \in Bez(D')$.

REFERENCES

[GS] Gohberg,I. and Semencul A.A., On inversion of finite-section Toeplitz matrices and their continuous analogues.(in Russian) Mat.Issled. 7,2 (1972),201-224.

[GSh] Gohberg,I. and Shalom,T., On Bezoutians of non-square matrix polynomials and inversion of matrices with non-square blocks. Linear Alg.Appl., to appear.

[HR1] Heinig,G. and Rost,K., Algebraic methods for Toeplitz-like matrices and operators. Akademie-Verlag, Berlin, and Birkhäuser Basel 1984.

[HR2] - " - , On the inverses of Toeplitz-plus-Hankel matrices. Linear Alg.Appl.106 (1988),39-52.

[HR3] - " - , Matrices with displacement structure, generalized Bezoutians, and Moebius transformations. Operator Theory: Advances and Appl., vol.40,1989, 203-230.

[HR4] - " - , Inversion of matrices with displacement structure. Integr.Equ. and Oper.Theory, to appear.

[HR5] - " - , On some classes of structured matrices, in preparation.

[I] Iohvidov,I.S., Hankel matrices and forms. Birkhäuser Basel 1982.

[K] Krein,M.G., Über eine Klasse von hermiteschen Formen und uber die Verallgemeinerung des trigonometrischen Momentenproblems. Izv.Akad.Nauk SSSR 9 (1933),1259-1275.

[LABK] Lev-Ari,H., Bistritz,Y. and Kailath,T., Generalized Bezoutians and families of efficient root location procedures. IEEE Trans.on Circ.and Syst., to appear.

[LT] Lerer,L. and Tismenetsky,M. , Generalized Bezoutians and the inversion problem for block matrices. Integr.Equ. and Oper.Theory 9,6 (1986), 790-819.

[MA] Marcellan,F. and Alfaro,M., Recurrence relations for orthogonal polynomials on algebraic curves. Portugalie Mathematica 42 (1984),41-52.

[MM] Marcellan,F. and Moral,L., Minimal recurrence formulas for orthogonal polynomials on Bernoulli's Lemniscate. Lecture Notes in Math. 1171 (1985),211-220.

[MP] Marcellan,F. and Perez-Grasa,I., The moment problem on equipotential curves. Nonlinear Numer.Meth. and Rational Appr., Reidel Publ.Co. 1988,229-238.

Georg Heinig
Karl-Marx-Universität Leipzig
Sektion Mathematik
Karl-Marx-Platz, Hauptgebäude
Leipzig
DDR-7010

Operator Theory:
Advances and Applications, Vol. 50

NUMERICAL RANGES FOR PARTIAL MATRICES

Charles R. Johnson[1] and Michael E. Lundquist[2]

The field of values (or numerical range) is an important notion in the generalization of facts about Hermitian matrices to general complex matrices. Completions of partial Hermitian matrices have now been studied in some depth, and it is time to study completions of more general complex partial matrices. Two natural definitions for the "field of values" of a square partial complex are suggested. The first "builds up" from the inside of any completion, or a specification of the free entries, of a given partial matrix, while the second "whittles down" from the classical field of values of any completion. In general, the resulting sets are different, though there is a universal containment. In this note we characterize those patterns of specified entries for which the two definitions are identical. Chordal graphs again play a key role, but in a somewhat different way from the classical positive definite Hermitian case.

1 INTRODUCTION

A *partial matrix* $A = [a_{ij}]$ is a rectangular array in which some of the entries are *specified* (known items from an indicated set), while the rest are *unspecified* (free to be chosen from the indicated set). A *completion* of a partial matrix is an allowed choice for each of the unspecified entries, resulting in a conventional matrix, i.e. a matrix $B = [b_{ij}]$ such that $b_{ij} = a_{ij}$ whenever a_{ij} is specified. The study of possible properties of completions of partial matrices has been the subject of active recent research. (See e.g. [1, 2, 3, 7, 8, 9].) One major focus of such work has been the case in which the partial matrix $A = [a_{ij}]$ over the complex numbers $\mathbf{C}$ is *partial Hermitian* (A is square and a_{ij} is specified whenever a_{ji} is, in which case $a_{ij} = \overline{a}_{ji}$) and only Hermitian completions are considered. For example, the existence of positive definite completions, the determinant-maximizing positive definite completions and possible inertias of completions have all been heavily studied. Often the *pattern* (or collective location) of the unspecified entries plays a key role in such results. It is typically assumed in this work that every diagonal entry a_{ii} of A is specified, and we

[1]The work of this author was supported in part by National Science Foundation grant DMS 88-02836 and a grant from the National Security Agency.

[2]The work of this author was supported in part by National Science Foundation grant DMS-88-02836 and Office of Naval Research contract N00014-88-K-0661.

maintain this assumption through the present work.

Historically, the field of values (or classical numerical range) has been an important object in understanding how to generalize facts about Hermitian matrices to more general matrices. The *field of values* of $B \in M_n$ is denoted and defined by

$$F(B) = \{x^*Bx : x \in \mathbf{C}^n,\ x^*x = 1\}.$$

Among the many important properties of F (see [5] for the most comprehensive survey) are the following:

(1) $F(B)$ is compact and convex for each $B \in M_n$;

(2) $F(\hat{B}) \subseteq F(B)$ whenever $\hat{B}$ is a principal submatrix of $B \in M_n$; and

(3) the spectrum $\sigma(B) \subseteq F(B)$ for each $B \in M_n$.

Two distinct notions of a numerical range (or field of values) for a square partial matrix naturally come to mind. The first of these builds up from the inside of the field of values of any completion in view of (1) and (2). If A is an $n \times n$ partial matrix over $\mathbf{C}$, let $A_1, \ldots, A_k$ be the maximal fully specified (i.e., consisting entirely of specified entries) principal submatrices of A and define

$$F_1(A) = \mathrm{Co}\Big(\bigcup_{i=1}^{k} F(A_i)\Big),$$

where $\mathrm{Co}(S)$ denotes the convex hull of a set S. Note that, since we assume all diagonal entries of A are specified, $k \geq 1$ (equality occurs only when A is already a fully specified matrix and then $F_1(A) = F(A)$), and $F_1(A)$ is well-defined. The second of these notions whittles down from the field of values of any completion of A. Define

$$F_2(A) = \bigcap F(B),$$

in which the intersection is taken over all completions B of A. Because of the compactness of the usual field of values, F_2 is well-defined. The set F_2 is convex because of (1) and each $F(A_i)$ is contained in each $F(B)$ because of (2); thus, it is clear that

(4) $$F_1(A) \subseteq F_2(A).$$

The sets in (4), however, are not in general identical. It is the principal goal of this note to characterize those patterns for the specified entries of A that insure equality in (4).

2 DEFINITIONS AND NOTATION

Let $D = (V, E(D))$ denote a loopless directed graph; that is, V is a finite set of vertices and the edge set $E(D)$ is a set of ordered pairs (x, y) of distinct vertices $x, y \in V$. Similarly, let $G = (V, E(G))$ denote a loopless *un*directed graph; here the edge set $E(G)$ is a set of unordered pairs $\{x, y\}$ of distinct vertices x, y. Throughout, we will assume $V = \{1, \ldots, n\}$ for some positive integer n. A directed (undirected) graph is *complete* if its edge set contains every ordered (unordered) pair of vertices. If D is a directed graph and $W \subseteq V$, the subgraph *induced* by W is the graph $D_W = (W, E(D_W))$, in which $E(D_W)$ consists of those edges in $E(D)$ whose endpoints are both in W. The corresponding notion for undirected graphs is identical. A *clique* is a set of vertices that induces a complete subgraph; a *maximal* clique is one that is not properly contained in any other clique. A *cycle* in an undirected graph G is a pairwise disjoint sequence of vertices $C = (v_1, \ldots, v_k)$ for which $\{v_1, v_2\}$, $\{v_2, v_3\}$, ..., $\{v_k, v_1\} \in E(G)$; the number k is the *length* of the cycle. A *chord* in a cycle C is an edge consisting of two nonconsecutive vertices in C, and a *chordal* graph is an undirected graph in which every cycle of length four or more has a chord.

Let A be an $n \times n$ partial matrix. The *directed graph* of the specified entries of A is the graph D whose edge set is

$$E(D) = \{(i, j) : a_{ij} \text{ is specified}\};$$

the *undirected graph* G of the specified entries of A is the graph whose edge set is

$$E(G) = \{\{i, j\} : a_{ij} \text{ and } a_{ji} \text{ are both specified}\}.$$

Note that even for partial matrices that are not combinatorially symmetric (that is, whose specified entries are not symmetrically placed) we define an associated undirected graph. In this case, we may regard an edge of an undirected graph as a pair of oppositely oriented directed edges, and then G may be thought of as the maximal undirected subgraph of D. Note that our definition of the undirected graph of a partial matrix is different from a common one for sparse matrices, in which the edge $\{i, j\}$ belongs to $E(G)$ in case *either* a_{ij} or a_{ji} is nonzero. If D is a directed graph a *D-partial* matrix is a partial matrix whose directed graph is D. Similarly, if G is an undirected graph a *G-partial* matrix is a combinatorially symmetric partial matrix whose graph is G. Now, if A is any partial matrix whose undirected graph is G, let A_G denote the G-partial matrix whose i, j entry is a_{ij} whenever $i, j \in E(G)$.

(Of course, if A is combinatorially symmetric, then $A_G = A$.) For example, if

$$A = \begin{bmatrix} 2 & 1 & ? \\ 3 & 2 & ? \\ 1 & ? & 5 \end{bmatrix},$$

in which the ?'s denote unspecified entries, then

$$A_G = \begin{bmatrix} 2 & 1 & ? \\ 3 & 2 & ? \\ ? & ? & 5 \end{bmatrix}.$$

Let A be a partial Hermitian matrix. Since every principal minor of a positive definite matrix must be positive, A may have a completion to a positive definite matrix only if each of its fully specified principal submatrices is positive definite; in this case we will call A *partial positive definite*. If each fully specified principal submatrix is positive semidefinite, then we say that A is *partial positive semidefinite*.

The question of which patterns insure the existence of positive definite completions of partial positive definite matrices was addressed in [3], where the following two results appeared. First, an undirected graph G is said to be *completable* (*nonnegative-completable*) if and only if every G-partial positive definite (semidefinite) matrix has a positive definite (semidefinite) completion.

THEOREM 1 ([3]). *An undirected graph G is completable if and only if G is nonnegative-completable.*

THEOREM 2 ([3]). *An undirected graph G is completable if and only if G is chordal.*

Let $H \in M_n$ be a Hermitian matrix, and denote the largest and smallest eigenvalues of H respectively by $\lambda_{max}(H)$ and $\lambda_{min}(H)$. Since λ_{max} and λ_{min} may be characterized as the the maximum and mimimum attained by the Hermitian form x^*Hx as x ranges over the unit sphere in $\mathbf{C}^n$, we have

$$F(H) = [\lambda_{min}(H), \lambda_{max}(H)],$$

which holds for any Hermitian matrix H.

It is well-known that any matrix $B \in M_n$ may be written uniquely as the sum of a Hermitian matrix and a skew-Hermitian matrix; by definition, the *Hermitian part* of B is the matrix $H(B) = \frac{1}{2}(B + B^*)$, and the *skew-Hermitian part* of B is the matrix $S(B) = \frac{1}{2}(B - B^*)$. Two useful facts that are easily verified are that for any $B \in M_n$

(see [5])

$$\operatorname{Re} F(B) = F(H(B)) \tag{5}$$

and

$$\operatorname{Im} F(B) = -iF(S(B)). \tag{6}$$

Finally, let us note the immediate fact that

$$F(zB) = zF(B)$$

for any complex number z.

3 MAIN RESULT

Our objective is to prove the following.

THEOREM 3. *Let G be an undirected graph on vertices $\{1, \dots, n\}$. Then $F_1(A) = F_2(A)$ for every partial matrix A whose undirected graph is G if and only if G is chordal.*

As a preliminary step we will examine the case in which A is partial Hermitian and only Hermitian completions are considered. Let A be a partial Hermitian matrix, let $A_1, \dots, A_k$ be the maximal fully specified principal submatrices of A, and define

$$\lambda_{min}(A) = \min_{1 \le i \le k} \lambda_{min}(A_i)$$

and

$$\lambda_{max}(A) = \max_{1 \le i \le k} \lambda_{max}(A_i).$$

The interlacing inequalities then imply that for any Hermitian completion B of A

$$\lambda_{min}(B) \le \lambda_{min}(A) \le \lambda_{max}(A) \le \lambda_{max}(B)$$

and therefore

$$[\lambda_{min}(A), \lambda_{max}(A)] \subseteq F(B).$$

One naturally wonders when we may expect to find Hermitian completions B_1 and B_2 satisfying

$$\lambda_{min}(B_1) = \lambda_{min}(A) \tag{7}$$

or

$$\lambda_{max}(B_2) = \lambda_{max}(A). \tag{8}$$

If (7) and (8) were to hold then in fact

$$F_1(A) = F_2(A). \tag{9}$$

Indeed, if B_1 and B_2 are Hermitian completions of A satisfying (7) and (8) then

$$\begin{aligned} F_2(A) &\subseteq F(B_1) \cap F(B_2) \\ &= [\lambda_{min}(A), \lambda_{max}(A)] \\ &= F_1(A). \end{aligned}$$

Since we always have $F_1(A) \subseteq F_2(A)$, equation (9) follows.

LEMMA 4. *Let G be an undirected graph. Then $F_1(A) = F_2(A)$ for every G-partial Hermitian matrix A if and only if G is chordal.*

Proof: In light of the preceding remarks it will suffice to show that Hermitian completions satisfying equations (7) and (8) exist for every G-partial Hermitian matrix if and only if G is chordal.

Suppose G is chordal, let A be any G-partial Hermitian matrix, and let $\lambda_0 = \lambda_{min}(A)$. Then

$$A_0 = A - \lambda_0 I$$

is a G-partial positive semidefinite matrix that, by Theorems 1 and 2, has a positive semidefinite completion B_0. The matrix

$$B_1 = B_0 + \lambda_0 I$$

is then a Hermitian completion of A satisfying (7). One may similarly construct a Hermitian completion B_2 satisfying (8) by using $-A$ in place of A.

Now suppose G is not chordal, and let A be any G-partial positive semidefinite matrix without positive definite completions; such a partial matrix exists by Theorem 2. Then, clearly, for every Hermitian completion B of A (or, equivalently, for the Hermitian part of an arbitrary completion) we have $\lambda_{min}(B) \leq 0 < \lambda_{min}(A)$; hence $F_1(A) \subsetneq F_2(A)$, which verifies the reverse implication. □

In proving Lemma 4 we showed, essentially, that for any nonchordal graph G there exists a partial Hermitian matrix A such that $\sup \lambda_{min}(B) < \lambda_{min}(A)$, where the supremum is taken with respect to all Hermitian completions B of A. It is not difficult to show that for any real number α the set of completions B satisfying $\lambda_{min}(B) \geq \alpha$ is compact (see, e.g., [4]). If, therefore, we restrict our attention to such a set we see, since the eigenvalues of a matrix are continuous functions of its entries, that λ_{min} actually attains a maximum; thus we are justified writing $\max \lambda_{min}(B) < \lambda_{min}(A)$. An interesting problem is then determining, for any partial Hermitian matrix, the maximum minimum eigenvalue over all its Hermitian completions. In the chordal case, as we have just proved, this value is simply $\lambda_{min}(A)$, but in the nonchordal case the question is open.

Proof of Theorem 3: If G is not chordal then by the Lemma there is a G-partial Hermitian matrix A such that $F_1(A) \subsetneq F_2(A)$.

Now suppose G is chordal, and choose $w \notin F_1(A)$. We will show that necessarily $w \notin F_2(A)$. First, consider the special case in which $\operatorname{Re} w > \operatorname{Re} z$ for all $z \in F_1(A)$, and let $x_0 = \max_{z \in F_1(A)} \operatorname{Re} z$. We claim that in fact $x_0 = \max_{z \in F_2(A)} \operatorname{Re} z$. Indeed, by the Lemma there exists a Hermitian completion $H_0 = [h^0_{ij}]$ of the G-partial Hermitian matrix $H(A_G)$ that satisfies

$$\lambda_{max}(H_0) = \lambda_{max}(H(A_G)) = x_0.$$

Let $B = [b_{ij}]$ be the completion of A formed by the following rule: If the (r,s) and (s,r) entries of A are both unspecified, then set $b_{rs} = h^0_{rs}$ and $b_{sr} = h^0_{sr}$; if the (s,r) entry is specified but the (r,s) entry is not, then set $b_{rs} = 2h^0_{rs} - \overline{a}_{sr}$. This insures that

$$H(B) = H_0,$$

which is verified by a simple calculation. We then have

$$\max_{z \in F(B)} \operatorname{Re} z = x_0 < \operatorname{Re} w,$$

whereby $w \notin F_2(A)$.

Now take any $w \notin F_1(A)$. Because $F_1(A)$ is convex, there is a suitable choice of θ so that $\operatorname{Re} e^{i\theta} w > \operatorname{Re} z$ for all $z \in F_1(e^{i\theta} A)$. The preceding argument shows that $e^{i\theta} w \notin F_2(e^{i\theta} A)$, so that $w \notin e^{-i\theta} F_2(e^{i\theta} A) = F_2(A)$. □

One may take an approach similar to that in the proof of Theorem 3 to show that, for any partial matrix A whose undirected graph is G, $F_2(A)$ depends only on A_G; that is,

$$F_2(A) = F_2(A_G).$$

We omit the details.

We have the following elementary consequences of Theorem 3.

COROLLARY 5. *Let G be an undirected graph. Then every partial matrix A whose undirected graph is G and for which $0 \notin F_1(A)$ has a completion B satisfying $0 \notin F(B)$ if and only if G is chordal.*

A matrix is *positive stable* if its spectrum lies in the open right half-plane of C, which we denote by RHP. By the spectral containment property (3), if the field of values of a matrix is contained in the open right half-plane then that matrix is necessarily positive stable.

COROLLARY 6. *Let G be an undirected graph. Every partial matrix A whose undirected graph is G and for which $F_1(A) \subseteq$ RHP has a completion B satisfying $F(B) \subseteq$ RHP if and only if G is chordal.*

Corollary 6 motivates the more general question of when, given a convex subset $\Omega \subseteq \mathbf{C}$, we may find a completion B of a given partial matrix whose field of values lies in Ω. In the Hermitian case this boils down to determining an interval $[a, b]$ for which there is a Hermitian completion B with $\lambda_{min}(B) \geq a$ and $\lambda_{max}(B) \leq b$. An open problem is determining the smallest such interval, or in other words minimizing $\lambda_{max} - \lambda_{min}$ over all Hermitian completions. The general question of when there is a positive stable completion seems also to be open.

In light of equations (7) and (8), then, it is natural to consider the general problem of determining a rectangle R in the complex plane and a completion B of a partial matrix A (not necessarily Hermitian) such that $F(B) \subseteq R$. Suppose that A is a combinatorially symmetric partial matrix, and define the Hermitian and skew-Hermitian partial matrices $H(A) = \frac{1}{2}(A + A^*)$ and $S(A) = \frac{1}{2}(A - A^*)$. If there is a completion H of $H(A)$ with $F(H) = [a, b]$ and a skew-Hermitian completion S of $S(A)$ with $F(S) = i[c, d]$, then the matrix $B = H + S$ is a completion of A with $\operatorname{Re} F(B) = [a, b]$ and $\operatorname{Im} F(B) = [c, d]$. That is, $F(B)$ lies within the closed rectangle $\{z : a \leq \operatorname{Re} z \leq b, c \leq \operatorname{Im} z \leq d\}$. Determining the smallest such rectangle is equivalent to determining the minimum difference of the largest and smallest eigenvalues in the Hermitian case. Now when A is not combinatorially symmetric we encounter a difficulty when trying to apply the preceding argument, due to the ambiguity of the notion of the Hermitian and skew-Hermitian parts of A; if we take these to mean $H(A_G)$ and $S(A_G)$, where G is the undirected graph of A, then of course the sum of arbitrary completions of of these partial matrices might not even be a completion of A. We must then complicate matters somewhat by considering only those pairs of completions H and S of $H(A_G)$ and $S(A_G)$ for which $h_{jk} + s_{jk} = a_{jk}$ whenever j and k are such that a_{jk} is defined but a_{kj} is not. One need only consider the two-by-two case to see that the combinatorially symmetric and unsymmetric "smallest rectangle" problems are different. In the symmetric case, the two off-diagonal entries are undefined, and the smallest rectangle is obtained by filling these entries with zeros; in the unsymmetric case, one off-diagonal entry is specified while the other is not, and it is unclear how, in general, to obtain the smallest rectangle.

REFERENCES

1. W. Barrett, C. R. Johnson and M. Lundquist, Determinantal formulae for matrix completions associated with chordal graphs, *Linear Algebra Appl.* **121** (1989), 265–289.

2. H. Dym and I. Gohberg, Extensions of band matrices with band inverses, *Linear Algebra Appl.* **36** (1981), 1–24.

3. R. Grone, C. R. Johnson, E. Marques de Sá and H. Wolkowicz, Positive definite completions of partial Hermitian matrices, *Linear Algebra Appl.* **58** (1984), 109–124.

4. R. Horn and C. R. Johnson, *Matrix Analysis*, Cambridge University Press, New York, 1985.

5. R. Horn and C. R. Johnson, *Topics in Matrix Analysis*, Cambridge University Press, to appear

6. C. R. Johnson, Numerical Ranges of principal submatrices, *Linear Algebra Appl.* **37** (1981), 23–34.

7. C. R. Johnson and L. Rodman, Inertia possibilities of completions of partial Hermitian matrices, *Linear and Multilinear Algebra* **16** (1984), 179–195.

8. C. R. Johnson and L. Rodman, Completions of partial matrices to contractions, *J. Functional Anal.* **69** (1986), 260–267.

9. H. Woerdemann, Minimal rank completions for partial block matrices, *Linear Algebra Appl.* **121** (1989), 105–122.

Charles R. Johnson
Department of Mathematics
College of William and Mary
Williamsburg, VA 23185

Michael Lundquist
Department of Mathematics
Brigham Young University
Provo, UT 84602

Operator Theory:
Advances and Applications, Vol. 50

ON THE RATIONAL HOMOGENEOUS MANIFOLD STRUCTURE OF THE SIMILARITY ORBITS OF JORDAN ELEMENTS IN OPERATOR ALGEBRAS

Kai Lorentz

Considering a topological algebra $\mathcal{B}$ with unit e, an open group of invertible elements $\mathcal{B}^{-1}$ and continuous inversion (e. g. $\mathcal{B}$ = Banach algebra, $\mathcal{B} = C^{\infty}(\Omega, M_n(\mathbb{C}))$ (Ω smooth manifold), $\mathcal{B}$ = special algebras of pseudo-differential operators), we are going to define the set of Jordan elements $\mathcal{J} \subset \mathcal{B}$ (such that $\mathcal{J}$ = Set of Jordan operators if $\mathcal{B} = L(H)$, H Hilbert space) and to construct rational local cross sections for the operation mapping

$$\mathcal{B}^{-1} \ni g \longmapsto gJg^{-1}$$

of $\mathcal{B}^{-1}$ on the similarity orbit $S(J) := \{gJg^{-1} : g \in \mathcal{B}^{-1}\}$, $J \in \mathcal{J}$.

We further show that $S(J)$ is a locally rational (esp. holomorphic or real analytic if $\mathcal{B}$ is Fréchet) manifold in the homogeneous topology. It turns out that this topology can be characterized as a special "gap"-topology in the case $\mathcal{B} = L(H)$.

1. INTRODUCTION

D. A. Herrero and L. A. Fialkow considered in 1977 the following situation: For a complex separable Hilbert space H they looked at the similarity orbit $S(T) := \{gTg^{-1} : g \in L(H)^{-1}\}$ of an operator T in the algebra $L(H)$ of bounded linear maps. They asked for which $T \in L(H)$ there exists a norm continuous local cross section for the natural operation mapping $\pi^T : L(H)^{-1} \longrightarrow S(T)$, $\pi^T(g) = gTg^{-1}$. The answer they gave in the following theorem (cf. [Apos, Th. 16.1] and [FiHe]):

THEOREM 1.1 *For $T \in L(H)$ there exists a norm continuous local cross section for π^T iff T is (similar to) a nice Jordan operator.*

Further, E. Andruchow and D. Stojanoff (cf. [AS 2]) considered the local structure of the similarity orbit of a nice Jordan operator, showing that the similarity orbit is in fact a holomorphic submanifold of $L(H)$.

We sharpen and generalize these results in two directions:

First: The manifolds considered by Herrero and Fialkow are even locally rational, this means in essential that the local cross sections are rational.

Second: Our proof is also valid for topological algebras with open group of invertible elements and continuous inversion. In connection with the theory of pseudo-differential operators, for this class of operator algebras a perturbation theory has been developed by B. Gramsch (see [Gr 1]). Especially the notion of Ψ^*-algebras fits many algebras of pseudo-differential operators as well as algebras connected with boundary value problems and algebras of C^∞-elements of C^*-dynamical systems (cf. [Bra], [Con], [Co 1], [Co 2], [Co 3], [Co S], [Gr 1, §5], [Schr], [Schu], [Ue], [Wag]) as well as the examples given in the abstract.

Our approach considers the similarity orbit as a manifold with the homogeneous topology induced by the group of all invertible operators and not as an embedded submanifold of the algebra. This has the advantage that we do not make use of the implicit function theorem, which is in general no longer valid in Fréchet algebras (cf. [Ha] and [Gr 1, §6]). In the Hilbert space case, we have by 1.1 the equivalence of these two topologies on the similarity orbit of a Jordan operator if and only if the Jordan operator is nice. The similarity orbit of a non-nice Jordan operator has no differentiable structure when it is considered in the norm-topology of the algebra (cf. [AS 2]). Our approach enables us to give to the similarity orbit of a general Jordan operator a very nice local structure in the topology inherited by the action of the group. Now the question arises how "far" from the norm-topology is the (clearly finer) homogeneous topology? As mentioned above there is no difference if the Jordan operator is nice but there has to be a difference if it is not nice. This note gives an answer to this question in functional-analytic terms. We show that the homogeneous topology on the similarity orbit of a general Jordan operator is always equivalent to a special kind of "gap"-topology (cf. Definition 3.2). To sum this up, we have in the Hilbert space case the following

THEOREM 1.2 *i) The similarity orbit of a Jordan operator is a locally $L(H)$-rational manifold in the homogeneous topology (for the definition of locally rational see section 1).*
ii) The similarity orbit of a nice Jordan operator, endowed with the norm topology of $L(H)$, is a locally $L(H)$-rational manifold.
iii) If there exist norm continuous local similarity cross sections for an operator $T \in L(H)$, then the similarity orbit of T is a locally $L(H)$-rational manifold in the topology of $L(H)$.
iv) If there exist norm continuous local similarity cross sections for an operator $T \in L(H)$, then there exists a rational (as a function of the group elements) local similarity cross section.

v) The similarity orbit of a Jordan operator is a locally $L(H)$-rational manifold with respect to the topology τ (for the definition of the topology τ see 3.2).

The result on the rational local structure of the similarity orbit and the existence of rational local cross sections for Jordan operators has been announced in [Lo 3] and a detailed proof is appearing in [Lo 2].

This note is organized as follows: In section 2 we give the definitions needed to state the result on the local structure of the similarity orbits and the local cross sections, namely 2.5, including a short sketch of the proof. In section 3 we specialize to the Hilbert space case and prove the above mentioned characterization of the homogeneous topology on the similarity orbit of a general Jordan operator as a "gap"-topology.

I want to thank Prof. B. Gramsch for suggesting this work and for many interesting and fruitful discussions on this subject as well as Prof. D. A. Herrero for focussing my attention on the papers [AS 1] and [AS 2].

The results are part of a dissertation project.

2. NOTATIONS

Throughout this note let $\mathcal{B}$ denote a topological algebra over $I\!K$ ($I\!K = \mathbb{C}$ or $I\!R$) with unit e and an open group $G := \mathcal{B}^{-1}$ of invertible elements and continuous inversion. First, we define the notion of Jordan elements in this setting.

DEFINITION 2.1 We call $J \in \mathcal{B}$ a Jordan element, if the following holds:

1. There exists a natural number n and non-zero projections $p^{(1)}, \dots, p^{(n)} \in \mathcal{B}$ such that:

   i) $p^{(1)} + \cdots + p^{(n)} = e$,

   ii) $p^{(i)}p^{(j)} = \delta_{ij}p^{(i)}$ $(i, j \in \{1, \dots, n\})$.

2. For every $j \in \{1, \dots, n\}$ there exists a $k_j \in I\!N$ and non-zero projections $p_1^{(j)}, \dots, p_{k_j}^{(j)} \in \mathcal{B}$ such that:

   i) $p_1^{(j)} + \cdots + p_{k_j}^{(j)} = p^{(j)}$,

   ii) $p_k^{(j)}p_l^{(j)} = \delta_{kl}p_k^{(j)}$ $(k, l \in \{1, \dots, k_j\})$.

3. For every $j \in \{1, \dots, n\}$ and $i \in \{1, \dots, k_j\}$ there exists an $n_i^{(j)} \in I\!N$ and non-zero projections $p_{i,1}^{(j)}, \dots, p_{i,n_i^{(j)}}^{(j)} \in \mathcal{B}$ such that:

   i) $1 \leq n_1^{(j)} < n_2^{(j)} < \cdots < n_{k_j}^{(j)}$

   ii) $p_{i,1}^{(j)} + \cdots + p_{i,n_i^{(j)}}^{(j)} = p_i^{(j)}$,

   iii) $p_{i,k}^{(j)}p_{i,l}^{(j)} = \delta_{kl}p_i^{(j)}$ $(k, l \in \{1, \dots, n_i^{(j)}\})$.

4. For every $j \in \{1, \dots, n\}$ and $i \in \{1, \dots, k_j\}$ there exist elements (matrix units in $C^\star$-algebras) $I_{r,s}^{j,i} \in p_{i,r}^{(j)}\mathcal{B}p_{i,s}^{(j)} \subset \mathcal{B}$ $(r, s \in \{1, \dots, n_i^{(j)}\})$ such that:

i) $I^{j,i}_{r,r} = p^{(j)}_{i,r}$,

ii) $I^{j,i}_{r,s} \circ I^{j,i}_{s,t} = I^{j,i}_{r,t}$ $(r, s, t \in \{1, \ldots, n^{(j)}_i\})$.

5. There exist n distinct complex numbers $\lambda_1, \ldots, \lambda_n$ such that the following relations hold:

i) $p^{(i)} J p^{(j)} = 0$ for $i, j \in \{1, \ldots, n\}$ such that $i \neq j$,

ii) $p^{(j)}_k J p^{(j)}_l = 0$ for $k \neq l$ $(j \in \{1, \ldots, n\}, k, l \in \{1, \ldots, k_j\})$,

iii) $p^{(j)}_{i,r} J p^{(j)}_{i,s} = 0$ for $r < s + 1$ and $r > s$ $(r, s \in \{1, \ldots, n^{(j)}_i\})$,

iv) $p^{(j)}_{i,r} J p^{(j)}_{i,r} = \lambda_j p^{(j)}_{i,r}$ $\forall r \in \{1, \ldots, n^{(j)}_i\}$,

v) $p^{(j)}_{i,r} J p^{(j)}_{i,r+1} = I^{j,i}_{r,r+1}$ $\forall r \in \{1, \ldots, n^{(j)}_i - 1\}$.

Let $\mathcal{J}$ denote the set of all Jordan elements in $\mathcal{B}$.

In the case $\mathcal{B} = L(H)$ (H Hilbert space), we call $J \in \mathcal{B}$ a nice Jordan element, if for every $j \in \{1, \ldots, n\}$ at most one of the projections $p^{(j)}_1, \ldots, p^{(j)}_{k_j}$ has infinite dimensional range.

REMARK 2.2 1. An equivalent Definition to 2.1 is the requirement that $J \in \mathcal{B}$ can be represented in the following ways:

$$\begin{aligned} J &= \sum_{j=1}^{n} \Big(\sum_{i=1}^{k_j} \Big(\sum_{r=1}^{n^{(j)}_i - 1} (\lambda_j p^{(j)}_{i,r} + I^{j,i}_{r,r+1}) \Big) + \lambda_j p^{(j)}_{i,n^{(j)}_i} \Big) \\ &= \sum_{j=1}^{n} \lambda_j p^{(j)} + \sum_{j=1}^{n} \sum_{i=1}^{k_j} \sum_{r=1}^{n^{(j)}_i - 1} I^{j,i}_{r,r+1}, \end{aligned}$$

where the projections $p^{(j)}_{i,r}$ and the $I^{j,i}_{r,s} \in \mathcal{B}$ satisfy the conditions 1. - 4. of 2.1.

2. $gJg^{-1} \in \mathcal{J}$ for $g \in \mathcal{B}^{-1}$, i. e. the group $\mathcal{B}^{-1}$ is acting on $\mathcal{J}$.

3. For $\mathcal{B} = L(H)$ (H a complex Hilbert space) the Jordan elements (resp. nice Jordan elements) of $\mathcal{B}$ are the operators similar to the Jordan operators (resp. nice Jordan operators) on H in the sense of [He].

DEFINITION 2.3 Let $J \in \mathcal{J}$ be a Jordan element. We define the similarity orbit of J by $S(J) := \{gJg^{-1} : g \in G\}$. The natural action of G on $S(J)$ will be denoted by π^J (i. e. $\pi^J(g) = gJg^{-1}$ for $g \in G$). We further call $A'(J) := \{b \in \mathcal{B} : bJ = Jb\}$ the commutant and $H_J := \{g \in G : gJg^{-1} = J\} = A'(J)^{-1} = (\pi^J)^{-1}(J)$ the stabilizer of $J \in \mathcal{J}$ with respect to the operation π^J. The set $S(J)$ will be endowed with the quotient topology of G/H_J. By a local cross section for π^J we mean a pair (U, s), such that $U \subset S(J)$ (U open) and $\pi^J \circ s = id_U$.

In order to state the Theorem on the local structure of the similarity orbit $S(J)$ of a Jordan element J, we need the following definition (cf. [Gr 1]):

DEFINITION 2.4 1. For a subset $X \neq \emptyset$ of $\mathcal{B}^n$ ($n \in I\!N$) and a set Ω of maps from X to $\mathcal{B}$ let

$$\Omega' := \{f + g, f \cdot g : f, g \in \Omega\} \cup \{f^{-1} : f \in \Omega \text{ s.t. } \forall x \in X \exists (f(x))^{-1} \in \mathcal{B}\}$$

(where $+$, $\cdot$, $^{-1}$ denotes pointwise addition, multiplication, and inversion (if possible) respectively within the ring $\mathcal{B}$).

2. Let X be as above, $pr_j (j = 1, \ldots, n)$ the projection of $\mathcal{B}^n$ onto the j-th component and $C_b : X \longrightarrow \mathcal{B}$ the constant map with value $b \in \mathcal{B}$. Further let $R_0 := \{pr_j : j = 1, \ldots, n\} \cup \{C_b : b \in \mathcal{B}\}$ and $R_{\nu+1} := R'_\nu$, $\nu = 0, 1, \ldots$. The set of $\mathcal{B}$-rational functions on X with values in $\mathcal{B}$ is:

$$\mathcal{B}r(X, \mathcal{B}) \; := \; \bigcup_{\nu=0}^{\infty} R_\nu .$$

The functions $f \in \mathcal{B}r(X, \mathcal{B})$ are continuous and even holomorphic if $\mathcal{B}$ is a locally convex space ([Gr 1, 1.7,1.8]). For $Y \subset \mathcal{B}^m$ ($m \in I\!N$) we define

$$\mathcal{B}r(X, Y) \; := \; \{f = (f_1, \ldots, f_m) : X \to Y : f_k \in \mathcal{B}r(X, \mathcal{B})\}.$$

We will call these maps $\mathcal{B}$-rational for short.

3. A topological space M will be called a locally $\mathcal{B}$-rational ($l\mathcal{B}r$ for short) manifold, if the following conditions hold:

0. There exist an open cover $\{U_\alpha : \alpha \in A\}$ of M; open subsets V_α of topological vector spaces $(T_\alpha, \tau(T_\alpha))$ and homeomorphisms $\phi_\alpha : U_\alpha \longrightarrow V_\alpha$ $(\alpha \in A)$.

1. T_α is a linear subspace of $\prod_\alpha := \mathcal{B}^{n_\alpha}$, $n_\alpha < \infty$, such that $\tau(T_\alpha)$ is finer than the topology of $\prod_\alpha$ on T_α.

2. All changes of coordinates

$$\phi_\alpha \phi_\beta^{-1} : \phi_\beta(U_\alpha \cap U_\beta) \longrightarrow \phi_\alpha(U_\alpha \cap U_\beta) \qquad \alpha, \beta \in A$$

are $\mathcal{B}$-rational maps.

Now we state the result on the local structure of the similarity orbits and the existence of local cross sections for Jordan elements:

THEOREM 2.5 *The homogeneous space $S(J)$ is a locally $\mathcal{B}$-rational manifold and there exist rational local cross sections which are continuous in the homogeneous topology.*

SKETCH OF THE PROOF: We give only a short sketch of the proof. The interested reader can find the details in [Lo 1] or [Lo 2].

Our proof of 2.5 consists of the following steps:

I) We determine the matrix structure of the group H_J and show the invertibility of all 'diagonal' matrix elements of $h \in H_J$.

II) We define a complementary subset $\mathcal{T}_J \subset \mathcal{B}$ (i. e. $H_J \cap \mathcal{T}_J = \{e\}$) such that:

III) The equation $g = ah$ (given $g \in G$, find $a \in \mathcal{T}_J$ and $h \in H_J$ such that $g = ah$) has a rational solution (that means $a = a(g)$ and $h = h(g)$ are rational expressions in g) for $g \in U_J$, U_J a neighborhood of the identity $e \in G$.

IV) The solutions of the above factorization equation $g = ah$ have the following uniqueness property: If an element $g \in \mathcal{B}$ can be written as $g = ah$ ($a \in \mathcal{T}_J$, $h \in H_J$), then a and h are rational expressions in g.

V) We define a topological vector space T_J and a homeomorphism γ_J between a neighborhood of $e \in \mathcal{T}_J$ and $0 \in T_J$.

Having done this, we proceed as follows:

By III) we have a rational factorization map $\zeta^J = (\tau^J, h^J) : U_J \longrightarrow \mathcal{T}_J \times H_J$, such that $\tau^J(g)h^J(g) = g \; \forall g \in U_J$. On the open neighborhood $\{gJg^{-1} : g \in U_J\} \subset S_G(J)$ we define the local cross section s_J by $s_J(gJg^{-1}) := \tau^J(g)$. From the uniqueness property IV) we see that this definition does not depend on the choice of g (even globally). By the rationality of the map τ^J we see that we have found a rational local cross section. As a chart around J we simply take $\gamma_J \circ s_J$. We have to check that this gives a homeomorphism (when restricting to a possible smaller neighborhood of $J \in S(J)$) and that the changes of coordinates are $\mathcal{B}$-rational maps. We are summing up the above construction in the following picture:

$$\begin{array}{ccc} G & \xrightarrow{\zeta^J=(\tau^J,h^J)} & \mathcal{T}_J \times H_J \\ \pi^J \downarrow\uparrow s_J & & \updownarrow \gamma_J \\ S(J) & & T_J \end{array}$$

(where the maps s_J and ζ^J are only defined locally).

3. THE HILBERT SPACE CASE

We let H be a complex, separable Hilbert space and $L(H)$ the algebra of bounded linear operators acting on H. Combining the results of C. Apostol, L. A. Fialkow, D. A. Herrero and D. Voiculescu with 2.5 we obtain the following (cf. 1.2,i)-iv)):

THEOREM 3.1 *i) The similarity orbit of a Jordan operator is a locally $L(H)$-rational manifold in the homogeneous topology.*

ii) The similarity orbit of a nice Jordan operator, endowed with the norm topology of $L(H)$, is a locally $L(H)$-rational manifold.

iii) If there exist norm continuous local similarity cross sections for an operator $T \in L(H)$, then the similarity orbit of T is a locally $L(H)$-rational manifold in the topology of $L(H)$.

iv) If there exist norm continuous local similarity cross sections for an operator $T \in L(H)$, then there exists a rational (as a function of the group elements) local similarity cross section.

PROOF: By [Apos, Th.16.1], we have for nice Jordan operators the equivalence of the quotient topology and the norm topology on the similarity orbit. 3.1 now follows from this equivalence, 1.1, 2.5 and our rational construction of the local similarity cross sections. □

CHARACTERIZATION OF THE HOMOGENEOUS TOPOLOGY ON $S(J)$:

In the last part of this note we want to give a functional analytic description of the homogeneous toplogy on $S(J)$, $J \in \mathcal{J}$, induced by the action of the group $L(H)^{-1}$ on $S(J)$. Our calculations will lead to a new way of defining cross sections on $S(J)$. Main parts of these observations can be found in [AS 1] and [AS 2]. First, we define a "gap"-topology on $S(J)$ by the following metric:

DEFINITION 3.2 Let $J = \sum_{j=1}^{n}\sum_{i=1}^{k_j}\sum_{r=1}^{n_i^{(j)}} \lambda_j p_{i,r}^{(j)} + \sum_{j=1}^{n}\sum_{i=1}^{k_j}\sum_{r=1}^{n_i^{(j)}-1} I_{r,r+1}^{j,i}$ be a Jordan operator in $L(H)$. For $S,T \in S(J)$ we define

$$d(S,T) := \|S-T\| + \sum_{j=1}^{n}\sum_{l=1}^{n_{k_j}-1} \|P_{\ker(\lambda_j I - S)^l} - P_{\ker(\lambda_j I - T)^l}\|,$$

where P_X denotes the orthogonal projection on X ($X \subset H$ closed subspace). Let τ denote the topology induced by the metric d.

THEOREM 3.3 *τ is equivalent to the homogeneous topology on $S(J)$.*

For the proof of 3.3 we need some preliminaries:

PROPOSITION 3.4 *Let $P \in L(H)$ be a projection, i. e. $P^2 = P$. Then the (uniquely determined) orthogonal projection $P_{R(P)}$ on $R(P)$ is given by the formulas*

$$P_{R(P)} = PP^*(I-(P-P^*)^2)^{-1} = (I-(P-P^*)^2)^{-1}PP^*.$$

PROOF: See [AS 1, (1.1)] or [Sa, 2.15]. □

REMARK 3.5 Let $\mathcal{P} := \{P \in L(H) : P^2 = P\}$ and $\mathcal{P}_\perp := \{P \in L(H) : P^2 = P = P^*\}$. From 3.4 we immediately see that the assignment $\mathcal{P} \ni P \longmapsto P_{R(P)} \in \mathcal{P}_\perp$ is continuous in $L(H)$.

As a consequence we can prove the (simpler) implication "the homogeneous topology is finer than τ" of 3.3:

It is clear that the assignment $L(H)^{-1} \ni g \longmapsto gJg^{-1} \in L(H)$ is continuous. Therefore it remains to show the continuity of

$$L(H)^{-1} \ni g \longmapsto P_{\ker(\lambda_j I - gJg^{-1})^l} \in L(H) \quad (j \in \{1,\ldots,n\}, l \in I\!N).$$

This follows from the observation that $\sum_{i=1}^{k_j}\sum_{r=1}^{\min(n_i^{(j)},l)} p_{i,r}^{(j)}$ is a projection on $\ker(\lambda_j I - J)^l$ as well as $\sum_{i=1}^{k_j}\sum_{r=1}^{\min(n_i^{(j)},l)} g p_{i,r}^{(j)} g^{-1}$ is a projection on $\ker(\lambda_j I - gJg^{-1})^l$ (which obviously depends continuously on $g \in L(H)^{-1}$). From 3.4 we get the desired implication. □

Now we turn our attention to the implication "τ is finer than the homogeneous topology":

REMARK 3.6 Let $J = \sum_{j=1}^{n}\sum_{i=1}^{k_j}\sum_{r=1}^{n_i^{(j)}} \lambda_j p_{i,r}^{(j)} + \sum_{j=1}^{n}\sum_{i=1}^{k_j}\sum_{r=1}^{n_i^{(j)}-1} I_{r,r+1}^{j,i}$ be a Jordan element in $L(H)$. For $T \in S(J)$ we denote by $Q_{\lambda_j}(T)$ the Riesz spectral projection associated to the spectral value λ_j of T (note that $\sigma(T) = \{\lambda_1, \ldots, \lambda_n\}$ for all $T \in S(J)$). From the functional calculus we know that $Q_{\lambda_j}(T)$ depends continuously (even holomorphically) on T. Since $\ker(\lambda_j I - T)^l$ is contained in $Q_{\lambda_j}(T)H$ ($l \in \mathbb{N}$) and $\ker(\lambda_j I - T)^l = Q_{\lambda_j}(T)H$ for $l \geq n_{k_j}^{(j)}$ we have the following algebraic identities:

1. $Q_{\lambda_j}(T) \cdot P_{\ker(\lambda_j I - T)^l} = P_{\ker(\lambda_j I - T)^l}$ for all $l \in \mathbb{N}$ and
2. $P_{\ker(\lambda_j I - T)^l} \cdot Q_{\lambda_j}(T) = Q_{\lambda_j}(T)$ for $l \geq n_{k_j}^{(j)}$.

From the first equation we see that $P_{\ker(\lambda_j I - T)^l} \cdot Q_{\lambda_j}(T)$ is always a projection (in fact it is the restriction of $P_{\ker(\lambda_j I - T)^l}$ on $Q_{\lambda_j}(T)H$).

DEFINITION 3.7 Let $J = \sum_{j=1}^n \sum_{i=1}^{k_j} \sum_{r=1}^{n_i^{(j)}} \lambda_j p_{i,r}^{(j)} + \sum_{j=1}^n \sum_{i=1}^{k_j} \sum_{r=1}^{n_i^{(j)}-1} I_{r,r+1}^{j,i}$ be a Jordan element in $L(H)$. Set $m_j := n_{k_j}^{(j)}$, $j \in \{1, \ldots, n\}$ (m_j is the order of nilpotency of $\lambda_j I - J$ on $p^{(j)}H = Q_{\lambda_j}(J)H$).
For $T \in S(J)$ we define:

$$\alpha(T) := \sum_{j=1}^{n} \sum_{i=1}^{m_j} (P_{\ker(\lambda_j I - T)^i} - P_{\ker(\lambda_j I - T)^{i-1}}) Q_{\lambda_j}(T) \cdot (P_{\ker(\lambda_j I - J)^i} - P_{\ker(\lambda_j I - J)^{i-1}}) Q_{\lambda_j}(J)$$

(see [AS 2, p. 14]).

REMARK 3.8 1. By the definition of τ we have the continuity of the map

$$(S(J), \tau) \ni T \longmapsto \alpha(T) \in L(H).$$

2.

$$\begin{aligned}
\alpha(J) &= \sum_{j=1}^{n} \sum_{i=1}^{m_j} [(P_{\ker(\lambda_j I - J)^i} - P_{\ker(\lambda_j I - J)^{i-1}}) \cdot Q_{\lambda_j}(J)]^2 \\
&= \sum_{j=1}^{n} [\sum_{i=1}^{m_j} (P_{\ker(\lambda_j I - J)^i} - P_{\ker(\lambda_j I - J)^{i-1}})] \cdot Q_{\lambda_j}(J) \\
&= \sum_{j=1}^{n} P_{\ker(\lambda_j I - J)^{m_j}} \cdot Q_{\lambda_j}(J) \\
&= \sum_{j=1}^{n} Q_{\lambda_j}(J) = I.
\end{aligned}$$

3. For $\nu \in \{1, \ldots, n\}$ and $l \in \mathbb{N}$ we have

$$\begin{aligned}
& P_{\ker(\lambda_\nu I - T)^l} \cdot Q_{\lambda_\nu}(T) \cdot \alpha(T) \\
= & \sum_{j=1}^{n} \sum_{i=1}^{m_j} P_{\ker(\lambda_\nu I - T)^l} \cdot Q_{\lambda_\nu}(T) \cdot (P_{\ker(\lambda_j I - T)^i} - P_{\ker(\lambda_j I - T)^{i-1}}) \cdot Q_{\lambda_j}(T) \cdot \\
& \cdot (P_{\ker(\lambda_j I - J)^i} - P_{\ker(\lambda_j I - J)^{i-1}}) \cdot Q_{\lambda_j}(J)
\end{aligned}$$

$$
\begin{aligned}
&= \sum_{j=1}^{n}\sum_{i=1}^{m_j} P_{\ker(\lambda_\nu I-T)^l}\cdot Q_{\lambda_\nu}(T)\cdot Q_{\lambda_j}(T)\cdot(P_{\ker(\lambda_j I-T)^i}-P_{\ker(\lambda_j I-T)^{i-1}})\cdot Q_{\lambda_j}(T)\cdot\\
&\quad\cdot(P_{\ker(\lambda_j I-J)^i}-P_{\ker(\lambda_j I-J)^{i-1}})\cdot Q_{\lambda_j}(J)\\
&= \sum_{i=1}^{m_\nu} P_{\ker(\lambda_\nu I-T)^l}\cdot Q_{\lambda_\nu}(T)\cdot(P_{\ker(\lambda_\nu I-T)^i}-P_{\ker(\lambda_\nu I-T)^{i-1}})\cdot Q_{\lambda_\nu}(T)\cdot\\
&\quad\cdot(P_{\ker(\lambda_\nu I-J)^i}-P_{\ker(\lambda_\nu I-J)^{i-1}})\cdot Q_{\lambda_\nu}(J)\\
&= \sum_{i=1}^{m_\nu} P_{\ker(\lambda_\nu I-T)^l}\cdot(P_{\ker(\lambda_\nu I-T)^i}-P_{\ker(\lambda_\nu I-T)^{i-1}})\cdot Q_{\lambda_\nu}(T)\cdot\\
&\quad\cdot(P_{\ker(\lambda_\nu I-J)^i}-P_{\ker(\lambda_\nu I-J)^{i-1}})\cdot Q_{\lambda_\nu}(J)\\
&= \sum_{i=1}^{\min\{m_\nu,l\}} (P_{\ker(\lambda_\nu I-T)^i}-P_{\ker(\lambda_\nu I-T)^{i-1}})\cdot Q_{\lambda_\nu}(T)\cdot\\
&\quad\cdot(P_{\ker(\lambda_\nu I-J)^i}-P_{\ker(\lambda_\nu I-J)^{i-1}})\cdot Q_{\lambda_\nu}(J)\\
&= \sum_{i=1}^{m_\nu}(P_{\ker(\lambda_\nu I-T)^i}-P_{\ker(\lambda_\nu I-T)^{i-1}})\cdot Q_{\lambda_\nu}(T)\cdot\\
&\quad\cdot(P_{\ker(\lambda_\nu I-J)^i}-P_{\ker(\lambda_\nu I-J)^{i-1}})\cdot P_{\ker(\lambda_\nu I-J)^l}\cdot Q_{\lambda_\nu}(J)\\
&= \sum_{i=1}^{m_\nu}(P_{\ker(\lambda_\nu I-T)^i}-P_{\ker(\lambda_\nu I-T)^{i-1}})\cdot Q_{\lambda_\nu}(T)\cdot\\
&\quad\cdot(P_{\ker(\lambda_\nu I-J)^i}-P_{\ker(\lambda_\nu I-J)^{i-1}})\cdot Q_{\lambda_\nu}(J)\cdot P_{\ker(\lambda_\nu I-J)^l}\cdot Q_{\lambda_\nu}(J)\\
&= \sum_{j=1}^{n}\sum_{i=1}^{m_j}(P_{\ker(\lambda_j I-T)^i}-P_{\ker(\lambda_j I-T)^{i-1}})\cdot Q_{\lambda_j}(T)\\
&\quad\cdot(P_{\ker(\lambda_j I-J)^i}-P_{\ker(\lambda_j I-J)^{i-1}})\cdot Q_{\lambda_j}(J)\cdot P_{\ker(\lambda_\nu I-J)^l}\cdot Q_{\lambda_\nu}(J)\\
&= \alpha(T)\cdot P_{\ker(\lambda_\nu I-J)^l}\cdot Q_{\lambda_\nu}(J).
\end{aligned}
$$

4. If in addition $\alpha(T)$ is invertible, we have

$$(\lambda_\nu I - S)^{n_i^{(\nu)}}\cdot p^{(\nu)}_{i,n_i^{(\nu)}} = 0 \text{ for } S=\alpha(T)^{-1}T\alpha(T)$$

$$(\nu\in\{1,\ldots,n\}, i\in\{1,\ldots,k_\nu\}).$$

PROOF: We have $p^{(\nu)}_{i,n_i^{(\nu)}} = P_{\ker(\lambda_\nu I-J)^{n_i^{(\nu)}}}\cdot p^{(\nu)}_{i,n_i^{(\nu)}}$ and $p^{(\nu)}_{i,n_i^{(\nu)}} = Q_{\lambda_\nu}(J)\cdot p^{(\nu)}_{i,n_i^{(\nu)}}$. Therefore

$$p^{(\nu)}_{i,n_i^{(\nu)}} = P_{\ker(\lambda_\nu I-J)^{n_i^{(\nu)}}}\cdot Q_{\lambda_\nu}(J)\cdot p^{(\nu)}_{i,n_i^{(\nu)}}.$$

From this equation and the above observation 3. we conclude

$$
\begin{aligned}
&(\lambda_\nu I-\alpha(T)^{-1}T\alpha(T))^{n_i^{(\nu)}}\cdot p^{(\nu)}_{i,n_i^{(\nu)}}\\
&= \alpha(T)^{-1}\cdot(\lambda_\nu I-T)^{n_i^{(\nu)}}\cdot\alpha(T)\cdot p^{(\nu)}_{i,n_i^{(\nu)}}\\
&= \alpha(T)^{-1}\cdot(\lambda_\nu I-T)^{n_i^{(\nu)}}\cdot\alpha(T)\cdot P_{\ker(\lambda_\nu I-J)^{n_i^{(\nu)}}}\cdot Q_{\lambda_\nu}(J)\cdot p^{(\nu)}_{i,n_i^{(\nu)}}\\
&= \alpha(T)^{-1}\cdot(\lambda_\nu I-T)^{n_i^{(\nu)}}\cdot P_{\ker(\lambda_\nu I-T)^{n_i^{(\nu)}}}\cdot Q_{\lambda_\nu}(T)\cdot\alpha(T)\cdot p^{(\nu)}_{i,n_i^{(\nu)}}\\
&= 0, \text{ since } (\lambda_\nu I-T)^{n_i^{(\nu)}}\cdot P_{\ker(\lambda_\nu I-T)^{n_i^{(\nu)}}} = 0.
\end{aligned}
$$

REMARK 3.9 We call an $A \in L(H)$ relatively invertible if there exists an $A^\sim \in L(H)$ such that

$$AA^\sim A = A \quad \text{and} \quad A^\sim AA^\sim = A^\sim.$$

$A^\sim$ is called a relative inverse of A. It is well known that $A \in L(H)$ is relatively invertible iff the range $R(A)$ is closed. It is easily seen that Jordan operators J are always relatively invertible and it is not hard to compute a canonical relative inverse of J by reflecting the entries of the matrix structure of J at the diagonal and inverting the non-zero diagonal entries of J. We will denote by $J^\sim$ this "canonical" relative inverse in the sequel. From this choice one can obtain the following identities (recall that also $(\lambda_j I - J)^l$ are Jordan operators for $j \in \{1, \ldots, n\}$ and $l \in I\!N$):

a) $(\lambda_j I - J) \cdot p_{i,r}^{(j)} = -I_{r-1,r}^{j,i} = p_{i,r-1}^{(j)} \cdot (\lambda_j I - J)\ \forall j, i\ \forall r \in \{2, \ldots, n_i^{(j)}\}$,

b) $p_{i,r}^{(j)} \cdot (\lambda_j I - J)^\sim = -I_{r,r-1}^{j,i} = (\lambda_j I - J)^\sim \cdot p_{i,r-1}^{(j)}\ \forall j, i\ \forall r \in \{2, \ldots, n_i^{(j)}\}$,

c) $p_{i,n_i^{(j)}-l}^{(j)} \cdot (\lambda_j I - J)^l((\lambda_j I - J)^l)^\sim = p_{i,n_i^{(j)}-l}^{(j)}\ \forall 0 \le l < n_i^{(j)}$,

d) $((\lambda_j I - J)^l)^\sim = ((\lambda_j I - J)^\sim)^l\ \forall l$.

DEFINITION 3.10 For $S \in S(J)$ we define

$$U(S) := \sum_{j=1}^{n} \sum_{i=1}^{k_j} \sum_{l=0}^{n_i^{(j)}-1} (\lambda_j I - S)^l \cdot p_{i,n_i^{(j)}}^{(j)} \cdot ((\lambda_j I - J)^\sim)^l.$$

PROPOSITION 3.11 *We have*

1. $U(J) = I$,
2. $S \cdot U(S) = U(S) \cdot J$ *for* $S \in S(J)$ *such that* $(\lambda_j I - S)^{n_i^{(j)}} \cdot p_{i,n_i^{(j)}}^{(j)} = 0$ $(j \in \{1, \ldots, n\})$.

PROOF: See also [AS 2, 3.3].

1. By definition $U(J) = \sum_{j=1}^{n} \sum_{i=1}^{k_j} \sum_{l=0}^{n_i^{(j)}-1} (\lambda_j I - J)^l \cdot p_{i,n_i^{(j)}}^{(j)} \cdot ((\lambda_j I - J)^\sim)^l$. Let $j \in \{1, \ldots, n\}$, $i \in \{1, \ldots, k_j\}$ and $l \in \{0, \ldots, n_i^{(j)} - 1\}$ be fixed.
   By the identities a), c) and d) of 3.9 we have

$$\begin{aligned} & (\lambda_j I - J)^l \cdot p_{i,n_i^{(j)}}^{(j)} \cdot ((\lambda_j I - J)^\sim)^l \\ = \ & p_{i,n_i^{(j)}-l}^{(j)} \cdot (\lambda_j I - J)^l((\lambda_j I - J)^\sim)^l \\ = \ & p_{i,n_i^{(j)}-l}^{(j)} \cdot (\lambda_j I - J)^l((\lambda_j I - J)^l)^\sim \\ = \ & p_{i,n_i^{(j)}-l}^{(j)}. \end{aligned}$$

   Therefore $U(J) = \sum_{j=1}^{n} \sum_{i=1}^{k_j} \sum_{l=0}^{n_i^{(j)}-1} p_{i,n_i^{(j)}-l}^{(j)} = I$.

2. Let $S \in S(J)$ such that $(\lambda_j I - S)^{n_i^{(j)}} \cdot p^{(j)}_{i,n_i^{(j)}} = 0$ for all $j \in \{1,\ldots,n\}$. For fixed $j \in \{1,\ldots,n\}$ we define $U_j(S) := \sum_{i=1}^{k_j} \sum_{l=0}^{n_i^{(j)}-1} (\lambda_j I - S)^l \cdot p^{(j)}_{i,n_i^{(j)}} \cdot ((\lambda_j I - J)^{\sim})^l$.
   As a first step we prove the relation

$$(\star) \qquad (\lambda_j I - S) \cdot U_j(S) = U_j(S) \cdot (\lambda_j I - J) \quad :$$

For the left-hand side of $(\star)$ we compute

$$\begin{aligned}
&(\lambda_j I - S) \cdot U_j(S) \\
= &\sum_{i=1}^{k_j} \sum_{l=0}^{n_i^{(j)}-1} (\lambda_j I - S)^{l+1} \cdot p^{(j)}_{i,n_i^{(j)}} \cdot ((\lambda_j I - J)^{\sim})^l \\
= &\sum_{i=1}^{k_j} \sum_{l=1}^{n_i^{(j)}} (\lambda_j I - S)^l \cdot p^{(j)}_{i,n_i^{(j)}} \cdot ((\lambda_j I - J)^{\sim})^{l-1} \\
= &\sum_{i=1}^{k_j} \sum_{l=1}^{n_i^{(j)}-1} (\lambda_j I - S)^l \cdot p^{(j)}_{i,n_i^{(j)}} \cdot ((\lambda_j I - J)^{\sim})^{l-1} \quad (\star\star)
\end{aligned}$$

(the last equality follows from the assumption $(\lambda_j I - S)^{n_i^{(j)}} \cdot p^{(j)}_{i,n_i^{(j)}} = 0$).
For the right-hand side of $(\star)$ we have

$$\begin{aligned}
&U_j(S) \cdot (\lambda_j I - J) \\
= &\sum_{i=1}^{k_j} \sum_{l=0}^{n_i^{(j)}-1} (\lambda_j I - S)^l \cdot p^{(j)}_{i,n_i^{(j)}} \cdot ((\lambda_j I - J)^{\sim})^l \cdot (\lambda_j I - J) \\
= &\sum_{i=1}^{k_j} \sum_{l=1}^{n_i^{(j)}-1} (\lambda_j I - S)^l \cdot p^{(j)}_{i,n_i^{(j)}} \cdot ((\lambda_j I - J)^{\sim})^l \cdot (\lambda_j I - J) \\
&(\text{since } p^{(j)}_{i,n_i^{(j)}} \cdot (\lambda_j I - J) = 0) \\
= &\sum_{i=1}^{k_j} \sum_{l=1}^{n_i^{(j)}-1} (\lambda_j I - S)^l \cdot p^{(j)}_{i,n_i^{(j)}} \cdot ((\lambda_j I - J)^{\sim})^{l-1} \cdot \\
&\cdot (\lambda_j I - J)^{\sim} (\lambda_j I - J)
\end{aligned}$$

From the identities a) - d) we obtain for $i \in \{1,\ldots,k_j\}$ and $l \in \{1,\ldots,n_i^{(j)}-1\}$:

$$\begin{aligned}
&p^{(j)}_{i,n_i^{(j)}} \cdot ((\lambda_j I - J)^{\sim})^{l-1} \cdot (\lambda_j I - J)^{\sim} (\lambda_j I - J) \\
= &((\lambda_j I - J)^{\sim})^{l-1} \cdot p^{(j)}_{i,n_i^{(j)}-l+1} \cdot (\lambda_j I - J)^{\sim} (\lambda_j I - J) \\
= &((\lambda_j I - J)^{\sim})^{l-1} \cdot p^{(j)}_{i,n_i^{(j)}-l+1} \cdot (I - \sum_{\gamma=1}^{k_j} p^{(j)}_{\gamma,1}) \\
= &((\lambda_j I - J)^{\sim})^{l-1} \cdot p^{(j)}_{i,n_i^{(j)}-l+1} - 0 \\
= &p^{(j)}_{i,n_i^{(j)}} \cdot ((\lambda_j I - J)^{\sim})^{l-1}
\end{aligned}$$

$$\Longrightarrow \quad U_j(S) \cdot (\lambda_j I - J) = (\star\star)$$
$$\Longrightarrow \quad \text{equation } (\star)$$

From $(\star)$ we conclude $S \cdot U_j(S) = U_j(S) \cdot J$ and since $U(S) = \sum_{j=1}^{n} U_j(S)$ we are done. □

SECOND CONSTRUCTION OF LOCAL CROSS SECTIONS AND FINAL PROOF OF 3.3:

DEFINITION 3.12 Let $V(J) \subset S(J)$ be an open neighborhood of J in $S(J)$ (with respect to τ) such that the following conditions are fulfilled for $T \in V(J)$:

1. $\alpha(T) \in L(H)^{-1}$
2. $U(\alpha(T)^{-1}T\alpha(T)) \in L(H)^{-1}$.

Define $\omega : V(J) \longrightarrow L(H)^{-1}$ by

$$\omega(T) := \alpha(T) \cdot U(\alpha(T)^{-1}T\alpha(T)) \in L(H)^{-1}.$$

REMARK 3.13 1. The map $\omega : (V(J), \tau) \longrightarrow L(H)^{-1}$ is continuous and satisfies $\omega(J) = I$.

2. For $T \in V(J)$ we have in view of 3.8 and 3.11

$$\begin{aligned} & \pi^J(\omega(T)) \\ = \; & \omega(T) \cdot J \cdot \omega(T)^{-1} \\ = \; & \alpha(T) \cdot U(\alpha(T)^{-1}T\alpha(T)) \cdot J \cdot \\ & \cdot U(\alpha(T)^{-1}T\alpha(T))^{-1} \cdot \alpha(T)^{-1} \\ = \; & \alpha(T) \cdot (\alpha(T)^{-1}T\alpha(T)) \cdot \alpha(T)^{-1} \\ = \; & T. \end{aligned}$$

This shows that $\omega : V(J) \longrightarrow L(H)^{-1}$ is a local cross section for π^J, which is continuous with respect to τ.

We therefore obtain the desired implication of 3.3.

From 3.3 and 3.1 we finally obtain (cf. 1.2, v)):

COROLLARY 3.14 The similarity orbit $S(J)$ of a Jordan operator J is a locally $L(H)$-rational manifold with respect to the topology τ.

References

[AFHS] Andruchow, E. ; Fialkow, L. A. ; Herrero, D. A. ; Pecuch-Herrero, M. B. ; Stojanoff, D. *Joint Similarity Orbits with Local Cross Sections*, Integral Equations and Operator Theory (to appear)

[Apos] Apostol, C.; Fialkow, L. A.; Herrero, D. A.; Voiculescu, D. *Approximation of Hilbert space Operators, volume II*, Pitman. Research Notes in Mathematics **102** (1984)

[AS 1] Andruchow, E. ; Stojanoff, D. *Nilpotent Operators and Systems of Projections*, J. Operator Theory **20** (1988), 359-374

[AS 2] Andruchow, E. ; Stojanoff, D. *Differentiable Structure of Similarity Orbits*, J. Operator Theory (to appear)

[Bra] Bratteli, O. *Derivations, Dissipations and Group Actions on $C^\star$-Algebras*, Springer LN **1229**, Berlin, Heidelberg, New York, London, Paris, Tokyo 1986

[Con] Connes, A. *$C^\star$-algèbres et géometrie differentielle*, C. R. Acad. Sci. Paris **290**, Ser. A, 599-604 (1980)

[Co 1] Cordes, H.O. *On Pseudodifferential Operators and Smoothness of Special Lie Group Representations*, Manuscripta math. **28**, 51-69 (1979)

[Co 2] Cordes, H.O. *On some $C^\star$-Algebras and Frechet*-Algebras of Pseudodifferential Operators*, AMS Proc. Symp. Pure Math. **43**,79-104 (1985)

[Co 3] Cordes, H.O. *Spectral Theory of Linear Partial Differential Operators and Comparison Algebras*, Cambridge University Press, London Math. Soc. Lecture Notes Series **76** (1987)

[Co S] Cordes, H.O.; Schrohe, E. *On the Symbol Homomorphism of a Certain Algebra of Singular Integral Operators*, Int. Eq. Op. Th. **8**, 641-649 (1985)

[CPR1] Corach, G. ; Porta, H. ; Recht, L. *The Geometry of Spaces of Projections in $C^\star$-Algebras*, Advances in Math. (to appear)

[CPR2] Corach, G. ; Porta, H. ; Recht, L. *Differential Geometry of Systems of Projectors in Banach Algebras*, Pac. J. Math. (to appear)

[Fi] Fialkow, L. A. *Similarity Cross Sections for Operators*, Indiana Univ. Math. J. **28**, 71-86 (1979)

[FiHe] Fialkow, L. A.; Herrero D. A. *Characterization of Operators with Local Similarity Cross Sections*, Notices Amer. Math. Soc. **232**, 205-220 (1977)

[Gr 1] Gramsch, B. *Relative Inversion in der Störungstheorie von Operatoren und Ψ-Algebren*, Math. Ann. **269**, 27-71 (1984)

[Gr 2] Gramsch, B. *Eine Klasse homogener Räume in der Operatorentheorie und Ψ-Algebren*, Funktionalanalysis Seminar. Universität Mainz, 114 S. (1981)

[Gr K] Gramsch, B.; Kalb, K. G. *Pseudo-locality and Hypoellipticity in Operator Algebras*, Semesterberichte Funktionalanalysis, 51-61, Tübingen 1985

[GrKa] Gramsch, B.; Kaballo, W. *Decompositions of Meromorphic Fredholm Resolvents and $\Psi^\star$-Algebras*, Integral Equations and Operator Theory **12**, 23-41 (1989)

[Ha] Hamilton, R. S. *The Inverse Function Theorem of Nash and Moser*, Bull. AMS **7**, 65-236 (1982)

[He] Herrero, D. A. *Approximation of Hilbert space Operators, volume I*, Pitman. Research Notes in Mathematics **72** (1982)

[He M] Herrero, M. P. *Global Cross Sections of Unitary and Similarity Orbits of Hilbert space Operators*, J. Operator Theory **12**, 265-283 (1984)

[Lo 1] Lorentz, K. *Lokale Struktur von Ähnlichkeitsbahnen im Hilbertraum* Diplomarbeit, Mainz 1987

[Lo 2] Lorentz, K. *On the Local Structure of the Similarity Orbits of Jordan Elements in Operator Algebras*, Annales Universitatis Saraviensis ,Series Mathematicae, Vol. 2, No. 3 (1989)

[Lo 3] Lorentz, K. *On the Structure of the Similarity Orbits of Jordan Operators as Analytic Homogeneous Manifols*, Integral Equations and Operator Theory **12**, 435-443 (1989)

[Rae] Raeburn, J. *The Relationship between a Commutative Banach Algebra and its Maximal Ideal Space*, J. Funct. Analysis **25**, 366-390 (1977)

[Sa] Salinas, N. *The Grassmann Manifold of a C^*-Algera and Hermitian Holomorphic Bundles*, Operator Theory: Advances and Applications, Vol. **28**, 1988 Birkhäuser Verlag Basel, 267-289.

[Schr] Schrohe, E. *The Symbols of an Algebra of Pseudodifferential Operators*, Pacific J. Math. **125**, 211-224 (1986)

[Schu] Schulze, B.-W. *Topologies and Invertibility in Operator Spaces with Symbolic Structure*, Proc. of the 9th Conf. on Problems and Methods in Math. Physics (9. TMP), Karl-Marx-Stadt, Teubner-Texte zur Mathematik, Bd. **111**, 257-270 (1989)

[Ue] Ueberberg, J. *Zur Spektralinvarianz von Algebren von Pseudodifferentialoperatoren in der L^p - Theorie*, Manuscripta Mathematica **61**, 459-475 (1988)

[Wag] Wagner, K. *Kommutatoren in der Theorie der Pseudodifferentialoperatoren mit Anwendungen auf die Submultiplikativität der Klassen $S^0_{1/2,1/2}$*, Diplomarbeit Mainz 1987

Fachbereich Mathematik

Johannes Gutenberg Universität

6500 Mainz

Fed. Rep. Germany

Operator Theory:
Advances and Applications, Vol. 50

PLANAR FACES OF OPERATOR SPACES IN L_p

E.M. Semenov and I.Ya. Shneiberg

The following statement was proved in [1]. Let $1<p<\infty$ and let M_p be the multiplier space in L_p spaces with respect to the exponent system. There exists $\varepsilon=\varepsilon(p)>0$ such that the inequality

$$\|\ldots\mu_{-1},0,\mu_1,\ldots\|_{M_p}\leq\varepsilon$$

implies

$$\|\ldots\mu_{-1},1,\mu_1,\ldots\|_{M_p}=1.$$

It is evident that $\varepsilon(2)=1$. This theorem has a geometrical interpretation. The unit sphere of the space M_p contains a planar face of codimension one in the point $e_0=(\ldots 0,1,0,\ldots)$, 1 in the 0–th position. It is clear that there exist such faces in the points $e_n=(\ldots 0,1,0,\ldots)$, 1 in the n–th position. There are no other planar faces on the unit sphere of the space M_p. This was proved by Y. Benjamini and P.–K. Lin [2].

The sequence e_0 generates the multiplier

$$S_1x(t)=\frac{1}{2\pi}\int_0^{2\pi}x(s)ds\ \mathbb{1}.$$

The following analogue of the Fefferman–Shapiro theorem was obtained in [3,4]. For given $p,q\in(1,\infty)$ a necessary and sufficient condition on the operator $T\in\mathcal{L}(L_q,L_p)$ was found in order that

$$\|S_1+\varepsilon T\|_{L_q\to L_p}=1$$

for sufficiently small ε. In this article we continue the investigation of this problem.

First of all we present some definitions and auxiliarly statements. We consider the spaces L_p on [0,1] with respect to Lebesgue measure. We denote the norm of an element x in L_p by $\|x\|_p$ and the norm of an operator T in $\mathcal{L}(L_q,L_p)$ by $\|T\|_{q,p}$. For a given positive function $a(t)$ we write

$$\|x\|_{p,a}=\Big(\int_0^1 |x(t)|^p a(t)dt\Big)^{1/p}.$$

LEMMA 1. *Let* $2\le p<\infty$. *There exist constants* $C_1(p)$, $C_2(p)$ *such that for* all *functions* $a,z\in L_p$ *satisfying the following condition*

$$\int_0^1 |a(t)|^{p-1}\mathrm{sign}\,a(t)z(t)dt=0 \tag{1}$$

the inequalities

$$\|a+z\|_p^p\le \|a\|_p^p+C_1(p)\big(\|z\|^2_{2,|a|^{p-2}}+\|z\|_p^p\big) \tag{2}$$

$$\|a+z\|_p^p\ge \|a\|_p^p+C_2(p)\big(\|z\|^2_{2,|a|^{p-2}}+\|z\|_p^p\big) \tag{3}$$

are satisfied.

PROOF. The following numerical inequality was proved in [1]. There exist constants $C_1(p)$ and $C_2(p)$ such that

$$1+pz+C_2(p)\big(|z|^2+|z|^p\big)\le|1+z|^p\le 1+pz+C_1(p)\big(|z|^2+|z|^p\big)$$

for all $z\in R^1$. Let $a\in R^1$, $a\ne0$. Substitute z by $\frac{z}{a}$ in these inequalities and multiply the resulting inequalities by $|a|^p$. Then we have

$$|a|^p+p|a|^{p-1}z\ \mathrm{sign}\,a+C_2(p)\big(|z|^2|a|^{p-2}+|z|^p\big)\le|a+z|^p\le$$

$$\le|a|^p+p|a|^{p-1}z\ \mathrm{sign}\,a+C_1(p)\big(|z|^2|a|^{p-2}+|z|^p\big).$$

By replacing formally the numbers a,z by the functions $a(t)$, $z(t)$, integrating the inequalities and using (1) we obtain (2) and (3).

The set of elements z satisfying condition (1) forms the orthogonal complement with respect to the norm of L_p. In this case

$$(4) \qquad \|a+\lambda z\|_p \geq \|a\|_p$$

for all $\lambda \in R^1$ (see [5], Ch. 2.). The converse statement is also true. If (4) holds for each $\lambda \in R^1$, then (1) is satisfied. If F_1, F_2 are closed subspaces of L_p, $L_p = F_1 \oplus F_2$ and (1) is satisfied for each $a \in F_1$, $z \in F_2$, then the subspace F_1 is called p–orthogonal to F_2. In that case we write $L_p = F_1 \oplus_p F_2$.

For given α, β, γ such that $\alpha, \beta, \gamma > 0$, $\alpha + \beta \geq 1$ the function

$$f(p) = (\alpha + \beta \gamma^p)^{1/p}$$

is decreasing ([6], 2.10). So

$$(5) \qquad (\alpha + \beta \gamma^p)^{\frac{1}{p}} \leq (\alpha + \beta \gamma^q)^{\frac{1}{q}}$$

if $q \leq p$.

For $S \in \mathcal{L}(L_q, L_p)$ and $\delta > 0$ we write

$$\Omega_\delta = \Omega_\delta(S) = \{a : a \in L_q, \|a\|_q = 1, \ \|Sa\|_p \geq \|S\|_{q,p} - \delta\}.$$

THEOREM 1. *Let* $2 \leq q \leq p < \infty$, $L_q = E_1 \oplus_q E_2$, $L_p = F_1 \oplus_p F_2$, $S \in \mathcal{L}(L_q, L_p)$, $SE_1 \subset F_1$, *and* $SE_2 = \{0\}$. *Then there exists* $\varepsilon > 0$ *such that for each operator* $T \in \mathcal{L}(L_q, L_p)$ *satisfying the conditions*

1) $TE_1 = \{0\}$, $TE_2 \subset F_2$
2) $\|T\|_{q,p} \leq \varepsilon$
3) *the estimate*

$$\|T\|_{L_{2,|a|^{q-2}} \to L_{2,|Sa|^{p-2}}} \leq \varepsilon$$

holds for some $\delta > 0$ *and for each* $a \in \Omega_\delta \cap E_1$

the following identity is valid:

$$\|S+T\|_{q,p} = \|S\|_{q,p}.$$

PROOF. Without loss of generality we may assume that $\|S\|_{q,p} = 1$. The inequality $\|S+T\|_{q,p} \geq 1$ is obvious. So we must prove only the inequality

$$\|(S+T)(a+z)\|_p = \|Sa+Tz\|_p \le \|a+z\|_q$$

for each $a \in E_1$, $\|a\|_q = 1$, $z \in E_2$.

By (2)

$$\|Sa+Tz\|_p \le \left[\|Sa\|_p^p + C_1(p)\left(\|Tz\|_{2,|Sa|^{p-2}} + \|Tz\|_p^p\right)\right]^{\frac{1}{p}}.$$

Now we assume that $a \in \Omega_\delta$, where δ is as in condition 3). Applying 2) and 3) and supposing that $\varepsilon < 1$, we obtain

$$\|Sa+Tz\|_p \le \left[1 + C_1(p)\left(\|Tz\|_{2,|Sa|^{p-2}} + \|Tz\|_p^p\right)\right]^{\frac{1}{p}} \le$$

$$\le \left[1 + C_1(p)\varepsilon^2\left(\|z\|^2_{2,|a|^{q-2}} + \|z\|_q^p\right)\right]^{\frac{1}{p}}.$$

Choose $\varepsilon > 0$ such that the following inequality

$$C_1(p)\varepsilon^2 \le C_2(q)$$

is fulfilled and apply (5). Then we have

$$\|Sa+Tz\|_p \le \left[1 + C_2(q)(\|z\|^2_{2,|a|^{q-2}} + \|z\|_q^p)\right]^{\frac{1}{p}} \le$$

$$\le \left[1 + C_2(q)\left(\|z\|^2_{2,|a|^{q-2}} + \|z\|_q^q\right)\right]^{\frac{1}{q}}.$$

By (3) we find

$$\|Sa+Tz\|_p \le \|a+z\|_q.$$

Now we consider the case $a \notin \Omega_\delta$. This means that $\|Sa\|_p \le 1-\delta$. If $\|z\|_q \le 2$ and $\|T\|_{q,p} \le \frac{\delta}{2}$ then, by (4) we get

$$\|Sa+Tz\|_p \le \|Sa\|_p + \|Tz\|_p \le 1 - \delta + 2\frac{\delta}{2} = 1 = \|a\|_q \le \|a+z\|_q.$$

If $\|T\|_{q,p} \leq \frac{\delta}{2}$ and $\|z\|_q \geq 2$ then

$$\|z\|_q \geq 2 = \frac{2-\delta}{1-\frac{\delta}{2}}.$$

Hence,

$$2-\delta \leq (1-\frac{\delta}{2})\|z\|_q$$

and

$$\|Sa+Tz\|_p \leq 1-\delta+\frac{\delta}{2}\|z\|_q \leq \|z\|_q - 1 = \|z\|_q - \|a\|_q \leq \|a+z\|_q.$$

Thus for

$$\varepsilon \leq \min\left(\frac{\delta}{2}, \left(\frac{C_2(q)}{C_1(p)}\right)^{\frac{1}{2}}\right),$$

we have

$$\|S+T\|_{p,q} = 1.$$

Denote by $\Omega_0 = \Omega_0(S)$ the set of elements $a \in L_q$, $\|a\|_q = 1$ on which the operator S attains its norm.

LEMMA 2. *Let* $1 < q, p < \infty$, $L_q = E_1 \oplus_q E_2$, $L_p = F_1 \oplus_p F_2$, $S, T \in \mathcal{L}(L_q, L_p)$, $SE_1 \subset F_1$, $SE_2 = \{0\}$ *and* $\|S+\varepsilon T\|_{q,p} = 1$ *for some* $\varepsilon \neq 0$.

1) *If* $a \in \Omega_0$ *then* $Ta = 0$.

2) Sa *and* Tz *are* p*-orthogonal for all* $a \in \Omega_0$, $z \in E_2$.

PROOF. Since $a \in \Omega_0$, we have

$$\|Sa \pm \varepsilon Ta\|_p \leq \|S \pm \varepsilon T\|_{q,p} = \|S\|_{q,p} = \|Sa\|_p.$$

The space L_p is strictly normed, so the last inequality is possible only for $Ta = 0$.

Let $a \in \Omega_0$, $z \in E_2$. Note that $\Omega_0 \subset E_1$. By [5, Ch. 2],

$$\|a + \alpha z\|_q - \|a\|_q = o(\alpha)$$

for $\alpha \to 0$. As $Ta = Sz = 0$,

$$\|Sa+\alpha\varepsilon Tz\|_p = \|(S+\alpha T)(a+\varepsilon z)\|_p \leq$$

$$\|S+\varepsilon T\|_{q,p}\cdot\|a+\alpha z\|_q = \|S\|_{q,p}\left(\|a\|_q+o(\alpha)\right) = \|Sa\|_p+o(\alpha)$$

for $\alpha\to 0$. Thus $S\Omega_0$ and TE_2 are p–orthogonal.

Lemma 2 says that condition 1) in Theorem 1 is close to being the necessary one. Consider from this point of view condition 3). For this it is necessary to make Lemma 1 more precise.

LEMMA 3. *Let $2\leq p<\infty$, $z\in L_p$, $\|a\|_p=1$, and let a and z be $p-$orthogonal. Then*

$$\|a+\lambda z\|_p-1=\frac{p-1}{2}\lambda^2\int_0^1 |a(t)|^{p-2}z^2(t)dt+o(\lambda^2)$$

for $\lambda\to 0$.

For the case $a(t)\equiv 1$ this lemma was proved in [3]. The proof for the general case is analogous. We omit it.

THEOREM 2. *Let $2\leq q\leq p<\infty$, $S,T\in\mathcal{L}(L_q,L_p)$, $L_q=E_1\oplus_q E_2$, $L_p=F_1\oplus_p F_2$, $SE_1\subset F_1$, $TE_2\subset F_2$, $SE_2=TE_1=\{0\}$, and $a\in\Omega_0$. If*

$$\|S+T\|_{q,p}=\|S\|_{q,p}=1,$$

then T may be extended to a bounded operator from $L_{2,|a|^{q-2}}$ into $L_{2,|Sa|^{p-2}}$ and

$$\|T\|_{L_{2,|a|^{q-2}}\to L_{2,|Sa|^{p-2}}}\leq\sqrt{\frac{q-1}{p-1}}. \tag{7}$$

PROOF. Since

$$\|S+\lambda T\|_{q,p}=1$$

for each $\lambda\in[0,1]$, the inequality

$$\|Sa+\lambda Tz\|_p-1\leq\|a+\lambda z\|_q-1$$

is valid for each $z\in E_2$. Applying Lemma 3 to both sides of this inequality, we have

$$\frac{p-1}{2}\int_0^1 |Sa(t)|^{p-2}|Tz(t)|^2dt \leq \frac{q-1}{2}\int_0^1 |a(t)|^{q-2}|z(t)|^2dt.$$

The estimate (7) follows from this inequality.

Let us consider the orthogonal projection

$$S_n x(t) = n \sum_{k=1}^{n} \int_{\frac{k-1}{n}}^{\frac{k}{n}} x(s)ds\, \mathcal{X}_{(\frac{k-1}{n},\frac{k}{n})}(t),$$

where n is a positive integer and $\mathcal{X}_e(t)$ is the characteristic function of a measurable set $e \subset [0,1]$. The operator S_n is bounded from L_q into L_p for all $q,p \in [1,\infty]$ and

$$\|S_n\|_{q,p} = \begin{cases} n^{\frac{1}{q}-\frac{1}{p}}, & q \leq p, \\ 1, & q \geq p. \end{cases}$$

For $k \in \{1,2,\ldots,n\}$, write

$$Q_k = \{x : x \in L_1,\ \text{supp}\ x \subset (\tfrac{k-1}{n},\tfrac{k}{n})\}$$

$$P = \Big\{x : x \in L_1, \int_0^{\frac{1}{n}} x(s)ds = \int_{\frac{1}{n}}^{\frac{2}{n}} x(s)ds = \ldots = \int_{1-\frac{1}{n}}^{1} x(s)ds = 0\Big\}.$$

COROLLARY 1. *Let $1 < q \leq p < 2$ or $2 < q \leq p < \infty$, and $T \in \mathcal{L}(L_q, L_p)$. The identity*

$$\|S_n + \varepsilon T\|_{q,p} = n^{\frac{1}{q}-\frac{1}{p}}$$

holds for sufficiently small ε if and only if the following conditions are met:

1) $T\mathcal{X}_{(\frac{k-1}{n},\frac{k}{n})} = 0,\ k = 1,\ldots,n$
2) $\text{Im}\ T \subset P$
3) $T \in \mathcal{L}(L_2, L_2)$
4) $T(L_q \cap Q_k) \subset Q_k$.

PROOF. For $n=1$ the statement was proved in [3]. Note that condition 4) is met in this case and may be omitted. Since the cases $1<q\leq p<2$ and $2<q\leq p<\infty$ are dual to each other we consider only the second case.
The necessity of condition 1) and 2) (resp. 3) and 4)) follows from Lemma 2 (resp. Theorem 2). Indeed, Ω_0 is contained the set of L_q-functions which are constant on each interval $(\frac{k-1}{n},\frac{k}{n})$, $1\leq k\leq n$. If $x\in L_q \cap Q_k$ we take

$$a(t)=n^{\frac{1}{q}}\, \mathcal{X}_{(\frac{j-1}{n},\frac{j}{n})}(t), \qquad (1\leq j\leq n, \quad j\neq k)$$

and apply inequality (7). In this way we obtain that $Tx\in Q_k$ and that T is a bounded operator in L_2. For the proof of the sufficiency we use Theorem 1. We take $E_1=F_1$ equal to the set of functions constant on each interval $(\frac{k-1}{n},\frac{k}{n})$, $1\leq k\leq n$ and $E_2(F_2)$ equal to the intersection of P and $L_q(L_p)$. It is easy to show that conditions 3), 4) from Corollary 1 and condition 3) from Theorem 1 are equivalent.

Generally speaking the concept of p-orthogonality depends on p. In Corollary 1 we find examples where pairs of elements are p-orthogonal for each $p\in(1,\infty)$.

Condition 3) in Theorem 1 is a very strong restriction on the operator T. There is an operator S such that only the zero operator satisfies the conditions of Theorem 1. This situation occurs when $q=p$ and S is the identity operator. This is not surprising. By [7] there exists for each $p\in[1,\infty]$, $p\neq 2$ a positive function $f=f_p$ such that

$$\|I+K\|_{p,p}\geq 1+f_p(\|K\|_{p,p})$$

for each compact operator in K in L_p.

Acknowledgement. The authors are grateful to Harm Bart and Philip Thijsse for their help in the preparation of this manuscript for publication.

REFERENCES

1. Fefferman, C. and H. Shapiro: A planar face on the unit sphere of the multiplier space M_p, $1<p<\infty$. Proc. Amer. Math. Soc., 36, (1972), 435–439.

2. Benyamini, Y. and Pei-Kee Lin: Norm one multipliers on $L^p(G)$.Arkiv for matematik, 24, (1986), 159–173.

3. Semenov, E.M. and I.Ya. Shneiberg: Geometrical properties of a unit sphere of the operator space L_p. The Gohberg Anniversary Collection. Vol. I. Operator Theory: Advances and Application. Vol. 41 (1988), 497–510.

4. Semenov, E.M. and I.Ya. Shneiberg: Hypercontracting operators and Khintchine's inequality. Funct. Anal. and its Appl. 22, (1988), 87–88 (Russian).

5. Diestel, J.: Geometry of Banach Spaces. Springer, Berlin–Heidelberg–New York (1975).

6. Hardy, G.H., J.E. Littlewood and G. Polya: Inequalities. Cambridge University Press, Cambridge (1934).

7. Benyamini, Y. and Pei-Kee Lin: An operator on L_p without best compact approximation. Israel J, Math. 51, (1985), 298–304.

E.N. Semenov
Department of Mathematics
Voronezh State University
Voronezh, 394693
U.S.S.R.

I.Ya Shneiberg
Voronezh Forest Technology Institute
Voronezh, 394043
U.S.S.R.

Operator Theory:
Advances and Applications, Vol. 50

SPECTRAL THEORY OF SELFADJOINT WIENER-HOPF OPERATORS WITH RATIONAL SYMBOLS

R. Vreugdenhil

Explicit formulas for the resolution of the identity of selfadjoint Wiener-Hopf operators with rational matrix symbol are constructed. The formulas are given in terms of a realization of the symbol.

0. INTRODUCTION

In this paper we study the Wiener-Hopf operator $K: L_2^m(\mathbb{R}_+) \to L_2^m(\mathbb{R}_+)$ defined by the formula

$$(Kf)(t) = \int_0^\infty k(t-s)f(s)\,ds, \quad 0 \leqq t < \infty.$$

Here the kernel k is an $m \times m$ matrix valued function with entries in $L_1(\mathbb{R})$. Throughout, we assume that the operator K is selfadjoint, and that its symbol

$$W(\lambda) = \int_{-\infty}^{\infty} k(t)\,e^{i\lambda t}\,dt, \quad -\infty < \lambda < \infty,$$

is a rational $m \times m$ matrix function. Since the symbol is rational, one can find (see, e.g., [Ka]) matrices A, B and C of sizes $l \times l$, $l \times m$ and $m \times l$, respectively, such that the matrix A has no real eigenvalue and

$$W(\lambda) = -C(\lambda - A)^{-1}B, \quad \lambda \in \mathbb{R}.$$

In what follows we shall assume that the size of A is as small as possible. In that case, because of the selfadjointness of K, the order l of A is even, say $l=2n$, and there exists (see, e.g., [R]) a unique invertible selfadjoint matrix $H: \mathbb{C}^{2n} \to \mathbb{C}^{2n}$ such that $HA = A^*H$, $HB = C^*$.

Our aim is to describe the resolution of the identity of K in terms of A, B, C and H. This requires an analysis of the Jordan structure of the matrices $A^\times(x) = A - \frac{1}{x}BC$, $x \neq 0$. The latter matrices are selfadjoint in the indefinite inner product defined by H, and the sign characteristic (see [GLR1]) of the pair $(A^\times(x), H)$ turns out to play an essential role. Before we state our main result, we introduce the subsets Θ and Σ:

$$\Theta = \{0\} \cup \{x \in \mathbb{R}\setminus\{0\} \mid A^\times(x) \textit{ has a real eigenvalue with a partial multiplicity greater than one}\}. \tag{0.1}$$

$$\mathbb{R}\setminus\Sigma = \{0 \neq 0 \mid \textit{all generalized eigenspaces of } A^\times(x) \textit{ corresponding to real eigenvalues are } H\textit{-definite}\}. \tag{0.2}$$

Here a subspace M is called H-definite if for every $m \in M$ $\langle Hm, m\rangle \geqq 0$ or $m \in M$ $\langle Hm, m\rangle \leqq 0$. The set Σ is finite, and $\Theta \subset \Sigma \subset \mathbb{R}$.

THEOREM. 0.1. *Let E be the resolution of the identity of the Wiener-Hopf operator K. If $[a, b]$ is a real interval that does not contain any point of Θ (see (0.1)), then for each $f \in (L_1^m \cap L_2^m)(\mathbb{R}_+)$*

$$(E([a, b])f)(t) = \int_0^\infty K_{a,b}(t, s) f(s)\, ds, \quad 0 \leqq t < \infty,$$

where

$$K_{a,b}(t, s) = -\frac{1}{2\pi} \int_a^b \frac{1}{x^2} B^* (e^{itA^\times(x)})^* (\Pi(x))^* H\, \Pi(x)\, e^{isA^\times(x)} B\, dx.$$

Here $A^\times(x) = A - \frac{1}{x}BC$, and the matrix $\Pi(x)$ is a projection, continuous on $\mathbb{R}\setminus\Theta$, and is defined as follows. For every $x \in \mathbb{R}\setminus\Sigma$ (see (0.2)) the kernel of $\Pi(x)$ is the *space spanned by all eigenvectors and generalized eigenvectors of the matrix A with respect to the eigenvalues in the open upper half plan, and the image of* $\Pi(x)$ *is the maximal* $A^{\times}(x)$*-invariant, H-nonpositive subspace* $M^{\times}(x)$ *of* $\mathbb{C}^{2n}$ *(i.e.,* $\langle Hm, m\rangle \leqq 0,\ m\in M^{\times}(x)$*), such that* $\sigma(A^{\times}(x)\,|_{M^{\times}(x)})$ *lies in the closed lower half plane.*

In the scalar case the sets Θ and Σ appearing in Theorem 0.1 are equal, and coinside with the set

$$\{0\}\cup\{x\in\mathbb{R} \mid \exists\mu\in\mathbb{R}: W(\mu)=x,\ \frac{d}{d\lambda}W(\lambda)\rfloor_{\lambda=\mu}=0\}.$$

We also prove that K may have an eigenvalue, but only at zero. This phenomenon does not occur in the scalar case (see [Ros1, RR, HW]). For the convolution operator acting on the space $L_2^m(\mathbb{R})$, corresponding to K, the existence of an eigenvalue at zero is equivalent to the condition that $\det W(\lambda)$ vanishes everywhere. For the Wiener-Hopf operator K this condition is only sufficient. In case $\det W(\lambda)$ is not identically zero we give a complete description for the kernel of K.

It is well known that a nontrivial Wiener-Hopf operator with a scalar symbol is absolutely continuous (see, e.g., [RR, Ros2]). For matrix-valued symbols this statement does not hold. However, using Theorem 0.1 we can prove that the operator K does not have a singular continuous component. For selfadjoint Toeplitz operators with a rational matrix symbol this has already been proved in [Rod].

For selfadjoint Toeplitz operators with scalar symbols (not necessary rational) a formula for the resolution of the identity is given in [Ros1] (see also [Ros3, RR]). These fomulas do not directly apply to matrix valued symbols, and are less explicit than the one given in Theorem 0.1.

The proof of Theorem 0.1 is based on a general formula for the resolution of the identity in terms of resolvents. To compute these resolvents we use the method for solving Wiener-Hopf equations appearing in [BGK2].

This paper consists of 6 sections. Some general facts about selfadjoint operators, such as the spectral theorem and a formula for the resolution of the identity, are gathered together in the first section. The first section also contains a survey of properties about indefinite scalar products spaces. The second section contains the spectral analysis of the symbol of K. Theorem 0.1 is proved in Section 3. In the latter section also some other spectral properties of K are derived. Differences between scalar and matrix case are discussed in the fourth section. In Section 5 we give an example, and in the last section we prove that K does not have a singular continuous component.

1. PRELIMINARIES

1.1. Selfadjoint Operators. In this section some, more or less well-known, results concerning selfadjoint operators on a Hilbert space are gathered together.

Let H be a complex Hilbert space, and let $T: H \to H$ be a bounded, selfadjoint operator. Put $m(T) = \inf_{\|x\|=1} \langle Tx, x\rangle$ and $M(T) = = \sup_{\|x\|=1} \langle Tx, x\rangle$.

A family $\{E(\lambda) \mid \lambda \in \mathbb{R}\}$ of orthogonal projections on H is called a *resolution of the identity for* T if

1. $E(\lambda) \leqq E(\mu)$, for $\lambda \leqq \mu$,
2. $\lim_{\lambda \downarrow \mu} E(\lambda)x = E(\mu)x$, for every $x \in H$,
3. $E(\lambda) = 0$, if $\lambda < m(T)$, $E(\lambda) = I$, if $\lambda \geqq M(T)$,
4. $T = \int_{\alpha}^{\beta} \lambda \, dE(\lambda)$ for any interval $[\alpha, \beta]$ such that $\alpha < m(T)$ and $\beta \geqq M(T)$.

Here the integral is a Riemann-Stieltjes integral. The spectral theorem says that each bounded selfadjoint operator T has a unique resolution of the identity corresponding to T (see, e.g., [BN, T]). For $\phi: [\alpha, \beta] \to \mathbb{C}$, continuous, with $\alpha < m(T)$ and $\beta \geqq M(T)$, define

$$\phi(T) = \int_{\alpha}^{\beta} \phi(\lambda) \, dE(\lambda). \tag{1.1}$$

The map $\phi \to \phi(T)$ has the usual properties of a functional calculus (see, e.g., [T]).

The set $P\sigma(T) = \{x \mid \text{Ker } (x - T) \neq (0)\}$ is called the *point spectrum* of T, and $C\sigma(T) = \{x \mid \text{Ker } (x - T) = 0, \text{ Im } (x - T) \neq \overline{\text{Im } (x - T)} = H\}$ is called the *continuous spectrum* of T. Since our operator T is selfadjoint, the spectrum $\sigma(T)$ is the union of the point and the continuous spectrum of T.

If λ_0 is a real number, then $\lambda_0 \in \varrho(T)$ if and only if $E(\lambda)$ is constant on a neighbourhood of λ_0. Furthermore, define $E_-(\lambda) = \lim_{\mu \uparrow \lambda} E(\mu)$. We have $\lambda \in P\sigma(T)$ if and only if $E(\lambda) \neq E_-(\lambda)$. The operator $E(\lambda) - E_-(\lambda)$ is the orthogonal projection on Ker $(\lambda - T)$.

Next, for intervals on the real line we define

$$E((a,b]) = E(b) - E(a), \quad E([a,b]) = E(b) - E_-(a),$$

$$E([a,b)) = E_-(b) - E_-(a), \quad E((a,b)) = E_-(b) - E(a),$$

and $E(\{a\}) = E(a) - E_-(a)$.

Theorem X 6.1 in [DS] gives a formula for the resolution of the identity on open intervals (a, b). In case a and b are not in the point spectrum of T the following somewhat stronger result holds.

THEOREM 1.1. *Let T be a bounded selfadjoint operator on a Hilbert space H, and let E be its corresponding resolution of the identity. Then for all $a, b \in \mathbb{R} \setminus P\sigma(T)$ $(a<b)$ we have that*

$$E([a, b]) = \lim_{y \downarrow 0} \frac{1}{2\pi i} \int_a^b [(\bar{z} - T)^{-1} - (z - T)^{-1}]\, dx, \tag{1.2}$$

in the strong operator topology, where $z = x + iy$ with $x, y \in \mathbb{R}$ and $y > 0$.

Proof. Let $z = x + iy$ with $x, y \in \mathbb{R}$ and $y > 0$. Take $a < b$ in $\mathbb{R} \setminus P\sigma(T)$. From the proof of Theorem X 6.1 in [DS] it follows that

$$\frac{1}{2\pi i} \int_a^b [(\bar{z} - T)^{-1} - (z - T)^{-1}]\, dx = \int_\alpha^\beta \frac{1}{\pi} [\arctan \frac{b - \mu}{y} - \arctan \frac{a - \mu}{y}]\, dE(\mu).$$

It is clear that

$$\frac{1}{\pi} \left| \arctan \frac{b - \mu}{y} - \arctan \frac{a - \mu}{y} \right| < 1, \quad a \leqq \mu \leqq b$$

and

$$\lim_{y\downarrow 0}\frac{1}{\pi}[\arctan\frac{b-\mu}{y}-\arctan\frac{a-\mu}{y}] = \begin{cases} 0, & \mu<a, \\ \frac{1}{2}, & \mu=a, \\ 1, & a<\mu<b, \\ \frac{1}{2}, & \mu=b, \\ 0, & \mu>b. \end{cases} \tag{1.3}$$

Let $h \in H$. Applying the dominated convergence theorem for vector valued measures (cf.[DS], Theorem IV 10.10) we get that the right-hand side of (1.2) exists, and is equal to

$$\int_{\alpha}^{\beta}\lim_{y\downarrow 0}\frac{1}{\pi}[\arctan\frac{b-\mu}{y}-\arctan\frac{a-\mu}{y}]dE(\mu)\,h.$$

Because E({a}) = 0 ($a \in C\sigma(T)\cup\varrho(T)$) and using (1.3), this integral is equal to

$$\int_{a}^{b}dE(\mu)\,h = E(b)\,h - E(a)\,h = E((a,b])\,h = E([a,b])\,h. \qquad \square$$

1.2. Indefinite scalar products. We introduce a matrix $H\colon \mathbb{C}^p \rightarrow \mathbb{C}^p$, which is invertible and selfadjoint. The indefinite scalar product $[\cdot,\cdot]$ associated with H is defined by

$$[x,y] = \langle Hx, y\rangle, \quad x,y \in \mathbb{C}^p,$$

where $H\colon \mathbb{C}^p \rightarrow \mathbb{C}^p$ is an invertible and selfadjoint matrix. We call the $p \times p$ matrix A *H-selfadjoint* if $HA = A^*H$ or, equivalentely, if A is selfadjoint with respect to $[\cdot,\cdot]$.

A subspace M of $\mathbb{C}^p$ will be called *H-nonnegative* (resp. *H-nonpositive, H-neutral, H-positive, H-negative*) if $\langle Hu, u\rangle \geqq 0$ (resp. $\langle Hu, u\rangle \leqq 0$, $\langle Hu, u\rangle = 0$, $\langle Hu, u\rangle > 0$ $(u \neq 0)$, $\langle Hu, u\rangle < 0$ $(u \neq 0)$) for every $u \in M$. We say that M is a *maximal H-nonnegative* subspace if there is no H-nonnegative subspace which contains M as a proper subspace. In a similar way one defines *maximal H-nonpositive* subspaces. A subspace M is called *hypermaximal H-neutral* whenever it is both maximal H-nonnegative and H-nonpositive. In this case

$HM = M^{\perp}$ (see, e.g., [GLR1]).

Given an arbitrary matrix T, we write M_+ for the linear span of all eigenvectors and generalized eigenvectors of T corresponding to the eigenvalues λ of T with Im $\lambda>0$. Similarly, M_- denotes the linear span of all eigenvectors and generalized eigenvectors of T corresponding to the eigenvalues λ of T with Im $\lambda<0$. Suppose that T has no real eigenvalue, and is a *H-dissipative* matrix, i.e., Im $[Tu, u] \geqq 0$ for all $u \in \mathbb{C}^p$, then the space M_+ (resp. M_-) is maximal H-nonnegative (resp. H-nonpositive) (see, e.g., Theorem 11.6 in [IKL]).

Let T be a H-selfadjoint matrix, then the pair (T, H) is *unitarily similar* to a pair (J, R) (i.e., for some invertible matrix S the equalities $T = S^{-1}JS$ and $H = S^*RS$ hold), where J is a Jordan matrix,

$$J = \operatorname{diag}\{J_1, \dots, J_q, J_{q+1}, \dots, J_{q+2r}\},$$

with $J_1, \dots, J_q$ elementary Jordan blocks with real eigenvalues, $J_{q+1}, \dots, J_{q+2r}$ elementary Jordan blocks with non-real eigenvalues, and $J_{q+2j} = \overline{J}_{q+2j-1}, j = 1, 2,\dots, r$ (the bar denotes the *complex conjugate* in every entry of the matrix), and where

$$R = \operatorname{diag}\left\{\epsilon_1 R_1, \dots, \epsilon_q R_q, \begin{pmatrix} 0 & R_{q+1} \\ R_{q+2} & 0 \end{pmatrix}, \dots, \begin{pmatrix} 0 & R_{q+2r-1} \\ R_{q+2r} & 0 \end{pmatrix}\right\}, \tag{1.4}$$

with

$$R_j = \begin{bmatrix} & & 1 \\ & \cdot^{\cdot^{\cdot}} & \\ 1 & & \end{bmatrix}, \quad j=1,\dots, q+2r,$$

the size of R_j coincides with the size of J_j (in particular $R_{q+2j-1} = R_{q+2j}, j=1,\dots,r$), and $\epsilon_j = \pm 1$, for $j = 1,\dots, q$. The pair (J, R) is uniquely determined (some trivial reorderings excluded) by (T, H), and is called the *canonical form* of (T, H). The set of signs $\epsilon_1, \dots, \epsilon_q$, which appear in equation (1.4), one sign for each Jordan block of J with a real eigenvalue, is called the *sign characteristic* of the pair (T, H). Thus, the sign characteristic of (T, H) prescribes a sign ± 1 to each partial multiplicity of the matrix T

corresponding to a real eigenvalue (see, e.g., [GLR1]).

Later, in Section 2, it will be convenient to use a slightly different ordering of the Jordan blocks in J, in fact, we shall replace J by:

$$J = \begin{pmatrix} J_+ & 0 & 0 \\ 0 & J_+ & 0 \\ 0 & 0 & J_- \end{pmatrix},$$

where

$$J_+ = \operatorname{diag}\{J_{q+1}, J_{q+3}, \ldots, J_{q+2r-1}\},$$

$$J_- = \operatorname{diag}\{J_{q+2}, J_{q+4}, \ldots, J_{q+2r}\},$$

$$J_0 = \operatorname{diag}\{J_1, J_2, \ldots, J_q\},$$

such that all eigenvalues of J_+ lie in the open upper half plane, and all eigenvalues of J_- lie in the open lower half plane. In this case, we have,

$$R = \begin{pmatrix} 0 & 0 & \tilde{R} \\ 0 & R_0 & 0 \\ \hat{R} & 0 & 0 \end{pmatrix},$$

where

$$\tilde{R} = \operatorname{diag}\{R_{q+1}, R_{q+3}, \ldots, R_{q+2r-1}\},$$

$$\hat{R} = \operatorname{diag}\{R_{q+2}, R_{q+4}, \ldots, R_{q+2r}\},$$

$$R_0 = \operatorname{diag}\{\epsilon_1 R_1, \ldots, \epsilon_q R_q\}.$$

2. SPECTRAL ANALYSIS OF SELFADJOINT RATIONAL SYMBOLS

2.1. Selfadjoint rational symbols. Throughout this section $W(\lambda)$ is an $m \times m$ matrix-valued function, which has no pole on the real line and is selfadjoint. The latter means $W(\lambda) = W(\bar{\lambda})^*$. We also assume that $W(\lambda)$ is given in the following form

$$W(\lambda) = -C(\lambda - A)^{-1}B, \quad \lambda \in \mathbb{R}, \tag{2.1}$$

where A is a $l \times l$ matrix, B is $l \times m$ and C is $m \times l$, such that A has no real eigenvalue. The representation (2.1) implies that $W(\cdot)$ is a rational matrix function whose entries vanish at infinity. It is well known (see, e.g., [Ka]) that any such function admits a representation of type (2.1). We refer to (2.1) as a *realization* of W.

We also assume the realization in (2.1) is *minimal*, which means that the order l of the matrix A is as small as possible. This is equivalent to saying that pair (A, B) is *controllable* (i.e., Im $(A\,|B) = \bigvee_{j=0}^{l-1} \operatorname{Im} A^jB = \mathbb{C}^l$) and the pair (C, A) is *observable* (i.e., Ker $(C\,|A) = \bigcap_{j=0}^{l-1} \operatorname{Ker} CA^j = (0)$) (see, e.g., [BGK2]).

Because the realization is minimal and selfadjoint, there exists a unique invertible selfadjoint operator $H\colon \mathbb{C}^l \to \mathbb{C}^l$ such that

$$HA = A^*H, \quad HB = C^*. \tag{2.2}$$

(see, e.g., [R]). Using equation (2.2) we obtain that $\lambda - A = H^{-1}(\bar{\lambda} - A)^*H$, and therefore $\lambda \in \sigma(A)$ if and only if $\bar{\lambda} \in \sigma(A)$. Moreover, the sizes of the Jordan blocks with eigenvalue λ are equal to sizes of the Jordan blocks with eigenvalue $\bar{\lambda}$ (see, e.g., Proposition 2.4 of part I in [GLR1]). Hence, l is an even number, and so we can write $2n$ instead of l. Note that $\det(\lambda - A) > 0$ for all $\lambda \in \mathbb{R}$.

Let P be the *Riesz projection* corresponding to the part of $\sigma(A)$ lying in the open upper half plane. Formula (2.2) implies

$$HP = (I - P^*)H. \tag{2.3}$$

From this identity it is easy to derive the next proposition (see [R]).

PROPOSITION 2.1. *The spaces* Im P *and* Ker P *are A-invariant, hypermaximal, H-neutral subspaces.*

2.2. The spectral projection $\mathbf{P}^\times(\mathbf{z})$. We continue to use the notation introduced in the previous section. We denote by Π_+ (resp. Π_-) the *open upper half plane* (resp. *open lower half plane*), and throughout this section $z = x + iy$ is a complex number, where $x, y \in \mathbb{R}$ and $x \neq 0$. For any complex number $z \neq 0$, we set $A^{\times}(z) = A - \frac{1}{z}BC$, and $P^{\times}(z)$ denotes the Riesz projection of $A^{\times}(z)$ corresponding to the part of $\sigma(A^{\times}(z))$ in Π_+.

Note, that $\lambda - A^{\times}(z) = H^{-1}(\bar{\lambda} - A^{\times}(\bar{z}))^* H$. Thus $\lambda \in \sigma(A^{\times}(z))$ if and only if $\bar{\lambda} \in \sigma(A^{\times}(\bar{z}))$. From (2.2) it follows that

$$HP^{\times}(z) = (I - P^{\times}(\bar{z}))^* H, \quad z \in \Pi_+. \tag{2.4}$$

PROPOSITION 2.2. *Put*

$$W(z,\lambda) = zI_m - W(\lambda) = zI_m + C(\lambda - A)^{-1}B, \quad \lambda \in \mathbb{R}. \tag{2.5}$$

Then the following hold.

(i) *For every* $\lambda \in \varrho(A)$ *and* $z \in \mathbb{C}\setminus\{0\}$ *we have*

$$\det W(z,\lambda) = z^m \frac{\det(\lambda - A^{\times}(z))}{\det(\lambda - A)}. \tag{2.6}$$

(ii) *For every* $z \in \mathbb{C}\setminus\mathbb{R}$ *the matrix* $A^{\times}(z)$ *has no real eigenvalue.*

(iii) *The matrix* $A^{\times}(z)$ *is H-dissipative for* $z \in \Pi_+$.

(iv) Let $z \in \Pi_+$. *The subspace* Im $P^{\times}(z)$ *is* $A^{\times}(z)$*-invariant, maximal H-nonnegative, and the subspace* Ker $P^{\times}(z)$ *is* $A^{\times}(z)$*-invariant, maximal H-nonpositive.*

Proof. (i) Let $z \in \mathbb{C}\setminus\{0\}$ and $\lambda \in \varrho(A)$, then

$$\det W(z,\lambda) = \det(z + C(\lambda - A)^{-1}B =$$

$$= z^m\, det(I + \frac{1}{z}C(\lambda - A)^{-1}B) = = z^m\, det(I + \frac{1}{z}BC(\lambda - A)^{-1}) =$$

$$= \frac{z^m \det(\lambda - A^{\times}(z))}{det(\lambda - A)}.$$

(ii) Let $z \in \mathbb{C}\setminus\mathbb{R}$. Because $W(\lambda)$ is selfadjoint for every $\lambda \in \mathbb{R}$ we have that $-z \in \varrho(W(\lambda))$ for all real λ. In other words, $W(z,\lambda)$ is invertible for any λ in $\mathbb{R}$. So, by using equation (2.6), we get that $\sigma(A^{\times}(z)) \cap \mathbb{R} = \varnothing$.

(iii) Let v be a vector in $\mathbb{C}^{2n}$ and $z \in \Pi_+$, then

$$[(A^{\times}(z)-A^{\times}(\bar{z}))v, v] = \langle H(A^{\times}(z)-A^{\times}(\bar{z}))v, v\rangle =$$

$$= (\frac{1}{\bar{z}}-\frac{1}{z})\langle HBCv, v\rangle = 2i \operatorname{Im} \frac{1}{\bar{z}} \|Cv\|^2.$$

Hence, (using $HA^{\times}(z) = A^{\times}(\bar{z})^* H$)

$$\operatorname{Im}[A^{\times}(z)v, v] = \frac{1}{2i}\{[A^{\times}(z)v, v]-[v, A^{\times}(z)v]\} =$$

$$=\frac{1}{2i}[(A^{\times}(z)-A^{\times}(\bar{z}))v, v] = \operatorname{Im} \frac{1}{\bar{z}} \|Cv\|^2 \geqq 0.$$

Thus the matrix $A^{\times}(z)$ is H-dissipative.

(iv) As $P^{\times}(z)$ is the Riesz projection of $A^{\times}(z)$ corresponding to its eigenvalues in Π_+ both $\operatorname{Im} P^{\times}(z)$ and $\operatorname{Ker} P^{\times}(z)$ are $A^{\times}(z)$-invariant. Furthermore, since $A^{\times}(z)$ has no real eigenvalue, the desired statement follows directly from the remark made the third paragraph of Subsection 1.2. □

For every $\lambda \in \mathbb{R}$ we may write

$$W(\lambda) = U(\lambda)^* \operatorname{diag}[d_1(\lambda),\ldots, d_m(\lambda)]U(\lambda),$$

where $U(\lambda)$ is a matrix-valued function, which is unitary for every λ in $\mathbb{R}_{\infty}$, $U(\lambda)$ is analytic on $\mathbb{R}_{\infty}$, and $d_1(\lambda),\ldots, d_m(\lambda)$ are real, analytic, scalar functions, for each $\lambda \in \mathbb{R}_{\infty}$ (see, e.g., Theorem S6.3 in [GLR2]). Moreover, $d_j(\infty) = 0$ $(j = 1,\ldots, m)$. From the above identity it follows that the following numbers are well-defined.

$$\begin{aligned} m_1 &= \min\{x \in \mathbb{R} \mid \det W(x, \lambda) = 0 \textit{ for some } \lambda \textit{ in } \mathbb{R}_{\infty}\}, \\ m_2 &= \max\{x \in \mathbb{R} \mid \det W(x, \lambda) = 0 \textit{ for some } \lambda \textit{ in } \mathbb{R}_{\infty}\}, \end{aligned} \tag{2.7}$$

Clearly, $-\infty < m_1 \leqq 0 \leqq m_2 < \infty$. Note that formula (2.6) implies that for $x \neq 0$ the matrix $A^{\times}(x)$ has a real eigenvalue if and only if $x \in [m_1, m_2]$.

PROPOSITION 2.3. (i) *The projection $P^{\times}(z)$ is a continuous function of z on $\mathbb{C}\setminus[m_1, m_2]$.*

(ii) *For every $z \in \mathbb{C}\setminus[m_1, m_2]$ we have*

$$\text{Im } P \oplus \text{Ker } P^{\times}(z) = \text{Ker } P \oplus \text{Im } P^{\times}(z) = \mathbb{C}^{2n}.$$

(iii) *The projection* $\Pi(z)$ *of* $\mathbb{C}^{2n}$ *onto* Ker $P^{\times}(z)$ *along* Im P *is also continuous on* $\mathbb{C}\setminus[m_1, m_2]$.

Proof. Property (i) is trivial, because the matrix $A^{\times}(z)$ has no real eigenvalue for $z \neq [m_1, m_2]$. Statement (ii) follows immediately from Theorem 3.3 in [R], Proposition 2.1 and Proposition 2.2 (iv), and (iii) is a consequence of (i) and (ii). □

LEMMA 2.4. *Let* $z \in \mathbb{C}\setminus[m_1, m_2]$. *Define* M *and* $N(z)$ *to be the* $2n \times n$ *matrices such that the columns of* M *form a basis for the space* Im P *and the columns of* $N(z)$ *form a basis for the space* Ker $P^{\times}(z)$. *Note, that the matrices* M *and* $N(z)$ *are injective. Let* $\Pi(z)$ *be the projection as defined in Proposition 2.3 (iii). Then the following conditions hold.*

$$(i) \quad \det(M^* HN(z)) \neq 0,$$

$$(ii) \quad \Pi(z) = N(z)(M^* HN(z))^{-1} M^* H,$$

$$(iii) \quad \Pi(z) = I_{2n} - H^{-1}\Pi(\bar{z})^* H,$$

$$(iv) \quad \Pi(z) = \begin{bmatrix} N(z) & \mathbf{0} \end{bmatrix} \begin{bmatrix} N(z) & M \end{bmatrix}^{-1},$$

$$(v) \quad \Pi(z) = (I - P - P^{\times}(z))^{-1}(I - P) =$$

$$= (I - P^{\times}(z)) + (I - P^{\times}(z))(I - P - P^{\times}(z))^{-1} P^{\times}(z).$$

$$(vi) \quad H(\Pi(z) - \Pi(\bar{z})) = (\Pi(z))^* H\Pi(z).$$

Proof. (i) Assume that $\det(M^* HN(z)) = 0$. Then there exists a nonzero vector $u \in \mathbb{C}^n$ such that

$$M^* HN(z)u = 0.$$

So, for all $v \in \mathbb{C}^n$

$$\langle HN(z)u, Mv \rangle = \langle M^* HN(z)u, v \rangle = 0.$$

Hence, (by using Proposition 2.1) we obtain that

$$N(z)u \in \operatorname{Im} P \bigcap \operatorname{Ker} P^{\times}(z).$$

Since $\operatorname{Im} P \oplus \operatorname{Ker} P^{\times}(z) = \mathbb{C}^{2n}$ (see Proposition 2.3) and $N(z)$ is an injective operator, we obtain that $u=0$, which contradicts our assumption.

(ii) From (i) we see that the right-hand side (which we shall denote as $\Xi(z)$) is well defined, and it is clearly a projection.

Let $v_1, \ldots, v_n$ be the columns of the matrix M, and $w_1, \ldots, w_n$ the columns of $N(z)$, then we get (by applying Proposition 2.1) that

$$v_j^* H v_i = 0. \quad (i,j=1,\ldots,n)$$

Hence, $\Xi(z)v_i=0$ $(i=1,\ldots,n)$. Thus, $\Xi(z)v=0$ for all $v \in \operatorname{Im} P$. Further, let $e_j=[0,\ldots,0,1,0,\ldots,0]^T$ $(j=1,\ldots,n)$, be the j-th standard basis vector of $\mathbb{C}^{2n}$. Then

$$\Xi(z)w_j = N(z)(M^*HN(z))^{-1}M^*HN(z)e_j = N(z)e_j = w_j. \quad (j=1,\ldots,n)$$

Hence, $\Xi(z)w=w$ for every $w \in \operatorname{Ker} P^{\times}(z)$. Thus, $\Xi(z) = \Pi(z)$ is the projection of $\mathbb{C}^{2n}$ onto $\operatorname{Ker} P^{\times}(z)$ along $\operatorname{Im} P$.

(iii) The following equations hold

$$H[\operatorname{Ker} \Pi(z)] = [\operatorname{Ker} \Pi(z)]^{\perp} = [\operatorname{Ker} \Pi(\bar{z})]^{\perp} = H[\operatorname{Ker} \Pi(\bar{z})],$$

$$H[\operatorname{Im} \Pi(z)] = [\operatorname{Im} \Pi(\bar{z})]^{\perp} = \operatorname{Ker}(\Pi(\bar{z})^*),$$

$$H[\operatorname{Im} \Pi(\bar{z})] = [\operatorname{Im} \Pi(z)]^{\perp}.$$

Thus, for any vector $u \in \mathbb{C}^{2n}$ there are vectors $u_1 \in \operatorname{Im} \Pi(z)$ and $u_2 \in \operatorname{Ker} \Pi(z)$ uniquely determined such that $u=u_1+u_2$. Now we get that

$$H\Pi(z)u = Hu_1 = (I-\Pi(\bar{z}))^*)Hu.$$

(iv) Let $u = \begin{bmatrix} u_1 \\ u_2 \end{bmatrix}$, such that $u_j \in \mathbb{C}^n$, $j=1,2$. Assume that

$\begin{bmatrix} N(z) & M \end{bmatrix} u = (0)$. Then $N(z)u_1 = -Mu_2 \in \operatorname{Im} P \cap \operatorname{Ker} P^{\times}(z) = (0)$. Since the matrices M and $N(z)$ are injective, we get that u is the zero vector. Hence, the matrix $\begin{bmatrix} N(z) & M \end{bmatrix}$ is invertible.

Proposition 2.1 implies that $M^* H M = 0$, and therefore an easy calculation shows that

$$\Pi(z) \begin{bmatrix} N(z) & M \end{bmatrix} = \begin{bmatrix} N(z) & \mathbf{0} \end{bmatrix}.$$

Hence,

$$\Pi(z) = \begin{bmatrix} N(z) & \mathbf{0} \end{bmatrix} \begin{bmatrix} N(z) & M \end{bmatrix}^{-1}.$$

(v) Let $(I - P - P^{\times}(z))u = 0$, then $Pu = (I - P^{\times}(z))u$ and $(I-P)u = P^{\times}(z)u$. Using Proposition 2.3 it follows that $u = 0$. Hence, the matrix $I - P - P^{\times}(z)$ is invertible. Next, observe that

$$(I - P - P^{\times}(z))^{-1}(I-P) \begin{bmatrix} N(z) & M \end{bmatrix}^{-1} = \begin{bmatrix} N(z) & \mathbf{0} \end{bmatrix}.$$

Now (iv) implies that

$$\Pi(z) = (I - P - P^{\times}(z))^{-1}(I-P).$$

The second identity follows from the fact that $P^{\times}(z)(z)\Pi(z) = 0$.

(vi) From (ii) it follows that $\Pi(\bar{z})\Pi(z) = \Pi(\bar{z})$. Hence, by using (iii) the desired statement follows. □

2.3. The projection Q(x). One of the results of Section 2.2 is that the projections $P^{\times}(z)$ and $\Pi(z)$ are continuous on $\mathbb{C}\setminus[m_1, m_2]$ (see Proposition 2.3). In this section we discuss what happens if we take the limit of these projections to the interval $[m_1, m_2]$.

Throughout this section, let $x \neq 0$ be a real number, and $z = x + iy$ $(y > 0)$. We define (in case it exists) the projection

$$Q(x) = \lim_{y \downarrow 0} P^{\times}(z).$$

We shall also introduce the projection

$$\tilde{Q}(x) = \lim_{y \downarrow 0} P^{\times}(\bar{z}),$$

which will exist if $Q(x)$ exists. Moreover, in that case it follows from (2.4) that $HQ(x) = (I - \tilde{Q}(x))^{*} H$. For $x \in \mathbb{R} \backslash [m_1, m_2]$ (with m_1, m_2 defined by (2.7)), the projection $Q(x)$ exists, and is equal to $P^{\times}(x)$. In this case $\tilde{Q}(x) = Q(x)$.

For $x \neq 0$ we define:

$$\sigma(x) = \{\lambda \in \sigma(A^{\times}(x)) \cap \mathbb{R} \mid \exists \textit{ a sequence } z_j = x + iy_j, y_j > 0, j = 1,2,\ldots,$$

$$y_j \downarrow 0, \textit{ and } \exists \lambda(z_j) \in \sigma(A^{\times}(z_j)) \cap \Pi_+, \mu(z_j) \in \sigma(A^{\times}(z_j)) \cap \Pi_- \textit{ such that}$$

$$\lambda = \lim_{j \to \infty} \lambda(z_j) = \lim_{y \to \infty} \mu(z_j) \}.$$

THEOREM 2.5. *Put* $\Sigma = \{x \in \mathbb{R} \backslash \{0\} \mid \sigma(x) \neq \varnothing\}$. *Then* Σ *is a finite set in* $[m_1, m_2]$. *Furthermore, if* $z = x + iy$, *with* $y > 0$ *and* $0 \neq x \in \mathbb{R} \backslash \Sigma$, *then the limit*

$$Q(x) = \lim_{y \downarrow 0} P^{\times}(z), \tag{2.8}$$

exists, and $\operatorname{Im} P \oplus \operatorname{Ker} Q(x) = \mathbb{C}^{2n}$.

Proof. Obviously, the matrix function $\mathscr{A}(z) = zA^{\times}(z) = zA - BC$, $z \in \mathbb{C}$, is analytic on $\mathbb{C}$. From Theorem V 7.1 in [B] we know that there exists an at most countable set Ω in $\mathbb{C}$, with ∞ as the only possible limit point, such that $\mathscr{A}(z)$ has fixed Jordan structure on $\mathbb{C} \backslash \Omega$ (i.e., the number of different eigenvalues of $\mathscr{A}(z)$ is independent of $z \in \mathbb{C} \backslash \Omega$, and for every pair $z_1, z_2 \in \mathbb{C} \backslash \Omega$ the different eigenvalues $\lambda_1(z_1), \ldots, \lambda_r(z_1)$ and $\lambda_1(z_2), \ldots, \lambda_r(z_2)$ of $\mathscr{A}(z_1)$ and $\mathscr{A}(z_2)$, respectively, can be enumerated so that the partial multiplicities of $\mathscr{A}$ at $\lambda_j(z_1)$ coincide with the partial multiplicities of $\mathscr{A}$ at $\lambda_j(z_2)$, for $j = 1, \ldots, r$). In particular, $\Omega \cap [m_1, m_2]$ is a finite set. The eigenvalues of the matrix $A^{\times}(z)$ are continuous functions on $\mathbb{C} \backslash \{0\}$ (see, e.g., [B, GLR3]). Hence, for $\lambda(x) \in \sigma(A^{\times}(x))$ there are $\lambda(z) \in \sigma(A^{\times}(z))$ $(z \in \Pi_+)$ such that $\lambda(z) \to \lambda(x)$ $(y \downarrow 0)$. Moreover, the set $\{\lambda(z) \mid y > 0\}$ lies completely in either Π_+ or Π_-. Since the matrix $\mathscr{A}(z)$ has the same Jordan structure as the matrix $A^{\times}(z)$, for any $z \in \mathbb{C} \backslash \{0\}$, it follows that $\Sigma \subset \Omega \cap [m_1, m_2]$. Hence, Σ is a finite set.

Since $\sigma(x) = \emptyset$, we can take a contour $\Gamma(x)$ around all eigenvalues $\lambda(x)$ of $A^{\times}(x)$ such that there are $\lambda(z) \in \sigma(A^{\times}(z)) \cap \Pi_+$ with $\lim_{y \downarrow 0} \lambda(z) = \lambda(x)$, and all other eigenvalues of $A^{\times}(x)$ lie outside $\Gamma(x)$. Now, for any $y > 0$ small enough we have that $P^{\times}(z) = \int_{\Gamma(x)} (\lambda - A^{\times}(z))^{-1} d\lambda$. Hence, it is obvious that the limit in (2.8) exists.

From Proposition 2.2 (iv) and equation (2.8) we obtain that Im $Q(x)$ is an $A^{\times}(x)$-invariant, maximal H-nonnegative subspace, and the space Ker $Q(x)$ is $A^{\times}(x)$-invariant, maximal H-nonpositive. Hence, the matching follows by using Theorem 3.3 in [R] and Proposition 2.1. □

THEOREM 2.6. *Put* $\Sigma = \{x \in \mathbb{R} \setminus \{0\} \mid \sigma(x) \neq \emptyset\}$, *and take* $0 \neq x \in \mathbb{R} \setminus \Sigma$. *Let* λ *be a real eigenvalue of* $A^{\times}(x)$, *and let* ϵ *be any sign in the sign characteristic of* $(A^{\times}(x), H)$ *corresponding to* λ. *Then, all partial multiplicities of* λ *are equal to one, and*

$$\epsilon = \begin{cases} 1, & \lambda \in \sigma(A^{\times}(x)|_{\text{Im } Q^{\times}(x)}), \\ -1, & \lambda \in \sigma(A^{\times}(x)|_{\text{Ker } Q^{\times}(x)}). \end{cases}$$

Furthermore, put

$$P(\lambda) = \frac{1}{2\pi i} \int_{\gamma} (\mu - A^{\times}(x))^{-1} d\mu,$$

where γ *is a circle with centre* λ *such that all other eigenvalues of* $A^{\times}(x)$ *lie outside* γ, *then* Im $P(\lambda)$ *is a* H*-positive subspace if* $\lambda \in \sigma(A^{\times}(x)|_{\text{Im } Q^{\times}(x)})$, *and* Im $P(\lambda)$ *is a* H*-negative subspace if* $\lambda \in \sigma(A^{\times}(x)|_{\text{Ker } Q^{\times}(x)})$.

Proof. Let $\lambda(x)$ be a real eigenvalue of the matrix $A^{\times}(x)$ and assume that $\lambda(x)$ has a partial multiplicity k greater than one. Since the matrix $A^{\times}(x)$ is H-selfadjoint there exists an invertible matrix $S(x)$ such that

$$A^{\times}(x) = S(x)^{-1} J(x) S(x),$$

$$H = S(x)^{*} R(x) S(x),$$

where the pair $(J(x), R(x))$ is the canonical form of the pair $(A^{\times}(x), H)$ (see, e.g., [GLR1]). Without loss of generality we can write

$$J(x) = \begin{bmatrix} J_1(x) & 0 \\ 0 & * \end{bmatrix}, \; R(x) = \begin{bmatrix} R_1(x) & 0 \\ 0 & * \end{bmatrix}, \; S(x)^{-1} = \begin{bmatrix} v_1 \ldots v_k & * \end{bmatrix},$$

where

$$J_1(x) = \begin{bmatrix} \lambda(x) & 1 & & 0 \\ & \ddots & \ddots & \\ & & \ddots & 1 \\ 0 & & & \lambda(x) \end{bmatrix}, \; R_1(x) = \begin{bmatrix} & & & \epsilon \\ & 0 & \cdot & \\ & \cdot & 0 & \\ \epsilon & & & \end{bmatrix}, \quad \epsilon = \pm 1,$$

both of size k, and the vectors $v_1, \ldots, v_k$ form a Jordan chain of $A^{\times}(x)$ corresponding to the eigenvalue $\lambda(x)$. Note that the vectors $v_1, \ldots, v_k$ all are elements of Im $Q(x)$ or they are element of Ker $Q(x)$. Now, obviously, we have

$$\langle H(v_1+v_k), v_1+v_k \rangle = 2\epsilon, \quad \langle H(v_1-v_k), v_1-v_k \rangle = -2\epsilon,$$

which is a contradiction with the definiteness of the spaces Im $Q(x)$ and Ker $Q(x)$. Hence, all partial multiplicities of the real eigenvalues of $A^{\times}(x)$ are one. Note that this implies directly the two other statements. □

REMARK. *We have the following alternative description for the set* Σ*:*

$$\mathbb{R}\backslash\Sigma = \{x \neq 0 \mid \textit{all generalized eigenspaces of } A^{\times}(x)$$
$$\textit{corresponding to real eigenvalues are } H\textit{-definite}\}.$$

2.4. The subset Θ.

Define the subset

$$\Theta = \{0\} \cup \{x \in \mathbb{R}\backslash\{0\} \mid A^{\times}(x) \textit{ has a real eigenvalue}$$
$$\textit{with a partial multiplicity greater than one}\}. \tag{2.10}$$

Note that Theorem 2.6 implies that $\Theta\backslash\{0\} \subset \Sigma$. Now we state two important theorems, which we will prove in the next subsection.

THEOREM 2.7. *Let $z = x + iy$ where x and y are real numbers. If $x \notin \Theta$, then the limit $Q^{\times}(x) = \lim_{y \downarrow 0} P^{\times}(z)$ exists.*

THEOREM 2.8. *Let $x \in \mathbb{R} \setminus \Theta$, then the following properties hold. First, we have the matching*

$$\operatorname{Im} P \oplus \operatorname{Ker} Q^{\times}(x) = \mathbb{C}^{2n}.$$

Next, the point x has a neighbourhood $\mathcal{B}$ such that

$$A^{\times}(z) = S(z)^{-1} \begin{pmatrix} J_1(z) & 0 & 0 & 0 \\ 0 & J_2(z) & 0 & 0 \\ 0 & 0 & J_3(z) & 0 \\ 0 & 0 & 0 & J_4(z) \end{pmatrix} S(z), \quad z \in \mathcal{B} \cap \overline{\Pi}_+, \tag{2.11}$$

where $S(z)$ and the Jordan matrices $J_1(z),\ldots,J_4(z)$ can be chosen such that the following holds. The size $n_j \times n_j$ of each $J_j(z)$ is independent of the choice of z, $n_1 = n_4$, $n_2 = n_3$, the matrices $J_2(z)$ and $J_3(z)$ are diagonal matrices and their elements are continuous on $\mathcal{B} \cap \overline{\Pi}_+$. Furthermore, for $z \in \mathcal{B} \cap \overline{\Pi}_+$, we have:

$$\sigma(J_j(z)) \subset \begin{cases} \Pi_+, & j=1, \text{ and } j=2 \text{ if } z \in \mathcal{B} \cap \Pi_+, \\ \Pi_-, & j=4, \text{ and } j=3 \text{ if } z \in \mathcal{B} \cap \Pi_+, \\ \mathbb{R}, & j=2,\,3 \text{ if } z \in \mathcal{B} \cap \mathbb{R}, \end{cases} \tag{2.12}$$

$\sigma(J_1(z)) \cap \sigma(J_2(z)) = \varnothing$, and $\sigma(J_3(z)) \cap \sigma(J_4(z)) = \varnothing$. Moreover, if

$$S(z)^{-1} = \begin{pmatrix} S_1(z) \ldots S_4(z) \end{pmatrix}, \quad S(z) = \begin{pmatrix} T_1(z) \\ \vdots \\ T_4(z) \end{pmatrix}, \tag{2.13}$$

are the partitionings of $S(z)$ and $S(z)^{-1}$ corresponding to the partitioning of the Jordan matrix, then the submatrices $\begin{pmatrix} S_2(z) & S_3(z) \end{pmatrix}$ and $\begin{pmatrix} T_2(z) \\ T_3(z) \end{pmatrix}$ are continuous on $\mathcal{B} \cap \overline{\Pi}_+$. For every $z = x \in \mathcal{B} \cap \mathbb{R}$ we have:

$$H = S(x)^{*} R(x) S(x),$$

where the pair $(J(x), R(x))$ is the canonical form of $(A^{\times}(x), H)$. The matrix $R(x)$

is given by:

$$R(x) = \begin{bmatrix} 0 & 0 & 0 & R_4(x) \\ 0 & R_2(x) & 0 & 0 \\ 0 & 0 & R_3(x) & 0 \\ R_1(x) & 0 & 0 & 0 \end{bmatrix},$$

where $R_1(x) = R_4^*(x)$, $R_2(x) \equiv I_2$, *and* $R_3(x) \equiv -I_3$. *This decomposition is also independent on the choice of* $x \in \mathscr{B} \cap \mathbb{R}$. *Finally, for every* $x \in \mathscr{B} \cap \mathbb{R}$ *the subspace* Im $S_2(x)$ *(resp.* Im $S_3(x)$ *) is* H*-positive (resp.* H*-negative).*

For every $z \in \mathscr{B} \cap \overline{\Pi}_+$ *put*

$$R^\times(z) = \begin{cases} P^\times(z), & z \in \Pi_+, \\ Q^\times(z), & z \in \mathbb{R}, \end{cases}$$

which is well-defined (see Theorem 2.7). The operator $R^\times(z)$ *is a projection, and, given (2.11) with the properties described in the previous paragraph, the following decomposition holds:*

$$R^\times(z) = P_1(z) + P_2(z), \quad I - R^\times(z) = P_3(z) + P_4(z),$$

where

$$P_2(z) = S_2(z)\,T_2(z), \quad P_3(z) = S_3(z)\,T_3(z).$$

Furthermore, by taking $\mathscr{B}$ *sufficiently small, we have:*

$$P_1(z) = \frac{1}{2\pi i}\int_{\gamma_1} (\lambda - A^\times(z))^{-1}\,d\lambda, \quad P_4(z) = \frac{1}{2\pi i}\int_{\gamma_4} (\lambda - A^\times(z))^{-1}\,d\lambda,$$

where γ_1 *is a contour in* Π_+ *(not depending on* z*) around the spectrum* $\sigma(J_1(z))$ *for every* $z \in \mathscr{B} \cap \overline{\Pi}_+$, *such that all other eigenvalues of* $A^\times(z)$ $(z \in \mathscr{B} \cap \overline{\Pi}_+)$ *lie outside* γ_1, *and* $\gamma_4 = -\overline{\gamma}_1$. *In particular, the projection* $R^\times(z)$ *is continuous on* $\mathscr{B} \cap \overline{\Pi}_+$.

We remark that the statements of the theorems also hold for $\overline{z}$ instead of z. In this case $\tilde{Q}(x) = \lim_{y \downarrow 0} P^\times(\overline{z}) = P_1(x) + P_3(x)$.

COROLLARY 2.9. *Let* $[a, b] \subset \mathbb{R} \backslash \Theta$. *Then the projections* $P_j(x)$, $j=1,\ldots,4$, *and the submatrices* $S_j(x)$ *and* $T_j(x)$, $j=1,2$, *(of Theorem 2.8) are continuous on* $[a, b]$.

THEOREM 2.10. *Let* $x \in \mathbb{R} \backslash \Sigma$. *Then there exists a unique* $A^{\times}(x)$*-invariant maximal* H*-nonpositive subspace* $M(x)$ *of* $\mathbb{C}^{2n}$, *such that* $\sigma(A^{\times}(x)|_{M(x)}) \subset \overline{\Pi}_-$. *Moreover,* $M(x)$ *has a continuous extension on* $\Sigma \backslash \Theta$.

Proof. Since the matrix $A^{\times}(x)$ is H selfadjoint, and $x \notin \Theta$, there exists an invertible matrix S such that $A^{\times}(x) = S^{-1}JS$ and $H = S^*RS$ such that (see Theorem 2.8) $J(x) = \operatorname{diag}\{J_1(x),\ldots,J_4(x)\}$, and

$$R = \begin{bmatrix} 0 & 0 & 0 & r \\ 0 & I_{n_2} & 0 & 0 \\ 0 & 0 & -I_{n_3} & 0 \\ r^* & 0 & 0 & 0 \end{bmatrix},$$

where the size of r is $n_1 \times n_1$ $(= n_4 \times n_4)$. Clearly, the subspace

$$M(x) = \operatorname{span}\{S^{-1}e_j \mid j=n+1,\ldots, 2n\},$$

is $A^{\times}(x)$-invariant maximal H-nonpositive, such that $\sigma(A^{\times}(x)|_{L(x)}) \subset \overline{\Pi}_-$. Moreover, by using Corollary 2.9, $M(x)$ is continuous on $\Sigma \backslash \Theta$. Finally, if $x \notin \Sigma$, then the pair $(A^{\times}(x), H)$ satisfies the sign condition (see, e.g., [RRod]). Now Theorem 2.2 in [RRod] implies that $M(x)$ is unique. □

COROLLARY 2.11. *Let* $x \in \mathbb{R} \backslash \Theta$, *and let* $M(x)$ *be the subspace of Theorem 2.10, then* Im $M(x)$ = Ker $Q^{\times}(x)$. *As before let* $\Pi(x)$ *be the projection of* $\mathbb{C}^{2n}$ *onto* Ker $Q^{\times}(x)$ *along* Im P, *then*

$$\Pi(x) = \lim_{y \downarrow 0} \Pi(z) = \begin{bmatrix} N(x) & \mathbf{0} \end{bmatrix} \begin{bmatrix} N(x) & M \end{bmatrix}^{-1},$$

where M *and* $N(x)$ *are matrices of size* $2n \times n$, *such that the columns of* M *form a basis for* Im P, *and the columns of* $N(x)$ *form a basis of* $M(x)$.

2.5. The proofs of Theorems 2.7 and 2.8.

The proofs are based on five lemmas, which follow below. Before we start with the lemmas we shall introduce several definitions. Note that the main difficulty in Theorems 2.7 and 2.8 concerns points in $[m_1, m_2]$ (where m_1, m_2 are defined in (2.8)). Therefore, fix $x_0 \in [m_1, m_2] \backslash \{0\}$, and define

$$\mathcal{B} = \{z \in \mathbb{C} \mid |z - x_0| < \epsilon\},$$

for $\epsilon > 0$ such that $\mathcal{B} \cap \Omega \subset \{x_0\}$, where Ω is the set introduced in the proof of Theorem 2.5. Hence, $\mathcal{B} \cap \Sigma \subset \{x_0\}$. Put

$$\mathcal{B}_0 = \mathcal{B} \backslash \{x_0\}, \quad \mathcal{U} = \mathcal{B} \cap \mathbb{R},$$

$$\mathcal{U}^+ = \mathcal{U} \cap [x_0, \infty), \quad \mathcal{U}^- = \mathcal{U} \cap (-\infty, x_0],$$

$$\mathcal{U}_0 = \mathcal{U} \backslash \{x_0\}, \quad \mathcal{U}_0^{\pm} = \mathcal{U}^{\pm} \backslash \{x_0\},$$

First we recall the following eigenvalue perturbation result (see, e.g., Theorem 19.1.1 in [GLR3], see also [B]). Let $\mu_1, \ldots, \mu_k$ be all the distinct eigenvalues of the matrix $A^{\times}(x_0)$, and let t_i $(i = 1, \ldots, k)$ be the geometric multiplicity of μ_i $(= \dim \operatorname{Ker} (\mu_i - A^{\times}(x_0))$) and p_{ij} $(j = 1, \ldots, t_i)$ the partial multiplicities of μ_i. Then all eigenvalues of the matrix $A^{\times}(z)$ for $z \in \mathcal{B}_0$ which converge to $\mu_1, \ldots, \mu_k$ are given by the fractional power series

$$\mu_{ij\sigma}(z) = \mu_i + \sum_{\alpha=1}^{\infty} a_{\alpha ij} [(z - x_0)_{\sigma}^{\frac{1}{m_{ij}}}]^{\alpha};$$

$$\sigma = 1, \ldots, m_{ij};\ j = 1, \ldots, s_i;\ i = 1, \ldots, k,$$

where $a_{\alpha ij} \in \mathbb{C}$ and for $\sigma = 1, \ldots, m_{ij}$

$$(z - x_0)_{\sigma}^{\frac{1}{m_{ij}}} = (z - x_0)^{\frac{1}{m_{ij}}} \left(\cos\left(\frac{2\pi\sigma}{m_{ij}}\right) + i \sin\left(\frac{2\pi\sigma}{m_{ij}}\right)\right),$$

(see, e.g., Theorem 19.1.1. in [GLR3], see also [B]). Moreover, all these eigenvalues are distinct, and the geometric multiplicity γ_{ij} as well as the partial multiplicities $m_{ij}^{(1)}, \ldots, m_{ij}^{(\gamma_{ij})}$ of the eigenvalues $\mu_{ij\sigma}(z)$ do not depend on σ and do not depend on z (for $z \in \mathcal{B}_0$). Hence,

$$\sum_{j=1}^{t_i} p_{ij} = \sum_{j=1}^{s_i} \sum_{\gamma=1}^{\gamma_{ij}} m_{ij} m_{ij}^{(\gamma)}, \quad (i=1,\ldots,k). \tag{2.14}$$

For each $i=1,\ldots,k$ and $j=1,\ldots s_i$ there exist vector-valued fractional power series converging for $z \in \mathcal{B}$ (for $\epsilon>0$ small enough):

$$v_{ij\sigma}^{(\gamma\delta)}(z) = \sum_{\alpha=0}^{\infty} v_{\alpha ij}^{(\gamma\delta)} [(z-x_0)_{\sigma}^{\frac{1}{m_{ij}}}]^{\alpha};$$

$$\delta=1,\ldots, m_{ij}^{(\gamma)};\ \gamma=1,\ldots,\gamma_{ij};\ \sigma=1,\ldots m_{ij}; \tag{2.15}$$

where $v_{\alpha ij}^{(\gamma\delta)} \in \mathbb{C}^{2n}$, such that for each $\gamma = 1,\ldots,\gamma_{ij}$ and each $z \in \mathcal{B}_0$ the vectors $v_{ij\sigma}^{(\gamma,1)}(z),\ldots, v_{ij\sigma}^{(\gamma,m_{ij}^{(\gamma)})}(z)$ form a Jordan chain of the matrix $A^{\times}(z)$ corresponding to the eigenvalue $\mu_{ij\sigma}(z)$, i.e.:

$$(A^{\times}(z)-\mu_{ij\sigma}(z))\, v_{ij\sigma}^{(\gamma\delta)}(z) = v_{ij\sigma}^{(\gamma,\delta-1)}(z);$$

$$\delta=1,\ldots, m_{ij}^{(\gamma)};\ \gamma=1,\ldots,\gamma_{ij};\ \sigma=1,\ldots m_{ij};$$

where by definition $v_{ij\sigma}^{(\gamma 0)}(z) = (0)$, $v_{ij\sigma}^{(\gamma 1)}(z) \neq (0)$. Finally, we remark that for every $z \in \mathcal{B}_0$ the vectors

$$v_{ij\sigma}^{(\gamma\delta)}(z);\ \delta=1,\ldots, m_{ij}^{(\gamma)};\ \gamma=1,\ldots,\gamma_{ij};\ \sigma=1,\ldots, m_{ij};$$

$$j=1,\ldots, s_i;\ i=1,\ldots,k;$$

form a linearly independent system.

LEMMA 2.12. *For every $i=1,\ldots,k$; $j=1,\ldots,s_i$; $\sigma=1,\ldots,m_{ij}$ there exists an $\epsilon>0$ such that $\mu_{ij\sigma}(x) \in \mathbb{R}$ (for all $x \in \mathcal{U}$) if and only if $m_{ij}=1$.*

Proof. Let $m_{ij}=1$, then $\sigma=1$, and the eigenvalue

$$\mu_{ij}(z) = \mu_{ij1}(z) = \mu_i + \sum_{\alpha=1}^{\infty} a_{\alpha ij}(z-x_0)^{\alpha}, \quad z \in \mathscr{B},$$

is an analytic function on $\mathscr{B}$ for a certain $\epsilon>0$. Now, assume there exists an $x \in \mathscr{U}$ such that $\mu_{ij}(x) \notin \mathbb{R}$. First we remark that it is not possible that $\mu_{ij}(x)$ lies in Π_+ for all $x \in \mathscr{U}_0$. Because Proposition 2.2 (ii) implies that $\mu_{ij}(z) \in \Pi_+$ for every $z \in \mathscr{B}_0$. Now we define the function

$$\phi(z) = \frac{z-i}{z+i}.$$

Then $\phi[\Pi_+] = \{z \mid |z|<1\}$ and $\phi[\mathbb{R}_\infty] = \{z \mid |z|=1\}$. Obviously, the function $\phi \circ \mu_{ij}$ is analytic on $\mathscr{B}$ and for each $z \in \mathscr{B}$ we obtain that

$$|\phi(\mu_i)| = |(\phi \circ \mu_{ij})(x_0)| \geqq |(\phi \circ \mu_{ij})(z)|.$$

Applying the Maximum Modulus Theorem (see, e.g., [C] Th. IV 3.11) it follows that the function $\phi \circ \mu_{ij}$ is a constant which is not possible, since neither ϕ or μ_{ij} are constant functions. Note that this argument also holds for Π_- instead of Π_+.

Choose an $\epsilon>0$ such that $\mu_{ij}(z)$ is analytic on $\mathscr{B}$, let $x_1 \in \mathscr{U}$ such that $\mu_{ij}(x_1) \notin \mathbb{R}$. Without loss of generality we assume that $\mu_{ij}(x_1) \in \Pi_+$. Furthermore, there is an $x_2 \in \mathscr{U}_0$ such that $\mu_{ij}(x_2) \in \overline{\Pi}_-$. In case $\mu_{ij}(x_2) \in \Pi_-$, we take a rectangle η, with vertices x_1, x_2, $x_2+i\delta$ and $x_1+i\delta$, where $\delta>0$ such that η lies in $\mathscr{B}$. Now we go along η and consider the behaviour of $\mu_{ij}(z)$ for $z \in \eta$. Starting at $x_1 \in \mathscr{U}_0$ we go via the real axis to the point $x_2 \in \mathscr{U}_0$, hence, from $\mu_{ij}(x_1)$ we arrive at $\mu_{ij}(x_2)$, where (by assumption) $\mu_{ij}(x_2) \in \Pi_-$. From this point x_2 we return to x_1 via $\eta \cap \Pi_+$. Since $\mu_{ij}(x_2) \in \Pi_-$ and $\sigma(A^{\times}(z)) \cap \mathbb{R} = \emptyset$ for every $z \in \Pi_-$ (see Proposition 2.2 (ii)) we conclude that $\mu_{ij}(z) \in \Pi_-$ for each $z \in \eta \cap \Pi_+$. Hence, $\lim_{\substack{z \to x_1 \\ z \in \eta \cap \Pi_+}} \mu_{ij}(z) \in \overline{\Pi}_-$. Because $\mu_{ij}(x_1)$ lies in Π_+ and $\mu_{ij}(z)$ is an analytic function on $\mathscr{B}$, this is not possible, and so we have that $\mu_{ij}(x_2) \in \mathbb{R}$. Repeating this argument for any $0<\epsilon<|x_0-x_2|$ we obtain a sequence $\{x_n\}_{n=2}^{\infty}$ in

$\mathcal{U}_0$ such that $\lim_{n\to\infty} x_n = x_0$ and $\mu_{ij}(x_n) \in \mathbb{R}$. Since $\mu_{ij}(z)$ is analytic on $\mathcal{B}$, we obtain that all the Taylor coefficients are real. Hence $\mu_{ij}(x)$ is real for every x in a certain real neighbourhood $\mathcal{U}$ of x_0.

Conversely, let $\mu_{ij\sigma}(x) \in \mathbb{R}$ for every $x \in \mathcal{U}$. Following the argument of Theorem S6.3 in [GLR2], let a_{mij} be the first nonzero coefficient (i.e., $a_{1ij} = a_{2ij} = \ldots = a_{m-1ij} = 0$, $a_{mij} \neq 0$). If $a_{\alpha ij} = 0$ for all $\alpha = 1,2,\ldots$, then our assertion is trivial. Now we obtain

$$a_{mij} = \lim_{x \downarrow x_0} \frac{\mu_{ij\sigma}(x) - \mu_i}{(x - x_0)^{\frac{m}{m_{ij}}}} \in \mathbb{R},$$

and

$$(-1)^{\frac{m}{m_{ij}}} a_{mij} = \lim_{x \uparrow x_0} \frac{\mu_{ij\sigma}(x) - \mu_i}{(x_0 - x)^{\frac{m}{m_{ij}}}} \in \mathbb{R},$$

Hence, $(-1)^{\frac{m}{m_{ij}}} \in \mathbb{R}$ and therefore m must be a multiple of m_{ij}. We repeat this argument for the function

$$\eta_{ij\sigma}(x) = \frac{\mu_{ij\sigma}(x) - \mu_i}{(x_0 - x)^{\frac{m}{m_{ij}}}} = a_{mij} + \sum_{\alpha=1}^{\infty} a_{m+\alpha ij} [(x - x_0)_\sigma^{\frac{1}{m_{ij}}}]^\alpha, \quad x \in \mathcal{U}.$$

Note that for every $x \in \mathcal{U}$ this function is real. Again, let a_{m+nij} be the first nonzero coefficient (after a_{mij}), and we obtain that n is also a multiple of m_{ij}. Hence,

$$\mu_{ij\sigma}(x) = \mu_i + \sum_{\substack{\alpha=1 \\ m_{ij} \mid \alpha}}^{\infty} a_{\alpha ij} [(z - x_0)_\sigma^{\frac{1}{m_{ij}}}]^\alpha, \quad x \in \mathcal{U},$$

which can be written as

$$\mu_{ij\sigma}(x) = \mu_i + \sum_{\alpha=1}^{\infty} \tilde{a}_{\alpha ij} [(z - x_0)]^\alpha, \quad x \in \mathcal{U},$$

where $\tilde{a}_{\alpha ij} = a_{\alpha m_{ij} ij}$. Hence, for each $x \in \mathcal{U}$:

$$\mu_{ij\sigma}(x) = \mu_{ij\delta}(x); \ \ \sigma,\delta = 1,\ldots, m_{ij}.$$

Since all the eigenvalues $\mu_{ij\sigma}(x)$ are distinct we must have that $m_{ij}=1$. In other words $\mu_{ij1}(z)$ is analytic on $\mathcal{B}$. □

Now we shall make some remarks about the real eigenvalues of the matrix function $W(x, \lambda)$. Let $x_0 \in \mathbb{R}\setminus\{0\}$ be a fixed number. Then from (2.7) we have

$$W(x_0, \lambda) = U(\lambda)^* \operatorname{diag}[x_0-d_1(\lambda),\ldots, x_0-d_m(\lambda)]\, U(\lambda), \ \ \lambda \in \mathbb{R}_{\infty}.$$

The real number λ_0 is called an eigenvalue of $W(x_0, \lambda)$ if $x_0=d_j(\lambda_0)$ for some j. Now, let λ_0 be a real eigenvalue of $W(x_0, \lambda)$. We may assume by rearranging rows of $U(\lambda)$ that $x_0=d_j(\lambda_0)$ for $j=1,\ldots,k$, where $1 \leqq k \leqq m$. Since $x_0-d_j(\lambda)$ is analytic on $\mathbb{R}_0$ we have that

$$x_0-d_j(\lambda) = x_0 - d_{j0} - \sum_{\alpha=1}^{\infty} d_{j\alpha}(\lambda-\lambda_0), \tag{2.16}$$

for λ in a real neighbourhood of λ_0. Note that $d_{j0}=x_0$ for $j=1,\ldots,k$. Put

$$m_j = \min\{d_{j\alpha} \neq 0 \mid \alpha \geqq 0\}, j=1,\ldots,m.$$

Then $m_j>0$ if $j=1,\ldots,k$ and $m_j=0$ for $j=k+1,\ldots,m$. The numbers $m_1,\ldots,m_k$ are called the partial multiplicities of λ_0 (see the comments below Theorem II.3.3 in [GLR1]). Moreover, λ_0 is a real eigenvalue of the matrix $A^{\times}(x_0)$ (see (2.6)). The partial multiplicities of λ_0 as eigenvalue of $A^{\times}(x_0)$ are precisely the numbers $m_1,\ldots,m_k$ (see Section II.3.2 in [GLR1]). We can write

$$x_0-d_j(\lambda) = (\lambda-\lambda_0)^{m_j}\phi_j(\lambda), \ \ j=1,\ldots,k,$$

where $\phi_j(\lambda)$ is analytic on a real neighbourhood of λ_0, and $\phi_j(\lambda_0) \neq 0$. The sign ϵ_j of $\phi_j(\lambda_0)$ is called the sign corresponding to m_j. The set of signs $\epsilon_1, \ldots, \epsilon_k$ is the part of the sign characteristic of $(A^{\times}(x_0), H)$ corresponding to λ_0 (see Theorem II.3.4 in [GLR1]).

LEMMA 2.13. *Define for each $j=1,\ldots, k$ the subset:*

$$\Theta_j = \{0\} \cup \{x \in \mathbb{R} \mid \exists \lambda \in \mathbb{R} : x = d_j(\lambda), \frac{d}{d\mu} d_j(\mu) \Big\rfloor_{\mu=\lambda} = 0\}.$$

Then $\Theta = \bigcup_{j=1}^{m} \Theta_j$.

Proof. Let $x_0 \neq 0$. Then

$$x_0 \in \bigcup_{j=1}^{m} \Theta_j \Leftrightarrow \exists j : x_0 \in \Theta_j \Leftrightarrow$$

$$\Leftrightarrow \exists j: \ \exists \lambda_0 \in \mathbb{R} : x_0 = d_j(\lambda), \frac{d}{d\mu} d_j(\mu) \Big\rfloor_{\mu=\lambda_0} = 0. \qquad (2.17)$$

Using (2.16) we have that (2.17) is equivalent to

$$\exists j: \ \exists \lambda_0 \in \mathbb{R} : x_0 = d_{j0}, \ d_{j1} = 0,$$

which is equivalent to $x_0 \in \Theta$. Hence, $\Theta = \bigcup_{j=1}^{m} \Theta_j$. □

Next we introduce the sets

$$J_j = \{d_j(\lambda) \mid \lambda \in \mathbb{R}_\infty\}.$$

Note that each J_j $(j=1,\ldots,m)$ is a closed bounded interval, or is equal to $\{0\}$. Moreover, $\Theta_j \subset J_j$ for every $j=1,\ldots,m$.

LEMMA 2.14. *Let $x_0 \in J_j \backslash \Theta_j$, and let $\lambda_0 \in \mathbb{R}$ such that $d_j(\lambda_0) = x_0$. Then there exists an $\epsilon > 0$ such that*

$$V = \{d_j(\lambda) \mid \lambda \in [\lambda_0 - \epsilon, \lambda_0 + \epsilon]\} \subset J_j \backslash \Theta_j.$$

Moreover, V is a closed interval.

Furthermore, put $U = [\lambda_0 - \epsilon, \lambda_0 + \epsilon]$, then the map

$$d_j \mid_U : U \rightarrow V, \qquad (2.18)$$

is injective.

Proof. Since $d_j[\mathbb{R}_\infty] = J_j$, there always exists a λ_0 such that $x_0 = d_j(\lambda_0)$ whenever $x_0 \in J_j \backslash \Theta_j$. The analyticity of $d_j(\lambda)$ on $\mathbb{R}_\infty$ implies that V is a closed interval. Note that J_j is not a singleton, as otherwise $d_j(\lambda) \equiv 0$. Since Θ_j is finite, there exists an $\epsilon > 0$ such that V lies in $J_j \backslash \Theta_j$. By using the mean value theorem we obtain that for any $a, b \in U$ with $a < b$ there exists a $\xi \in (a, b)$ such that

$$d_j(b) - d_j(a) = (b-a) \frac{d}{d\mu} d_j(\mu) \Big\rfloor_{\mu=\xi} .$$

Hence, $d_j(b) \neq d_j(a)$. In other words, the map in (2.18) is injective. $\square$

We introduce the set Φ, which is defined as follows. The point $x_0 \in \mathbb{R} \backslash \Phi$ if and only if $x_0 \neq 0$, and one of the following two properties hold:

(*i*) $\sigma(A^\times(x_0)) \cap \mathbb{R} = \varnothing$,

(*ii*) *if* $\mu \in \sigma(A^\times(x_0)) \cap \mathbb{R}$, *then* $\exists \lambda(z) \in \sigma(A^\times(z))$, λ *is analytic in a neighbourhood of* x_0 *and* $\lambda(x_0) = \mu$.

Note that $\Phi \subset [m_1, m_2]$, where m_1, m_2 are defined in (2.8). One of our aims is to prove that $\Theta = \Phi$.

Next, take $x_0 \notin \Phi$. Since all m_{ij} are equal to one we shall skip for this case the index "σ". Put for each $i = 1, \ldots, k$; $j = 1, \ldots, s_i$:

$$\mathcal{M}_{ij}(z) = \operatorname{Ker}\, [A^\times(z) - \mu_{ij}(z)], \quad \mathcal{M}_i(z) = \dot{+}_{j=1}^{s_i} \mathcal{M}_{ij}(z), \quad z \in \mathcal{B}_0,$$

$$\mathcal{M}_i = \operatorname{Ker}\, [A^\times(x_0) - \mu_i],$$

Furthermore, define for every $i = 1, \ldots, k$; $j = 1, \ldots, s_i$; $\gamma = 1, \ldots, \gamma_{ij}$:

$$l(i, j, \gamma) = \min \{\alpha \in \mathbb{N} \cup \{0\} \mid v_{\alpha ij} \neq (0)\},$$

$$w_{ij}^{(\gamma)}(z) = (z - z_0)^{-l(i,j,\gamma)} v_{ij}^{(\gamma)}(z), \quad z \in \mathcal{B}_0,$$

and

$$w_{ij}^{(\gamma)}(x_0) = \lim_{z \to x_0} w_{ij}^{(\gamma)}(z).$$

Then for every $z \in \mathscr{B}$ the function $w_{ij}^{(\gamma)}(z)$ is an eigenvector of $A^{\times}(z)$ corresponding to the eigenvalue $\mu_{ij}(z)$. Moreover, for each $z \in \mathscr{B}_0$ the vectors $w_{ij}^{(\gamma)}(z)$ $(i=1,\ldots,k;\ j=1,\ldots,s_i;\ \gamma=1,\ldots,\gamma_{ij})$ are linearly independent. Hence, for all $z \in \mathscr{B}_0$ we have that

$$\mathscr{M}_{ij}(z) = \operatorname{span}\{w_{ij}^{(\gamma)}(z) \mid \gamma=1,\ldots,\gamma_{ij}\},\ \ i=1,\ldots,k;\ j=1,\ldots s_i.$$

Using Theorem 18.2.1 in [GLR3] on the family of analytic transformations $A^{\times}(z)-\mu_{ij}(z)$ on $\mathscr{B}$, we obtain the existence of analytic vector functions on $\mathscr{B}$

$$y_{ij}^{(\gamma)}(z);\quad \gamma=1,\ldots,\gamma_{ij};\ j=1,\ldots s_i;\ i=1,\ldots,k, \tag{2.19}$$

such that for each $z \in \mathscr{B}$; $i=1,\ldots,k;\ j=1,\ldots s_i;$ the vectors $y_{ij}^{(1)}(z),\ldots,y_{ij}^{(\gamma_{ij})}(z)$ are linearly independent, and

$$\mathscr{M}_{ij}(z) = \operatorname{span}\{y_{ij}^{(\gamma)}(z) \mid \gamma=1,\ldots,\gamma_{ij}\},\ \ z\in\mathscr{B}_0;\ \ i=1,\ldots,k;\ j=1,\ldots s_i.$$

Next, define

$$\mathscr{M}_{ij}(x_0) = \operatorname{span}\{y_{ij}^{(\gamma)}(x_0) \mid \gamma=1,\ldots,\gamma_{ij}\};\ \ i=1,\ldots,k;\ j=1,\ldots s_i;$$

and

$$\mathscr{M}_i(x_0) = \mathop{\dot{+}}_{j=1}^{s_i} \mathscr{M}_{ij}(x_0),\quad i=1,\ldots,k.$$

Then $\mathscr{M}_i(x_0) \subset \mathscr{M}_i$, $i=1,\ldots,k$.

Define for each $z \in \mathscr{B}$ and $i=1,\ldots,k$

$$Q_i(z) = \frac{1}{2\pi i}\int_{\gamma_i}(\lambda-A^{\times}(z))^{-1}\,d\lambda, \tag{2.20}$$

where γ_i ($i=1,\ldots,k$) is a circle with centre μ_i such that all other eigenvalues of $A^{\times}(x_0)$ lie outside γ_i. Then for every $i=1,\ldots,k$ (and $\epsilon>0$ small enough) it follows that all the eigenvalues $\mu_{ij}(z)$ (for every $z\in\mathscr{B}$; $j=1,\ldots,s_i$) of $A^{\times}(z)$ lie inside γ_i and all other eigenvalues of $A^{\times}(z)$ lie outside γ_i. Note that for every $i=1,\ldots,k$ the function $Q_i(z)$ is continuous on $\mathscr{B}$.

LEMMA 2.15. *We have* $\Theta = \Phi$.

Proof. Let $x_0 \notin \Phi$. Then the real eigenvalues of $A^{\times}(x)$ are analytic on a neighbourhood of x_0, i.e., $m_{ij} = 1$ for all i, j. From Lemma 2.12 we know that for every $x \in \mathcal{U}$; $i=1,\ldots, k$; $j=1,\ldots, s_i$ the eigenvalues $\mu_{ij}(x)$ are real. Since $\mathcal{U}_0 \cap \Sigma = \emptyset$, we have (by using Theorem 2.6) that all the partial multiplicities $m_{ij}^{(\gamma)}$ $(\gamma=1,\ldots,\gamma_{ij})$ of $\mu_{ij}(x)$ $(x \in \mathcal{U}_0)$ are equal to one (for every $i=1,\ldots, k$; $j=1,\ldots, s_i$). Hence, for each $x \in \mathcal{U}_0$

$$\mathcal{M}_i(x) = \operatorname{Im} Q_i(x), \quad i=1,\ldots, k.$$

But now we have the following relation:

$$\operatorname{Im} Q_i(x_0) = \lim_{x \to x_0} \operatorname{Im} Q_i(x) = \lim_{x \to x_0} \mathcal{M}_i(x) =$$

$$= \mathcal{M}_i(x_0) \subset \mathcal{M}_i \subset \operatorname{Im} Q_i(x_0); \quad i=1,\ldots, k,$$

and by taking the dimensions:

$$\sum_{j=1}^{t_i} p_{ij} = \dim \operatorname{Im} Q_i(x_0) = t_i, \quad i=1,\ldots, k.$$

Hence, $p_{ij} = 1$ (for every $i=1,\ldots, k$; $j=1,\ldots, s_i$). In other words $x_0 \notin \Theta$.

Conversely, let $x \in \mathbb{R}\setminus\{0\}$, and denote $n(x)$ as the sum of the algebraic multiplicities of all the real eigenvalues of $A^{\times}(x)$. Next, put

$$J(x) = \{ j \in \{1,\ldots,m\} \mid x \in J_j \}.$$

For each $j \in J(x)$, let

$$\lambda_{j1}(x),\ldots,\lambda_{j s_j(x)}(x),$$

be all distinct eigenvalues of $A^{\times}(x)$ such that $d_j(\lambda_{ji}(x)) \in J_j$, $i=1,\ldots,s_j(x)$. Moreover, let $m_{ji}(x)$ be the partial multiplicity of $\lambda_{ji}(x))$, $i=1,\ldots,s_j(x)$. Then

$$n(x) = \sum_{j \in J(x)} \sum_{i=1}^{s_j(x)} m_{ji}(x).$$

Now, assume $x_0 \in \mathbb{R} \setminus \Theta$. Then $n(x_0) = \sum_{j \in J(x)} s_j(x)$. Furthermore, let λ_0 be a real eigenvalue of $A^\times(x_0)$. Since $x_0 \notin \Theta_j$ by Lemma 2.13, Lemma 2.14 implies that for $x - x_0$ (x real) small enough there is a real $\lambda(x)$ close to λ_0 such that $d_j(\lambda) = x$ for each $j \in J(x_0)$. Hence, for $x - x_0$ (x real) small enough $J(x) = J(x_0)$. Note that $\lambda(x)$ is an eigenvalue of $A^\times(x)$. Moreover, the partial multiplicities of $\lambda(x)$ as an eigenvalue of $A^\times(x)$ are all one, as $\frac{d}{d\mu} d_j(\mu) \rfloor_{\mu=\lambda} \neq 0$ by Lemma 2.14. Hence, $N(x_0) = n(x)$ for $x - x_0$ small enough. So the matrix $A^\times(x)$ has precisely $n(x)$ eigenvalues $\lambda(x)$, all real, such that $\lim_{x \to x_0} \lambda(x)$ is an eigenvalue of $A^\times(x_0)$, and in this way all real eigenvalues of $A^\times(x_0)$ are obtained. Then Lemma 2.12 implies that $x_0 \notin \Phi$. So we conclude $\Theta = \Phi$ □

LEMMA 2.16. *Let $x_0 \in \mathbb{R} \setminus \{0\}$. Define for z close to x:*

$$R^\times(z) = \begin{cases} P^\times(z), & z \in \Pi_+, \\ Q^\times(x), & z = x \in \mathbb{R}. \end{cases}$$

Let $\{z_n\}_n$ be a sequence in $\overline{\Pi}_+$ such that $\lim_{n \to \infty} z_n = x_0$. If

$$\lim_{n \to \infty} R^\times(z_n) = Q^\times(x_0), \tag{2.21}$$

then $\lim_{\substack{z \to x_0 \\ z \in \overline{\Pi}_+}} R^\times(z) = Q^\times(x_0)$.

Proof. There is a neighbourhood $\mathscr{B}$ of x_0 such that $A^\times(z) = S(z)^{-1} J(z) S(z)$ for every $z \in \mathscr{B}$, where $J(z)$ is the Jordan form of $A^\times(z)$. Moreover, by taking $\mathscr{B}$ small enough, the Jordan structure of $A^\times(z)$ is fixed on $\mathscr{B}_0 = \mathscr{B} \setminus \{0\}$ (see the proof of Theorem 2.5). Hence, we can write $R^\times(z) = S(z)^{-1} \left(\begin{smallmatrix} I & 0 \\ 0 & 0 \end{smallmatrix}\right) S(z)$ for every $z \in \mathscr{B}_0$. We know (see the beginning of this section) that eigenvectors and generalized eigenvectors can be expressed by vector-valued fractional power series. So, $R^\times(z)$, $z \in \mathscr{B}_0$, is a combination fractional power series. Adding and/or multiplying of fractional power series is again a fractional power series. Furthermore, if we divided one by a fractional power series we get a fractional power series, but with possibly negative powers. Hence,

$$R^{\times}(z) = \sum_{j=l}^{\infty} R_j (z-x_0)^{\frac{j}{r}}, \quad z \in \mathscr{B}_0,$$

where $r \in \mathbb{N}$, $l \in \mathbb{Z}$ and $R_l \neq 0$.

Assume that (2.21) holds. We shall prove that $l \geqq 0$, which implies that $R^{\times}(z)$ is continuous on $\mathscr{B}$. So, suppose $l<0$, and consider the following:

$$\lim_{n\to\infty} (z_n - x_0)^{-\frac{l}{r}} R^{\times}(z_n) = \lim_{n\to\infty} \sum_{j=0}^{\infty} R_{j+l}(z-x_0)^{\frac{j}{r}} = R_l.$$

On the other hand we have:

$$\lim_{n\to\infty} R^{\times}(z_n) = Q^{\times}(x_0), \quad \lim_{n\to\infty} (z_n - x_0)^{-\frac{l}{r}} = 0.$$

Hence $R_l = 0$, which contradicts our assumption. □

Proof of Theorems 2.7 and 2.8. Let $x_0 \notin \Theta$. Then from the first part of theproof of Lemma 2.15 we obtain that $\mathscr{M}_i(x_0) = \mathscr{M}_i$ $(i=1,\ldots,k)$. Hence, the vectors $y_{ij}^{(\gamma)}(x_0)$ $(\gamma=1,\ldots;\ \gamma_{ij};\ j=1,\ldots,s_i)$ (obtained at (2.16)) are eigenvectors of the matrix $A^{\times}(x_0)$ and form a basis for the subspace $\mathscr{M}_i$ (for $i=1,\ldots,k$).

For fixed i and j we know (from the Lemma's 2.12 and 2.14) that the eigenvalue $\mu_{ij}(x)$ is real for each $x \in \mathscr{U}$, and (from Theorem 2.6) that all the signs of $\mu_{ij}(x)$ $(x \in \mathscr{U}_0)$ are equal. Now, let

$\omega = \{(i,j) \mid i=1,\ldots,k;\ j=1,\ldots,s_i$ *such that*

the signs of $\mu_{ij}(x)$ $(x \in \mathscr{U}_0)$ *are equal to* $+1\}$.

Applying Theorem A.10 in [GLR1] on the subspace $\mathscr{M}_{ij}(x)$ we obtain the existence of the vectors $u_{ij}^{(\gamma)}(x)$ (for $x \in \mathscr{U}$; $i=1,\ldots,k$; $j=1,\ldots,s_i$; $\gamma=1,\ldots,\gamma_{ij}$) such that these vectors are continuous functions on $\mathscr{U}$, and for each i and j the vectors $u_{ij}^{(1)}(x),\ldots,u_{ij}^{(\gamma_{ij})}(x)$ form a basis for the subspace $\mathscr{M}_{ij}(x)$, and moreover, for every $x \in \mathcal{U}$

$$\langle H u_{ij}^{(\gamma)}(x), u_{ij}^{(\delta)}(x)\rangle = \begin{cases} +1, & \gamma=\delta \text{ and } (i,j)\in\omega, \\ -1, & \gamma=\delta \text{ and } (i,j)\notin\omega, \\ 0, & \gamma\neq\delta. \end{cases}$$

For every $x \in \mathcal{U}$ there is an invertible matrix $S(x)$ such that $A^{\times}(x) = S(x)^{-1}J(x)S(x)$ and $H = S(x)^{*}R(x)S(x)$, where $(J(x), R(x))$ is the canonical form of the pair $(A^{\times}(x), H)$. Moreover, we can write the Jordan form $J(x)$ in the following decomposition:

$$J(x) = \operatorname{diag}\{J_1(x),\ldots,J_4(x)\}, \tag{2.22}$$

where $\sigma(J_1(x))\subset\Pi_+$, $\sigma(J_4(x))\subset\Pi_-$, $J_2(x)$ (resp. $J_3(x)$) is a diagonal matrix with only the eigenvalues $\mu_{ij}(x)$ such that $(i,j)\in\omega$ (resp. $(i,j)\notin\omega$) on its diagonal. With respect to the decomposition of equation (2.22) we write:

$$S(x)^{-1} = \begin{bmatrix} S_1(x)\ldots S_4(x) \end{bmatrix},$$

and

$$S(x) = \begin{bmatrix} T_1(x) \\ \cdot \\ \cdot \\ T_4(x) \end{bmatrix}.$$

Then it is clear that we can take for the columns of the submatrices $S_2(x)$ (resp. $S_3(x)$) all the vectors $u_{ij}^{(\gamma)}(x)$; $(i=1,\ldots,k;\ j=1,\ldots,s_i;\ \gamma=1,\ldots,\gamma_{ij})$ such that $(i,j)\in\omega$ (resp. $(i,j)\notin\omega$). Hence, the submatrices

$$\begin{bmatrix} S_2(x)\ S_3(x) \end{bmatrix}, \quad \begin{bmatrix} T_2(x) \\ T_3(x) \end{bmatrix} = \begin{bmatrix} S_2(x)^{*} \\ -S_3(x)^{*} \end{bmatrix} H,$$

are continuous functions on $\mathcal{U}$.

For every $x \in \mathcal{U}_0$ the projection $Q^{\times}(x)$ exists (see Theorem 2.5) and

$$Q^{\times}(x) = P_1(x) + P_2(x),$$

where

$$P_1(x) = \frac{1}{2\pi i} \int_{\gamma_1} (\lambda - A^{\times}(x))^{-1} d\lambda, \quad P_2(x) = S_2(x) T_2(x),$$

such that γ_1 is a contour in Π_+ around $\sigma(A^{\times}(x)) \cap \Pi_+$ for all $x \in \mathcal{U}$. Note that for every $x \in \mathcal{U}_0$ we have that $P_1(x) P_2(x) = P_2(x) P_1(x) = 0$. Now, we define

$$P_1(x_0) = \frac{1}{2\pi i} \int_{\gamma_1} (\lambda - A^{\times}(x_0))^{-1} d\lambda,$$

$$P_2(x_0) = \lim_{x \to x_0} S_2(x) T_2(x) = S_2(x_0) T_2(x_0).$$

Hence, the limit

$$\lim_{x \to x_0} Q^{\times}(x) = P_1(x_0) + P_2(x_0),$$

exists. Using Proposition 2.3, Theorem 2.5 and Lemma 2.16 we get the existence of the projection $Q^{\times}(x_0)$. Moreover:

$$Q^{\times}(x_0) = P_1(x_0) + P_2(x_0).$$

We know from the proof of Theorem 2.5 that the Jordan structure of $A^{\times}(z)$ is fixed on $\mathcal{B}_0$. The continuity of the submatrices

$$\begin{bmatrix} S_2(\cdot) & S_3(\cdot) \end{bmatrix}, \quad \begin{bmatrix} T_2(\cdot) \\ T_3(\cdot) \end{bmatrix},$$

follows by the same argument as used in the proof of Lemma 2.16. Hence, the decomposition above also holds on $\mathcal{B}$. Finally, for every $x_0 \notin \Theta$ the matching

$$\operatorname{Im} P \oplus \operatorname{Ker} Q^{\times}(x_0) = \mathbb{C}^{2n},$$

follows from the same argument as used in the proof of Theorem 2.5. □

As an application of the previous results we have the following. Let $x_0 \in \mathbb{R} \setminus \Theta$, and let λ_0 be a real eigenvalue of $A^{\times}(x_0)$ with eigenvector v_0 and sign ϵ_0. We may assume that $\epsilon_0 = \langle H v_0, v_0 \rangle$. There exists a $j \in J(x_0)$ such that

$x_0 = d_j(\lambda_0)$ and $\frac{d}{d\mu} d_j(\mu) \Big\rfloor_{\mu=\lambda_0} = -\epsilon_0$. Put $z = x_0 + iy$ with $y > 0$. For $y > 0$ small enough we have $A^\times(z)v(z) = \lambda(z)V(z)$ such that $\lambda(z) \to \lambda_0$ and $v(z) \to v_0$ as $y \downarrow 0$. Furthermore, for $y > 0$ small enough:

$$\operatorname{span}\{v(z)\} \subset \begin{cases} \operatorname{Im} P^\times(z), & \lambda(z) \in \Pi_+, \\ \operatorname{Ker} P^\times(z), & \lambda(z) \in \Pi_-. \end{cases}$$

Hence,

$$\epsilon_0 = \begin{cases} +1, & \lambda(z) \in \Pi_+, \\ -1, & \lambda(z) \in \Pi_-. \end{cases}$$

Furthermore,

$$\epsilon_0 = \begin{cases} +1, & d_j(\lambda) \textit{ is strictly decreasing in a real neighbourhood of } \lambda_0, \\ -1, & d_j(\lambda) \textit{ is strictly increasing in a real neighbourhood of } \lambda_0. \end{cases}$$

3. SELFADJOINT WIENER-HOPF INTEGRAL OPERATORS WITH RATIONAL SYMBOLS

3.1. Preliminaries. Consider the *Wiener-Hopf operator*

$$K: L_2^m(\mathbb{R}_+) \to L_2^m(\mathbb{R}_+) \quad (Kf)(t) = \int_0^\infty k(t-s) f(s)\, ds, \quad t \geqq 0. \tag{3.1}$$

Here the *kernel* k is an $m \times m$ matrix function with entries in $L_1(\mathbb{R})$. We assume throughout that the *symbol* of equation (3.1)

$$W(\lambda) = \int_{-\infty}^{\infty} k(t)\, e^{i\lambda t}\, dt, \quad -\infty < \lambda < \infty, \tag{3.2}$$

is a rational $m \times m$ matrix function, given in the form

$$W(\lambda) = -C(\lambda - A)^{-1} B, \quad \lambda \in \mathbb{R}, \tag{3.3}$$

where A, B and C are matrices, of sizes $l \times l$, $l \times m$ and $m \times l$, respectively, such that A has no real eigenvalues. Then the kernel k admits the following exponential representation

$$k(t) = \begin{cases} iCe^{-itA}(I-P)B, & 0<t<\infty, \\ -iCe^{-itA}PB, & -\infty<t<0. \end{cases}$$

Here P is the Riesz projection corresponding to the part of $\sigma(A)$ lying in the open upper half plane. Conversely, if the kernel k can be written in this form, then the symbol W can be described as in (3.3) (see [BGK2]).

Assume that the matrix $A^{\times} = A - BC$ has no real eigenvalue, and let $P^{\times}$ be the Riesz projection corresponding to the part of $\sigma(A^{\times})$ in the upper half plane. Then (cf. Theorem I 3.4 in [BGK2]) the Wiener-Hopf equation

$$((I-K)\phi)(t) = f(t), \quad 0 \leqq t < \infty, \tag{3.4}$$

has a unique solution in $L_2^m(\mathbb{R}_+)$ for each f in $L_2^m(\mathbb{R}_+)$ if and only if

$$\mathbb{C}^{2n} = \operatorname{Im} P \oplus \operatorname{Ker} P^{\times}.$$

In that case, for a given f in $L_2^m(\mathbb{R}_+)$, the solution of equation (3.4) is given by

$$\phi(t) = f(t) + \int_0^{\infty} \gamma(t,s) f(s)\, ds, \quad 0 \leqq t < \infty,$$

where

$$\gamma(t,s) = \begin{cases} iCe^{-itA^{\times}} \Pi e^{isA^{\times}} B, & 0 \leqq s < t < \infty, \\ -iCe^{-itA^{\times}}(I-\Pi) e^{isA^{\times}} B, & 0 \leqq t < s < \infty, \end{cases}$$

and Π is the projection of $\mathbb{C}^{2n}$ onto $\operatorname{Ker} P^{\times}$ along $\operatorname{Im} P$.

3.2. Some spectral properties of selfadjoint Wiener-Hopf operators. We shall assume that the operator K is selfadjoint, which means that its symbol is selfadjoint with respect to the real line, i.e., $W(\lambda) = W(\bar{\lambda})^*$.

We remark that the spectral properties of $W(\lambda)$ are discussed in Section 2. Therefore, we shall use throughout this section all notations and results of the previous sections.

Note that $W(z,\lambda)$ as defined in (2.5) is the symbol of the Wiener-Hopf equation

$$z\phi(t) - (K\phi)(t) = f(t), \quad 0 \leqq t < \infty. \tag{3.5}$$

THEOREM 3.1. *(i) For each real* $x \neq 0$, *the following are equivalent:*

(a) $x \in \sigma(K)$,

(b) *there is a* $\lambda \in \mathbb{R}$ *such that* $\det W(x,\lambda) = 0$,

(c) $A^\times(x)$ *has a real eigenvalue,*

(ii) $\sigma(K) = [m_1, m_2]$ *(where* m_1 *and* m_2 *are defined in (2.7)).*

Proof. (i) The equivalence between (b) and (c) follows from (2.6). Next, take $x \in \varrho(K) \cap \mathbb{R}$. From Theorem 4.10 in [BGK1], it follows among other things, that $\det W(x,\lambda) \neq 0$ for all real λ.

Conversely, let x in $\sigma(K)\backslash\{0\}$, and assume that $\det W(x,\lambda) \neq 0$ for all $\lambda \in \mathbb{R}$. From equation (2.6) we know that the matrix $A^\times(x)$ has no real eigenvalues. Hence, by applying Propositions 2.2 and 2.3,

$$\operatorname{Im} P \oplus \operatorname{Ker} P^\times(x) = \mathbb{C}^{2n}.$$

Use again Theorem 4.10 in [BGK1] to conclude that $x \in \varrho(K)$. But this contradicts our assumption. Hence, there exists a real λ such that $\det W(x,\lambda) = 0$.

Finally, (ii) follows immediately from the definitions of m_1 and m_2, Proposition 2.2 and (i). □

Because K is selfadjoint, the Wiener-Hopf equation (3.5) has a unique solution ϕ in $L_2^m(\mathbb{R}_+)$ for each $f \in L_2^m(\mathbb{R}_+)$ whenever $z \in \Pi_+$. From (3.1),

(3.2) and (3.3) it is obvious that the solution ϕ can be written as

$$\phi(t) = \frac{1}{z} f(t) + \int_0^\infty \gamma_z(t,s) f(s)\, ds, \quad 0 \leqq t \infty, \tag{3.6}$$

where

$$\gamma_z(t,s) = \begin{cases} \frac{i}{z^2} C e^{-itA^\times(z)} \Pi(z) e^{isA^\times(z)} B, & 0 \leqq s < t < \infty, \\ -\frac{i}{z^2} C e^{-itA^\times(z)} (I - \Pi(z)) e^{isA^\times(z)} B, & 0 \leqq t < s < \infty. \end{cases} \tag{3.7}$$

Here $\Pi(z)$ is the projection of $\mathbb{C}^{2n}$ along Im P onto Ker $P^\times(z)$.

THEOREM 3.2. *(i) The nonzero part of the spectrum of* K *is continuous, i.e.,* $\sigma(K)\backslash\{0\} = C\sigma(K)\backslash\{0\}$.

(ii) If the determinant of the symbol is identically zero, then $0 \in P\sigma(K)$.

PROOF (i) Fix $0 \neq x \in \sigma(K)$, and assume there is a nonzero vector $f \in L_2^m(\mathbb{R}_+)$ such that

$$((x-K)f)(t) = 0. \quad 0 \leqq t < \infty.$$

From Corollary 2.3 in [BGK2] we obtain that

$$f(t) = \frac{1}{x} C e^{-itA^\times(x)} v, \quad 0 \leqq t < \infty,$$

where $Pv = v$ and $\lim_{t \to \infty} e^{itA} P e^{-itA^\times(x)} v = 0$. Furthermore, from Theorem 3.1 in [BGK2] we obtain a unique $\varrho \in L_2^{2n}(\mathbb{R}_+)$ such that

$$\begin{cases} \dot{\varrho}(t) = -iA\,\varrho(t) + iB\,f(t), & 0 \leqq t < \infty, \\ f(t) = \frac{1}{x} C\,\varrho(t), & 0 \leqq t < \infty, \\ (I-P)\,\varrho(0) = 0. \end{cases}$$

Next, put $F(t) = e^{-itA^\times(x)} v$, then a simple calculation shows that $\varrho - F \in$ Ker

$(C \mid A) = (0)$. Hence, F lies in $L_2^{2n}(\mathbb{R}_+)$. (It is also possible to obtain the equality $\varrho(t) = e^{-itA^\times(x)}v$ directely from [BGK1]).

Define:

$$P(x) = \frac{1}{2\pi i} \int_\gamma (\lambda - A^\times(x))^{-1} d\lambda,$$

where γ is a contour in Π_- around $\sigma(A^\times(x)) \cap \Pi_-$. Since the matrix $A^\times(x)$ is H-selfadjoint one easily checks that Im $P(x)$ is an H-neutral subspace.

Now, put

$$g(t) = e^{-itA^\times(x)}P(x)v, \quad 0 \leqq t < \infty,$$

then $g \in L_2^{2n}(\mathbb{R}_+)$ (because $\sigma(A^\times(x) \mid_{\mathrm{Im}\ P(x)}) \subset \Pi_-$). Hence,

$$F - g = e^{-i(\cdot)A^\times(x)}(I - P(x))v \in L_2^{2n}(\mathbb{R}_+).$$

Since $\sigma(A^\times(x) \mid_{\mathrm{Ker}\ P(x)}) \subset \overline{\Pi}_+$ we must have that $F - g = 0$. Hence, $(I - P(x))v = (0)$. In other words $P(x)v = v = Pv$, and because of the H-neutrality of the subspaces Im P and Im $P(x)$ we get that

$$\langle HAu, u \rangle = \langle HA^\times(x)u, u \rangle = 0,$$

for every $u \in \mathrm{Im}\ P \cap \mathrm{Im}\ P(x)$. Then, by using equation (2.2),

$$\frac{1}{x} \| Cv \|^2 = \frac{1}{x} \langle HBCv, v \rangle = \langle H(A - A^\times(x))v, v \rangle = 0.$$

Hence, $Cv = 0$. But then $A^\times(x)v = Av$ and so Im $P \cap$ Im $P(x)$ is A-invariant. Hence,

$$CA^k v = 0, \quad k \geqq 0.$$

Since the pair (C, A) is observable it follows that $v = 0$ and so $f = 0$ which is a contradiction with our assumption. Hence, Ker $(x - K) = (0)$.

(ii) Assume $\det W(\lambda) \equiv 0$. Since the symbol is a rational function we can write

$$W(\lambda) = \frac{1}{q(\lambda)} R(\lambda),$$

where $q(\lambda)$ is a scalar polynomial and $R(\lambda)$ is a $m \times m$ matrix polynomial. Moreover, $R(\lambda) = F_1(\lambda) D(\lambda) F_2(\lambda)$, where $F_j(\lambda)$ $(j=1, 2)$ is a $m \times m$ matrix polynomial with constant nonzero determinant, and

$$D(\lambda) = \operatorname{diag}\{d_1(\lambda), \ldots, d_r(\lambda), 0, \ldots, 0\},$$

such that the diagonal elements $d_j(\lambda)$ $(j=1,\ldots,r)$ are scalar polynomials. The matrix $D(\lambda)$ is called the Smith form of $R(\lambda)$ (see, e.g., [GLR2, GLR3]). Because $\det W(\lambda) \equiv 0$ we have that $r<m$.

The last column $f_m(\lambda)$ of $F_2(\lambda)^{-1}$ is a nonzero vector-polynomial which we can write as

$$f_m(\lambda) = \sum_{j=0}^{p} (\lambda+i)^j g_{mj},$$

for some vectors $g_{m1}, \ldots, g_{mp}$ in $\mathbb{C}^m$. Since

$$\int_0^\infty t^j e^{-t} e^{i\lambda t}\, dt = \frac{i^{j+1} j!}{(\lambda+i)^{j+1}}, \quad j=0,1,2,\ldots .$$

we can rewrite $f_m(\lambda)$ as

$$f_m(\lambda) = (\lambda+i)^{p+1} \int_0^\infty P_m(t) e^{-t} e^{i\lambda t}\, dt,$$

where $P_m(t)$ is a certain nonzero vector-valued polynomial of degree p. Now we define

$$h(t) = \begin{cases} P_m(t) e^{-t}, & 0 \leqq t < \infty, \\ 0, & -\infty < t < 0. \end{cases}$$

Then h is a nonzero vector in $L_2^m(\mathbb{R})$, and $W(\lambda) \hat{h}(\lambda) \equiv (0)$, where $\hat{h}(\lambda)$ is the Fourier transform of h, i.e.,

$$\hat{h}(\lambda) = \int_{-\infty}^{\infty} e^{i\lambda t} h(t)\, dt = \frac{1}{(\lambda+i)^{p+1}} f_m(\lambda).$$

Put

$$v(t) = (Lh)(t) = \int_{-\infty}^{\infty} k(t-s)h(s)\, ds, \quad -\infty<t<\infty.$$

We obtain for the Fourier transform of v that $\hat{v}(\lambda) = W(\lambda)\hat{h}(\lambda)=0$, for every $\lambda\in\mathbb{R}$. Hence, by applying Plancherel's theorem (see, e.g., [Ru]) we get that $v = 0$, and so h is a nonzero vector in Ker L. In particular, for $t>0$ we have that $h\in L_2^m(\mathbb{R}_+)$ and $((Kh)(t) = (Lh)(t) = 0$. □

REMARK 3.3. *If the matrix B is not injective, then there is a nonzero vector in the kernel of the operator K.*

The following theorem will not play a role in the sequel, and therefore its proof is omitted.

THEOREM 3.4. *If* $\det W(\lambda) \not\equiv 0$, *then*

$$\text{Ker } K = \{f\in L_2^m(\mathbb{R}_+) \mid f(t) = \hat{R}e^{-it\tilde{R}}u,\ 0<t<\infty,$$

$$u\in \text{Im } P \cap \text{Im } R\},$$

where

$$R = \frac{1}{2\pi i}\int_{\gamma} \begin{pmatrix} 0 & I_{2n} \end{pmatrix} \begin{pmatrix} 0 & C \\ B & \lambda-A \end{pmatrix}^{-1} \begin{pmatrix} 0 \\ I_{2n} \end{pmatrix} d\lambda,$$

$$\tilde{R} = \frac{1}{2\pi i}\int_{\gamma} \lambda \begin{pmatrix} 0 & I_{2n} \end{pmatrix} \begin{pmatrix} 0 & C \\ B & \lambda-A \end{pmatrix}^{-1} \begin{pmatrix} 0 \\ I_{2n} \end{pmatrix} d\lambda,$$

$$\hat{R} = \frac{1}{2\pi i}\int_{\gamma} \begin{pmatrix} I_m & 0 \end{pmatrix} \begin{pmatrix} 0 & C \\ B & \lambda-A \end{pmatrix}^{-1} \begin{pmatrix} 0 \\ I_{2n} \end{pmatrix} d\lambda,$$

and γ is a contour in Π_- such that the set

$$\Phi = \{\lambda\in\Pi_- \mid \det \begin{pmatrix} 0 & C \\ B & \lambda-A \end{pmatrix} = 0\},$$

lies inside γ.

3.3. The proof of Theorem 0.1. In this section we shall prove our main result, Theorem 0.1.

LEMMA 3.5. *Let* $[a, b]$ *be a real interval such that* $[a, b] \cap \Theta = \emptyset$, *and put*

$$\mathscr{X} = \{z = x + iy \mid x \in [a, b], y \in [0, 1]\}. \tag{3.8}$$

For every $z \in \mathscr{X}$ *we define the projection* $R(z)$ *as follows.*

$$R(z) = \begin{cases} P^{\times}(z), & z \in \Pi_+, \\ Q(z), & z \in \mathbb{R}. \end{cases} \tag{3.9}$$

Then for all $t > 0$ *and* $\phi \in L_1(\mathbb{R}_+)$ *the map*

$$z \mapsto \int_t^\infty R(z) e^{isA^{\times}(z)} \phi(s)\, ds,$$

is continuous on $\mathscr{X}$.

Proof. Let z_0 be any number in $\mathscr{X}$ and $\{z_n\}_{n=1}^\infty$ a sequence in $\mathscr{X}$ such that $\lim_{n\to\infty} z_n = z_0$. We shall consider two cases.

(i) $z_0 \notin \sigma(K)$.

Since the matrix $A^{\times}(z_0)$ has no real spectrum we can take a contour Γ in Π_+ around $\sigma(A^{\times}(z_0)) \cap \Pi_+$. Then for n sufficiently large (say $n \geqq N$ for some $N \in \mathbb{N}$), all the eigenvalues of the matrix $A^{\times}(z_n)$ in Π_+ are completely inside Γ. Hence, for every $n \geqq N$ and $s \geqq 0$ we get that

$$\|R(z_n) e^{isA^{\times}(z_n)} - R(z_0) e^{isA^{\times}(z_0)}\| =$$

$$= \| \frac{1}{2\pi i} \int_\Gamma e^{is\lambda} [(\lambda - A^{\times}(z_n))^{-1} - (\lambda - A^{\times}(z_0))^{-1}] d\lambda \| \leqq$$

$$\leqq \frac{1}{2\pi} \ell(\Gamma) \max_{\lambda \in \Gamma} |e^{is\lambda}| \, \| [(\lambda - A^{\times}(z_n))^{-1} - (\lambda - A^{\times}(z_0))^{-1}] \|, \tag{3.10}$$

where $\ell(\Gamma)$ is the length of the contour Γ. Since $|e^{is\lambda}| \leqq 1$ for every $s \geqq 0$ and $\lambda \in \Gamma$, the right-hand side of (3.10) tends to zero as $n \to \infty$ (uniformly on $[0, \infty)$). Hence, by applying the dominated convergence theorem we get that

$$\lim_{n\to\infty} \int_t^\infty R(z_n)\, e^{isA^\times(z_n)} \phi(s)\, ds = \int_t^\infty R(z_0)\, e^{isA^\times(z_0)} \phi(s)\, ds,$$

for $0<t<\infty$.

(ii) $z_0 \in [a, b]$, where $[a, b] \subset \sigma(K) \cap \mathscr{X}$.

Using Theorem 2.8 we can write the projection $R(z_0) = P_1(z_0) + P_2(z_0)$ and for n sufficiently large (say $n \geqq N$) we have that $R(z_n) = P_1(z_n) + P_2(z_n)$, such that $P_j(z_0) = \lim_{n\to\infty} P_j(z_n)$, for $j=1, 2$. Now, the argument of case (i) implies that

$$\lim_{n\to\infty} \int_t^\infty P_1(z_n)\, e^{isA^\times(z_n)} \phi(s)\, ds = \int_t^\infty P_1(z_0)\, e^{isA^\times(z_0)} \phi(s)\, ds,$$

for $0<t<\infty$. Furthermore, for every $s \geqq 0$ the following inequality holds:

$$\sup_{n \geqq N} \| P_2(z_n)\, e^{isA^\times(z_n)} \| \leqq \sup_{n \geqq N} \| P_2(z_n) \| \, \| S_2(z_n) \| \, \| T_2(z_n) \| < \infty.$$

Note that $\| e^{isJ_2(z_n)} \| \leqq 1$ $(n \geqq N, s \geqq 0)$. Again, we apply the dominated convergence theorem to obtain that

$$\lim_{n\to\infty} \int_t^\infty P_2(z_n)\, e^{isA^\times(z_n)} \phi(s)\, ds = \int_t^\infty P_2(z_0)\, e^{isA^\times(z_0)} \phi(s)\, ds,$$

for $0<t<\infty$. Hence,

$$\int_t^\infty R(z)\, e^{isA^\times(z)} \phi(s)\, ds,$$

is continuous on $\mathscr{X}$. □

REMARK 3.6. *Let $\mathcal{X}$ be the set defined in Lemma 3.5. Then the matrix functions $A^{\times}(z)$, $R(z)$ (see (3.9)), and $\Pi(z)$ are continuous on $\mathcal{X}$, and Lemma 3.6 also holds for $\bar{z}$ instead of z.*

Let $[a, b]$ be an interval in $\mathbb{R}$ such that $[a, b] \cap \Theta = \varnothing$ (where Θ is defined in (2.9)). From equation (3.7) we derive, for every $x \in \mathbb{R}$, $y \in (0, \infty)$ and $s, t \in [0, \infty)$, the following function

$$\gamma(x, y, t, s) = \frac{1}{2\pi i}\left[\gamma_{\bar{z}}(t, s) - \gamma_z(t, s)\right],$$

where $z = x + iy$. Using equation (3.6) we obtain that, for any real interval $[a, b]$ and $f \in L_2^m(\mathbb{R}_+)$, the following holds:

$$\int_a^b ([(\bar{z}-K)^{-1} - (z-K)^{-1}]f)(t)\,dx = \int_a^b \left(\frac{1}{\bar{z}} - \frac{1}{z}\right) dx\, f(t) +$$

$$+ \int_a^b \int_0^\infty \gamma(x, y, t, s) f(s)\, ds\, dx \quad 0 \leqq t < \infty. \tag{3.11}$$

Note that whenever $0 \notin [a, b]$, we have $\lim_{y \downarrow 0} \int_a^b \left(\frac{1}{\bar{z}} - \frac{1}{z}\right) dx = 0$.

If we take $x \in [a, b] \subset \sigma(K) \backslash \Theta$ (where Θ is defined in (2.9)), then it follows from Theorem 2.8 that the limit

$$\gamma(x, t, s) = \lim_{y \downarrow 0} \gamma(x, y, t, s), \quad 0 \leqq s, t < \infty,$$

exists. Using (2.2) and Theorem 2.4:

$$\gamma(x, t, s) = -\frac{1}{2\pi x^2} B^* (e^{itA^{\times}(x)})^* (\Pi(x))^* H\, \Pi(x)\, e^{isA^{\times}(x)} B.$$

Using Proposition 2.3, it follows that $\gamma(x, t, s) = 0$ for every $x \in \varrho(K) \cap \mathbb{R}$.

LEMMA 3.7. *Let $\phi \in (L_1 \cap L_2)(\mathbb{R}_+)$, $z \in \mathcal{X}$ and $0 < t < \infty$ (where $\mathcal{X}$ is as in (3.8)). Then the function*

$$g_{\phi,t}(z) =$$

$$=\frac{1}{2\pi}\int_0^t[\frac{1}{\bar{z}^2}e^{-itA^\times(\bar{z})}\Pi(\bar{z})e^{isA^\times(\bar{z})}-\frac{1}{z^2}e^{-itA^\times(z)}\Pi(z)e^{isA^\times(z)}]\phi(s)ds-$$

$$-\frac{1}{2\pi}\int_t^\infty[\frac{1}{\bar{z}^2}e^{-itA^\times(\bar{z})}(I-\Pi(\bar{z}))e^{isA^\times(\bar{z})}-\frac{1}{z^2}e^{-itA^\times(z)}(I-\Pi(z))e^{isA^\times(z)}]\phi(s)ds,$$

is continuous on $\mathscr{X}$.

Proof. Fix $t>0$. From Remark 3.6 we have that the integrands of the two integrals are continuous on $\mathscr{X}$. Since the first integral is finite and $f\in L_1(\mathbb{R}_+)$ it is clear that the first integral is continuous on $\mathscr{X}$. From the definition of $\Pi(z)$ (see Proposition 2.3 (iii)) we have that

$$I-\Pi(z) = (I-\Pi(z))P^\times(z),$$

for all $z\in\mathbb{C}\backslash\mathbb{R}$. Now Lemma 3.5 implies the continuity on $\mathscr{X}$ of the second integral. □

Proof of Theorem 0.1. Let $\{e_1, \ldots, e_m\}$ be the standard basis of $\mathbb{C}^m$, then any function $f\in(L_1^m\cap L_2^m)(\mathbb{R}_+)$ can be written as:

$$f = \begin{bmatrix} f_1 \\ \cdot \\ \cdot \\ f_m \end{bmatrix} = \sum_{j=1}^m e_j f_j,$$

where $f_j\in(L_1\cap L_2)(\mathbb{R}_+)$. Hence, $Bf = \sum_{j=1}^m f_j B e_j$. Next, let $z=x+iy$ with $x\in[a, b]$ and $y\in[0, 1]$, and consider the function

$$\varrho_t(z) = \int_0^\infty \gamma(x,y,t,s)f(s)\,ds = \sum_{j=1}^m Cg_{f_j,t}B e_j, \quad 0<t<\infty.$$

For every $t>0$ the function ϱ_t is continuous on $\mathscr{X}$ (see Corollary 3.7). But then $c_t = \max_{z\in\mathscr{X}} \|\varrho_t(z)\|$ exists and is finite. Hence,

$$\max_{x\in\mathscr{X}} \|\int_a^b \varrho_t(z)\,dx\| \leqq (b-a)c_t.$$

So, by applying the dominated convergence theorem, we get the equality:

$$\int_a^b \int_0^\infty \gamma(x,t,s) f(s)\, ds\, dx = \lim_{y \downarrow 0} \int_a^b \int_0^\infty \gamma(x,y,t,s) f(s)\, ds\, dx, \tag{3.12}$$

for $0<t<\infty$. From Theorem 2.8, Corollary 2.9 and Theorem 2.4 (vi) the following holds,

$$(\Pi(x))^* H\, \Pi(x) P_4(x) = 0, \quad x \notin \Theta,$$

and there exists a $\delta>0$ such that

$$\max\{ \|\gamma(x, t, s)\|_{\mathbb{C}^m} \mid x \in [a, b];\ s, t \geqq 0 \} \leqq \delta.$$

Hence, by applying Fubini's theorem,

$$\int_0^\infty K_{a,b}(t,s) f(s)\, ds\, dx = \lim_{y \downarrow 0} \int_a^b \int_0^\infty \gamma(x,y,t,s) f(s)\, ds\, dx, \quad 0<t<\infty.$$

Now, let $\{z_n\}_{n=1}^{\infty}$ be a sequence in $\mathscr{X}$ (see (3.8)) such that $z_n = x+iy_n$, $y_n>0$ and $\lim_{n\to\infty} y_n = 0$. For convenience we define the functions:

$$F(t) = \int_0^\infty K_{a,b}(t,s) f(s)\, ds\, dx, \quad 0<t<\infty,$$

$$F_n(t) = \frac{1}{2\pi i} \int_a^b \{[(\bar{z}_n - K)^{-1} - (z_n - K)^{-1}] f\}(t)\, dx, \quad 0<t<\infty,$$

$$\hat{E}(t) = (E([a, b]) f)(t), \quad 0<t<\infty.$$

Then, from Theorem 1.1 we get that

$$\| \hat{E} - F_n \|_2 \to 0, \quad n\to\infty, \tag{3.13}$$

and so

$$\| F_n \|_2^2 \to \| \hat{E} \|_2^2, \quad n\to\infty. \tag{3.14}$$

Moreover, using (3.11), equation (3.12) implies that

$$\lim_{n\to\infty} \| F_n(t) \|^2 = \| F(t) \|^2, \quad -\infty<t<\infty \tag{3.15}$$

Then, using equations (3.14) and (3.15) and Fatou's lemma the following holds:

$$\| F \|_2^2 = \int_0^\infty \| F(t) \|^2 dt = \int_0^\infty \liminf_{n\to\infty} \| F_n(t) \|^2 dt \leqq$$

$$\leqq \liminf_{n\to\infty} \int_0^\infty \| F_n(t) \|^2 dt = \lim_{n\to\infty} \| F_n \|_2^2 =$$

$$= \| \hat{E} \|_2^2 \leqq \| f \|_2^2 < \infty.$$

Hence, $F \in L_2^m(\mathbb{R}_+)$.

We remark that $\| g(t) \| < \infty$ a.e. on $\mathbb{R}_+$ for every $g \in L_2^m(\mathbb{R}_+)$. Thus,

$$\lim_{n\to\infty} \| \hat{E}(t) - F_n(t) \|^2 = \| \hat{E}(t) - F(t) \|^2, \quad \text{a.e. on } \mathbb{R}. \tag{3.16}$$

Then, by applying Fatou's lemma and equations (3.13) and (3.16) we obtain that

$$\| \hat{E} - F \|_2^2 = \int_0^\infty \liminf_{n\to\infty} \| \hat{E}(t) - F_n(t) \|^2 dt \leqq$$

$$\leqq \liminf_{n\to\infty} \| \hat{E} - F_n \|_2^2 = 0.$$

Hence, $\hat{E} = F$. In other words,

$$(E([a,b])f)(t) = \int_0^\infty K_{a,b}(t,s) f(s)\, ds, \quad \text{a.e. on } \mathbb{R}_+. \quad \square$$

4. THE SCALAR CASE

In this section we specify the results from Sections 2 and 3 for the case when the symbol $W(\lambda)$ is a scalar function (i.e., $m=1$). The formula for the resolution of the identity is given for intervals outside the exceptional Θ (see (2.10)). We shall show that there exists a direct relation between the sets Θ and Σ (see (2.10) and Theorem 2.5), and the local extremal points of the symbol $W(\lambda)$.

Throughout we assume that the symbol $W(\lambda)$ is not identically zero, in other words, we assume that the Wiener-Hopf operator K is not the zero operator. In this case the spectrum of K is continuous (see, e.g., [RR, HW, Ros1]).

We can write the symbol $W(\lambda)$ as

$$W(\lambda) = -\frac{p(\lambda)}{q(\lambda)}, \quad \lambda \in \mathbb{R},$$

where p and q are polynomials such that $gcd(p, q) = 1$. The condition at infinity (i.e. $W(\infty) = 0$) implies that $\deg p < \deg q$ (where $\deg p$ stands for the degree of the polynomial p).

Let $p(\lambda) = \sum_{j=0}^{2n-1} c_j \lambda^j$ and $q(\lambda) = \lambda^{2n} + \sum_{j=0}^{2n-1} a_j \lambda^j$. Put

$$A = \begin{pmatrix} 0 & 1 & & & \\ & & \cdot & \cdot & \\ & & & \cdot & \cdot & \\ & & & \cdot & 1 \\ -a_0 & -a_1 & \cdot & \cdot & -a_{2n-1} \end{pmatrix}, \quad B = \begin{pmatrix} 0 \\ \cdot \\ \cdot \\ 0 \\ 1 \end{pmatrix}, \quad C = \begin{pmatrix} c_0 \dots c_{2n-1} \end{pmatrix}, \tag{4.1}$$

where $C \neq 0$. Then $W(\lambda) = -C(\lambda - A)^{-1}B$ and (A, B, C) is a minimal realization (see, e.g., [BGK1]). Moreover, $p(\lambda) \in \mathbb{R}$ and $q(\lambda) = \det(\lambda - A) > 0$ for all $\lambda \in \mathbb{R}$ (so, in particular, $a_0 > 0$). Note that A is a matrix in first companion form, and that the matrix $A^{\times}(z) = A - \frac{1}{z}BC$ has the same form. All matrices of this type have only eigenvalues with geometric multiplicity one. Moreover, if μ is an eigenvalue of the matrix A with partial multplicity m, then the vectors $v_1(\mu), \dots, v_m(\mu)$ of $\mathbb{C}^{2n}$ defined by:

$$(v_j(\mu))_i = \begin{cases} 0, & i = 1, \dots, j-1, \\ \binom{i-1}{i-j} \mu^{i-j}, & i = j, \dots, 2n, \end{cases}$$

form a Jordan chain of the matrix A corresponding to the eigenvalue μ, and they form a basis of the spectral subspace of A at μ. Similar statements holds for $A^{\times}(z)$.

Let x be a real number, then the graph of the symbol

$W(x, \lambda) = x + C(\lambda - A)^{-1}B$ (of the operator $x - K$) can be seen as a continuous curve in $\mathbb{R}^2$. Note, that the shape of this curve does not depend on the choice of x.

PROPOSITION 4.1. *In the scalar case, we have $\Theta = \Sigma$ (where Θ and Σ are as in (2.10) and Theorem 2.5).*

Proof. We already know that $\Theta \subset \Sigma$ (see Theorem 2.6) and that all geometric multiplicities of the eigenvalues of $A^{\times}(z)$ $(z \neq 0)$ are equal to one. Hence, equation (2.14) becomes,

$$p_{i1} = \sum_{j=1}^{s_i} m_{ij} m_{ij}^{(1)}, \quad i = 1, \ldots, k.$$

Now, let $x_0 \notin \Theta$, i.e., $p_{i1} = \ldots = p_{ik} = 1$, and so each s_i, m_{ij} and $m_{ij}^{(1)}$ are equal to one. This means that for every real eigenvalue μ of the matrix $A^{\times}(x_0)$ there is precisely one eigenvalue $\mu(z)$ $(z \in \Pi_+)$ of $A^{\times}(z)$ which converges to μ as z tends to x_0. Hence, the set $\sigma(x_0)$ (defined above Theorem 2.5) is empty, and therefore $x_0 \notin \Sigma$. $\square$

THEOREM 4.2. *Let $x_0 \neq 0$, and let λ_0 be any real eigenvalue of the matrix $A^{\times}(x_0)$ with partial multiplicity m and sign ϵ. Then the following statements hold:*

(1) *$m > 1$ if and only if λ_0 is a local extremal point of $W(\lambda)$, i.e.,*

$$\frac{d}{d\lambda} W(\lambda) \Big|_{\lambda = \lambda_0} = 0.$$

(2) *Let $m=1$, and let $\lambda(z)$ $(z = x_0 + iy,\ y > 0)$ be an eigenvalue of the matrix $A^{\times}(z)$ such that $\lim_{y \downarrow 0} \lambda(z) = \lambda_0$. Then the following are equivalent:*

 (i) $\epsilon = +1$,

 (ii) $\lambda(z) \in \Pi_+$,

 (iii) *$W(\lambda)$ is strictly increasing on a neighbourhood of λ_0.*

(3) *Let m be an even number, then $\epsilon = -1$ (resp., $+1$) if λ_0 is a local maximum (resp., minimum) of $W(\lambda)$.*

(4) *Let $m > 1$, m odd, and let λ_0 be a decreasing (resp., increasing) inflexion*

point of $W(\lambda)$ *(i.e.,* $\frac{d}{d\lambda} W(\lambda) \rfloor_{\lambda=\lambda_0} = 0$ *and* $W(\lambda)$ *is a decreasing (resp., increasing) function on a neighbourhood of* λ_0,*) then* $\epsilon = -1$ *(resp.,* $+1$*).*

The prooof follows directly from the remarks at the end of Subsection 2.5.

COROLLARY 4.3. *Let* Θ *and* Σ *be the sets as defined in (2.10) and Theorem 2.5, then*

$$\Theta = \Sigma = \{0\} \cup \{x \in \mathbb{R} \mid \text{there exists a } \lambda_0 \text{ in } \mathbb{R} \text{ such that } W(\lambda_0) = x, \text{ and } \frac{d}{d\lambda} W(\lambda) \rfloor_{\lambda=\lambda_0} = 0\}.$$

5. AN EXAMPLE

Consider the kernel $k(t) = e^{-|t|}$, $t \in \mathbb{R}$. Then (from equation (3.2)) the symbol is given by

$$W(\lambda) = \frac{2}{1+\lambda^2}.$$

Using (4.1) we get the following matrices in a minimal realization $W(\lambda) = -C(\lambda - A)^{-1}B$:

$$A = \begin{pmatrix} 0 & 1 \\ -1 & 0 \end{pmatrix}, \quad B = \begin{pmatrix} 0 \\ 1 \end{pmatrix}, \quad C = \begin{pmatrix} -2 & 0 \end{pmatrix}, \quad H = \begin{pmatrix} 0 & -2 \\ -2 & 0 \end{pmatrix}.$$

If $x \neq 0$, then:

$$A^{\times}(x) = \begin{pmatrix} 0 & 1 \\ -1+\frac{2}{x} & 0 \end{pmatrix}.$$

From Theorem 3.1 and Corollary 4.3 we get that $\sigma(K) = C\sigma(K) = [0, 2]$, and $\Theta = \{0, 2\}$. Next, define for $0 < x < 2$: $p = \sqrt{\frac{2-x}{x}}$. Then p and $-p$ are the eigenvalues of the matrix $A^{\times}(x)$ with signs -1 and 1, respectively. For $0 < x < 2$ we compute the following matrices:

$$A^{\times}(x) = \frac{1}{2p} \begin{pmatrix} 1 & 1 \\ p & -p \end{pmatrix} \begin{pmatrix} p & 0 \\ 0 & -p \end{pmatrix} \begin{pmatrix} p & 1 \\ p & -1 \end{pmatrix},$$

$$P = \frac{1}{2}\begin{pmatrix} 1 & -i \\ i & 1 \end{pmatrix},$$

$$P^{\times}(x) = I - \tilde{P}^{\times}(x) = \frac{1}{2}\begin{pmatrix} 1 & -\frac{1}{p} \\ -p & 1 \end{pmatrix},$$

$$\Pi(x) = \frac{1}{i-p}\begin{pmatrix} i & -1 \\ pi & -p \end{pmatrix},$$

Now, we compute the formula for the resolution of the identity E (where we have changed the integration variable x to $p = \sqrt{\frac{2-x}{x}}$):

$$(E([a,b])f)(t) = \int_0^\infty \{ \frac{1}{2\pi} \int_{\sqrt{\frac{2-b}{b}}}^{\sqrt{\frac{2-a}{a}}} [e^{(t-s)pi} + e^{(s-t)pi} +$$

$$+ \frac{p-i}{p+i} e^{(t+s)pi} + \frac{p+i}{p-i} e^{-(t+s)pi}] f(s)\, ds \}\, dp, \quad 0 \leqq t < \infty, \quad 0 < a, b < 2,$$

where f is a function in $(L_1 \cap L_2)(\mathbb{R}_+)$. Since $E(2) = I_{\ell_2}$, we can compute $E(\lambda)$, for $0<\lambda<2$, as follows:

$$(E(\lambda)f)(t) = f(t) - \int_0^\infty \{ \frac{1}{2\pi} \int_0^{\sqrt{\frac{2-\lambda}{\lambda}}} [e^{(t-s)pi} + e^{(s-t)pi} +$$

$$+ \frac{p-i}{p+i} e^{(t+s)pi} + \frac{p+i}{p-i} e^{-(t+s)pi}] f(s)\, dp \}\, ds.$$

6. THE SINGULAR CONTINUOUS COMPONENT

Let $\{E(\lambda) \mid \lambda \in \mathbb{R}\}$ be the resolution of the identity for T (see Section 1). There exists a unique function $\hat{E}$, defined on the Borel sets Δ of $\sigma(T)$, such that the following holds:

$$\hat{E}(\emptyset) = 0, \quad \hat{E}(\sigma(T)) = I,$$

$$\hat{E}(\Delta_1 \cap \Delta_2) = \hat{E}(\Delta_1)\, \hat{E}(\Delta_2),$$

$$\hat{E}(\Delta_1 \cup \Delta_2) = \hat{E}(\Delta_1) + \hat{E}(\Delta_2), \quad \text{if } \Delta_1 \cap \Delta_2 = \emptyset,$$

for every $x, y \in H$ *the function* $\hat{E}_{x,y}(\Delta) = \langle \hat{E}(\Delta)x, y \rangle$ *is a regular complex Borel measure on* $\sigma(T)$,

and on intervals $\hat{E} = E$. This unique function is called the *spectral measure* of T (see, e.g., [K]). In what follows, we shall write E instead of $\hat{E}$.

Let $H_p = \overline{\bigvee_{\lambda \in \mathbb{R}} \operatorname{Im} E(\{\lambda\})}$ and $H_c = H_p^{\perp}$ be the *discontinuous* and *continuous parts* of H. Furthemore, let

$H_{ac} = \{u \in H \mid$ *if* Δ *is a Borel set of Lebesgue measure zero, then* $E(\Delta)u = 0\}$,

$H_s = H_{ac}^{\perp} = \{u \in H \mid$ *there exists a Borel set* Δ *of Lebesgue measure zero such that* $E(\Delta)u = u\}$,

and $H_{sc} = H_c \ominus H_{ac}$ as the *absolute continuous, singular,* and the *singular continuous subspaces* of H, respectively (see, e.g., [K]).

Let T be the Wiener-Hopf operator K of Section 3. Then we can write the resolution of the identity as follows:

$$(E([a,b])f)(t) = \int_a^b \varrho_{f,t}(x)\,dx, \tag{6.1}$$

for functions f in $(L_1^m \cap L_2^m)(\mathbb{R}_+)$, where $[a,b] \cap \Theta = \varnothing$ (see (2.10)). For fixed f and t we have that $\varrho_{f,t}$ is a continuos function on $[a,b]$. Now we extend (6.1) by setting:

$$(E(\Delta)f)(t) = \int_{\Delta} \varrho_{f,t}(x)\,dx, \tag{6.2}$$

where Δ is a Borel set in $[a,b]$.

THEOREM 6.1. *Let* Δ *be a Borel set of* $\sigma(K)$ *of Lebesgue measure zero, then*

$$E(\Delta) = \begin{cases} E(\{0\}), & 0 \in \Delta, \\ 0, & 0 \notin \Delta. \end{cases}$$

Proof. Since the set Θ is finite, we may assume that $\Delta \subset [c, d]$ such that $|\Theta \cap [c, d]| \leqq 1$. If $\Theta \cap [c, d] = \emptyset$, it follows from (6.2) that $E(\Delta) f = 0$ for all $f \in (L_1^m \cap L_2^m)(\mathbb{R}_+)$. Because $(L_1^m \cap L_2^m)(\mathbb{R}_+)$ is dense in $L_2^m(\mathbb{R}_+)$ we obtain $E(\Delta) = 0$. Next, assume $\Theta \cap [c, d] = \{x\} \subset \Delta$. Then there is an $\epsilon_0 > 0$ such that $[x - \epsilon_0, x + \epsilon_0] \setminus \{0\} \subset C\sigma(T)$. Now define for $0 < \epsilon \leqq \epsilon_0$ the Borel set $\Delta_\epsilon = \Delta \setminus (x - \epsilon, x + \epsilon)$. Note that Δ_ϵ has Lebesgue measure zero and that (6.2) holds for Δ_ϵ (instead of Δ), therefore $E(\Delta_\epsilon) = 0$. From one of the properties of the spectral measure we get that

$$\begin{aligned} E(\Delta \cup (x - \epsilon, x + \epsilon)) &= E(\Delta_\epsilon) + E((x - \epsilon, x + \epsilon)) = \\ &= E(x + \epsilon) - E(x - \epsilon). \end{aligned} \tag{6.3}$$

Because $E_{f,f}$ is a positive measure (for each $f \in L_2^m(\mathbb{R}_+)$), we obtain (in the strong topology)

$$E(\Delta) = \lim_{\epsilon \downarrow 0} E(\Delta \cup (x - \epsilon, x + \epsilon)).$$

Hence, with (6.3) we get the equality $E(\Delta) = E(\{x\})$. Finally, we remark that

$$E(\{x\}) = \begin{cases} E(\{0\}), & x = 0, \\ 0, & x \neq 0, \end{cases}$$

because $P\sigma(K) \subset \{0\}$. □

THEOREM 6.2. *A first kind selfadjoint Wiener-Hopf operator, with a rational symbol, has a trivial singular continuous subspace.*

Proof. From Theorem 6.1 we have that $H_c = H_p^\perp = \text{Im } E(\{0\})^\perp = H_s^\perp = H_{ac}$. Hence, $H_{sc} = (0)$. □

REFERENCES

[B] Baumgärtel, H.: Analytic perturbation theory for matrices and operators. Operator Theory: Advances and Applications. Vol.15, Birkhäuser Verlag (Basel) 1985.

[BGK1] Bart, H., Gohberg, I., Kaashoek, M.A.: Minimal factorization of matrix and operator functions. Operator Theory: Advances and Applications. Vol.1, Birkhäuser Verlag (Basel) 1979.

[BGK2] Bart, H., Gohberg, I., Kaashoek, M.A.: Wiener-Hopf integral equations, Toeplitz matrices and linear systems. In Toeplitz Centennial (ed. I. Gohberg), Operator Theory: Advances and Applications. Vol.4, Birkhäuser Verlag (Basel) 1982, 85-135.

[BN] Bachman, G., Narici, L.: Functional analysis. Academic Press (New York, etc.) 1966.

[C] Conway, J.B.: Functions of one complex variable. Springer-Verlag, New York, 1978.

[DS] Dunford, N., Schwartz, J.T.: Linear operators, part II: Spectral theory. New York: Interscience Publishers 1963.

[GLR1] Gohberg, I., Lancaster, P., Rodman, R.: Matrices and indefinite scalar products. Operator Theory: Advances and Applications. Vol.8, Birkhäuser Verlag (Basel) 1983.

[GLR2] Gohberg, I., Lancaster, P., Rodman, R.: Matrix polynomials. Academic Press (New York, etc.) 1982.

[GLR3] Gohberg, I., Lancaster, P., Rodman, R.: Invariant subspaces of matrices with applications. Canadian mathematical society of monographs and advanced texts, John Wiley and Sons. New York, etc., 1986.

[HW] Hartman, P., Wintner, A.: The spectra of Toeplitz matrices. Amer. Journal of Math., 76 (1954), 867-882.

[IKL] Iohvidov, I., Krein, M.G., Langer, H.: Introduction to spectral theory of operators in spaces with an indefinite metric. Math. Research, Vol. 9, Akademie-Verlag, Berlin, 1982.

[K] Kato, T.: Perturbation theory for linear operators. Springer-Verlag, Berlin, 1966.

[Ka] Kailath, T.: Linear systems. Prentice-Hall, Englewood Cliffs, NJ, 1980.

[R] Ran, A.C.M.: Minimal factorization of selfadjoint rational matrix functions. Integral Equations and Operator Theory, 5 (1982), 850-869.

[Rod] Rodman, L.: On the structure of selfadjoint Toeplitz operators with rational symbol. Proc. Amer. Math. Soc., Vol. 92 (1984), 487-494.

[Ros1] Rosenblum, M.: Selfadjoint Toeplitz operators and associated orthonormal functions. Proc. Amer. Math. Soc., Vol. 13 (1962), 590-595.

[Ros2] Rosenblum, M.: The absolute contonuity of Toeplitz matrices. Pacific J. Math. 10 (1960), 987-996.

[Ros3] Rosenblum, M.: A concrete spectral theory for selfadjoint Toeplitz operators. Amer. J. Math. 87 (1965), 709-718.

[Ru] Rudin, W.: Real and complex analysis, third edition. McGraw-Hill Book Company, New York, etc., 1986.

[RR] Rosenblum, M., Rovnyak, J.: Hardy classes and operator theory. Oxford mathematical monographs, Oxford Unversity Press, New York, etc., 1985.

[T] Taylor, A.E.: Introduction to functional analysis. John Wiley and Sons. New York, etc., 1958.

R. Vreugdenhil,
Faculteit Wiskunde en Informatica,
Vrije Universiteit,
De Boelelaan 1081,
1081 HV Amsterdam,
The Netherlands

WORKSHOP PROGRAM

Monday, June 26, **1989**

09.30	***Welcome*** by Prof.dr. A.H.G. Rinnooy Kan, Rector Magnificus
09.40	***Opening*** by I. Gohberg

Plenary Sessions

09.50 - 10.30	***L. de Branges*** (Lafayette) *A construction of Krein spaces of analytic functions*
10.35 - 11.15	***T.J. Laffey*** (Dublin) *Some remarks on commuting matrices*

Parallel Sessions

11.45 - 12.15	***A. Dijksma*** (Groningen) *Spectral properties of Hamiltonian systems with boundary conditions containing the eigenvalue parameter*
12.20 - 12.50	***C.S. Sadosky*** (Washington) *A Nehari theorem for Hankel forms bounded by non-Toeplitz norms*
11.45 - 12.15	***R. Loewy*** (Haifa) *Cones of nonnegative matrices having given left and right Perron eigenvectors*
12.20 - 12.50	***B.S. Mitiagin*** (Ohio) *Littlewood inequality Bennett megalemma.*
14.10 - 14.40	***W. Kaballo*** (Quezon City) *Decomposition of meromorphic Fredholm resolvents and* Ψ^**-algebras*
14.45 - 15.15	***R. Mennicken*** (Regensburg) *Expansions of analytic fuctions in Berson series and Carlitz series*
14.10 - 14.40	***L. Lerer*** (Haifa) *The matrix quadratic equation and factorization of matrix polynomials*
14.45 - 15.15	***A. Ben-Artzi*** (San Diego) *Band matrices and dichotomy* *(joint work with I. Gohberg).*

Plenary Sessions

15.45 – 16.25 **F.-O. Speck** and **E. Meister** (Darmstadt)
Modern Wiener–Hopf methods in diffraction theory

16.30 – 17.10 **P. Dewilde** (Delft)
On a new class of maximum entropy approximations to positive operators on ℓ^2

17.30 **Reception** by invitation of the Econometric Institute

Tuesday, June 27, 1989

Plenary Sessions

08.45 – 09.25 **T. Ando** (Sapporo)
Some extremal problems related to majorization

09.30 – 10.10 **M.S. Livsic** (Beer Sheva)
Commuting nonselfadjoint operators and the concept of corpuscle in systems theory

Parallel Sessions

10.40 – 11.10 **Qiang Ye** (Calgary)
Rayleigh–Ritz methods using Krylov subspaces for symmetric matrix pencils

11.15 – 11.45 **P. Bruinsma** (Groningen)
The Nevanlinna–Pick interpolation problem

11.50 – 12.20 **D. Alpay** (Blacksburg)
Structured matrices, De Branges pairs of spaces and a generalization of Iohvidov laws

12.25 – 12.55 **D. Hershkowitz** (Haifa)
Level and height characteristics and their relations

10.40 – 11.10 **K. Lorentz** (Mainz)
On the local structure of the similarity orbits of Jordan elements in topological algebras

11.15 – 11.45 **F. Mantlik** (Dortmund)
On linear vector function equations

11.50 – 12.20 **I. Zaballa** (Vitoria)
Completing a pair of matrices with fixed principal part

12.25 – 12.55 **F. van Schagen** (Amsterdam)
Operator blocks and completions of submatrices

Plenary Sessions

14.15 – 14.55	***H. Schneider*** (Madison) *Perron–Frobenius theory of (highly) reducible non-negative matrices.*
15.00 – 15.40	***V.I. Matsaev*** *On the factorization of non analytic selfadjoint operator functions and the principal parts of their resolvents*
16.10 – 16.50	***H. Dym*** (Rehovot) *Structured reproducing kernel spaces and matrix equations*
16.55 – 17.35	***L. de Branges*** (Lafayette) *The Riemann hypothesis.*

Wednesday, June 28, 1989

Plenary Sessions

08.45 – 09.25	***G. Heinig*** (Karl–Marx–Stadt) *Matrices with displacement structure, generalized Bezoutians, and Moebius transformations*
09.30 – 10.10	***E.M. Semenov*** (Voronezh) *Planar faces of the operator space* L_p

Parallel Sessions

10.40 – 11.10	***D.C. Lay*** and ***R.L. Ellis*** (Maryland) *Rank–preserving extensions of band matrices*
11.15 – 11.45	***T. Shalom*** (Tel–Aviv) *On Bezoutians and inverses of generalized block Toeplitz matrices*
11.50 – 12.20	***N. Young*** (Lancaster) *A Schur–Cohn theorem for matrix polynomials*
10.40 – 11.10	***H.S.V. de Snoo*** (Groningen) *Generalized resolvents, Weyl coefficients and their kernels*
11.15 – 11.45	***H.J. Woerdeman*** (Amsterdam) *A maximum entropy principle in the general framework of the band method*

11.50 - 12.20 **C.V.M. van der Mee** (Newark)

A Riemann-Hilbert problem arising from the inverse scattering problem for the 3-D Schrödinger equation

13.30 ***Excursion to Gouda***

18.00 ***Workshop Dinner***

Thursday, June 29, 1989

Plenary Sessions

08.45 - 09.25 **N.K. Nikolskii** (Leningrad)

Quasi-orthogonal Hilbert space decompositions, with an application to a proof of De Branges theorem (Bieberbach conjecture)

09.30 - 10.10 **J.A. Ball** (Blacksburg)

De Branges-Rovynak operator models and systems theory: a survey

Parallel Sessions

10.40 - 11.10 **F.Ali Mehmeti** (Mainz)

Interaction problems

11.15 - 11.45 **E. Schrohe** (Mainz)

Normal solvability for elliptic boundary value problems on asymptotically flat manifolds

11.50 - 12.20 **A.C.M. Ran** (Amsterdam)

Regular matrix polynomials with prescribed zero structure in the finite complex plane

12.25 - 12.55 **L. Rodman** (Williamsburg)

Positive definite completions of matrices over C^*-algebras

10.40 - 11.10 **R. Vreugdenhil** (Amsterdam)

Spectral theory of selfadjoint Wiener-Hopf operators with rational symbol

11.15 - 11.45 **S. Rubinstein** (Tel-Aviv)

Minimal rational symplectic orbits of contractions

11.50 - 12.20 **L. Frank** (Nijmegen)

Elliptic singularly perturbed eigenvalue problems and applications

12.25 - 12.55 **M. Dominguez** (Caracas)

Systems of Toeplitz operators and matricial weighted inequalities

Plenary Sessions

14.15 - 14.55 **R. Duduchava** (Tbilisi)

Singular integral equations on plane domains and the factorization of rational matrix functions

15.00 - 15.40 **C. Foias** (Bloomington)

A strong Parrott lemma and applications

16.10 - 16.50 **C.R. Johnson** (Williamsburg)

Vanishing minor conditions, inverse zero patterns and determinants

16.55 - 17.35 **J.W. Helton** (San Diego)

Factorization of non-linear operators

17.40 ***Closing Remarks by T. Ando, H. Bart and M.A. Kaashoek***

LIST OF PARTICIPANTS

Ali Mehmeti, F.A., Mainz, GERMANY
Alpay, D., Blacksburg, Virginia, U.S.A.
Ando, T., Sapporo, JAPAN
Ball, J.A., Blacksburg, Virginia, U.S.A.
Bart, H., Rotterdam, THE NETHERLANDS
Ben-Artzi, A., La Jolla, California, U.S.A.
Branges, L. de, Lafayette, Indiana, U.S.A.
Bruinsma, P., Groningen, THE NETHERLANDS
Cohen, N., Rehovot, ISRAEL
Dewilde, P., Delft, THE NETHERLANDS
Dominguez, M., Caracas, VENEZUELA
Duduchava, R., Tbilisi, U.S.S.R.
Dijksma, A., Groningen, THE NETHERLANDS
Dym, H., Rehovot, ISRAEL
Ellis, R.L., College Park, Maryland, U.S.A.
Foias, C., Bloomington, Indiana, U.S.A.
Frank, L., Nijmegen, THE NETHERLANDS
Gohberg, I., Ramat-Aviv, ISRAEL
Groenewald, G., Amsterdam, THE NETHERLANDS
Heinig, G., Chemnitz, GERMANY
Helton, J.W., La Jolla, California, U.S.A.
Hershkowitz, D., Haifa, ISRAEL
Hoogland, H.J.T., Rotterdam, THE NETHERLANDS
Huijsmans, C.B., Leiden, THE NETHERLANDS
Johnson, C.R., Williamsburg, Virginia, U.S.A.
Kaashoek, M.A., Amsterdam, THE NETHERLANDS
Kaballo, W., Dortmund, GERMANY
Khvedelidze, B.V., Tbilisi, U.S.S.R.
Korevaar, J., Amsterdam, THE NETHERLANDS
Kuijper, A.-B., Amsterdam, THE NETHERLANDS
Laffey, T.J., Dublin, IRELAND
Lay, D.C., College Park, Maryland, U.S.A.
Lerer, L., Haifa, ISRAEL

Livsic, M.S., Beer Sheva, ISRAEL

Loewy, R., Haifa, ISRAEL

Lorentz, K., Mainz, GERMANY

Mantlik, F., Dortmund 50, GERMANY

Matsaev, V.I., Chernogolovka, USSR

Mee, C.V.M. van der, Newark, U.S.A.

Meister, E., Darmstadt, GERMANY

Mennicken, R., Regensburg, GERMANY

Mitiagin, B.S., Columbus, Ohio, U.S.A.

Nikolskii, N.K., Leningrad, U.S.S.R.

Pagter, B. de, Delft, THE NETHERLANDS

Qiang, Ye, Calgary, Alberta, CANADA

Ran, A.C.M., Amsterdam, THE NETHERLANDS

Rodman, L., Williamsburg, Virginia, U.S.A.

Rubinstein, S., Ramat-Aviv, ISRAEL

Sadosky, C.S., Washington, DC, U.S.A.

Schagen, F. van, Amsterdam, THE NETHERLANDS

Schneider, H., Madison, Wisconsin, U.S.A.

Schrohe, E., Mainz, GERMANY

Semenov, E.M., Voronezh, U.S.S.R.

Shalom, T., Ramat-Aviv, ISRAEL

Snoo, H.S.V. de, Groningen, THE NETHERLANDS

Speck, F.-O., Darmstadt, GERMANY

Thijsse, G.Ph.A., Rotterdam, THE NETHERLANDS

Vreugdenhil, R., Amsterdam, THE NETHERLANDS

Woerdeman, H.J., Amsterdam, THE NETHERLANDS

Young, N., Lancaster, ENGLAND

Zaballa, I., Vitoria-Gasteiz, SPAIN

Zuidwijk, R., Leiden, THE NETHERLANDS

Titles previously published in the series
OPERATOR THEORY: ADVANCES AND APPLICATIONS
BIRKHÄUSER VERLAG

13. **G. Heinig, K. Rost:** Algebraic Methods for Toeplitz-like Matrices and Operators, 1984, (3-7643-1643-8)
14. **H. Helson, B. Sz.-Nagy, F.-H. Vasilescu, D.Voiculescu, Gr. Arsene** (Eds.): Spectral Theory of Linear Operators and Related Topics, 1984, (3-7643-1642-X)
15. **H. Baumgärtel:** Analytic Perturbation Theory for Matrices and Operators, 1984 (3-7643-1664-0)
16. **H. König:** Eigenvalue Distribution of Compact Operators, 1986, (3-7643-1755-8)
17. **R.G. Douglas, C.M. Pearcy, B. Sz.-Nagy, F.-H. Vasilescu, D. Voiculescu, Gr. Arsene** (Eds.): Advances in Invariant Subspaces and Other Results of Operator Theory, 1986, (3-7643-1763-9)
18. **I. Gohberg** (Ed.): I. Schur Methods in Operator Theory and Signal Processing, 1986, (3-7643-1776-0)
19. **H. Bart, I. Gohberg, M.A. Kaashoek** (Eds.): Operator Theory and Systems, 1986, (3-7643-1783-3)
20. **D. Amir:** Isometric characterization of Inner Product Spaces, 1986, (3-7643-1774-4)
21. **I. Gohberg, M.A. Kaashoek** (Eds.): Constructive Methods of Wiener-Hopf Factorization, 1986, (3-7643-1826-0)
22. **V.A. Marchenko:** Sturm-Liouville Operators and Applications, 1986, (3-7643-1794-9)
23. **W. Greenberg, C. van der Mee, V. Protopopescu:** Boundary Value Problems in Abstract Kinetic Theory, 1987, (3-7643-1765-5)
24. **H. Helson, B. Sz.-Nagy, F.-H. Vasilescu, D. Voiculescu, Gr. Arsene** (Eds.): Operators in Indefinite Metric Spaces, Scattering Theory and Other Topics, 1987, (3-7643-1843-0)
25. **G.S. Litvinchuk, I.M. Spitkovskii:** Factorization of Measurable Matrix Functions, 1987, (3-7643-1843-X)
26. **N.Y. Krupnik:** Banach Algebras with Symbol and Singular Integral Operators, 1987, (3-7643-1836-8)
27. **A. Bultheel:** Laurent Series and their Pade Approximation, 1987, (3-7643-1940-2)
28. **H. Helson, C.M. Pearcy, F.-H. Vasilescu, D. Voiculescu, Gr. Arsene** (Eds.): Special Classes of Linear Operators and Other Topics, 1988, (3-7643-1970-4)
29. **I. Gohberg** (Ed.): Topics in Operator Theory and Interpolation, 1988, (3-7634-1960-7)

30. **Yu.I. Lyubich:** Introduction to the Theory of Banach Representations of Groups, 1988, (3-7643-2207-1)
31. **E.M. Polishchuk:** Continual Means and Boundary Value Problems in Function Spaces, 1988, (3-7643-2217-9)
32. **I. Gohberg** (Ed.): Topics in Operator Theory. Constantin Apostol Memorial Issue, 1988, (3-7643-2232-2)
33. **I. Gohberg** (Ed.): Topics in Interplation Theory of Rational Matrix-Valued Functions, 1988, (3-7643-2233-0)
34. **I. Gohberg** (Ed.): Orthogonal Matrix-Valued Polynomials and Applications, 1988, (3-7643-2242-X)
35. **I. Gohberg, J.W. Helton, L. Rodman** (Eds.): Contributions to Operator Theory and its Applications, 1988, (3-7643-2221-7)
36. **G.R. Belitskii, Yu.I. Lyubich:** Matrix Norms and their Applications, 1988, (3-7643-2220-9)
37. **K. Schmüdgen:** Unbounded Operator Algebras and Representation Theory, 1990, (3-7643-2321-3)
38. **L. Rodman:** An Introduction to Operator Polynomials, 1989, (3-7643-2324-8)
39. **M. Martin, M. Putinar:** Lectures on Hyponormal Operators, 1989, (3-7643-2329-9)
40. **H. Dym, S. Goldberg, P. Lancaster, M.A. Kaashoek** (Eds.): The Gohberg Anniversary Collection, Volume I, 1989, (3-7643-2307-8)
41. **H. Dym, S. Goldberg, P. Lancaster, M.A. Kaashoek** (Eds.): The Gohberg Anniversary Collection, Volume II, 1989, (3-7643-2308-6)
42. **N.K. Nikolskii** (Ed.): Toeplitz Operators and Spectral Function Theory, 1989, (3-7643-2344-2)
43. **H. Helson, B. Sz.-Nagy, F.-H. Vasilescu, Gr. Arsene** (Eds.): Linear Operators in Function Spaces, 1990, (3-7643-2343-4)
44. **C. Foias, A. Frazho:** The Commutant Lifting Approach to Interpolation Problems, 1990, (3-7643-2461-9)
45. **J.A. Ball, I. Gohberg, L. Rodman:** Interpolation of Rational Matrix Functions, 1990, (3-7643-2476-7)
46. **P. Exner, H. Neidhardt** (Eds.): Order, Disorder and Chaos in Quantum Systems, 1990, (3-7643-2492-9)
47. **I. Gohberg** (Ed.): Extension and Interpolation of Linear Operators and Matrix Functions, 1990, (3-7643-2530-5)
48. **L. de Branges, I. Gohberg, J. Rovnyak** (Eds.): Topics in Operator Theory. Ernst D. Hellinger Memorial Volume, 1990, (3-7643-2532-1)
49. **I. Gohberg, S. Goldberg, M.A. Kaashoek:** Classes of Linear Operators, Volume I, 1990, (3-7643-2531-3)